Academy of Sciences of the USSR
Institute of Microbiology

N. A. Krasil'nikov

SOIL MICROORGANISMS AND HIGHER PLANTS

Published by the Academy of Sciences of the USSR
Moscow 1958

Published for
THE NATIONAL SCIENCE FOUNDATION, WASHINGTON, D.C.
and
THE DEPARTMENT OF AGRICULTURE, USA by
THE ISRAEL PROGRAM FOR SCIENTIFIC TRANSLATIONS
1961
Title of Russian Original: Mikroorganizmy pochvy i vysshie rasteniva

Translated by: Dr. Y. Halperin

Printed in Jerusalem by: S. Monson

Table of Contents

INTRODUCTION 5

Part I – FUNDAMENTALS OF THE STRUCTURE AND DEVELOPMENT OF MICROORGANISMS, AND
THEIR CLASSIFICATION 9

STRUCTURE OF BACTERIAL CELLS..9

Cell Wall 9

Flagella...12

Protoplast 14

Nucleus and Nucleoli..15

Gram-staining of Bacteria..22

GROWTH AND MULTIPLICATION OF MICROORGANISMS 25

Growth of Cells and Development of Cultures...25

The Sexual Process in Bacteria..38

VARIABILITY OF MICROORGANISMS 53

Individual Variability or Polymorphism of Cells...54

Species Variability of Microorganisms..59

Adaptive and Directed Variability 63

"Induced" Variation in Microorganisms 68

Variability of Soil Microorganisms 77

Bacterial Variability under the Influence of Plants...81

THE PRINCIPLES OF CLASSIFICATION OF MICROORGANISMS 85

The Species Concept in Microorganisms 86

SUBDIVISION OF MICROORGANISMS INTO MAIN GROUPS 93

Actinomycetes 93

Bacteria 98

Phages..103

Part II – THE SOIL AS AN ENVIRONMENT FOR MICROORGANISMS 105

SOIL STRUCTURE 106

SOIL POROSITY 110

The liquid phase of soils 112

The soil solution..116

The gaseous phase of soil 119

SOIL RESPIRATION..123

THERMAL REGIME OF THE SOIL 124

SOIL INSOLATION 125

ORGANIC MATTER OF THE SOIL 126

RADIOACTIVE SUBSTANCES OF THE SOIL 133

THE ADSORPTION CAPACITY OF SOILS 135

Adsorption of bacteria by soil 136

Adsorption of products of microbial metabolism by soil 141

THE MICROFLORA OF THE SOIL 143

DISTRIBUTION OF MICROORGANISMS IN SOIL 152

The morphology of microbes in soil 158

Ecological and geographical distribution of microorganisms in soils 162

Part III – BIOLOGICAL FACTORS OF SOIL FERTILITY 168

PLANT NUTRITION 168

HETEROTROPHISM OF PLANTS 173

The effect of humus on the growth of plants 178

The effect of humus on the vitamin content of plants 184

BIOTIC SUBSTANCES OF THE SOIL 186

The origin of the biotic substances of soil 191

Microorganisms synthesizing biotic substances 193

The effect of biotic substances on plants 201

The assimilation of biotic and other substances by plants 207

EFFECT OF BACTERIA ON THE ASSIMILATION OF NUTRIENTS BY PLANTS 212

Part IV – INTERACTION BETWEEN SOIL MICROORGANISMS AND PLANTS 219

EFFECT OF PLANTS ON THE SOIL MICROFLORA 219

Root mass of plants 219

Root excretions 222

Root residue 229

Rhizosphere 230

The microflora of the rhizosphere 232

The specificity of the microflora of the root zone 244

Microflora of decomposing roots 257

EFFECT OF SOIL MICROORGANISMS ON PLANTS 260

Microbial activators 260

Distribution of microbial activators in the soil 271

Microbial inhibitors and their action on plants 274

TOXICOSIS OF SOILS AND BIOLOGICAL FACTORS CAUSING IT 286

Formation of toxic substances by plants 292

MICROBIAL ANTAGONISTS 299

The protective role of microbial antagonists 306

Significance of antagonists in plant immunity 311

Formation and accumulation of antibiotics in soil 312

3

Entry of antibiotics into plants..319

Antibiotic substances as a therapeutic means in plant cultivation 323

CONCERNING THE EPIPHYTIC MICROFLORA 341

This book is devoted to the problem of the interaction between soil microorganisms and higher plants. The material presented includes basic information on the structure, development, variability and classification of bacteria, actinomycetes and fungi in the light of recent scientific achievements, as well as information on the importance of microorganisms in plant nutrition, the role of micro-activities in the complementary nutrition of plants, the effect of microbes on the vitamin content of plants, their importance in plant development and their influence on soil fertility. In addition, data are given on the importance of antibiotics as a means of therapy and prevention of diseases in agricultural practice.

The book is designed for the use of microbiologists. plant physiologists, soil specialists, phytopathologists, mycologists, agrobilologists, and agronomists. It may also serve as a textbook for students In biological faculties of universities or agricultural and forestry institutes.

INTRODUCTION

The scientific basis of soil science as a natural science was established by the classical works of Dokuchaev. Previously, soil had been considered a product of physicochemical transformations of rocks, a dead substrate from which plants derive nutritious mineral elements. Soil and bedrock were in fact equated.

Dokuchaev considers the soil as a natural body having its own genesis and its own history of development, a body with complex and multiform processes taking place within it. The soil is considered as different from bedrock. The latter becomes soil under the influence of a series of soil-forming factors--climate, vegetation, country, relief and age. According to him, soil should be called the "daily" or outward horizons of rocks regardless of the type; they are changed naturally by the common effect of water, air and various kinds of living and dead organisms.

The outstanding soil-science specialist, P.A. Kostychev (1892), in developing the theory of soil, attaches special importance to biological factors. He considers the soil as a botanist and not as a geologist. On the question of the origin of the chernozem, Kostychev attributes the most essential role to plants and microbes. He writes that chernozem formation is involved with the geography and physiology of higher plants as well an that of lower ones which perform the decomposition of organic matter. The accumulation of soil humus depends on the intensity and completeness of the decomposition of plant residues, the roots and the parts which are above the ground. In these processes Kostychev identifies the most important role with microscopic creatures--fungi and bacteria. Being an excellent microbiologist he carried out interesting experiments on the decomposition of organic matter and the formation of humus. The experiments showed that in different cases the decay of plant residues began in a different way. Sometimes bacteria inhabit decaying matter first, and sometimes fungi emerge first. Various parts of the same decaying matter decompose differently, in one part one organism multiplies and near it an entirely different organism may be found. It is further pointed out that various forms of decomposition change consecutively according to changes in the properties of the decaying substance.

Kostychev (1889) for the first time established that humus is formed by soil fungi, Moreover, the discovery of the regularity of the relationship between carbon and nitrogen (C:N) in soil and of its importance In the development of plants and microbes must be attributed to him. Kostychev revealed the essence of the enrichment of soil humus by nitrogen. During the process of the decomposition of plant residues, which are known to contain no more than 1.5-2% nitrogen, humus with 4-5% of nitrogen is obtained. This transformation of organic matter as may be seen from Kostychev's data, occurs with the aid of microorganisms (see Kostychev, 1951).

With his works Kostychev laid the foundation of soil microbiology. He also outlined a vast program of investigations, the problems of which are at present being solved by Soviet microbiologists.

Vernadskii (1927), who expounded the theory of "bio-inert" natural bodies, wrote that the entire soil is a characteristic "bio-inert" body. All physicochemical properties of the soil would appear considerably different if the living substance in it were not taken into, account. In this way Vernadskii formulated the dependence of the fundamental property of soil upon the organisms lodged in it.

Vil'yams (Russian soil scientist (1863-1939) whose name is thus spelled when transliterated from the Russian, though it may appear as Williams in foreign publications) developing the teachings of Dokuchaev and Kostychev. introduced many new principal elements into the science of soil.

He considers the soil a natural body and means of agricultural production, an attributes to it the new qualitative property of fertility, i. e., the ability to produce plant crops. According to him ,the notion of soil and its fertility are inseparable. He considers fertility an essential property, a qualitative indication of soil independent of the degree of its quantitative expression. We oppose the idea of fertile soil to the idea of sterile stone, in other words to the notion of massive rock (Vil'yams, 1949).

It must be noted, that rocks also possess fertility to some degree. Investigations show that even the hardest rock massifs a are inhabited by various organisms. Lichens develop on their surfaces and often cover large spaces on mountain tops. The upper layer of rock massifs, the so-called weathering crust, is saturated with bacteria and algae, as well as fungi, actinomycetes, protozoa and other organisms. According to our data, in the upper layer of basalt rocks, the bacteria in one grain of substrate number from some tens of thousands to millions. Similar data are also given by other investigators (Novogrudskii, 1950; Glazovskaya, 1950; Parfenova, 1955; Krasil'nikov, 1949 and others).

Rocks differ in their fertility. Some of them are densely overgrown with lichens and microorganisms, others are sparsely covered. There are rocks, or more properly parts of one and the same rock, which are not overgrown with lichens, but contain only certain microbial forms--bacteria, actinomycetes and fungi (Krasil'nikov, 1949b).

Even under arctic conditions, on the islands of the northern Arctic Ocean (Franz Josef Land, Novaya Zemlya, Severnaya Zemlya and other areas) rocks and loose calcareous soils contain a considerable number of microorganisms. Tens of thousands to hundreds of millions were counted in one gram. Moreover, these organisms live actively and carry out biochemical and chemical transformations (Krasil'nikov and Artarnonova, 1958).

In order to study the properties which determine the crop capacity of soil and to increase the soil's fertility by exercising an influence on it, "it is first of all necessary to know these properties, to enumerate them, and to choose from the large number of properties and qualities of soil just those which determine the capacity of soil to produce the products necessary for mankind" (Vil'yams, 1949, p 138). Consequently, Vil'yams inseparably related the theory of soil with the theory of its fertility and its crop capacity.

The principal property of the soil fertility is determined by biological factors, mainly by microorganisms. The development of life in soil endows it with the property of fertility. "The notion of soil is inseparable from the notion of the development of living organisms in it". Soil is created by microorganisms. "Were this life dead or stopped, the former soil would become an object of geology" (Vi'lyams, 1950, p 204). Kostychev and Vil'yams transferred the science of soil from the chapter of geology to the chapter of biology.

The new understanding of the biological essence of the soil-creating processes, which was established by Kostychev and Vil'yams, has given the majority of Soviet microbiologists the principal leading landmarks in their investigations. Microbiology also contributed considerably to the development of this new direction in the science of soil, successfully solving many essential questions of soil fertility.

During the last two decades, microbiology has shown that life processes taking place in soil, were larger and deeper than it had been assumed earlier. Whereas earlier, hundreds of thousands and millions of microbial cells were counted in one gram of soil, at present, with more improved methods of investigation, hundreds of millions and billions are determined. The total bacterial mass on one hectare of the surface layer of fertile soils amounts to five to seven tons. This mass to composed of single cells which live, develop and multiply.

Along with bacteria, a very large number of fungi, actinomycetes, algae, ultramicrobes, phages, protozoa, insects, worms and other living creatures. inhabit the soil. Fungi and actinomycetes are counted in hundreds of thousands and millions in one gram, algae--in thousands and tens of thousands. Quite often their number even amounts to a hundred thousand in one gram of soil. The total mass of these organisms in the upper surface layer of the soil may amount to two-three tons in one hectare.

Analyses show that there is in the soil a great number of phages--actinophages and bacteriophages, which have very intense activity.

No less important is the soil fauna. According to different authors, amoebae, ciliata, and other protozoa are numbered in tens and hundreds of thousands in one gram (Brodskii, 1935, Nikolyuk, 1949; Dogel', 1951). In one square meter of the surface layer of soil some tens to hundreds of larger invertebrates may be found--earth worms, myriapods, larvae of various beetles, etc. The population of small nonmicroscopic arthropods (ticks, Collembola, and others) is numbered in tens and hundreds of thousands in 1 m^2 of cultivated soil layer, and in forest soils their number often amounts to a million individuals.

The number of nematodes is sometimes counted in millions per 1 m^2. According to the counts of Gilyarov (1949, 1953), the total mass of this fauna comprises several tons (3-4) per hectare of soil.

As seen from the data cited, every particle of soil is saturated with living creatures. The enormous mass of these creatures is in a state of continuous activity during the whole vegetative period. Separate individuals or cells of simpler creatures multiply rapidly and attain astronomical numbers. In the process of life activity the whole soil population carries out work of cosmic importance. It transforms enormous masses of organic and mineral compounds, and continuously synthesizes new organic and inorganic substances.

Various active biocatalysts--metabolites of microbes and other living creatures are found in the soil, including enzymes, vitamins, auxins, antibiotics, toxins and many other compounds. All these substances together with living organisms lend particular properties to the soil, differentiating it from a geological body or a mineral rock.

The biogeny of soil is the most significant indicator of its fertility. As soon as the activity of a microbial population begins in a rock, the first signs of fertility are manifested. The degree of soil fertility is determined by the intensity of the life processes of the microbial population.

Quantitative manifestations of biological processes are diverse and depend upon climatic, and geographic al or topographical conditions, as well as the seasons of the year and other external factors.

The knowledge of life in soil, investigation of biological and biochemical processes taking place in it, are inseparably connected with the knowledge of living organisms inhabiting the soil. Consequently, investigation of the microbial world of the soil is one of the basic problems of microbiology and agrobiology. The knowledge of biological processes, caused by the living population of the soil, should be one of the most important problems of pedology and agriculture.

This work deals with present knowledge of microorganisms, mainly of bacteria, actinomycetes and partly of fungi, inhabiting the soil, of their relationships with higher plants, of the importance of different groups and microbial species in the life of higher plants, and of the effect of metabolic products on growth and yield of agricultural crops.

It should be pointed out that in recent years, in the investigation of the problem of soil fertility. more and more attention in paid to biological factors. A growing interest in microbial population is displayed by plant growers as well as by soil scientists. This to not surprising, since it is impossible to solve problems of pedology, not to speak of agriculture and plant growing, without taking into account the microflora of soil.

The importance of microorganisms in the life of plants, as shown by present data, in very great but still little investigated. A lesser influence on the development and life activity of microbes in the soil is manifested by plants.

The interaction of soil microorganisms with higher plants is very complex and multiform. The effect of the former on the latter may be positive or negative. Depending upon the plant cover on the same soil under equal external climatic conditions, the composition of the microflora changes sharply. Plants are a very strong ecological factor. selecting certain species of bacteria, fungi, actinomycetes and other inhabitants of soil. As a result of wrong agricultural practices and crop rotation, the soil becomes infested with harmful microbial forms. By use of suitable plants in the crop rotation, one may change the microflora of soil in the desired direction, and eliminate harmful organisms, in other words--restore the health of soil.

The influence of soil microorganisms on growth and development of higher plants is of great diversity. The role of microbes is not by any means limited to the mineralization of substances in nature. At present, microorganisms should not be considered as only a link in the circulation of substances, as agents delivering sources of mineral nutrition to plants. Microorganisms of soil display a direct, very essential influence on plants, a positive or negative one depending upon the species and external conditions.

The positive role of symbiotic root-nodule bacteria, mycorhizal fungi and others is well known though little investigated. Great attention in paid to free-living nitrogen-fixing organisms--*Azotobacter, Clostridium* and others. Fertilizers comprising bacteria and fungi are largely used (azotogen, nitrogen, silicogen, phosphorobacterin and others).

A large chapter of science is devoted to the biology of phytopathogenic bacteria and fungi.

Little investigated and elucidated is the role of the free-living microflora, particularly that which inhabits the root zone. Investigation of the rhizosphere microflora has begun comparatively recently. The data obtained show that its effect on plants to very considerable. Among the rhizosphere microflora there is a large number of species which affect the growth of plants with products of their metabolism. Some microbial species are active producers of various biotic substances vitamins, auxins, amino acids and other substances essential to the growth of plants. Other species are antagonists of phytopathogenic bacteria, fungi and protozoa. Microorganisms of these species strengthen the immunization properties of tissues, thus protecting the plants from infections.

Among the soil microflora there are many organisms which produce toxic substances and suppress the plants; they inhibit their growth and development. It must be assumed that there are in the soil other microbial forms, as yet unknown, exerting a positive or negative effect on plants.

Representatives of useful and harmful microflora inhabit the rhizosphere. Their quantitative ratio in various soils and under various conditions of treatment is different. Consequently, the total effect of the activity of the root zone microflora will not be the same. It may be positive or negative, depending on what microbial species predominates.

In this work the latest information on the importance of this microflora for plants is given.

It is evident that in investigations of the activity of soil microflora exact knowledge of its species composition is necessary, as are exact conceptions of the soil climate and the ecological conditions of its development and accumulation. At present. investigations of soil biology in general, and questions of interaction of microorganisms with higher plants in particular, cannot be resolved with a merely quantitative account of the general composition of the microbes. It is necessary to establish the degree of dissemination of the different species, to investigate their biology and specificity of interaction with the plant, and their interaction with other species of soil microflora. The latter should be an important problem of soil microbiology. Microbial antagonists represent an important ecological factor of development and formation of microbial associations in general and dissemination of individual species in particular.

It is important to point out that the determination of the microbial species and particularly of bacteria is very difficult. The similarity of structure and the lack of external marks of identification does not allow the investigator to recognize, immediately or precisely, the forms and species he is faced with in the analysis of soil or other natural substrate.

We shall not be able to reveal the specificity of the root-zone microflora until we learn to identify and differentiate bacterial species. Without this it is impossible to determine the specificity of biochemical transformations in soil, and, consequently, the site of its fertility is also indeterminable.

Various representatives of microorganisms inhabit the soil--bacteria, actinomycetes, fungi, algae, protozoa and more highly organized animals. There are also various ultramicrobes--phages and others. Our knowledge of the structure and development of soil microorganisms is obtained by observation of their growth on artificial nutrient media, but little is known about the form in which they inhabit the soil and what their dimensions, structure, growth, and multiplication are. Does our knowledge, obtained by investigation of growth of microorganisms on artificial media, reflect the state in which they occur in soil? Many microbes are known only because they occur in soil. From general observations of bacterial growth on nutrient media we know that the structure and form of their cells is relatively simple and monotypic. Three types of cell structure are recognized--spherical. rod-shaped and spiral form. According to this, bacteria are divided into groups: cocci, bacilli, and spirilla. In every group there are subgroups.

The cell structure of these organisms is monotypic. As in higher organisms the bacterial cells possess a cell wall and a protoplast. The internal structure of the protoplast and cell wall differ from the structure of higher forms.

STRUCTURE OF BACTERIAL CELLS

Cell Wall

The cells of bacteria, fungi, actinomycetes, and yeasts, like other organisms of plant origin have a cell wall. Very few bacteria (myxobacteria) do not possess one. It forms the external skeleton of bacteria, actinomycetes and fungi, and determines the cell shape of these organisms. Cells having a cell wall do not change their shape while moving; they are rigid. The cell wall itself has some elasticity, due to its physical and chemical structure.

There are data showing that bacterial cell walls have a complex structure differing in various representatives. In some gram-negative bacteria *(Shigella* and others), the cell wall is composed of polysaccharides, proteins and residual phospholipids. Immunological properties of intact cells, separation of the "O" antigen and its components from them, show that a polysaccharide is the basic substance of the bacterial cell surface. The protein component of the "O" antigen as well as the phospholipid part of the cell wall is evidently located deeper.

In gram-positive bacteria the inner layer of the cell wall, close to the plasma membrane (the upper layer of plasma or cytolemma), is composed of protein molecules which contain a considerable number of basic amino acids and sulfhydryl groups. In this layer there are compounds which condition the gram-positive staining. The layer in question is covered on the outside with a second layer composed of magnesium ribonucleate, which is in the gram-positive complex.

On the outside of this second layer desoxyribonucleoproteins are located. With their aid specific polysaccharides are produced by the smooth (S) bacteria forms on their surfaces. This three-component layer of nucleoproteins represents the outer system of various enzymes and coenzymes. On the one hand the first phases of metabolism, synthesis and resynthesis of molecules from elements entering from the environment take place here, on the other hand, the decomposition of complex molecules leaving the organism also takes place here. In smooth forms of bacteria (S-forms) on the surface, the outer layer of desoxyribonucleic proteins produces specific polysaccharides which enter into the composition of the capsules. This synthesis is carried out by an enzyme, containing magnesium. In the absence of magnesium in the medium, cells appear, as a result of magnesium starvation, that do not contain the gram-positive substance and do not synthesize the capsular polysaccharide. In such cases the cells are gram-negative. having properties of rough forms and grow as R-forms (Stacey, 1949, Peshkov, 1955).

With the aid of electron microscopy it was revealed that the surface of the bacterial cell wall often has a fibrous structure. Separate submicroscopic fibers of a surface mucous substance, mucomucin, extend beyond the cell wall, in the form of slender filaments sometimes imitating flagella.

In this way we imagine the structure of the cell wall as a complex multi-layer formation, covering the living substance of the cell and having a sufficiently loose structure to let various compounds through and to secure an exchange of substances of the cell protoplasm with the environment.

The mechanical strength of the bacterial cell wall is relatively great. This is proved by the fact that upon disruption of the cells, the cell walls are often completely preserved. One may achieve a complete destruction of the cell, fully preserving the cell walls. For instance by shaking the bacterial cells with microscopically small glass beads the cells are fractionated, yet the cell walls remain intact. The cell wall is disrupted in one spot and the plasma leaks out through the opening. Upon rupture, the broken edges of the cell walls may be seen. This destruction with rupture of cell walls is also observed after ultrasonic and other mechanical action.

The cell walls of bacteria as well as those of higher plants are formed by the membrane of the periplast or cytolemma.

As was mentioned above, this bacterial surface excretes a mucous substance, composing the mucous capsule. When this substance is formed in large amounts, the capsule becomes thick, well outlined under the microscope, and the whole culture assumes a mucous character. In cases of scanty excretion of the mucous substance, the capsule around the cell wall is very small and not often visible by ordinary microscopy.

Mucous capsules of various bacteria differ in their chemical composition. Thus, for instance, the capsule of the acetic bacteria--*Acetobacter xylinum* consists of pure cellulose, the micelles of which were successfully revealed and photographed by electron microscopy, (Van Iterson, 1949). In the butyric acid bacteria *Clostridium pasterianum,* the mucous capsule consists of hemicellulose and may be stained blue with iodine (according to Peshkov, 1955).

The mucous capsular substance may be mechanically or chemically separated from the cell. It may be dissolved In aqueous, alkaline, or buffer solutions. From acid solutions it may be precipitated by alcohol. Upon acid hydrolyasis and subsequent chemical treatment, one succeeds in differentiating two groups of compounds in the capsular mucous substance. One group does not contain nitrogenous substances or contains them in negligible quantities. The substances of the other are composed essentially of nitrogenous compounds Knaysi et al., 1950).

The capsular substance of the first group in some root-nodule bacteria decomposes during hydrolysis with the formation of glucose, in pneumonia bacteria--with formation of galactose. Beside these two sugars, in the mucous capsular substance fructose and arabinose, or a mixture of these substances were found. The capsular substance of the second group of compounds found in the sporeforming bacteria, *Bacillus anthracis* and others, contains 7.4 - 8.0% nitrogen and is considered as a glucoprotein of the pseudomucin group (Kramer, 1921). A large amount of nitrogen was found in the mucous capsule of *Rhizobium leguminosarum* and in some lactic acid bacteria. In the capsules of bacteria, amino acids, ribonucleic and desoxyribonucletc acids are found. (Catlin, 1956, Smith et al., 1957).

The study of the capsules of many bacteria and primarily of pneumococci, is of special interest due to the presence in them of substances which condition virulence and the gram-positive staining. The chemical composition of the capsule determines the specific serological and antigenic reactions. The capsular antigen is connected with the phenomenon of gram-staining. An autolyzed culture of pneumococci, having lost the ability to stain gram-positively, becomes unable to provoke the formation of specific precipitins for the capsular polysaccharide when injected, subcutaneously, into a rabbit. Upon autolysis, 4-10% of the dry weight of the cells is obtained and this substance consists mainly of ribonucleoproteins and ribonucleic acid Dubos, 1948).

The ability of bacterial cells in the animal body to provoke the formation of precipitins for the capsular polyeaccharides is inseparably connected with the integrity of the gram-positive complex. This property in attributed to smooth forms (S). The rough forms (R) having completely lost the ability to form capsules, do not stain gram-positively.

Some investigators (Bisset, 1950) consider the cell wall a dead skeleton, a product of secretion of the cell protoplast. The data cited prove the contrary. The cell wall represents an organ of a living organism able, not only to carry out the function of a skeleton, but also to perform a series of purely biochemical processes of great importance in the metabolism of the cell, as well as in the exchange of substances between the inner and outer medium, between the organism and the substrate. The cell wall, together with the protoplast, comprises an entity--the bacterial cell.

The cell membrane plays a great role in the multiplication of the cell. It forms two transverse, protoplasmic, parallelly located threads, with a true system between them. This system appears as a result of secretion of transverse membranes and in fact consists of two thin septa which part and perform the division.

On reproduction of cells by formation of a septum, the cell membrane at the site of division is constricted inside the protoplast to complete closure or a small area in the form of a canal is left. In both cases the membrane forms the cell wall simultaneously with constriction inside the protoplast.

The transverse membrane in some bacteria in strongly thickened at the ends of the cell. Since it stains strongly with basic dyes, due to its basophilic character, in such cases it takes the form of caps or even polar bodies, assumed by some investigators to be nuclear elements (see Imshenetskii, 1950, Peshkov, 1955).

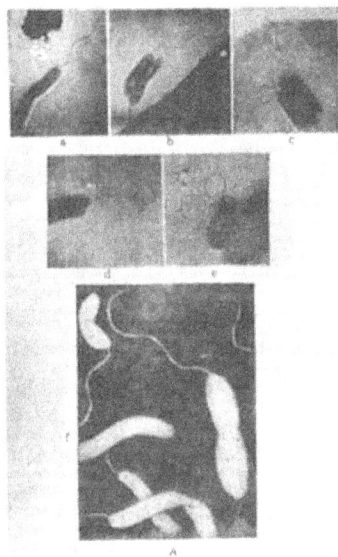

Figure 1. Polar flagella in bacteria: A. Monotrichous a) *Pseudomonas fluorescens* (1:3,000); b) *Pseudomonas malvacearum* (Azerbaijan strain, 1:9,000) ; c) *Pseudomonas malvacearum* (Fergana strain. 1:9,000), d) *Rhizobium trifolii* (1:3,000); e)*Rhizobium meliloti* (1:3,000); f) *Vibrio metchnikovii* (1:18,000, after Iterson, 1949).

Figure 1. (continuation) B. Lophotrichous: *Spirillum serpens* (after Iterson, 1949, 1:18,000).

Flagella in bacteria were revealed for the first time by Ehrenberg (1838); later, they were studied by many investigators in various bacterial species--sporiferic and nonsporiferic, spiral forms, some cocci and others.

According to the location of flagella, bacteria are divided into: monotrichous forms--those which have a single polar flagellum; amphitrichous forms--a single flagellum at both ends of the cell; lophotrichous forms--with tufts of flagella at the ends of the cell (Figure 1A, B); peritrichous forms--with flagella distributed over the whole cell surface (Figure 2). In monotypes one may frequently observe the flagellum located not on the end of the cell but at the side and sometimes in the middle of the cell. Such an anomaly appears in root-nodule bacteria, vibrio and others. The cause of this has not been clarified: it is not known if this is an anomaly caused by a pathological development or is an accidental event in the development and formation of the flagellum an a result of internal disturbances in the protoplasm or cell membrane.

According to our observations, the lateral flagella are formed by a dislocation of the polar flagella during the growth of the cell. The latter grow in length with the end attached at the point of growth. Sometimes the site of attachment of the flagellum is moved by some event from the side of the growth point and, as the cell lengthens, it moves farther away from the end (Krasil'nikov. 1932, 1935).

Flagella are formed by the protoplast, they are organically connected with the membrane and obtain impulses for movement from it. Contemporary methods of investigation elucidated that at the base of flagella there are granules located directly under the membrane. These granules are similar to the basal bodies of cilia in protozoa.

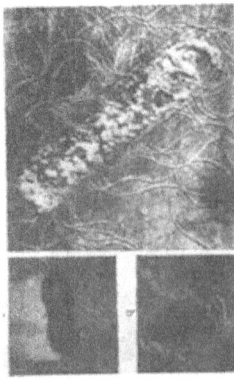

Figure 2. Bacteria with peritrichous flagella: a) *Bact. proteus* (1:17,900, after Iterson, 1949); b) *Azotobacter chroococcum* (Moscow strain); c) *Azotobacter chroococcum* (Central Asia strain, Vakhsha Valley, 1:8,000).

Loeffler (1889) established that flagella have a spiral-like form. They often stick together into tufts and locks which may be seen by ordinary microscopy without having been specially stained. As shown by recent studies, the separate flagella are of a more complex structure than was earlier assumed. In spiral bacteria the flagellum consists of many very thin elements, visible only under electron microscopy. There are 17-20 in one flagellum. A similar flagellum structure was described in protozoa.

The structure of flagella is different in sporeforming bacteria. According to observations of Roberts and Franchini (1950), the cells of *Bac. cereus* have flagella, consisting of a spirally-twisted axial thread and an outer layer. The length of the spiral twist equals 80 Å. The spiral consists of two very thin inter-woven threads. These spirals appear to be the motion apparatus of the flagellum.

The length of flagella differ in various bacteria. In some species the flagella are long, sometimes being more than a hundred times the length of the cell. In other species they are short, distributed on the surface of the cell in the form of bristles. The width of flagella is several Å or millimicrons, the length attains several tens of microns.

The development of flagella proceeds consecutively. At first, in young, developing cells they are very thin and short. Then, as the cell grows, they become longer and larger. The growth of flagella begins from the periplast, directly from the basal or kinetoblast body (Erikson, 1949).

With the aid of flagella the bacterial cells move actively in the liquid medium. The movement of bacteria is a progressive motion. The cells move rapidly or slowly forward, backward or sideways. More often than not, the movement is uneven, sometimes rapid, sometimes slow with sudden halts. The cells move by jumping.

There are indications in the literature that in some bacterial species flagella occur, but the cells themselves are nonmotile. Some authors assume that in such bacteria the flagella are paralyzed.

The movement may be stopped artificially by placing the cells in strong solutions of salts or sugars, by lowering the pH of the medium, by strong illumination by increasing or lowering temperature and by other means. Frequently, the flagella movement is stopped as a result of abundant slime formation. If they are grown on unsuitable media, bacteria may not display any motility. Often motility may only be seen in young cultures. as for instance observed in the hay bacillus *Bac. subtilis,* When the cells form long filaments, they become nonmotile. This in observed in *Bac. mycoides, Bac. megatherium, Bac. mesentericus, Bac. cereus* and other bacteria. Migula (1892) observed the formation of generations in *Bac. subtilis,* which, although they possessed flagella, remained nonmotile. The same was noted in *Micrococcus agilis, Sarcina mobilis* and other bacteria.

On the basis of the data cited, some investigators are inclined to assume that there are no bacteria in nature without flagella (Meyer, 1912). Some investigators tried to prove this assumption experimentally. Thus, for instance. Kobblmüller (1934, 1937) observed motility in lactic-acid streptococci considered to be motionless; Clark and Carr (1951) found that mycobacteria and corynebacteria--*Mycob. phlei, Mycob. fimi,* are motile. They claimed that they observed flagella in these organisms through electron microscopy. These data are as yet not confirmed and it to doubtful whether they are correct. Mycobacteria according to their nature are related not to bacteria but to the group of actinomycetes. One must assume, that the above investigators dealt with root-nodule bacteria or with Mycoplana. or they observed mixed (impure) cultures. Perhaps artifacts which were formed upon the treatment of the external cell-wall layer with reagents were taken to be flagella as Ptjper (1931, 1949) observed.

In many works Ptjper tried to show that bacteria do not possess flagella at all. and what in considered to be flagella are artifacts in the form of thin threads obtained an treatment, at the expense of the mucus of the cell capsule.

The theory of Ptjper was not confirmed by other investigators. Bolsches (1948. 1949) disproved Ptjper's statements experimentally. Weiball's a data (1948, 1949) on the protein nature of flagella and the works of Fleming et al., (1950), of Van Iterson (1949) and many others are in variance with Ptjper's a observations. The structure of the flagella, their subsequent development, and the connection with the membrane and basal bodies, an was mentioned above, all speak against the theory of Ptjper.

There are data in the literature, with the aid of which some investigators try, to prove that, in general, bacterial motility is caused not by flagella but by another mechanism, In particular, Ptjper explains the motility of cells by an outflow of mucus, an it taken place in myxobacteria and blue algae. The motility of microbial cells without flagella has for a long time attracted the attention of investigators. An yet, there is no full and clear idea on the mechanism of motility in these microorganisms. It in assumed to be essentially similar to that of diatoms and bluegreen algae, i.e. , of the reactive type. Myxobacteria evidently move by contraction of the whole cell peroplast. Wavelike contractions, accompanied by a longitudinal shrinkage and stretching of the protoplast. brings the cell to a sliding movement (Peshkov, 1955). On moving, jets of mucus are excreted. The cell moves according to the principle of recoil i. e , in the direction opposite to the direction of ejection of the mucus.

The chemical composition of flagella differs from that of the cell wall. The basic chemical component of flagella is protein; polysaccharides are not present.

On the basis of data of physicochemical analysis and study of X-ray spectrum. an individual flagellum is considered to be a gigantic muscle macromolecule capable of rhythmic movements. However, cystine, characteristic of muscle protein--myosin, was not detected in flagella. The protein nature of flagella structure is also indicated by immunological reactions. These reactions also reveal differences between the protein of flagella and that of the cell protoplasm.

Flagella protein is the basis of the H-antigen which provokes formation of specific antibodies in the animal body. The presence of these antibodies in serum causes the agglutination of the flagella of bacterial cells, and differs from O agglutination, during which the agglutination of the cells themselves takes place.

Protoplast

The bacterial cell contains plasma, which, with its inclusions is called the protoplast; in the immediate vicinity of the cell wall it is covered with a condensed plasma layer which is called the membrane. The plasma, as in all other cells, consists of a living substance of a very complex structure--proteins. polysaccharides, fats and other compounds.

The structure of bacterial plasma is also very complex and of rather great diversity, depending upon the bacterial species, age and the conditions of growth.

The plasma of young cells is optically homogeneous. there are no inclusions, no fat or volutin; they are without vacuoles. As the culture grows older, small granules of a different nature, and vacuoles appear in the plasma. In old cultures the plasma of the cells becomes fine-grained, strongly vacuolized with a larger or smaller content of various granules and bodies--volutin, chromatin; fat droplets and other inclusions stained with various dyes appear in it.

The detailed structure of bacterial protoplasm has not been investigated. It in known that protoplasm has great viscosity which varies greatly in various species. The lowest viscosity of the protoplasm, according to data of Gostev (1951), exceeds that of water approximately 3-4 times; in the majority of cases it exceeds the viscosity of water 800-8, 000 times. The viscosity of plasma depends directly upon the condition of the culture, its age, nutritional conditions, etc. As in other organisms, the viscosity of bacterial plasma changes strongly in response to external factors (temperature, mechanical damage), under the effect of radiant energy, chemical reagents and other factors.

The internal osmotic pressure of bacterial cells equals on the average 3-6 atmospheres. In some bacterial species it amounts to considerable values--300 atmospheres and more (Mishustin, 1947). The isoelectrIc point of bacterial protoplasm in the majority of species to of the range of pH = 3.0-4.0. The point of acid agglutination of bacteria also lies within these limits of pH. In smooth variants the point of acid agglutination lies a somewhat lower (pH from 3.0 to 4.0) than in rough forms (pH from 4.0 to 4.5). The specific gravity of bacterial plasma is 1.055.

The chemical composition of bacterial protoplasm is hardly known. More detailed investigations of the chemical composition have only begun in recent years. General data on the total chemical composition of bacterial plasma are known. For instance. the plasma of higher organisms consists essentially of protein substances, ribonucleotides, lipides, polysaccharides, fats and water. The latter constitutes 90-95% on the average.

Data of chemical analysis of the bacterial cell. obtained by Peshkov (1955), Belozerskii (1941) and others, show that the quantitative composition of the mentioned substances in the plasma changes, depending on the age of the cell. For instance, in a 5-hour-old culture of *B. coli,* the content of nucleic acids in the cell plasma is greater than that in cells of a 40-hour-old culture (respectively 22, 30 and 9.66%); on the contrary, the general amount of proteins increases with the age of the culture: in 5 hours--57.0 %, in 40 hours-70.4 % of the dry weight. The same relationships are also found in *Shigella.* The basophilic character also changes simultaneously with the change in content of these substances, and with it the stainability of the plasma. In young 5-hour-old cultures, the plasma absorbs aniline dyes better than that of 40-hour-old cultures. In young cells, just beginning to develop, the ribonucleic acids are firmly bound to the proteins. With the aging of the culture this tie becomes weaker (Belozerskii, 1941).

Detailed data on the chemical composition of the bacterial plasma are given in the books of Gubarev (1952), Gostev (1951), Kuzin (1946), Model (1952) and others.

Nucleus and Nucleoli

Different microorganisms have different nuclear structures. In protozoa as well as in fungi and yeasts there is a fully formed distinct nucleus with a characteristic internal structure and development. attributed to that of higher organisms. In blue-green algae it is represented by a primitive structure in the central part of the cell called the "central body".

The "central body" occupies the greatest part of the cell. It consists of a thin reticulum with distinct granules of chromatin distributed in the loops. This part stains well with basic dyes. Located on the periphery, close to the cell wall, is a thin layer of protoplasm. There are often granules of cyanophycin. Before the cell divides. the "central body" splits into two portions by the aid of a transverse septum (Figure 3).

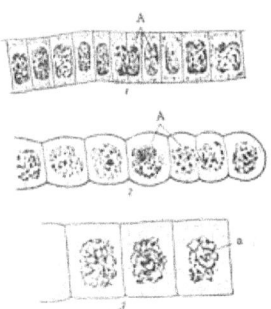

Figure 3. Central body in blue-green algae: 1--*oscillatoria;* 2--*Nostoc;* 3--*Oscillatoria;* A--central body; a--chromatin net.

In fungi and yeast the nuclear aparatus was thoroughly studied. It does not differ essentially from that of higher plants. As in the latter, the nucleus of yeasts and fungi have a vesicular structure and contain nucleoli; when dividing, chromosomes are formed in the nucleus with characteristic structural figures according to the phases of development, the division of the nucleus proceeds mitotically. Detailed investigations of the nucleus in yeast and fungi were performed by Guilliermond (1920). Until his investigations, the yeast cells were regarded as being without a nucleus; the distinct bodies found inside the cells were considered as nucleus-like

15

formations, but not as true nuclei (Kursanov. 1940; Nadson, 1935; Guilliermond, 1941). The conceptions of the nucleus of bacteria and actinomycetes are not as clear.

For a long time bacteria were regarded as organisms without nuclei. At the end of the nineteenth century, Büchili (1880) suggested that bacteria, like all other organisms, possess a nucleus. Ten years later, after a very careful study of the protoplast, he came to the conclusion that bacteria do in fact possess a nucleus. but not like that of higher organisms. According to his data, the nucleus in bacteria constitutes the central part of the protoplast and is constructed like the nuclei of blue-green algae. As in the latter, the central body or the prototype of the nucleus is surrounded by a thin layer of protoplasm directly adjacent to the cell wall.

This opinion was shared by many other investigators of that time--Weigert (1887), Tsetnov (1891), Frenzel (1892), Ruzhichka (1909), Mitrofanov (1093). Shevyakov (1893), and others (see Peshkov, 1955).

Fisher (1902) developed another point of view on the question of the existence of a nucleus in bacteria. According to his data, there is no nucleus in bacteria, or more precisely, the whole protoplast of the bacterial cell constitutes the nucleus. The author subjected the cells to microscopic analysis, but did not find any inclusions which resembled a nucleus in the plasma. According to his data, the separate bodies and granules mentioned did not have anything in common with it. Only some granules like the chromatin, stained with aniline dyes.

This opinion was also shared by Migula (1892). Like Fisher. he found only small granules which upon staining were similar to the nuclear substance. According to these authors, the nucleus or nuclear substance in bacteria exists in a diffused or fragmented state.

In developing the theory of the diffused nucleus, Hertwig (1902), proceeded from this analogy, with the formation of the so-called chromidia upon decomposition of the nucleus in some protozoa *(Heliozoa).* According to his opinion, bacteria do not possess an individual nucleus, but a nuclear substance, which is in the form of tiny granules or threads, and is distributed as a net (chromidial net) in the protoplast throughout the cell. The cbromidial net may occupy the whole cell or a greater part of it. In the latter case the plasma is located on the periphery. Schaudin (1902), conducted extensive investigations on the nuclear apparatus in the gigantic sporeforming bacillus--*Bac. bütchlii,* which he found; he confirmed what has been stated above. This microbe is 80 μ long and 3.5 u wide and is very motile; it is peritrichous. The protoplast of the cells consists of two parts: a peripheral portion, in the form of a light rim, and a central portion. According to the author's observations, the first represents a thin layer of plasma, the second--the central body, or a primitive nucleus. It has a vesicular structure, staining well with basic dyes, and small granules--the chromidia--are embedded in it. Prior to spore formation the granules of the central body move to the center of the cell and form a thread along the longer axis of the cell. This zigzag-like chromatin thread extends from one end of the cell to the other, stains strongly, and strongly refracts light. After some time the granules of the thread begin to move to the poles where they assemble and form one large body of round or oval form. The spore is later formed from this body. A central body was noted by Schaudin In the sporeforming bacillus--*Bac. Sporonema.*

The theory of diffused nucleus was also elaborated by Guilliermond, Swellengrebel (1909) and other investigators. At present this is accepted by many physiologists (see Imshenetskii, 1940). Swellengrebel described the chromatin thread, formed from chromidia in the sporeforming bacillus *Bac. maximus buccalis.* In other bacteria he found nuclear threads which were formed either from round nuclear bodies or from chromidia.

Guilliermond, an outstanding specialist in cytology, did much work investigating the protoplast of protist organisms--algae. fungi, yeasts and bacteria. In bacteria he did not find nuclei. but found a chromatin substance which was present in the protoplasm in a soluble or fragmented state. During spore formation Guilliermond observed precipitation of chromatin in the form of separate granules.

Contrary to the opinions on the diffuse structure of the nucleus just cited , there are in the literature statements as well-founded, on the presence of a well-defined, fully formed nucleus in bacteria.

This point of view was expressed for the first time by Meyer (1897). According to him, bacteria possess a true nucleus similar to that of higher organisms. He based his conclusions on data from analysis of fungal organisms in which the nuclei are very well defined and manifest themselves clearly. The author assumed that fungi and bacteria are organisms phylogenetically close to each other. If nuclei occur in fungi they should occur also in bacteria. As an object of his investigations, Meyer chose the large sporeforming bacillus *Bac. asterosporus.* He established the presence of defined bodies, which he regarded an nuclei inside this bacillus. He counted from 3 to 4 such bodies of a diameter of 0.3 µ in the cell.

The followers of these opinions extensively developed the theory of a defined nucleus in bacteria. At present this theory is the most popular one.

Data recently obtained by electron microscopy are of interest. By introducing some improvements, it was possible to disclose and differentiate very small separate cell structures.

It is known that the most characteristic substances in the composition of the nucleus are nucleic acids, namely thymonucleic or desoxyribonucleic acid in the nucleus, and ribonucleic acid or plasma acid in the protoplasm.

The penetrability, i. e., transparency to a beam of electrons, of the nuclear or chromatin substance which consists essentially of thymonucleic acid, differs from that of plasma nucleic substances (ribonucleic acid). Owing to this it is possible to differentiate and to discern nuclear from nonnuclear elements.

In the study of the nucleus and nuclear substance of bacterial cells great importance is attached to chemical methods. By chemical reactions one succeeds in revealing the separate components of the nucleus, and in establishing their chemical nature. Among the chemical reactions, the Feulgen reaction is noteworthy. It is based on the hydrolysis of thymonucleic acid by hydrochloric acid. During hydrolysis guanine and adenine, as well as the polysaccharide fraction, are released. The latter, when treated with fuchsin sulfate acquires a bright pink-violet color. This stain is characteristic of aldehydes, and, consequently, shows that the polysaccharide part of thymonucleic acid consists of aldehydes.

Feulgen (1926) applied this reaction in order to reveal thymonucloic acid in protozoa and obtained a positive result. He did not find this substance in bacteria and yeasts, on which he based the erroneous conclusion that in bacteria the essential nuclear substance was absent.

Subsequent investigations of many other authors showed that this conclusion was at fault.

By this method thymonucleic acid was found in bacteria of various groups and species, in actinomycetes, yeasts, fungi and other representatives of microorganisms (Imshenetskii, 1940; Dubos, 1948; Guillermond, 1941 and others).

The studies showed that the distribution of thymonucleic acid in bacterial cells as established by the Feulgen method fully corresponds to the localization of structural formations revealed by microscopic methods. Depending on the bacterial species, their age and growth conditions, thymonucleic acid in distributed either diffusely or is concentrated in the form of small bodies and masses of various dimensions and configurations.

A valuable microchemical method for the establishment of the composition of nuclear substance is the enzymatic treatment. Pepsin with hydrochloric acid (stomach juice) dissolves bacterial cells and proteins of various composition, but not the chromatin substance. Neither does it dissolve nucleic acids. Peshkov (1955) applied ferments of nuclease to discern nuclear substances in basophilic bodies of bacteria.

A macrochemical method is also applied for the investigation of the nuclear substance. Belozerskii (1944, 1945) extracted nucleoproteins from the bacterial mass with alkali and separated them into two fractions. nuclear and protoplasmatic. The first contains a typical thymonucleic or desoxyribonucleic acid. The second - only the plasma or ribonucleic acid.

Recently another very sensitive method of recognizing nuclear substances was suggested--the method of spectroscopy. It is based on the ability of the chromatin substance or nucleoli and other nuclear structures to absorb certain parts of the spectrum of ultraviolet light. By this method it has been successfully   shown that the nucleoli, and, in general, the nuclear substance of bacteria, has exactly the same absorption spectrum (in the region of 2,600Å) as that characteristic of the chromatin of higher organisms (Peshkov, 1955).

As a result of the application of all these methods in the cytological analysis of the microbial cell, one may obtain quite convincing date on the presence of a nucleus in bacteria. At present one can definitely say that bacteria possess a nuclear substance, not differing in its composition from the nuclear substance of higher plants. However, in contradistinction to the latter, the nuclear substance of chromatin in bacteria in not always distributed in the form of defined bodies or organelles. It occurs in various forms depending on the species, age of cells, and growth conditions of the culture. The chromatin may occur in bacterial cells in a diffuse state or in the form of tiny bodies and granules distributed throughout the protoplast of the cell. At given stages of development the nuclear substance is present in bacterial cells in the form of defined structural formations, having dimensions, forms and patterns characteristic of the true nuclei of other organisms--fungi, yeasts and others.

Such distinct structural formations of chromatin, in contradistinction to true nuclei are called bacterial nuclei or nucleoids. The most characteristic and convincing proof that these formations are cell organelles and, not some other inclusion, is their ability to reproduce themselves during the process of the multiplication of the cells. This property is known to be characteristic of every cell organelle and primarily of the nucleus. This distinguishes it from superficially similar nonliving inclusions as for instance reserve food substances, various metabolites, etc.

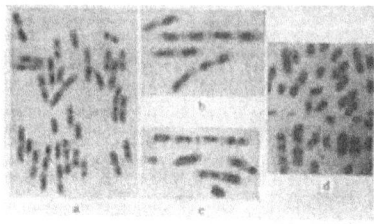

Figure 4. Division of nucleoli in bacteria (after Robinow, 1942): a) *Bac. cereus;* b) *Bac. mycoides;* c) *Bac. mesentericus;* d) *Bact. proteus.*

As shown by many recent investigations, nucleoids multiply by simple division or constriction, in other words, by amitosis. The round body extends and becomes rodlike; a furrow appears in the center, with the aid of which the splitting into two bodies occurs (Figure 4). Sometimes one may observe a splintering of the nucleoids, i. e. , division of the chromatin mass into several pieces simultaneously (Figure 5). The nuclear substance of the nucleoids may reproduce by budding. A protrusion appears on the surface of the body which gradually increases in size and after reaching certain dimensions is pinched off from the parent body. From one body several daughter nucleoids may be pinched off.

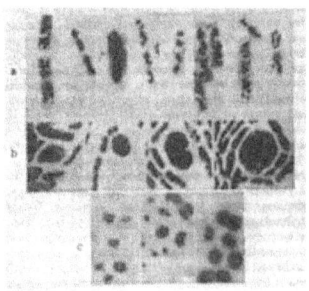

Figure 5. Splintering of nucleoids in bacteria: a) *Bac. megatherium* (De Lamater, 1951); b) *Bac. proteus* (after Stempen and Hutchinson, 1951); c) *Microc. cryophilus* (De Lamater, 1951).

Budding and splintering of the chromatin mass of nucleoids is usually observed in distended, so-called involution, cells. Such forms of reproduction of nucleoids are very often accompanied by formation of special spores, reproductive or regenerative bodies, inside the cell. These formations were observed by us in *Azotobacter,* root-nodule bacteria and actinomycetes (Krasil'nikov, 1932 d, 1954 e).

Following the division of nucleoids there is cell division. In bacteria a plural cell reproduction is often observed, multiplication by fragmentation when the cell divides simultaneously into several daughter cells. Such division is observed in micrococci, mycrobacteria, sporeforming bacteria--*Bac. megatherium, Bac. mesentericus* and others, in *Azotobacter* and in certain filamentous bacteria. In these cases in every daughter cell one small chromatin body or nucleoid may be observed (Peshkov, 1955; Robinow, 1942; Knaysi, 1950; Bisset, 1950; Pickarski, 1937, and others).

Nucleoids are always formed during spore formation. The fragmented chromatin in the form of tiny granules concentrates in small bodies in a part of the cell, more often in the part where the spore is formed. These bodies become rounded; a thin rim of plasma appears and then the cell wall is formed. A ripe spore is obtained. In all species of sporeforming bacteria, mycobacteria, actinomycetes, and, one should assume, in all other microorganisms which produce spores internally or have other reproductive bodies, there are always nucleoids in the mature spore.

The internal structure of the "bacterial nucleus" is not apparent; the whole body represents a homogeneous mass, strongly stained with basic dyes. Inside such bodies neither chromosomes nor nuclear grains are found.

Some investigators try to show the presence of defined structures inside the "bacterial nucleus", they describe chromosomes and various bodies, and are of the opinion that these bodies are like the chromidial net of true nuclei. By making comparisons with nuclei of higher organisms, mitosis with its various phases was described in bacteria. For instance. De Lamater (1951-1952) notes in *Bac. Megatherium,* a prophase, metaphase, anaphase, telophase, then formation of a typical spindle with centrioles. He observed this picture of nuclear division in micrococci, *Micr. cryophilus* and in *Bac. coli.* The author indicates that various bacterial species contain a different number of chromosomes in the nucleus. Analagous data on structure and development of the nucleus in bacteria are presented by Lindegren (1950) and some other investigators.

The material cited by these investigators is not convincing. The structural changes of the chromatin accumulation bodies described by them, are of a quite diverse character. Chromatin or chromatin-like substances in bacteria quite often assume various and rather indefinite configurations and dimensions. These changes are without any regularity and connection with the process of cell division. Among the various formations of chromatin accumulation one may always find fortuitous figures somehow reminding one of the forms of one or another phase of nuclear division.

Bisset (1953) could not confirm the data of De Lamater. He considers his classification of the structures observed inside the cell during the process of division as erroneous. What De Lamater assumed to be centrioles, proved in fact to be rudiments of transverse septa. Neither could Bisset find mitochondria in mycobacteria as did De Lamater.

The nuclear substance in the form of nucleoids occurs in the cells for a short time, usually only in the period of their early development. As the culture ages the chromatin substance in the calls increases noticeably. It often occupies a considerable part of the cell, almost filling it completely. In such cases the chromatin substance, as a rule, has indefinite patterns and configurations. Large masses of chromatin stain densely with nuclear dyes, giving the Feulgen reaction, and acquiring a loose vesicular structure, often the whole mass is broken into separate parts and small lumps or separate small pieces. Some authors consider these enormous accumulations as one large nucleus, and consider its fragmentation as cell division (Peshkov, 1955, Bisset, 1950, Robinow, 1951, Pickarski, 1937 and others). However, it is quite impossible to agree with this. As a rule, the enormous chromatin aggregations are observed during the abnormal development of a culture, when the cells are in a state of involution. They are always observed during unfavorable growth conditions, under the effect of increased temperature, irradiation with ultraviolet, radium X-rays, etc. An active and rapid formation of chromatin substance occurs under the influence of phages on bacteria and actinomycetes. According to observations of Gerchik (1945). a penetration of the phage tail into the cell body suffices to provoke in it, all the indicated changes. After several minutes of contact between the protoplast of bacteria or actinomycetes with the phage, large lumps of chromatin or chromatin-like substance are formed. The cells swell and assume unusual forms and dimensions.

As a rule the accumulation of chromatin substance in cells is accompanied by a decrease in their viability.

Formation of such a large amount of nuclear substance can hardly be regarded as a normal function of development and cell multiplication. It must be assumed that this process reflects an abnormal development and disturbance of certain biochemical reactions of metabolism. The chromatin accumulations in the form of various masses represent a result of an unnatural metabolism and in no case, a phase in the development of the nucleus. In these small masses distinct rudiments of germs, the so-called regenerative bodies, may be formed.

Nucleoids have been described in bacteria of the coli group - *Bac. coli, Bac. typhi, Bac. dysenteriae, Bact. proteus* and other species.

We have observed nucleoids in actinomycetes, as a rule in young individuals. They are particularly noticeable when the organisms are cultivated in a liquid synthetic medium. In the threads of the mycelium separate bodies which are located far away from each other are revealed. One may often see two nucleoids closely located or even fused together. We regard this phenomenon as the process of division of these formations. In old cultures the threads do not possess definite nucleoids, instead granules, or irregular small blocks and large aggregations of chromatin are found. They are dispersed in a disorderly manner over the entire mycelium or in separate threads only. These chromatin aggregations are not organelles of normal cell development, but represent aggregations resulting from a degenerative change of the whole or part of the protoplast, namely that connected with the nucleoproteins.

It was shown experimentally that with the disorderly accumulation of chromatin masses in the cells, biochemical processes noticeably change as well as the nature of exchange and formation of metabolic products. This may easily be seen particularly in actinomycetes during their process of formation of antibiotic substances. The indicated changes in chromatin structures of the actinomycete mycelium are quite constant and regular during culture development and are often used in antibiotic practice as an indication of the current state of the process. It must be assumed that chromatin may possibly be one of the potent factors of metabolism regulation. Unfortunately we know very little of the relationship between the biochemical processes and the formation and aggregation of chromatin substance in the cell.

There are bacteria in which the nuclear substance is distributed in the form of a central body in the cells as in blue-green algae. Such nuclei have been described by us in *Pontothrix longissima,* in *Oscillospira guilliermondii,* and in *Anabaeniolum sp.* These bacteria consist of threadlike individuals of various lengths. Every individual or thread is divided by transverse septa in a series of short cells, whose length usually does not exceed the width and is more often less than the latter. The internal structure of such cells reminds one of the structure of the protoplasts of blue-green algae. A great part of the cell is occupied by the central body. On the periphery the plasma is distributed in the form of a thin layer. The central body stains well with aniline dyes and is easily found by ordinary magnification of the microscope owing to its dimensions (Figure 6).

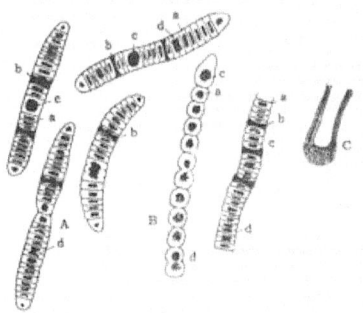

Figure 6. Central body in A--*Oscillospira guilliermondii,* B--*Anabaeniolum langeroni* and in C--*Pontothrix longissima* a--central body; b-nucleoids; c-prospore; d--division of central bodies.

Such a distribution of chromatin has also been described in *Caryophanon* by Sall and Mudd (1955).

In many bacteria so-called polar bodies or small grains at the ends of the cells are noted. Like chromatin, they stain densely with stains. The role of these bodies is not exactly known. Some authors regard them as nuclear structures, some others as volutin. There are indications that polar bodies represent a specific enlargement of the cell wall. Recently many investigators are inclined to regard these enlargements as points of growth of cells, directing the process of elongation of the individuals.

Thus, one may conclude that the nuclear apparatus in bacteria and actinomycetes is of a peculiar nature. It differs strongly from the nuclear apparatus of fungi, yeasts, protozoa and higher plants or animals.

The nucleus in bacteria and actinomycetes is primitive. It does not have a constant, established, structural formation. It occurs either in a diffuse dissolved state, or in the form of small grains and bodies which are dispersed all over the protoplast of the cell, or in the form of an organized intracellular organoid-nucleoid. In the latter case, such a body, according to its external appearance, form or, dimension, and patterns, calls to mind the true nucleus of fungi or yeasts. However, in contradistinction to the latter, the nucleoids of bacteria are structurally undifferentiated, the characteristic formations of true nuclei as for instance nucleoli, chromosomes or other structures, are not found in them. The most convincing proof of the nuclear nature of nucleoids is their ability to reproduce during the process of cell division.

It is a characteristic property of nucleoids that they are often reformed from diffusely dissolved or dispersed chromatin during spore formation in sporiferous bacteria, mycobacteria and actinomycetes; they are also organized inside the vegetative cells at certain stages of their growth.

It should be noted that some forms of manifestation of the bacterial nucleus are attributed also to the nuclear apparatus of higher organisms.

Unusual methods of cell multiplication and division of the nucleus have been noted long ago in the cytology of the plant cell. Under certain conditions the nuclei of cells of some tissues begin to multiply in a way not peculiar to ordinary cells of higher organisms; this multiplication is neither by complex development and formation of various phases, nor by mitosis, but by a simple division or splitting of the nuclear mass, or amitosis. In amitosis the nuclei multiply by division, constriction, fragmentation and budding. Amitosis, as shown by recent investigations, is widespread among the representatives of the vegetable kingdom. It in found in cells of animal organisms (Polyakov, 1949; Usov, 1924; Ellenhorn, 1951; Glushchenko et al. 1953 and others).

Amitotic division of the nucleus is found in the cells of callus, cicatrization, regeneration and in tissues formed anew. In these cases the cells undergo a series of cytochemical and morphological changes. In the nucleus processes of direct division or fragmentation go on and several daughter nuclei are formed consecutively; sometimes the nucleus splits simultaneously into several daughter nuclei. The latter separate and are distributed in various parts of the plasma, where they become the centers of formation of daughter cells. According to Glushchenko et al, (1953) the budding of nuclei was established in the cells of cicatricating tissue of the potato fiber. Such nuclei undergo deformations, swell. acquire various patterns, and oarlike or budlike protrusions appear on their surface. The latter gradually become rounded, pinch off from the maternal nucleus and transform into separate daughter nuclei.

Ellenhorn and Zhironkin (1953) showed that in certain cases, upon the development of the primordial root, its cells possess no nuclei at all. These cells, lacking nuclei, multiply by fragmentation. In the subsequent development of the rootlet the cells lacking nuclei form primitively organized nuclei or "protokaryons". The "protokaryon" multiplies by simple division or binary fission. Mitosis is absent. Only afterward when the rootlet develops sufficiently do the primitive nuclei begin to multiply mitotically in their cells.

Cells without nuclei were found by Glushchenko in tissues of lentils of black currants, where they form rootlets, and later in cells of cicatricating Neder's tissue. These calls multiply by fragmentation in the early state of rootlet formation without any indication of the presence of a nucleus. Also, according to the experiments of Ellenhorn, after some time primitive nuclei are formed anew in the cells. They multiply by simple division, binary fission. In these nuclei--"protokaryons", neither nucleoli nor chromosomes are present.

In these instances (there are many in the literature) some similarity of nuclear formation and structure in plants and bacteria or actinomycetes has been shown. As in the latter. the nuclei of plant cells may be formed anew from chromatin substance diffusely distributed in the protoplast. Such a way of formation and development of primitive nuclei in plants evidently reflects the picture of the early stage of evolution and formation of cells.

Gram-staining of Bacteria

During the elaboration of methods for differentiating bacterial cells in tissues of animal organisms, Christian Gram (1884) and his collaborators suggested a special method of staining. By this method the division of all bacteria into two groups, gram-positive and gram-negative bacteria. has been established.

This staining method afterward underwent some changes and improvements, but in principle remained the same. The technique of staining is an follows: the bacterial cells are stained with a slightly alkaline basic stain such as crystal-violet; then they are treated with mordant iodine in the form of potassium iodide, or with picric acid. The stained preparations are washed with water and neutral alcohol or acetone. By this treatment the stain in removed in some bacterial species and retained in others. The former are called gram-negative, the latter, gram-positive bacteria.

The fact that different bacteria retain the stain in a different manner is important, not only from the point of view of the problem of staining, but also in its broader significance, since it indicates a chemical difference in the cells of these microorganisms. The gram stain is an indication of the biological or hereditary properties and state, in other words, of the nature of the organism. These properties should be used, not only as a diagnostic, but also as a systematic indication in bacterial taxonomy.

In clinical laboratory practice, this indication is widely used and yields good results. However, it is not taken into account in classification or only considered to a small degree since it in undoubtedly underestimated here.

The ability to absorb and to retain the stain according to Gram is characteristic of many organisms of the bacterial class of almost all actinomycetes and mycobacteria, fungi and yeasts. The majority of the cells of higher plants and animals are gram-negative. However, separate inclusions in the cells of these organisms, in particular the nucleus, the nuclear hyalin and the nuclear substance, etc, are gram-positive. Viral proteins, bacteriophages and actinophages are gram-positive.

Many other bacterial properties are also related to gram staining. Thus, for instance, gram-positive bacteria are more resistant to the lysing effect of alkalies and the proteolytic effect of enzymes--trypsin, pepsin, pancreatic juice; they are more sensitive to many inhibitors--to antibiotics, aniline, phenol, ethanol, toluene, benzene, xylene, chloroform, ether, iodine, to basic dyes and other substances. Concentrated in the cells of these bacteria are such amino acids as arginine, glutamic acid, histidine, lysine and tyrosine. Upon staining with methylene-blue and eosin, they become more sensitive to light, etc, (see Bartholomew and Mittwer, 1952).

The property of gram staining may change to a considerable degree even in the same species, depending on the age of the culture, the nutrient medium and other external conditions. For instance, on unfavorable media, bacteria often lose the ability to stain gram-positively or they stain weakly. The hay bacillus--*Bac. subtilis* becomes gram-negative when grown in a medium with immune serum (Simonini, 1914). Some authors observed gram- positiveness in *Bac. coli*, grown in a liquid protein medium with a high content of glucose and the salts $MgSO_4$, NaCl. The sporeforming bacillus--*Bac. cerus*, after having been in distilled water or tap water, stains gram-positively rather weakly, and many cells become completely gram-negative. How water acts is unknown. Whether the gram-positive substance in used as a nutrient during starvation, or if this substance is dissolved as a result of autolytic processes which proceed quite intensely under these conditions was not elucidated.

Knaysi et al. (1950) established that small doses of benzimidazole added to the Dubos medium transforms the tubercle bacillus--*Mycob. tuberculosis,* avium type - from a gram-positive, acid fast form into a gram-negative, nonacid fast form. The authors assume that benzimidazole inhibits the synthesis of ribonucleic acid--the essential ingredient in the composition of the gram-positive substance.

As seen from the previously-mentioned data, the medium is of great importance in connection with the gram stainability of cells. A medium not sufficiently well chosen may mislead the investigator in his differentiation of bacterial species. Therefore, in all doubtful cases of uncertain staining, the use of several nutrient media in recommended. The gram stainability is also greatly affected by the age of the culture. As a rule, young cells stain more strongly than old ones. A 24-hour culture is more gram-positive than a two-to three-day culture and a five-to six-day-old culture even more so. In some cases old cultures are more gram-positive than young cultures. In the sporeforming bacteria--*Bac. mesentericus, Bac. subtilis* and other species; the gram-positive substance occurs in cells in the sporulation phase and even after the spore had already been formed, i.e. , in the residual plasma (epiplasma). This substance is not decolorized on treatment for 10 minutes with 95% ethanol.

The better gram stainability of young cells is evidently caused by the basophily of the protoplast and its ability to strongly absorb stains in general.

The stainability of cells depends to a considerable degree upon their individual features. Two cells of the same age placed side by side often stain with a different intensity. The degree of the stainability changes depending upon the duration of the decolorization, the fixation method and, in general, upon the preparation of the cells and reagents. Thick smears take longer to decolorize than thin ones.

Gram stainability is connected with the species characteristics of organisms. Certain species, for example gonococci and certain mycobacteria easily change their staining characteristics when grown under different conditions, while others preserve those characteristics in a more or less stable fashion. Consequently, the gram staining depends not only on the staining technique but also on the species characteristics of the culture and the properties of the cell substance. It is obvious, that when gram stainability is being determined one should adhere to certain standards in the methods of making the preparations, as well an in the following procedures--fixation, staining. decolorization and others.

The essence of gram stainability in spite of numerous investigations performed, has not yet been elucidated, Various opinions and theories have been expressed. They can all be reduced to three basic ideas: a chemical theory, an isoelectric theory, and that of cell permeability.

The chemical theory of the gram-positive staining of bacteria in based on the particular composition of the cell plasma, According to this theory, there are in the cells of gram-positive bacteria particular substances which retain the stain and do not release it, or release it with difficulty upon washing. Some authors relate the gram stain to the presence of fatty acids of lecithin or lipoproteins in the plasma. These substances strongly combine with the stain and iodine and do not decolorize on treatment with ethanol, Schumacher (1920 isolated fatty acids from the cells of yeasts; afer this treatment the cells lost the ability to stain gram-positively. However, the same cells regained their gram stainability after treatment with fatty acids. According to the works of Peterson (1955) and of other authors, the role played by unsaturated fatty acids in gram stainability was refuted In connection with the fact that gram-negative bacteria do not contain fewer of these acids than gram-positive bacteria,

Recently gram stainability has been attributed to a particular nucleoprotein gram-positive substance, ribonucloic acid, more precisely, to magnesium ribonucleate. It was established that by hydrolysis or treatment with bile salts one can deprive the cells of the gram-positive substance and transform them into gram-negative (Denesen, 1948; Stacey et al., 1940; Stacey, 1949). Bartholomew and Umbrait (1944) removed the magnesium ribonucleate by crystalline ribonuclease and transformed the cells into gram-negative ones. The artificially removed magnesium ribonucleate may be returned to the cells.

Although this theory sounded convincing. it proved to be groundless. It was revealed that when the gram-positive substance is separated from the cells, it does not transform naturally gram-negative bacteria. In the latter there is no less ribornucleic acid and magnesium ribonucleate. These substances isolated from bacteria of the coli group do not transform gram-positive cells (after previous ,separation of the gram-positive substance) for instance *Clostridium* cells--*Clostridium welchii* (Jones, Mugglestone and Stacey, 1950).

Henry, Stacey and others came to the conclusion that the gram stain depends upon the ratio of nucleoproteins to ribonucloic and desoxyribonucleic acids. In streptococci and in *Clostridium--Clostridium welchii* this ratio is 8: 1. and, in gram-negative bacteria--1: 3 (Stacey, 1949; Henry, Stacey, 1943). Other authors did not confirm the regularity of these relationships.

Mitchell and Moyle (1950) attribute greatest importance to a particular substance, "XP", of unknown nature and containing phosphorus which is combined with the ribonucleate of the gram-positive substance. According to their data. the substance "XP" is always found in the cells of gram-positive bacteria. However, subsequent studies showed that this substance also occurs in large quantities in gram-negative bacteria and in some of them- -*Bac. coli, Bac. aerogenes, Nelsseria catarrhalis*--even more is found than in the gram-positive bacillus--*Bac. subtilis* and in brewer's yeasts. Shugar and Baranowska (1954) came to the conclusion that a decisive role in the gram stain belongs not to, ribonucloic acid but to the protein of the cells.

The theory of the isoelectric point, suggested by Stearn and Stearn (1926-1931) is based on the fact that the isoelectric state of the plasma in gram-positive bacteria differ from that in gram-negative bacteria. The authors determine this state of the plasma by the call's ability to stain or to adsorb acid and basic dyes, at various values of pH. If one determines the pH of the cell protoplast at a stage when it absorbs basic and acidic dyes to the same degree, then it may be noted that In gram-positive bacteria this state occurs at pH 2 and in gram-negative at pH 5. Bacteria whose gram stainability is not completely clear occupy an intermediate position (according to Dubos, 1948).

It should be noted that the establishment of the isoelectric point is carried out in the cell treated with iodine and, if it conditions the result of the gram stain, it does so in nonliving cells. The applied fixation method by iodine causes oxidation of plasma. Oxidizing agents are also such substances as bromine, picric acid, potassium bichromate, tri-nitrobenzene, etc. These substances oxidize some components of the protoplasts, render it more acid and shift the isoelectric zone. As shown by Stearn and Stearn (1931), the displacement occurs in all bacteria. but is expressed more strongly in gram-positive bacteria than in gram-negative ones. Due to this fact, the difference in the degree of acidity in given groups of bacteria increases considerably under the action of iodine and other mordants. This fact, if true, only shows that the biochemical condition of the plasma of gram-positive bacteria differs from that of gram-negative bacteria. In other words, the living substance in the former and the latter organisms is different.

The given theory of gram stain of bacteria has not been sufficiently confirmed by a series of experiments. First, some investigators did not confirm that iodine as an oxidizing agent may be replaced by other reagents in the same treatment of cells. Second, if the theory is true, then iodine should also produce this effect before staining, but this does not occur (Bartholomew and Mittwer, 1952). Further, if the theory in correct, then decomposed cells or the content of decomposed cells should be stained by this staining procedure as well an the plasma of whole cells, however, in fact, this is not observed; the protoplast of the disrupted cells is not gram-stainable.

At the basis of the permeability theory is the different ability of the cell wall and cell membrane to be permeated by stains and mordants. A suggestion was put forward that, in the cell, insoluble precipitates of the stain and iodine are formed, which are not washed out by ethanol. It in known that iodine does indeed form a colloidal complex with methylviolet which in not soluble in water. In favor of this theory are some other facts; for instance, that decomposed cells are gram-negative. It is sufficient to destroy the integrity of the cell wall by some method (grinding with sand, autolysis, lysis, etc) to make the cell lose its ability to stain gram-positively. Benians (1920) divides bacteria according to their stainability into three groups.

The first group comprises bacteria whose cell wall allows stains and iodine to permeate but the molecules of iodine and stain combined inside the cell are retained. As a result. the whole protoplast remains stained after washing with ethanol. The second group of bacteria possess cell walls which do not allow stains to permeate and. due to this, are easily decolorized upon gram staining. In the third group of bacteria, the cell wall allows the stain and iodine to permeate, but the complex formed inside the cell is freely released upon decolorization.

Stearn and Stearn refute Benians' point of view on the basis that iodine and the stain do not form such large particles that they cannot penetrate into the cell. Besides this, the addition product of iodine with the stain dissociates in ethanol. Consequently, on washing the preparation, the obtained complex should also be dissolved and go out through the cell wall.

It was suggested that the permeability, for iodine itself, of the cell walls of gram-positive and gram-negative bacteria is different. For instance, in order to prevent the exit of the stain from the cell of gram-negative bacteria a strong concentration of 0. 01% or less of iodine in methanol is sufficient (Mitchell, Bartholomew. Kallman, 1950).

Summing up, it may be stated that the essence of the gram stain has not been solved until now. No single hypothesis or theory explains the diversity of cell stainability in various species and bacterial groups.

It must be assumed that the ability of gram staining is determined by the property of the whole protoplast and cell wall together and not by any one part of the cell. The degree of stainability or stability of the combination of the plasma with the stain depends upon external growth conditions of the culture, its age and other causes.

GROWTH AND MULTIPLICATION OF MICROORGANISMS

Microorganisms, like all other living creatures, grow, develop, multiply, undergo changes, age and die. They have their own cycle of development. One must differentiate between the growth and development of separate individuals or microbial cells and the growth and development of their cultures. The latter consist of an enormous number of separate individuals and represent large populations or associations of various individuals.

It is quite natural that the character and regularities of growth of separate cells and whole associations of the same microbial species sharply differ.

Growth of Cells and Development of Cultures

The growth and development of separate bacterial individuals are very simple in their external manifestations. When they are observed under laboratory conditions of growth on nutrient media, the following may be noted. The cells elongate, reach definite dimensions and start division. The daughter cells repeat the same cycle and this continues until the nutrient substances become exhausted, the medium and environmental conditions change.

In sporeforming bacteria spore formation occurs at a definite phase of development. Upon inoculation of a fresh nutrient substrate, the spores grow into rodlike individuals, which begin the cycle anew.

The growth process of the microbial cell is characterized by an increase in its volume. The separate cells increase in length; the growth occurs at the ends. It was established that in bacteria there are accumulations (polar bodies), at the ends of the cell, which are of a particular substance, and stain a dark color with aniline dyes. This substance is often regarded as chromatin or metachromatin. The studies of Bielig et. al., (1949), Bergersen (1953) and Bisset (1953) proved that these bodies are related to the growth of the cell. According to their observations, intense biochemical processes take place in this area and the body of the cell increases there. Because of its properties, this part of the cell may be regarded as a growth point. Such distal cell growth occurs in actinomycetes, protoactinomycetes and mycobacteria. The elongation of the separate hyphae at the growth point in easily noticed upon observation of a hanging drop. With the growth of the mycelium the length increases just at the distal part, from the youngest apical branch to the end of the hyphae; the distance between consecutive branches of the same sprig remains unchanged (Figure 7).

We observed growth at the end of cells in *Azotobacter,* in purple sulfur bacteria (Krasil'nilkov, 1932, 1935) and in some other bacteria. Streshinskii (1955, 1956) observed such growth of cells in sporeforming bacteria-- *Bac. subtilis* and *Bac. megatherium.*

In spherical bacteria (family Coccaceae) the growth point may occur at various areas on the periphery of the cell. Respectively their division takes place in various planes, in sarcina the growth is orientated in three perpendicular directions, in streptococci in one direction, and in micrococci--in all directions.

A multilateral increase of the volume of cells occurs during the growth of the culture in yeast, which proliferate by budding. As known many yeasts multiply by budding. On the surface of the cell a small knob is formed--the bud. The dimensions of the bud increase, its growth proceeds uniformly in all directions, or unequally. In the former case spherical cells are obtained and in the latter-oval or oblong ones.

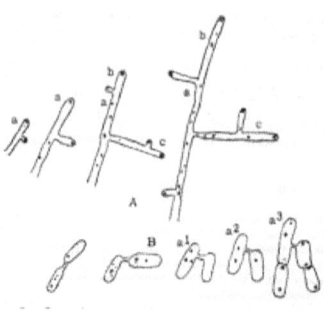

Figure 7. Growth of microbial cells. Body increase at the distal part of the cell, at the "growth point" A. *Actinomyces streptomycini:* a, b, c--cell increase from the site of sprig formation; B. *Azotobacter chroococcum* growth of the distal part of the cell: a^1. a^2 and a^3--increase of the cell at the "growth point" from the site of attachment of the septum.

Under conditions of normal growth the cells of microorganisms start to multiply when they reach certain dimensions. At this stage of life various cytochemical transformations occur inside the cell, which lead to the maturation of the individual. The state of its nuclear apparatus changes, and a redistribution of chromatin and other structural substances occur. These processes have been little studied in bacteria, but one must assume that they are no less complicated than in higher organisms.

The cells under conditions of normal growth are of definite dimensions. For many species or groups of microbes this value is quite constant. For instance in representatives of the *Pseudomonas* bacteria the dimensions of the cell are on the average 3-5 x 0.7 μ, while the representatives of sporeforming bacteria of the group *Bac. subtilis* and *Bac. mesentericus* consist of larger cells--5-7 x 1.0μ. In representatives of the *Clostridium* bacteria the cells are of still larger dimensions 5-10 x 1.5μ. The representatives of *Azotobacter,* etc, have quite large cells.

The dimensions of cells are in general constant and characteristic for separate microbial groups or even their different species. Under unfavorable growth conditions or under the action of special agents as well as upon the aging of the culture the cells acquire different and variable sizes. As a rule they increase greatly in volume, assume various forms and patterns.

In these cases there is a lack of correspondence between the growth of the cells and proliferation. The cells continue to grow, whereas the multiplication slows down or stops completely. Strongly elongated individuals appear with a specific internal structure of the protoplast and nuclear substance. Some investigators observed in such cells formation of many nuclei or their analogues--nucleoids--and assumed them to be polynuclear cells (Peshkov, 1955, De Lamater and others, 1955). Under unfavorable growth conditions the cell may grow only in length without a noticeable change in width. Elongated, threadlike cells are obtained. In all cases of such an abnormal growth the internal regulation of reproduction is disturbed, the system may be artificially evoked by means of external factors.

Upon development of cultures on artificial nutrient media, definite and well-expressed. regularly proceeding phases are observed. They follow consecutively.

Four phases are observed: the lag phase, phase of logarithmic growth, phase of stationary growth, and phase of reverse development. In the first phase--the lag phase--the cells do not proliferate. After inoculation of a culture of bacteria, yeasts, or other microbes on a fresh nutrient substrate, the cells act for some time as if they are at rest, their number does not increase. However. one should not regard this period as a resting state. Investigations show that in this phase an intense preparatory activity of cells proceeds. The cells increase in volume, become elongated and enlarge; the plasma acquires a more strongly expressed basophily, and is optically homogeneous, without granular inclusions. The cells remain in this phase from 2 to 10 hours and more, depending upon the microbial species, the composition of medium and the environmental conditions. During this period the cells undergo reorganization, they adapt themselves to the new conditions of life. For an old culture the fresh nutrient substrate becomes a new medium. Upon each new transfer the process of adaptation of the cells to the fresh medium proceeds very slowly; the older the transferred culture, the slower the process. When young cells, for instance those 5-6 hours old are transferred, the lag phase is considerably shorter than upon transfer of old, (5-7 days) cultures. The number of surviving cells is greater in the former case than in the latter.

Upon the inoculation of bacteria. actinomycetes, yeasts or other microorganisms in a fresh nutrient medium, not all the cells adapt themselves. Some remain in a resting state without any apparent development, and finally die. Separate cells grow after a great delay, while the neighboring individuals may already have produced several generations. A delayed development is observed in weak cells.

They evidently lack the growth factors, whose ability to be formed is lost. When all those substances are produced in a sufficient amount by the neighboring, normal cells, then the weak and the abnormal cells develop.

In microbes the cells are polymorphous. They differ from each other in many respects, for example, in viability. Some of them reveal a high degree of adaptability, growth and proliferation, in others the life functions are weak. As the culture becomes old the number of weakened cells increases. In *Azotobacter,* when an old 7-10 day culture is transferred to a fresh Ashby medium approximately 10-30% of the cells grow; upon transfer of a young 24-hour culture 70-90% grow (observations were carried out in a hanging drop).

The number of growing cells of an old culture may be increased by selection of suitable conditions, or by a change of the composition of the medium. When a little of the filtrate of the old *Azotobacter* or yeast culture and often of other microbes, is added to the medium, the percentage of growing cells increases.

The phase of logarithmic growth is characterized by a rapid growth and proliferation of cells. Their size increases, and they assume the usual dimensions characteristic of the respective species. When they reach the limiting size, the cells multiply. The daughter cells formed grow to their limiting size and soon also begin to multiply. The plasma of the cells in this period of growth is less basophilic and less homogeneous, small granules and various other structural formations appear in it, but in a very small quantity, without changing the homogeneity of the protoplast.

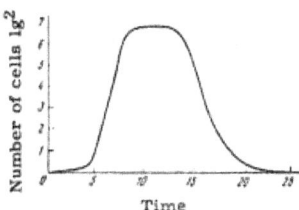

Figure 8. Typical growth curve of bacterial culture; $\lg^2$ of growth as a function of time (after Stephenson, 1951)

In the phase of logarithmic growth the number of cells increases exponentially, if the logarithms of the survivors are plotted on the ordinate and the time on the abscissa (Figure 8). Such an increase of the cell number in the culture lasts until a definite value is reached, then the intensity of growth decreases, the proliferation slows down and the culture passes into the third phase--the stationary phase. In this phase the number of living cells does not increase, the formation of new cells equalling the number of dying ones. The culture reaches its limiting age. The plasma loses the basophily, becomes granular and has various inclusions: chromatin, metachromatin, fat droplets and other substances. The growth curve in this period is parallel to the abscissa.

The fourth phase is a reversed development of the culture or phase of accelerated death. Its main feature is that the number of dying cells proves to be greater than the newly formed ones. The multiplication of cells in this period slows down. The culture becomes old, senile and dies.

In this phase of senescence and degeneration the cells undergo considerable cytomorphological changes. While in the second phase the cells assumed their usual form and size for the given species, and the plasma showed a somewhat weaker basophily in comparison with the first phase, in the third phase the cells become more polymorphous. In this period of development the cultures reach maximal variation in size and form, as well as in the internal structure of the protoplast; they also differ biochemically.

Beside normal cells, various deviating forms are found in varying quantities. As a rule, in old cultures there are many strongly fractionated germ formations almost invisible by optical microscopy and of a size smaller than the resolving power. In these cultures one may also find many large specimens, often reaching enormous dimensions, ten times greater than those of normal ones.

These swollen cells differ from normal ones by unusually diverse forms. The deformed cells also differ in internal structure, structure of protoplast, in the quantity and formation of chromatin and other granular formations.

The physiology and fermentative properties of cells in the lag phase differ from those in the logarithmic phase of growth; cells of a later growth period differ from those of the two preceding phases (Ierusalimskii, 1949).

Enlarged cells of the lag phase with a more basophilic plasma, strongly absorb stains. A shift of the isoelectric point of the protoplasm to the acid side may be noted in them. In young cells passing into the second phase or already being in this phase, the metabolism is much more strongly expressed than in the preceding and in the following phases. The oxygen uptake, the release of carbonic acid, and the heat formation increase: upon the decomposition of proteins, the release of ammonia and other decomposition products is increased. Intensification of the process is caused not by the increase of the cell mass or total volume of individual cells but by the condition and properties of the living substance of the individual cells. This may be seen from the data in Table 1.

Table 1

Intensity of cell metabolism depending upon the phase of development of a *Bac. coli* culture. Observations in a peptone-glucose medium (Hungington and Winslow, 1937)

Time, hours	Release of CO_2	Volume of cells μ^3	Rate of multiplication (number of cells formed during one hour)
0	--	0.41	--
1	78	0.77	0.13
2	86	1.03	1.04
3	79	0.93	1.41
4	59	0.91	0.64
5	31	0.81	0.42
6	24	0.84	0.04
7	19	0.89	0.18
23	--	0.76	--
25	7	0.75	0.05

Note. Calculation of CO_2 in mg x 10^{-11} per 1 μ^3 viable cells during one hour.

The rate of metabolism (from CO_2) per unity of living substance in the lag phase is considerably higher.

According to the statements of some authors, cells of the lag phase are more sensitive to environmental factors. They are less resistant to higher temperature and salt concentrations, to various chemicals, stains, antibiotics, etc, (Peahkov, 1955).

Proliferation of Bacteria

Bacterial cells proliferate in various ways--by division, constriction and sometimes by budding and fragmentation.

Mostly they multiply by division. This process proceeds in the following manner. According to recent data, the division of the cell is preceded by a differentiation of the protoplast. The amount of chromatin or nucleoproteins increases in it. Various inclusions of reserve food substances and others are formed. Consecutively an intracellular separation of the protoplast into two daughter units follows. Each forms its own membrane by which they are separated from one another. The membrane of each daughter protoplast forms a transverse septum on its exterior. Afterward, division of the cell into two daughter cells occurs (Figure 9).

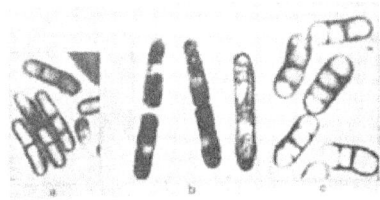

Figure 9. Division of cells by means of transverse septa: a) *Bac. anthracis;* b and c) *Bac. megatherium* (after Robinow from Dubos, 1948),

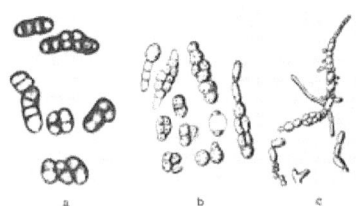

Figure 10. Fragmentation of bacterial cells: a) *Bac. megatherium* (after Robinow from Dubos, 1948); b) *Bac. megatherium* (after Kudryavtsev, 1932); c) *Act. globisporus* (after Krasil'nikov, 1938).

In separate cases under unfavorable growth conditions a simultaneous division of the cell into several daughter cells is observed in bacteria. The cell splits or undergoes fragmentation to three, four, five and more small individuals (Figure 10). Such splitting is observed in sporeforming bacteria--*Bac. megatherium* (Kudryavtsev, 1932), in *Bac. mesentericus, Bac. mycoides* in lactic bacteria (Krasil'nikov, 1954 B), in filamentous bacteria-- *Pontothrix longissima* (Krasil'nikov, 1932a) and in many others (see Dubos, 1948; Bisset, 1950; Robinow, 1942, 1951; Malek, 1955 and others). During the process of splitting, septa are formed in various directions-- transverse. longitudinal and oblique, like those formed in micrococci.

Constriction is less often observed than division. It may, however, be seen quite often in various specimens of bacteria and mycobacteria. This process proceeds in the following manner. In the central part of the cell an almost invisible constriction is formed, it deepens gradually separating the cell into two halves. Sometimes the constriction is not complete and leaves a small bridge in the form of a copulation canal between the formed daughter halves. Finally this bridge breaks and the daughter cells part (Figure 11).

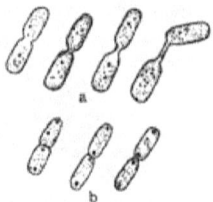

Figure 11. Multiplication of bacterial cells by constriction: a--*Azotobacter chroococcum* (the author's observations); b--*Rhizobium trifolii* (the author's observations).

Such a multiplication is observed in *Azotobacter* during growth on mustagar, In mycobacterium, in purple sulfur bacteria, *Chromatium* and other bacteria.

Often a mixed type of proliferation is noted in bacteria, division is combined with constriction. At first a small annular depression appears--a constriction--then a transverse septum is formed and the cell divides. The process begins with constriction and ends with division (Figure 12). A similar type of proliferation is observed in Azotobacter, in various specimens of sporeforming and other bacteria.

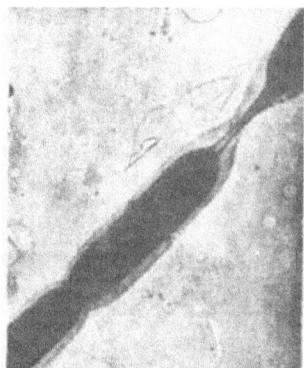

Figure 12. Multiplication of bacterial cells by constriction: *Bac. cereus* (after Johnson, 1944).

Sometimes under unfavorable conditions, small particles in the form of cocci, detach themselves from the end of the cell. They split off from the maternal cell, and under favorable conditions, grow into new organisms. Such coccuslike germs, formed by splitting off, are found in many bacterial species, and in mycobacteria: in *Bac. mycoides, Bac. megatherium, Bac. cereus,* and others (Kudrayvtsev, 1932; Krasil'nikov. 1932 and others). They may often be seen in filamentous bacteria - *Pontothrix* and others Krasil'nikov, 1932a, 1945 B),

One and the same organism may proliferate by division, constriction and some times by fragmentation (splitting into small fragments) depending upon the growth conditions of the culture.

Budding in bacteria is not observed under normal growth conditions. It is found in old cultures or in cultures subjected to the action of unfavorable factors. It is expressed in the following way. On the surface of the cell, on any part of it, a tiny body appears, which gradually increases, reaches definite dimensions and acquires a contour in the form of a more or less distinct membrane or cell wall. Such a body has the form of a budlike cell (Figure 13).

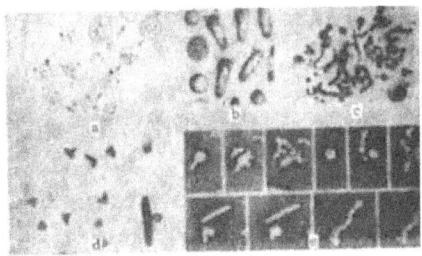

Figure 13. Formation of budlike bodies in bacteria: a) *Azotobacter chroococcum* Krasil'nikov, 1931); b) *Bac. megatherium* (after Kudryavtsev, 1932); c) *Bact. vulgaris* (after Peshkov, 1955); d) *Bac. mesentericus* (after Robinow from Dubos, 1948); e)*Bact. proteus* (after Stempen and Hutchinson, 1951).

As a rule under conditions of the hanging drop, where the formation of buds proceeds, they do not develop. Only in rare cases does one succeed in following their further development. A great many of these buds become swollen and after some time burst. In single cases one may observe buds growing into normal rod-like cells, while still affixed to the maternal cell. Such a formation of reproductive forms was observed by us in *Azotobacter,*in root-nodule bacteria, in some sporeforming, and nonsporeforming bacteria, *Acetobacter* and others (Krasil'nikov. 1954 B).

In mycobacteria, mycococci, actinomycetes and other specimens of Actinomycetales, budding as a form of multiplication is often observed, the process of budding hereby proceeds exactly as that in the yeastlike fungi. At first a small protrusion appears on the surface of the cell wall, then it enlarges. reaches a definite size and then either splits off from the maternal cell and continues to grow and develop,or it grows without separation from the initial cell (Figure 14). In Actinomycetales, the process of budding should be regarded as the normal way of proliferation. The buds formed in them possess a completely normal dense protoplasm which strongly refracts light and stains intensely. A small granule of chromatin may be differentiated inside the bud (Krasil'nikov, 1938 b).

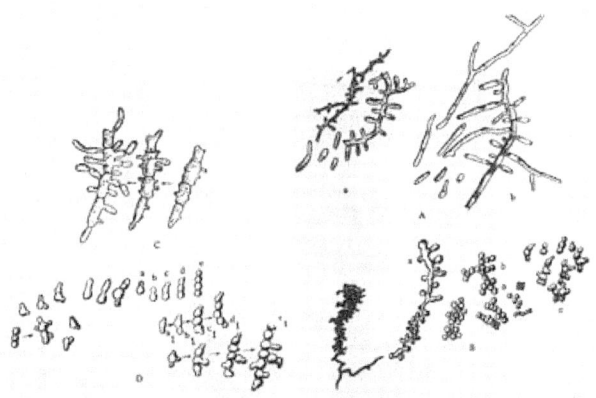

Figure 14. Budding in Actinomycetales A--*Act. candidus:* a--filaments of mycelium with short rodlike sprigs--buds, b--germination of rodlike appendages. B--*Proact ruber:* a--chain of rodlike cells, b--cells with buds, c--process of bud formation. C--*Mycobac. nigrum,* D--*Mycoc. ruber:* a--formation of the bud on the surface and further development b, c, d and e. a_1-- e_1--the same. The arrows show the sequence of development.

The reproduction of bacterial cells may proceed by the formation of special germ cells, the so-called regenerative bodies inside the cells. These bodies are usually formed during the degenerative process, in old cultures or under unfavorable growth conditions of the culture. The cells undergo deformation, the plasma changes noticeably and granules of chromatin and other structural bodies appear. In such degenerative protoplasma, separate particles or granules of chromatin become centers of formation of very small germ cells.

Around the chromatin granule a zone of plasma concentration is formed, on the surface of which there is a thin, hardly visible membrane. A tiny germ cell is obtained, which is usually scarcely noticeable in the protoplasm. Refraction of light of this germ cell hardly differs from that of the cell protoplasm. The diameter of the whole germ does not exceed 0. 5μ, more often 0.1-0. 2μ (Figure 15). When the cells undergo autolysis their cell wall disintegrates, the germ cells are released, and under favorable conditions they may grow and produce a normal generation (Krasil'nikov, 1954B).

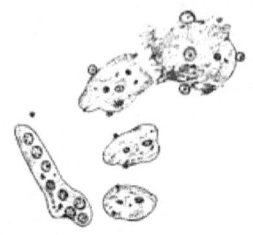

Figure 15. Formation of regenerative bodies inside swollen and dying cells of *Azotobacter*

Such regenerative bodies have nothing in common with endogenous spores in sporeforming bacteria. They are closer to the regenerative forms occurring an budding described above. Their characteristic feature is a low viability. Under laboratory condition they do not, as a rule. grow, or very seldom, there by producing only several generations.

The formation of buds, and regenerative bodies inside the cells, then the fragmentation of the cells into tiny germ elements and the branching off of small particles from the extremity of the cell--all these methods of reproduction take place in particular pathological states in organisms under unfavorable growth conditions. Evidently, these forms of reproduction constitute a biological adaptation of the species. The probability is not excluded. that in many bacterial species they are the most frequent under natural conditions. In the soil the cells of bacteria and actinomycetes occur in other forms and states, and, consequently, the ways of reproduction may sharply differ from those observed by us in artificial laboratory media.

Under conditions of the normal growth and development of microbes a definite and quite constant relationship between the growth of cells and their proliferation is observed. Multiplication of cells begins after the latter reach a certain size. Growth and multiplication of cells occur at a definite rate which differs in various species.

If growth and multiplication is suppressed for some reason, various kinds of formation disturbances are observed. Under the influence of environmental factors the process of multiplication of microbes may be suppressed or stopped, while the growth function is preserved or only slightly suppressed. The opposite may also occur--the growth is stopped, but the process of multiplication proceeds normally or only slowly.

In the former case, when the function of multiplication is suppressed, the cells increase, reach enormous sizes, undergo deformation and are transformed into so-called involution forms (see below). Such forms occur in old cultures which are influenced by their own metabolic products. They can be obtained by the action of penicillin, lithium chloride and other substances.

In those cases where the growth function is slowed down or suppressed and the process of proliferation continues, small cells are produced. Smaller cells are formed by each division until ultramicroscopic cells are formed (Figure 16). This decreasing cell size is observed in many bacteria, mycobacteria and actinomycetes.

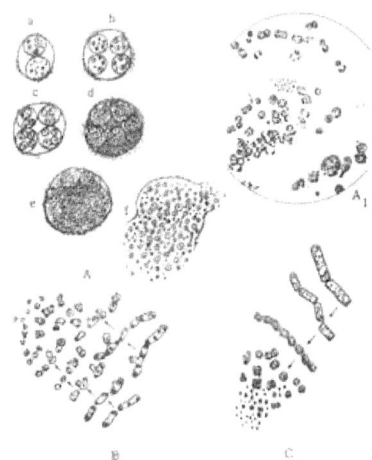

Figure 16. Decrease of bacterial cell size in the process of consecutive divisions during slowed growth A) *Azotobacter chroococcum;* inside the capsula consecutive division of cells leads to formation of small elements, like camocytes in algae (a - f); A₁) the same--microphotography; B) *Bac. mycoides;* C)*Mycob. rubrum;* arrows show consecutive transformations of cells.

The viability of such forms decreases or vanishes completely at a determined stage of this process during both a strong increase and a strong decrease in size. The cells of decreased. size stop growing on nutrient laboratory media. Particularly the viability of ultramicroscopic elements decreased when they were passed through fine filters.

On studying the process of bacterial proliferation a biologically important question arises: are the daughter cells formed equivalent? It is usually assumed that during proliferation bacterial cells divide into two equal parts; on division the maternal cell is transformed into two identical "sister" cells. On this basis some investigators altogether negate the development of bacterial cells. The latter grow and increase in length, but do not develop; qualitatively they do not change. According to these opinions the newly formed daughter cells do not differ in their properties from the initial maternal cell. They are only shorter. Consequently, an ontogenetic development does not occur in bacteria.

If it is indeed so, then after each division of the cell, the daughter cells which are formed are identical in nature with the initial cell before the division. In other words bacterial cells are invariable.

This opinion is not correct. It has been formed under the influence of observations which are primarily of a morphological nature. Modern improved methods of investigation show that in the cells, formation processes proceed de novo and that the protoplast of the cell is not equivalent in all its parts. The daughter cells formed are not identical in their physiological and biochemical properties.

It was assumed earlier that the viability of the "sister" cells is the same, but it was recently clearly established that this is not always the came. Developing cultures of bacteria, even in the most active growth phase (logarithmic growth phase), under the most favorable conditions of nutrition and respiration, contain a considerable number of dead cells. These cells die a natural death as a result of exhaustion after a consecutive series of multiplications (Malek, 1954, 1955; Streshinskii, 1955, 1956).

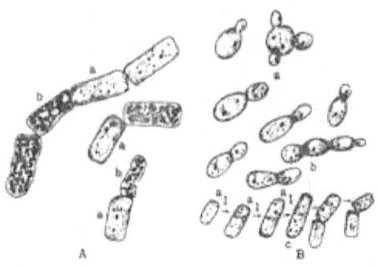

Figure 17. A. Aging and death of cells in yeasts *Schizosaccharomyces octosporus:* a--young viable cells; b--dying and dead old maternal cells B. Enlargement of cells in yeasts: a--*Saccharomyces cerevisiae;* growth and multiplication of cells in all directions; b--*Saccharomycodes ludwigie;* growth of cells in one direction along the long axis; c cells multiply by budding; the formed buds branch off by transverse septa; c--*Schizosaccharomyces octosporus:* polar growth from one end (a_1); multiplication by division.

We observe senescence and the death of cells in yeast organisms, budding--*Saccharomyces cerevisiae* and proliferating *Schizosaccharomyces octosporus.* In the maternal cell, upon senescence, the process of bud formation is slowed down, the plasma becomes more granular, and fat inclusions appear. Soon after, such a cell stops budding and perishes. If at the onset, the maternal cell was similar to the daughter cells, it now becomes very different, not only physiologically, but also cytochemically. The latter difference is not always well expressed. Sometimes the dying maternal cell cannot be distinguished externally from the young daughter cells. Only cessation of growth and multiplication indicates its death.

A similar aging of cells was noted in proliferating yeasts *Schizosaccharomyces octosporus.* This organism multiplies like bacteria. Its cells on reaching a definite form a transverse septum and separate. The daughter cells prove to be externally identical, as in bacteria it is hard to see any difference whatsoever between "sister" cells, when the culture is still young. However, after a prolonged series of division one of the daughter cells becomes weaker, lags in growth, and its division slows down or ceases altogether. Soon its protoplast noticeably also changes, the plasma becomes coarse-grained with a fatty degeneration; the permeability of the cell wall to, stains, increases considerably. After a successive division, one may often see that one of the cell a die immediately, and the other continues to develop and to multiply normally. In two daughter cells which are still attached to each other, one is often dead and the other alive. This is well seen upon cytochemical analysis (Figure 17A).

In budding yeasts--Saccharomyces the formation of daughter cells proceeds at various places on the periphery of the cell. The growth of the protoplast in them is evidently homogeneous. The buds in the early growth phase stand out sharply in form and size as well as in the state of the plasma. When they reach maturity. they are hardly distinguishable from the not yet aged maternal cell.

In some yeasts *(Saccharomycodes)* growth of the protoplast proceeds in one direction, along the longer axis of the cell. Cells of these organisms multiply by budding in the following manner. At the end of the cell a bud is formed, which grows and reaches a definite size, afterward a transverse septum is formed at the site of the constriction and branching off from the maternal cell occurs (Figure 17B). This type of multiplication is intermediary between budding and typical division. Upon division of yeast organisms *(Schizosaccharomyces)* growth of the protoplast proceeds at one end at the point of growth. In this case constriction does not occur. The cells divide into two equal parts by means of a transverse septum.

In bacteria as in *Saccharomycetes,* the growth of the protoplast proceeds from one end of the cell. Their content undergoes differentiation as growth and at the same time, different cells are formed on division. One of them has an older type of structure, and is less viable than the other. Often one cell ceases to grow and multiply while the other divides intensely. The two cells formed as a result of the division are, in fact, not "sisters": one of them in the maternal cell and the other is a daughter cell. Bisset (1950, 1951) showed that the young growing part of the bacterial cell does not possess flagella. The latter are formed later, when the transverse septum appears and the daughter cell splits off (Figure 18).

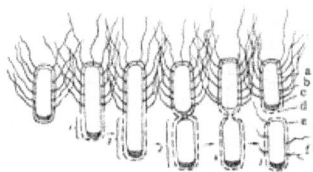

Figure 18. Enlargement of cells in bacteria (diagram after Bisset, 1950): 1-5 polar growth of cell at the "growing point": a--external cell wall of maternal cell; b--cytolemma; c--"growing point"; d--external cell wall in "growing point"; e--cell wall of daughter cell; f--flagella in state of formation.

As seen from the afore-mentioned, the protoplasm of the bacterial cell varies in quality during the growth process. There is in the cell, an older and a younger part. The latter is principally connected with the formation of daughter cells. Malek (1955) showed that formation of the daughter cell proceeds in the maternal cell. Before starting division, the cell increases in mass of living substance and length and also undergoes qualitative changes. A definite cycle of development takes place--formation of a qualitatively different portion of protoplasm occurs in it. This is indispensable for the formation of the daughter cell. Consequently, in bacterial cells an ontogenetic development takes place.

At a definite stage of development, many microorganisms begin to form spores. This process proceeds in various microbial groups in a different manner. Spares may be formed exogeneously as for instance, in actinomycetes, and endogeneously--in bacteria (sporeformers). Both the manner in which they are formed and the properties of spores in various specimens vary.

In sporeforming bacteria endogenous spores are formed in the following way. As was indicated above, at first a chromatin substance in the form of a distinct body appears inside mature cells. This body is regarded as a rudimental spore. Around it, the protoplasm concentrates into a large round formation, the prospore. The prospore soon becomes dense, decreases in volume, the contour becomes clearer ,and a thin membrane, the intima, appears on its surface. This membrane is in turn covered by an external, thicker membrane, the enzyme, In this way, the prospore matures and is transformed into a spore.

Upon formation of the spore the chromatin body disappears; it is, not found, in the prospore but in the mature spore it appears once more in the form of a definitely formed, nucleus-like formation (Figure 19).

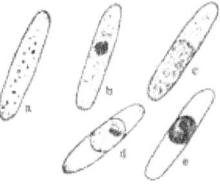

Figure 19. Spore formation in bacteria *Bacillus* sp.: a--vegetative mature cell with chromatin and metachromatin granules inside it, b--chromatin aggregated into a separate body--nucleoid, around which plasma is concentrated; c--formation of prospore, nucleoid is absent; d--maturation of prospore, plasma becomes dense, chromatin appears in the form of nucleoid; e--mature spore, chromatin in the form of nucleoid, membrane is clearly visible.

Upon spore formation the plasma changes its staining properties. The plasma which is concentrated around the chromatin body, stains more intensely than normal protoplasm; the prospore stains most densely of all. The latter is also well seen without staining. Due to its great refraction of light under the microscope, the content of the prospore appears to be glistening. Mature spores lose the ability to stain. Only after treatment with a weak solution of hydrochloric acid at 60°C does the plasma of the spore take on stain. The stained spores are decolorized by acid more slowly than vegetative cells. The method of differentiation and recognition of spores is based on this fact. The inability of the mature spore to stain by usual methods is ascribed to the relative lack of permeability of the membrane. However, analysis shows that isolated membranes stain well, but the protoplasm of the spore itself does not stain. Consequently, the plasma of the cell undergoes essential changes during the process of spore formation. The plasma of mature spores has physicochemical properties which differ from those of the plasma of vegetative cells, although the composition of their ash is the same.

Free enzymes are not found in mature spores. It is assumed that they are present in a bound form, such that its active groups are not destroyed upon heating.

Spores, as is known, are resistant to heating. The mechanisms and causes for this thermostability of spores are not known. Some authors ascribe it to a lowered water content of the plasma, but this was not confirmed by investigations; it was found that the amount of water in the plasma of the spore and vegetative cells is equivalent.

Recently, the resistance of spores has been related to an increased concentration of calcium. Attempts have been made to explain the thermoresistance of spores by their content of lipides and other factors. However, neither hypothesis was confirmed experimentally.

Formation of spores in bacteria proceeds under various conditions. It is observed in both deficient and enriched media. In the former, spores appear earlier.

The factors which condition spore formation are not clear. There is no basis for explaining spore formation by a deficiency of nutrition. In a rich medium the total quantity of spores in the culture is always greater than in a deficient medium. Neither aeration, nor temperature nor other factors constitute by themselves a direct cause of spore formation; they only create conditions affecting the given process.

The process of spore formation in bacteria is subject to the same rules, as those which are noted for spore formation or offspring production in other groups of microorganisms--actinomycetes, fungi, yeasts. and others. Nutritional deficiency is an accelerating factor under conditions of spore formation. The resistance of the spores to unfavorable life conditions is regarded as a biological adaptation of bacteria for preserving the species.

We assume that the biological essence of spore formation consists not solely in the preservation of the species but it has perhaps another biologically essential purpose

It is not always conducive to species preservation. In the majority of the soil bacteria, an a rule, it does not occur under many unfavorable conditions. According to our observations, it does not occur in bacteria of the temperate zone at increased temperatures (36-38°C), or at a low temperature 3-5°C). We did not succeed in obtaining spores in nutrient media in the presence of many antibiotics and some chemicals.

Neither does spore formation occur under many natural conditions. Cells with spores are seldom found during microscopic soil analysis. If a young culture of sporeforming bacteria is introduced into the soil, at a time when spores have not yet been formed, the latter do not appear. We did not succeed in obtaining spores in soil (podsol), from *Bac. mycoides, Bac. mesentericus* and *Bac. megatherium.*

Under conditions of the Far North. on the islands of the Northern Arctic Ocean, (islands of Franz-Josef, Severnaya Zemlya and others) sporeforming bacteria even lose the ability to form spores. In our investigations and those of Sushkina and Ryzhkova (1955) the majority of these bacteria neither form spores in artificial nutrient media nor in the soil itself. We tested many media and grew cultures under various conditions; but the majority of them did not form spores. Only some, under particular growth conditions at an increased temperature (36-38°C). started spore formation. but they were not many. In some organisms the process is incomplete, only prospores being formed, i.e. , not fully mature spores, without membranes. Spores, as shown by daily observations, are not particularly resistant to unfavorable environmental factors. They often die with the same rapidity as vegetative cells. For instance, we observed their death simultaneously with that of vegetative cells under the action of some chemical antiseptics. They are not always resistant to high temperature. There are species whose spores die, like the vegetative cells, at a temperature of 80-100°C. Frequently, nonsporeforming bacteria occur, which are as resistant or even more so, to unfavorable factors, than species of sporeforming bacteria.

We assume that spore formation in bacteria is a biological form of renewal of the organism a means of increasing the viability of the cells and, consequently, of the whole species. The spore may be regarded as a zygote cell, formed after fusion of various parts of the protoplast, as it takes place upon autogamous copulation.

The bacterial cells reach a definite size, and after a series of consecutive divisions begin spore formation. which is accompanied by a complex picture of microscopic transformations in the protoplast, leading to the fusion of separate chromatin elements and other parts into a compact body--the center of spore formation (see sexual process).

Some authors (Sorokin, 1890; Gibson, 1935; Starkey. 1938; Rubenchik. 1953) indicate that the described formation of endogenous spores in characteristic, not only of sporeforming bacteria, but also of spirilla and vibriones. According to their observations, a large oval body is formed inside the cells of spirilla whose external form resembles spores in sporeforming bacteria. For their formation the whole protoplast or a considerable part of it in used. Sometimes two or three such sporelike bodies are formed simultaneously in the cell. We observed the formation of similar bodies in *Spirillum voluntans* and other bacteria (Krasil'nikov, 1949 B), by placing them in an artificial synthetic medium or in a drop of the same water as that from which they were obtained. The same could be observed in a hanging drop of this medium. The motility of the spirillum slows down after some time the protoplast begins fragmentation into separate parts which become round and assume the forms and sizes of large spores. The number of such sporelike bodies varies from one to four or more (Figure 20.)

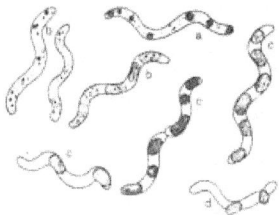

Figure 20. Formation of sporelike bodies (fragmentation spores) in *Spirillum volutans* (after Krasil'nikov, 1949) a, b, c, d--consecutive stages of spore formation.

According to our observations. these formations are not like the true spores of sporeforming bacteria. They represent fragments of protoplasts, divided into parts as a result of these or other causes. Their formation rather resembles fragmentation of the filaments of mycelium in actinomycetes, mycobacteria and in some filamentous and other bacteria.

In actinomycetes, as will be shown further, spore formation proceeds in two ways--endogenously and exogenously. Spores or reproductive elements are formed on special branches of the aerial mycelium. and inside vegetative filaments of the substrate mycelium. They are formed by segmentation or fragmentation.

Spore formation by fragmentation is observed in some species of mycobacteria and lactic acid bacteria (Krasil'nikov, 1938a, 1952d). As in actinomycetes, the protoplast of the cell splits into separate parts, which become round bodies--spores. The number of such spores in the cell varies from 2 to 6, or more. The quantitative regularity so characteristic of yeasts and fungi is absent.

In all cases of multiple spore formation in actinomycetes, mycobacteria or lactic acid bacteria a concentration of chromatin substance as a rudiment or center for their formation is noted.

One of the most important expressions of life of organisms is the sexual process. In higher organisms this process is morphologically well expressed and has been investigated in detail. It is also well expressed and has been investigated in lower organisms--algae, fungi, and protozoa.

The question of a sexual process in bacteria and actinomycetes remains undetermined. Does it occur in these organisms? This question has been vividly discussed in the literature of recent years. Former views on bacteria as primitive organisms gradually changed under the pressure of accumulated factual material. A search for complicated cycles of development in bacteria as well as for copulation processes was started. In the period of enthusiasm for theories of a complicated cycle, various forms of sexual process in various specimens of bacteria were described. Many authors assumed that all bacteria reproduce by copulation, which is in general peculiar to all other higher and lower organisms. This process takes place in various specimens of bacteria, although in a primitive manner, externally manifested in various ways. It may proceed, according to the opinions of some authors, in the form of autogamy, hologamy, or oögamy (see Krasil'nikov, 1932 b, c).

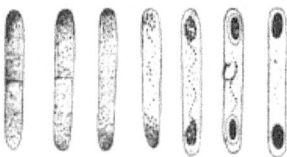

Figure 21. Phenomenon of autogamy in *Bac. bütchlii*

A primitive type of copulation in bacteria deserving great attention was described by Schaudin (1902). This investigator observed fusion of a divided cell in the relatively large bacillus which he discovered--*Bac. bütchlii.* This microbe lives in the intestine of the black beetle, however, apparently, only sometimes , since many investigators could not find it there.

According to Schaudin's data, copulation in *Bac. bütchlii* does not occur between separate organisms, but between parts of one and the same cell. The whole process is described by the author as follows: after isolation of the coarse-grained bacillus in a drop of the intestinal juice of the beetle, glistening granules, surrounded by a clear halo are visible in the center of the cell after 30 minutes. Exactly as in division the granules are distributed in a row and after 20-40 minutes they form a transverse septum, which does not differ in living or in fixed and stained cells from the septum observed upon ordinary cell division. The bacillus remains in this phase approximately 1-2 hours, after which the transverse septum becomes pale and thinner, the granules distributed in a row along the septum disappear and the cell of the bacillus looks as it did before the formation of the septum (Figure 21). Before the dissolution of the septum multiple granules of chromatin appear in the plasma, they are distributed in a long line and form a peculiar chromatin thread. This thread disintegrates into granules upon spore formation, the majority of granules moving to the poles of the cell and forming large bodies there. The plasma is concentrated around these bodies and a prospore is formed, and later, a mature spore. There the cell that underwent division fuses again, the chromatin of the two cellular parts fuses into one body and then divides in two. This is followed by division of the protoplast into two parts, from which spores are formed.

The described way of autogamous copulation closely resembles a similar process in some algae--in different species of *Spyrogyra condensata, Sp. sprellana* and other forms.

An analogous picture was observed by us (1928) in the bacteria *Oscillospira* which live in the intestine of the guinea pig. In its internal and external structure this organism is similar to the filamentous blue-green algae of the genus *Oscillaria,* but in contradistinction to them, it does not possess a pigment. Each organism represents a filament of greater or shorter length, divided by transverse septa into short cells.

During the course of its development Oscillospira forms spores. Before sporulation two and sometimes three adjacent cells of one filament fuse; the septa separating these cells dissolve and disappear. and the protoplasts of cells fuse into one body, which, by an appropriate reorganization, transforms into a spore. Upon fusion of the protoplasts, the central bodies and chromatin granules distributed into them fuse into one large body (Figure 22).

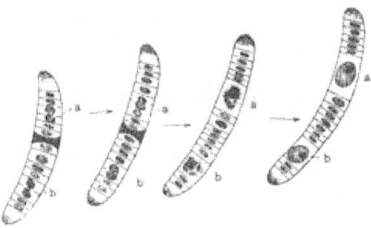

Figure 22. Copulation of cells in bacteria *Oscillospira guilliermondii* before spore formation: a--fusion of three adjacent cells; b--fusion of two cells. The arrow shows consecutive stages of development.

We observed the fusion of parts of a divided protoplast in the yeasts--*Saccharomyces paradoxus*. As is known, in this organism the protoplast of the cell divides into 3-4 parts before sporulation. First the division of the nucleus takes place, and is followed by that of the plasma. From each part of the protoplast a spore is formed. Mature spores released from the maternal cell germinate into vegetative cells (Bachinskaya, 1914).

Often before the spores become permanent vegetative organisms, they combine into pairs and copulate. Copulation of spores may proceed inside the capsule. Often 2-4 spores copulate. This kind of copulation is observed in many yeasts (see Krasil'nikov. 1935) and is regarded as a usual form. However in the yeast organism mentioned, aside from this form, an abortive copulation which does not proceed to completion in noted.

After the fragmentation of the protoplast into separate parts, spore formation begins, and small bits of protoplasm become rounded and dense with the cell walls outlines. Afterward maturation of spores stops and the process proceeds in the opposite direction. The contours of the outlined cell wall disappear and the protoplasm becomes less dense, and undergoes vacuolation. Separate not organized parts appear; very often the residual cell walls may be seen among them (Figure 23). Then a fusion of the parts and the nuclei occurs. As a result, the cell capsule takes on the initial appearance of a vegetative cell, and soon processes of growth and multiplication start again.

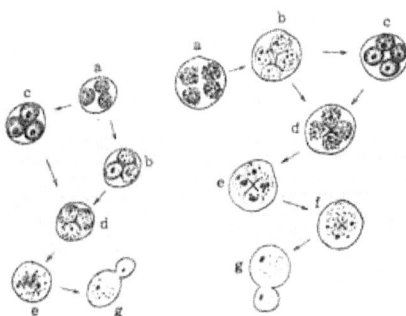

Figure 23. Autogamous copulation in the yeasts *Saccharomyces paradoxus* Batschin. Consecutive phases of sporulation:-b--protoplast divides into 3 -4 parts; c--onset of wall formation around the prospore; d--the wall disappears; e--content of prospores fuses; f--cell is transformed anew into vegetative cell; g--budding starts.

This, process was regarded by us some time ago as a reversed development of the organism (Nadson and Krasil'nikov. 1920). Afterward we came to another conclusion, and related thin phenomenon to autogamous copulation, somehow similar to that which takes place in *Bac. bütchlii* or in protozoal organisms.

The process of autogamy may evidently take place between various parts of the cell, which did not undergo preliminary division. The expression of extreme autogamy was noted by many authors. Data on observations of fusion of chromatin granules into one body before sporulation and sometimes before division have already been cited above. We observed such a fusion of cellular parts in certain large bacilli, which were isolated from the intestine of a guinea pig. The chromatin granules distributed throughout the protoplast combine into one large body. After some time the body either divides into two and is followed by the cell division, or it becomes the center of spore formation. In some microbes the fusion of chromatin is accompanied by formation of long threadlike fibers, which are straight or spirally bent from one pole of the cell to the other. Such formations were observed by us (1928), in a peculiar microbe--*Metabacterium octosporus*. This microbe lives in the intestine of the guinea pig; it is of a large size, 10-20 x 3-5 µ, nonmotile, and forms from one to eight spores. The cells multiply by division. The protoplasm in young cells is homogeneous, dense, and stains a dark color with basic dyes. As the cell grows, granules of chromatin and metachromatin appear in the plasma.

Before spore formation, chromatin aggregates into a spirally bent thread which is distributed along the longer axis in the central part of the cell. Afterward this thread divides into parts, usually into four to eight, sometimes into two to three Each part becomes a center of organization and spore formation.

Copulation of the hologamic type in various specimens of bacteria was described by many authors. In essence, this manner of copulation is as follows: after a series of consecutive divisions, two isolated cells combine in various ways and exchange their contents or fuse into one organism. The fusion of protoplasts proceeds through copulation canals and small bridges. Formation of bridges between two cells was noted by many investigators. Rindfleisch (1872) described a similar combination in bacteria and considered it as a process of copulation. Klebs (1896) and Albrecht (1881) observed the fusion of cells by means of canals in spirochaeta, Fuhrman (1906)--in coccoid bacteria, Forster (1892)--in purple bacteria, etc.

The observations of Potthoff (1924) are most widespread. He described the formation of small bridges in the purple sulfur bacteria--*Chromatium okenii, Ch. weissii, Ch. violaceus* and in the *Spirillum photometricum.* According to his observations sometimes three or more cells are consecutively interconnected by canals, which in his opinion are copulation canals. Potthoff even observed fusion of chromatin granules (Figure 24.

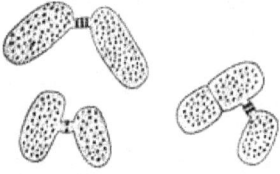

Figure 24. Joining of cells by means of "canals" in *Chromatium okenii* (after Potthoff, 1924)

Löhnis (1921) described such a process of cell fusion in *Azotobacter* and some other bacteria. According to his observations the cells combine by bridges, through which the fusion of protoplasts occurs. A similar type of copulation was described by Lieske (1926) in bacteria of the coli group which were isolated from a tobacco extract. Cunningham (1931) found cells combined by means of a canal in 20 strains of *Bac. saccharobutiricus.*

Mellon (1925) described spherical elements in a culture of *Bact. coli.* In his opinion they represented zygospores formed upon fusion of two rodlike cells. Stoughton (1932) noted similar formations in *Bact. malvacearum.* He regarded a part of the spherical bodies as zygotes, and part as budding germs.

Smith (1944) described fusion of cells in *Pseudobacter funduliformis;* Lindegren and Mellon (1932) in the tubercle bacillus *(Mycob. tuberculosis,* var. avium). Klieneberger-Nobel (1949) observed the fusion of nuclear bodies in bacteria before the formation of L-forms. According to her observations, rodlike cells are fragmented into small parts, each possessing one chromatin body. These small parts then combine. As a result, the unusual. so-called L-forms are formed.

Stempen and Hutchinson (1951) observed development of cells in *Bact. Proteus* (OX-19) by means of microphotography. They noticed formation of strongly swollen spherical cells. These spheres are combined with one to three rodlike cells. As the growth of the spherical cell proceeds, the rods become shorter, paler and disappear completely; their substance appears to have entered into the spherical elements. On this basis the authors concluded that these spheres represented zygote cells which undergo disintegration with the formation of either fine-grained elements or are transformed into rodlike elements by consecutive divisions. Under favorable conditions the granules may be transformed into regenerative elements and develop into vegetative cells (Hutchinson and Stempen 1954).

Analogous formations were observed by Dienes (1930, 1943) in *Bact. proteus, Bact. moniliformans* and in *Pseudobacterium (Bacteriides)* (from Dubos, 1948).

Examining the microphotographs obtained by Stempen and Hutchinson we can not agree with their interpretation. In our opinion the spherical cells observed by them represent not zygotes but involution forms. The forms described by them were observed by us in *Azotobacter, Mycobacteria* and other microorganisms. They are often combined with rodlike elements.

The cited data on the fusion of cells are also not convincing. They mostly represent a result of deductions based on the study of fixed and stained preparations. Direct microscopic observations on development and fusion of cells were not carried out by them.

Our study of a large number of bacteria did not enable us to establish any external signs of cell fusion, which resembled a sexual process. In some species peculiar small bridges between the cells are observed. However, a detailed investigation of such formations showed that they represent the result of an unfinished constriction of cells. As was indicated above, when cells of *Azotobacter* multiply by constriction, long plasmodesms are often formed, which resemble copulation canals (Figure 25). These plasmodesms sometimes move to a lateral surface of the cell, and even possess a small swelling with a tiny granule inside it, in the central part, which resembles a sexual process even more. We did not observe any indications of fusion of unseparated cells.

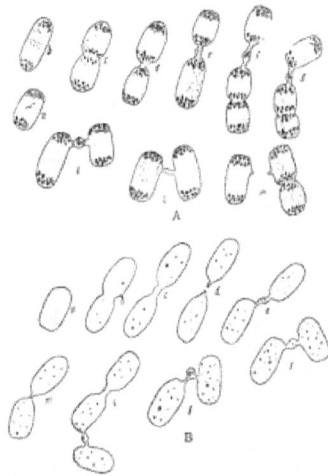

Figure 25. Joining of cells by plasmodesms (incomplete division): (A) Chromatium warmingii and (B) Azotobacter chroococcum: a. b, c, d, e, f. g, i, l, m--consecutive stages of cell division.

The same may be said in relation to fusion of cells in *Chromatium* described by Potthoff (1924). We had the opportunity to follow in detail the formation of bridges in *Chromatium okenii* and *Chromatium warmingii* in a hanging drop. As in *Azotobacter,* the cells of these bacteria under special conditions , multiply by constriction and form long canals (Figure 25 B). The canals are often displaced to a lateral surface. This displacement, as indicated above, takes place because the cell grown at the end, in the growing point, and thus shifts the site of attachment of the canal sideward (Krasil'nikov 1932b, c).

Löhnis and some other investigators regard the extracellular fusion of the protoplast of two and more cells after the disintegration of their cell walls as a sexual process. This fusion of the substance of the disintegrated cells was named by Löhnis "symplasm". It is observed on the aging of cells, and under unfavorable growth conditions. The cells, often strongly swollen, disintegrate, their cell wall disrupts and undergoes lysis and the cell contents leak out and fuse with the contents of other cells. A protoplasmic extracellular mass or symplasm is obtained. According to Löhnis processes characteristic of copulation proceed, there. After some time, new organisms having the nature of a zygote are formed in the symplasm.

According to our observations, the described symplasm sometimes does indeed contain germs or regenerative bodies. However they are usually formed in the cell before its disintegration and do not constitute a fusion product of two or more protoplasts. Symplasm represents a post-mortem formation of cells, a mixture of protoplasm of already dead or moribund cells. It to possible that in single cases. regenerative units are formed In such a mixture of still living protoplasts (Krasil'nikov 1932 d 1954 b).

Several investigators describe combinations of cells of bacteria which strongly resemble the conjugation of protozoal organisms. Rodlike cells unite at their ends into groups, several cells in each group, and form peculiar groups. Inside the cells a shift of the nuclear inclusions takes place. A similar fusion of cells is observed in *Pseudomonas radiobacter, Ps. tumefaciens, Bac. stellatus* and other forms.

In cultures of these bacteria, starlike groups of cells which are joined by their ends may often be seen. The cause of these formations and their meaning are not clear. Stapp and others (1931, 1949, 1955) subjected these cells to careful study and came to the conclusion that the formation of cells into starlike groups represents a sexual process during which a combination and fusion of hereditary material proceeds. They follow the formation of these groups, the fusion of the cells and further events, step by step. It was established that the cells indeed combine by their ends into firm groups. The nuclear substance, chromatin granules and nucleoids move toward the site of fusion, the cell wall dissolves and these granules and nucleoids of the combined cells fuse into one large body. The latter remains in one of the cells, which starts to develop and gives rise to a new generation. The body gradually becomes looser. decreases in size and begins division. The division of nucleoids is followed by the division of the cell. Two. three, four, five and more cells may take part in the process of union described (Figure 26).

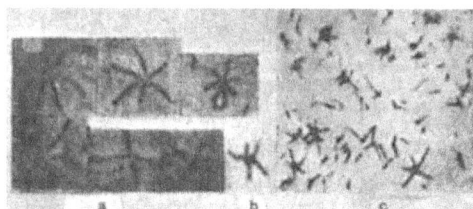

Figure 26. Union of cells into starlike complexes (after Stapp, 1956): a--*Ps. Radiobacter;* b--*Ps. tumefaciens;* c--*Bac. stellatus.*

A detailed picture of cell fusion was observed by us in some strains of root-nodule bacteria of the pea, bean, vetch, and other legumes. When these bacteria were cultivated on synthetic media (CPI. Capelu or "chkapeka" in Russian) or on extracts of leguminous plants, starlike groups of cells were often seen. By a careful cytochemical analysis and observations of their development in a hanging drop it was established that the cells are joined quite firmly by the ends and are not disrupted under pressure. This connection is not mechanical, but organic, although a dissolution of the cell wall at the sites of the attachment of the cell was not observed by us. At the end of each cell one chromatin granule (nucleoid) is distinctly seen. After some time these nucleoids undergo dissolution or become invisible and then the cells begin to grow and multiply. The plasma of such reorganized cells is optically homogeneous; granular inclusions and nucleoids appear later.

As a rule. the cells in starlike groups after transfer to a fresh nutrient substrate do not develop immediately. A long time passes before they begin growth and multiplication. Until that moment they remain in the state of reorganization of the protoplasts. We assume that the picture of cell combination into starlike groups described in root-nodule bacteria is connected with some metabolic process. The cells mutually exchange the products of their metabolism, one cell assimilates some substances released by the other.

Recently papers were published in which the sexual process in bacteria is regarded as an exchange of substances between one cell and another upon direct contact without any special copulation structure and without disintegration of the cell wall. It is suggested that specific cell substances--carriers of hereditary properties--penetrate through the cell walls of the contiguous cells and in this way the process of fertilization is achieved. In the author' a opinion, in each culture of bacteria there are unisexual and bisexual cells. In the latter, fertilization takes place by the autogamy described above. The unisexual cells constitute a small percentage and are formed in special cases of metabolic disturbances, under the action of some environmental factors [such as] "ultraviolet light", and chemical agents; they also occur in old cultures. They lack the ability to perform certain functions, which are restored upon contact with other bacteria carrying the opposite features (Braun. 1953 ; Lederberg and Tatum. 1954. and others).

This method of mutual fertilization was described in several strains of *Bact. coli* variant K = 12 by Tatum and Lederberg (1947, 1954); Hayes (1953), Catcheside (1951), and others. They indicate that on the mixing of cells of various strains of *Bact. coli* after their subsequent plating on an agar medium, one obtains new strains with properties of the parent cells (see chapter on variation).

Some investigators assume that bacteria perform a sexual process of the organic type. According to observation of Ferran (1885), cells of *Vibrio cholerae* form male gametes--antheridia, and female gametes--archegonia. The archegonium, according to him, is fertilized by the antheridium and transforms into a zygospore. From the latter a vegetative cell develops under favorable conditions.

Enderlein (1925) made the course of a developmental cycle of bacteria in general and of the sexual process in particular very complicated. According to his views, at a determined stage of development cells form special bodies--gonites or gonidia, subjected to meiosis. The gonites develop either into spermites or into oites. The former are small, rodlike, straight or slightly bent, very motile with a flagellum on one extremity; the latter are larger. of spherical form. nonmotile. An oite fertilized by a spermite transforms into a mycete, initiating normal vegetative cells. The sexual cells--oite and spermite represent haploid forms, as a result of fertilization, a diploid--the mycete--form is created.

Enderlein has no factual material to confirm his hypothesis. Since he was not a microbiologist he mechanically transferred data from zoology into the microbial world.

In spite of lack of objective evidence, this hypothesis had many followers among microbiologists. There were and still are followers who try to find a basis for it. Data from bacterial life are given by them in order to confirm Enderlein's views (Broadhurst, Mariyama Pease, 1931, Almquist, 1925; Mellon, 1925; De Lamater, 1951, and others).

Almquist (1925) describing the sexual process between differentiated sexual cells in bacteria, assumes that this process may proceed not only between cells of one species but also between those of different species. producing hybrid offspring.

The latter were obtained by him upon mixing cultures of *Bac. typhi* and *Bac. dysenteriae*. The hybrids differed from the initial cultures and at the same time had features common to both.

Almquist and other followers of Enderlein's views bring data which is not very convincing as evidence. The so-called antheridia and oögonia or sporogonia observed by them have nothing in common with the differentiated sexual cells of bacteria. In cultures of the latter particularly, as is known to microbiologists, in old cultures, there are always greatly enlarged organisms, and very small forms, which, if one cares to do so, may be regarded as female and male sexual cells. However, none of these authors showed the essential importance of these cells in the process of fusion under direct and constant observation in a hanging drop. All statements are based on analogies with known facts of sexual processes in other organisms.

In actinomycetes the question of a sexual process remains to be elucidated. There are only a few casually stated opinions. For instance, Kober (1929) expressed the view that in actinomycetes fusion of cells occurs by direct contact.

According to our observations, the fusion of cells in actinomycetes occurs in two ways: a) by a combination of outgrowths of the spores, b) by combination of mycelial filaments by means of anastomoses. Both are found under usual growth conditions (Krasil'nikov, 1938 a).

Fusion of cells may be observed under direct constant observation in a hanging drop in many species of actinomycetes. The process proceeds as follows: upon germination of spores, small appendages shaped like small tubes are formed (outgrowths); these tubes come into close contact at the ends, at the site of contact the cell walls dissolve and a canal is formed. Through this canal the contents of the two germinating spores combine and fuse into one protoplast . The combined outgrowths produce a common sprout which extends into a long filament and then grows, becoming a mycelium (Figure 27).

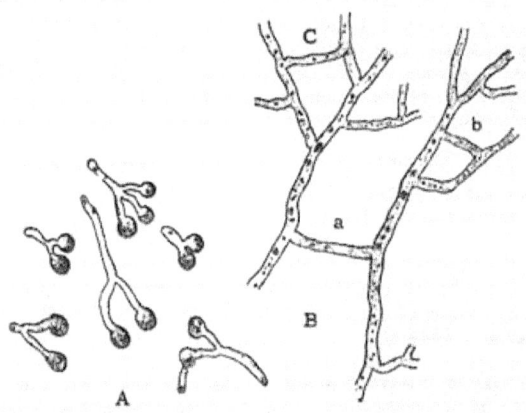

Figure 27. Fusion of cells in actinomycetes: A--Joining of germinating spores in *Act. chromogenes.* B--union of mycelial filaments by means of anastomoses: a--joining of filaments genetically distant from one another; b--joining of branches of daughter and maternal hyphae genetically related; c--joining of genetically related sister branches.

This fusion of spores was observed by us in different species of actinomycetes (1938a), Later (1950) this process of fusion of germinating spores was noted in many species of actinomycetes. By its external manifestation the described fusion of spores did not differ from similar fusions in many yeasts (Krasil'nikov. 1935; Kudryavtsev, 1954; Guilliermond, 1920; 1941, Gäumann, 1949).

As in other yeasts, fusion of chromatin, nucleoids and chromatin granules occurs upon fusion of the growth tubes. If in yeasts a similar act is regarded as a sexual process, there is no basis for not accepting it in Actinomycetes, though we have not yet sufficiently investigated the cytochemical changes in the protoplast of conjugated cells.

The joining of cells by means of anastomoses is observed in actinomycetes considerably more often than the fusion of spores. This may be seen in any culture on various nutrient media, liquid or agar. Externally, anastomoses in actinomycetes do not differ from those in fungi. As in the latter, hyphae are formed between the mycelial filaments of actinomycetes. These are small bridges in the form of a canal connecting two more or less distant filaments. This canal may be quite long when it connects two filaments located at a great distance from one another. The canal may occur between branches located side by side and closely related, originatng from one hypha, as well as between distant branches originating from various hyphae of the same mycelium. (Figure 27 B).

Anastomoses have no septa, and because of this they do indeed represent canals, through which the protoplasts of filaments fuse. This may be seen directly under the microscope in a preparation of a living culture. Separate granules, which are in constant Brownian movement, pass from the filament into the canal, from there these granular inclusions move into the second filament. An exchange of protoplasmatic substance occurs between the hyphae and fusion of substances, such as chromatin.

Consequently, between the hyphae of actinomycetes a process takes place which may be qualified as an autogamous copulation.

Formation of anastomoses occurs not only between filaments of one and the same culture but also between filaments of various strains of one and the same species: we observed anastomoses between various strains of*Actinomyces streptomycini,* producers of streptomycin, isolated from various soils of different areas of the Soviet Union.

This form of fusion of mycelial filaments is not observed between cultures belonging to different species or even to different varieties. We carried out observations under the microscope. in a drop of a semiliquid nutrient medium. Spores of two actinomycetal species which were inoculated at different sites of the drops germinated and gave rise to hyphae, which grew out and after some time came close together. At the site where the filaments touched, the formation of anastomoses could be observed.

The process of formation of anastomoses proceeds in the following manner: a side branch of the mycelial filament grows in the direction of a neighboring filament, touches it, and attaches itself by its end. After some time, the walls of the filaments dissolve at the site of attachment and a canal is formed, through which fusion of protoplasts occurs.

Probably not all branches form anastomoses. Many of them, which came in contact with neighboring hyphae did not attach themselves and did not form anastomoses. It to possible that anastomoses are formed by branches which have some sexual property, and are different from those of usual mycelial hyphae. If it is so, then the whole culture of mycelium should be regarded as a heterogeneous system, where separate filaments and branches have different sexual qualities.

In general, it should be noted that, while solving the problem of a sexual process in bacteria and actinomycetes, it is necessary to clarify the meaning of the terms "sexual process" and how it is regarded in biology in general and in lower organisms in particular.

There are various theories regarding the biology of the sexual process. They may be reduced to two basic ones: the theory of plasma mixing and the theory of rejuvenation.

The theory of the mixing of plasma was first presented by Weismann in the 1880's. Fusion of two cells and mixing of germ plasm or amphimixis, was primarily considered by Weismann, in relation to heredity and formation of species. Such a point of view in still widespread among biologists at present. Developing his views, Weismann stated that the germ plasm was immortal and is passed from generation to generation in an unmodified state. This view of a "potential immortality" of the germ plasm was widely developed in his studies on protozoa. He stated that unicellular organisms were endowed with the ability to proliferate indefinitely. With each division, the two cells formed anew are equivalent. This point of view in refuted at present.

The hypothesis of rejuvenation puts forward the problem: do cells become old upon prolonged asexual multiplication? Bütchlii (1880, 1887) and then Maupas (1888, 1889) tried to solve this problem experimentally on infusoria. Many lengthy observations were carried out in this direction by Metal'nikov, Hertwig (1902) and others. The investigations led the authors to reject the indicated hypothesis. Numerous observations made by contemporary investigators on bacteria and yeasts also show the erroneousness of this point of view.

The explanation of the essence of the fertilization process, given by the chromosomal theory of heredity, is built entirely on the assumption that gametes contain particles of a special hereditary substance. According to this theory, factors responsible for the hereditary transmission of properties are located in the chromosomes of cells. They are connected with the chromatin substance of the latter and, according to modern ideas, with desoxyribonucleic acid, which constitutes the fundamental component of the nuclear substance.

Upon fusion of the copulating cells an entirely mechanical combination of cytoplasm and nuclei occurs. The substances of chromosomes are not dissolved in the protoplast, they do not disappear, and they preserve their peculiarities as well as their characteristic features of heredity. The chromosomes of the female and male cells preserve their individuality in the chemical as well as in the biological sense.

This invariable part of the nuclear substance assures the continuity of inheritance. The material carriers of heredity, the genes, which are concentrated in it, perform the transmission of properties and traits from parents to progeny.

These concepts on the essence of the sexual process are refuted as groundless by modern biology and particularly by Soviet specialists, followers of the school of Michurin.

An entirely new interpretation of the question of fertilization to given by the academician Lysenko. He is of the opinion that the essence of the sexual process consists of the combination of the chromosomal sets of the gametes but not of exchange of substances between these gametes. According to the author, fertilization represents a peculiar process where the copulating cells mutually assimilate substances. The protoplast of one cell is assimilated by the protoplast of the second.

According to Lysenko, the essential difference in fertilization from all other biological assimilation processes consists of the following: "In any physiological process one part is the assimilating one and the other the assimilated one. . . . . The substances which are assimilated are used as building material for the assimilating component. In the sexual process, when two seemingly equivalent cells combine, there is a mutual assimilation. Each builds itself from the substance of the other, according to its own pattern. Finally, neither of these cells remains, but instead of the two a third new cell is produced" (Lysenko, 1948. p 383).

The peculiarity of this exchange or assimilation consists in that there is not one but two cells which assimilate. During the process of fertilization the two combined cells--the maternal and the paternal--are equivalent; upon fusion neither preserves its previous individuality. A new cell is obtained, dissimilar from both the paternal and maternal ones. This new cell--zygote--initiates a new organism combining the paternal and maternal traits. The living substance of the zygote is of dual quality, it contains elements of both the paternal and maternal organisms, these elements are there, not in the form of special corpuscles or genes, concentrated in chromosomes. but exist throughout the whole living organism. This heterogeneous quality of the newly formed zygotic cells insures the formation of an organism with new properties inherited from the parent couple. In this way the continuity of inheritable traits and the evolution of the species is secured. According to Lysenko's conclusion, the biological importance of the process of fertilization consists in an increase of viability of the organisms. As a result of fertilization, organisms with a dual heredity, maternal and paternal, are obtained. "Dual heredity conditions a high vitality (in the direct sense of the word) of organisms and great adaptability to variable conditions of life" (Lysenko, 1948, p 381).

In the majority of cases prolonged self-fertilization in animals and plants, as well as coupling of closely related animals, leads to the extinguishment of life. As a rule, normal viable organisms occur only in cases when plants and animals which differ at least slightly one from the other are coupled. Normal internal life contradictions, and, consequently, also life impulses are created mainly by means of crossing and breeding.

The above-mentioned principles of the biological importance of the fertilization process. elaborated by academician Lysenko on the basis of Darwin's and Michurin's theories. reflect one of the laws of adaptability and evolution of the living population.

The usefulness of the sexual process was stressed by Darwin and subsequently by many other investigators. Maupas (1888-89), summarizing the results of numerous observations and experiments, came to the conclusion that conjugation in Infusoria constitutes a process which is indispensable to the renewal of viability. According to his opinion, for every species of Protozoa there exists only a certain number of asexual generations, the cells multiply vegetatively to a certain limit, then age, degenerate and inevitably die. The sexual process leads to restoration of the vital activity of cells. It rejuvenates the Infusoria and initiates a new series of sexual generations.

Calkins came to these conclusions after ten years of research on conjugations in the infusorium *Uroleptus mobilis.* He considers the "wearing out " of plasma and organoids as the cause of aging and degenerative senility of cells. Conjugation is indispensable for the restoration of viability. In this process reconstruction of the whole living substance occurs (according to Dogel', 1951).

In recent years the concept of the biological usefulness of the sexual process was well confirmed by the work of Cleveland. This author studied the phenomenon of fusion of gametes in the flagellates Polymastigida and Hypermastigida in detail. In one of them--Trichonymphs--the gametes are morphologically identical. Upon copulation the male cell penetrates into the female cell and dissolves there, or more precisely, the protoplasts of both gametes mix completely and form a new organism. In Oxymonas the fused gametes remain in the form of a double organism for a few days and then the fusion of important organoids follows. Not only nuclei and chromosomes, but also resistant formations of axostyles which do not have any relation to the protoplast fuse. In the zygote a double, larger axostyle is formed (according to Dogel', 1951).

The decisive importance of metabolism in the sexual process is confirmed in numerous works of Hartmann, Moewus and other investigators; their subjects of study were algae and protozoa. As will be shown further, for a successful fusion of gametes and formation of healthy progeny, at least a small difference in the chemical composition of their living substance is necessary.

The process of fertilization in lower organisms, protozoa, algae and fungi, often proceeds in the same way as that of higher forms i. e.. by the fusion of sexual cells in which the nucleus, nucleolus and other important inclusions of the cell always take part.

Externally the sexual process expressed itself in different ways in various specimens of microorganisms. Besides the true oögamy i.e. , fertilization occurring between highly differentiated sexual cells as in some yeasts and fungi, copulation often takes place between individuals externally similar to one another. Gametes do not often differ from the vegetative cells. There are microbial forms, in which the sexual process occurs not between cells but between parts of one and the same cell, i.e., autogamously.

In some yeast organisms, upon combination of cells, only fusion of the plasma takes place, the nuclei do not fuse, and in some yeasts a combination of copulation of protrusions is noted, but there is no fusion of cells (Guilliermond, 1920, Krasil'nikov, 1935).

As was shown above, there is in the literature factual material which constitutes a basis for the assumption that in bacteria and actinomycetes processes occur between cells or parts of cells which are consistent with the notion of fertilization.

Proceeding from the conjecture that bacteria. (at least some of them) represent not a primitive, ancestral cell but a quite complex organism whose functions of growth, development and life activity are stabilized, in an evolutionary way, and that many of them are evidently degenerative forms of more organized creatures, one must think, that bacterial cells also have sexual rudiments in a potential form which, in a more developed form, are characteristic of higher organisms.

According to contemporary concepts, any sexually differentiated individual (female or male) and any sexually differentiated gamete concurrently contains all the rudiments necessary for development of the opposite sex. As a result of the increased development of one of the rudiments, and the suppressed development of the other, the male or female tendency of the cell is expressed. The sexual tendency becomes expressed under the influence of various kinds of environmental and internal factors.

Even in higher organisms the expression of sexual tendency to often conditioned by environmental influences. For instances in corn the sexual tendency changes upon change of nutrition. If in the early period of growth the plant does not get sufficient nitrogen, female flowers develop mostly; on the other hand, if there is a lack of potassium. more male flowers develop.

In cucumbers and melons, when there is a deficiency of nitrogen during the embryonic period of development when reproductive organs are initiated, formation of female flowers is mainly observed. The market gardeners of Klin have long employed the smoking of cucumber plantations in order to obtain female flowers.

Milliard (1898) and Schafner (1927) obtained 100% formation of female flowers in hemp, by regulating the duration of daylight (according to Sabinin, 1940). There are many other examples of similar changes of the sexual tendency in higher plants.

Kuhn (1941), Zhukovskii and Medvedev (1948) and others, assume that the formation of specific substances, sex-"determinants" are determined by light stimuli, short-wave rays of the solar spectrum. It to assumed that sexual tendencies are connected with the photochemical reactions of specific substances. According to some authors, in the expression of sexual tendency. pigments, especially carotenoids play an important role (Lebedev, 1953).

In our opinion Sabinin in right in indicating that the sexual tendency is determined not by specific individual compounds but by the whole living substance of the cell.

In lower plants--algae, the sexual tendency is evidently subjected to more variability than in higher plants.

At present, there is voluminous data in the literature on the hermaphroditism of species and variants, as well as on experimentally obtained variants in protozoan organisms. Changes of sexual tendency have been studied in many specimens of protozoa and especially in flagellar algae--*Chlamydomonas, Polytoma, Stephanosphaera* and in some species of *Ectocarpus, Dasycladus, Tetraspora* and others.

Among the known species of *Chlamydomonas*, there are dioecious organisms, hermaphrodites, and organisms where the sex cannot be clearly distinguished. Among the latter, the Hissen group (from the town Hissen) of*Chalmydomonas pseudoparadoxa,* is of special interest. In this group there is no copulation between the cells. but after treatment with filtrates obtained from cultures of another dioecious form of *Chlamydomonas,* the sexual process between the gametes proceeds in a normal way. Thereby one of the "Hissen" clones is activated only by filtrates of one type of gamete, and other clones--only by filtrates of another type of gamete of the dioecious group (Moewus 1933, 1935, 1950: Lewin, 1954, Smith, 1951; Hartmann, 1923, 1943, 1955). Moewus performed crossings between dioecious groups, and also between the clones of *Chlamydomonas paradoxa* and *Chlamydomonas pseudoparadoxa* which are identical in respect to sex. He established that female clones copulate with other females. and male clones with other males, gametes of one species of algae copulate with gametes of another species. These data confirm earlier observations described by Hartmann (1923, 1943), which lead to him conclusion on the relativity of the sexual tendency and served as the foundation of his general theory of sexuality. In 1925 Hartmann showed that the existing unisexual male and female groups of algae *Chlamydomonas paupera* were of different potency or valence, hence, cells of one and the same sexual potency may copulate as gametes of different sexes. Hartmann, and subsequently other investigators, observed a relativity of sexual tendency in the dioecious green algae *Spyrogyra quinina.* In these algae the separate filaments consist either of only female cells, or only of male cells.

Under special conditions, cells of one and the same filament sometimes react as female cells, and other times as male cells. This is well demonstrated when three or more adjacent filaments take part in the sexual process. Cells of the middle filament copulate with those of one of the neighboring filaments as gametes of male traits (Figure 28).

Figure 28. Schematic representation of triple copulation in the dioecious alga *Spirogyra quinina:* Middle filament B functions as a male relative to filament A and a female relative to filament C (after Hartmann, 1943)

Detailed investigations were performed by Moewus on the alga--*Chlamydomonas eugametos*. In experiments with this alga not only facts pertinent to relative sexual tendency in gametes of one and the same sexuality were established but also facts relative to its physiological and biochemical differences. It was shown, that sex changes depended on growth conditions, the composition of the nutrient medium, light sources and other factors.

It was previously known that dioecious gametes have distinct physiological properties, the two sexes secrete different substances into the medium. The presence of these substances stimulates the gametes to copulate. The formation and presence of these two differentiated substances in the medium were investigated in detail by Kuhn, Moewus and others and by collaborators of Hartmann, Forster and Wiese (1954) and others.

It was shown that in the formation of gametes and the determination of sex, as well as during fertilization itself, a complexity of variously named substances took part. In three of the most minutely investigated species of algae *Chlamydomonas eugametos, Chl. dreadenais* and *Chl. braunii* the following substances were found.

1. *A motility substance,* stimulating the gametes to motility. It is formed under the influence of light and then secreted into the medium. If a filtrate of *Chlamydomonas* is added to a culture with nonmottle gametes. the latter acquire flagella and become motile. In the absence of light and filtrate the algae are nonmotile and do not copulate. The motility substance affects cells of its own species more strongly than those of foreign species. Consequently, it is specific to some degree.

The chemical nature of the motility substance was investigated on dense concentrates of culture filtrates of *Chlamydomonas,* 16 ml of a bright yellow concentrate was obtained from 300 liters after evaporation. The presence of a carotenoid very close to crocin was established. Crocin ($C_{44}H_{64}O_{24}$) obtained from saffron, proved, on investigation, to have the same effect on cells of algae of *Chlamydomonas* as filtrates of the substance obtained from them. The sensitivity of algae cells to the latter to very high. Even at a dilution of 1:250 billions, crocin as well as the substance of filtrates activates the cells of *Chlamydomonas eugametos* in the dark. After 4 - 5 minutes the gametes become motile and ready for the sexual process.

2. *Fertilization substance or gamones.* Investigations showed that in order to evoke copulation of alga cells of *Chl. eugametos,* the presence of the "motility" substance alone is not enough. Still other elements, the so-called gamones are needed. Copulation occurs when a filtrate of culture of *Chlamydomonas,* kept under light, is added to the medium. It was shown that in the filtrate two different substances connected with sex occur in the gamones: one affects female gametes, the other affects male gametes. Gamones affecting female gametes are formed by them, and gamones affecting gametes of the opposite sex are formed by male gametes under the influence of the violet and blue parts of the solar spectrum. Therefore, the male cells need a more prolonged radiation than the female cells. When the same cells are subjected to the effect of light, (after 24-26 min) the female gamone appears first in the filtrate and the male gamone appears later (after 74-76 min). The female gamone was named gynogamone, the male androgamone.

Gynogamones and androgamones consist of a mixture of two chemically established substances: trans-dimethyl crocetin and cis-dimethyl crocetin. At first the latter to formed under conditions of the culture, and then under the effect of violet and blue light it transforms into trans-dimethyl crocetin.

The quantitative relation of trans-and cis-dimethyl crocetin determines the sexual tendency of the gametes of algae. The composition of gynogamones of *Chlamydomonas eugametos* form a simplex, corresponds to three parts of cis- and one part of trans-dimethyl crocetin ester. In other groups and variants the relation of these two gamones is different and may change within the limits of 10%.

Studying various groups and variants of *Chlamydomonas eugametos,* Moewus found that they have different sexual valency, i. e., have different ability to start the copulation process. Expressing this valency by four numbers one may state, that the weakest valency (male [1] and female [1]), is observed in *Chl. eugametos* forma synoica; it is expressed somewhat stronger in *Chl. eugametos* forma simplex (male [2] and female [2]), The valency is higher in *Chl. eugametos* forma typica (male [3] and female [3]) Finally. it is manifested most strongly in the anisogamic type *Chl. braunii* (male [4] and female [4]). In experiments on crossing these variants, various combinations were obtained; in the first group gametes with a valency of 1 did not copulate or copulated in single cases; the experimental cells of the second group with a valency 2 produced about 20 copulating pairs; in the third group 100 and more pairs were obtained out of the total number of organisms tested.

The biological effect of gamones consists in that, in the presence of these substances, the gametes are attracted to each other, stick together in groups and copulate. They also cause agglutination of gametes. Moewus assumed that the adherance of algae cells which he observed was caused by an admixture of bacteria. Forster and Wiese (1954) found agglutination of cells in sterile cultures of algae in complete absence of bacteria. These authors established that the process of sex determination in algae to neither conditioned by carotenoid nor crocin but by other specific substances of protein nature--glycoproteins. Hartmann is also of this opinion (1955).

As was indicated above, at an appropriate degree of development of only female or only male tendency, the gamete cells may react with each other as organisms of different sexes. The more the degree or valency of sexual potency is expressed, and the stronger the difference in the potency of the reacting cells, the greater the intensity in which copulation proceeds. Moewus cited six combinations for copulation of gametes of the same sex. Combinations male [4] x female [1] give the highest percentage of copulating pairs, the combinations male [4] x female [3]--least. Gametes male [4] and female [4] do not copulate at all. It was shown that, the greater the difference in the content of cis- and trans-substance in copulating gametes, the more vigorous is their fusion. If the difference in the content of these substances is small (not exceeding 20%), copulation does not occur.

In 1923 Hartmann stressed the quantitative character of alterations of gamones on the basis of his theory of relativity of sexual tendency. He did not regard male and female cells as absolutely male and female, but as an expression of quantitative relationships of gamones (Figure 29).

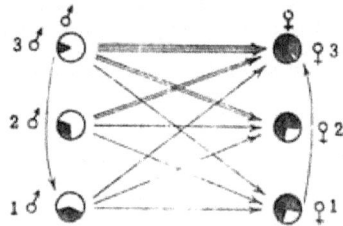

Figure 29. Scheme explaining experiments on relative sexuality (after Hartmann, 1931): white color--male substance in gametes; black--female substance; left (1, 2, 3)--various kinds of male gametes, right (1. 2. 3)--females gametes. Arrows show positive reactions: triple arrows--strong reactions; double--average reactions; single--weak reactions. Gametes with arrows directed to them behave as female gametes with respect to gametes from which the arrows start.

3. *Termones* or substances determining sex in algae were found by Iollos in 1926. Afterward the existence of these substances was also revealed by Moewus. Upon the addition of filtrate of dioecious species*(Dasycladus)* or flagellates *(Chlamdomonas)* of algae to gametes of monoecious algae *Chl. eugametos* forma synoica, the whole population acquires a monosexual character, namely, from filtrates of male gametes--a male tendency, and from filtrates of female ones--a female tendency. A specific pH of the medium should be preserved, for the first alkaline, for the second acid.

According to Moewus (1950), the indicated effect of filtrates is not conditioned by gamones, but by special substances--termones (the female--quinotermone and the male--androtermone). These substances have not been isolated and investigated. It is assumed that they are related to the bitter substance of saffron; gynotermone to picrocrocin, androtermone, to safranal.

The effect of termones in manifested only on hermaphrodite forms, the dicecious gametes do not react.

The nature of mutual relations between chemical substances taking part in copulation processes of algae, according to Moewus (1950), are as follows: they are all chemically close to the bitter, fragrant substance of saffron, largely distributed in the plant kingdom and derived from protocrocin. Protocrocin occurs in all cells growing in the dark. Two substances--cis-crocin and trans-crocin, and also crocin, are formed from it. The latter is formed by splitting of protocrocin under the influence of a special enzyme. Consequently, picrocrocin also appears. Crocin in sexually oriented cells, transforms into cis- and trans-crocetin dimethyl ester, i.e., in one type of gamete--into androgamone, in the others into gynogamone. The latter brings the cells to copulate. Picrocrocin under the influence of cellular ferments trans-forms into safranal, i.e., into termone, and the termone transforms the bisexual cells into cells of different sexes--either males or females.

The relativity of sexual tendency is noted in some protozoan organisms. In the infusorian *Paramecium aurelia* various groups and types have been revealed whose cells copulate in certain combinations. Lines of Infusoria are described, living in various geographical parts, in which several types of cell coupling have been noted. On these grounds the mentioned organisms were subdivided into independent species (Nanney, 1954).

Substances were also found in Infusoria which activate the sexual process. Kimball (1939) established that filtrates from cultures of *Euplotes patella* of one type of coupling evokes conjugation in organisms of a second type of coupling. Thereby types of one line induce cell copulation in all other types, except its own. Types of another category evoke conjugation only in certain types of distant relation. This type of induction resembles the effect of andro- and gynogamones of algae mentioned above.

Special substances of the type of antibiotics formed by separate types or line's of Infusoria display an effect on processes of copulation in Infusoria. In a series of investigations it was shown that some line a of Infusoria formed the so-called paramecins which suppress life processes in other lines. Some paramecins evoke a strong swelling and vacuolization of cells, others attack the motility organs, and still others affect metabolism. All of them display a great influence on copulation processes in various ways.

Of special interest are substances, formed by cells of *Paramecium* and appearing in the protoplast in the form of small granules or bodies, called "kappa". These particles, too small to be seen by eye, are determined by indirect methods. Upon division of cells they also multiply and pass from the parent individual to the daughter cell through a great number of generations. Their number in the cell may reach considerable values (250-450) and remain on this level if the rate of cell division in not too high. Upon accelerated multiplication of cells the size of the "kappa particles decreases, their number becomes smaller and they may disappear completely, In this way one may obtain a line entirely deprived of these particles. However, if one single "kappa" particle remains in the cell, an increase of their number to the limits mentioned above occurs, if the multiplication of individuals becomes slower.

The process of paramecin formation and lack of cell sensitivity to it is closely connected with the formation of "kappa" particles. The process of formation of the antibiotic slows down with a decrease of the number of the "kappa" particles in the cell. Lack of sensitivity to foreign paramecin is preserved till at least one "kappa" particle remains in the cell. The cells become sensitive to paramecin after the disappearance of these particles. The "kappa" particles are inactivated at 36°. Maximal formation is observed at 27°; at 10° this process slows down, and at 30° it stops. Numerous attempts have been made to explain the nature of this mysterious substance. Some investigators (Altenburg, 1946) regard the "kappa" particles as symbionts, others as viruses (Lindegren. 1945). However the majority of authors reject these views and do not find an explanation for this phenomenon (according to Dogel'l, 1951).

The process of conjugation and autogamy in Infusoria may be induced by the addition of killed cells to living ones. Metz (1947) and others added about 1,500 to 3,000 cells killed by formalin to each 800-1,200 living cells of a culture of *Paramecium aurelia*. In this mixture agglutination of living with dead cells occurred. After 60-90 minutes the living cells part from the dead and begin coupling and conjugating. This coupling is assumed by the authors to occur during agglutination when the cells are in the aggregates (Metz, 1954; Tyler, 1948).

The described process of conjugation of paramecia is apparently affected by some chemical substances formed in the dead cells. By its action it resembles the gamones of algae.

The relativity of sexual tendency in protozoa is well manifested in experiments with various clones. By suitable cultivation one may obtain male or female gametes at will. Lerkhe grew a large number of clones on a medium of normal composition and on media deficient in nitrogen and phosphorus. Cells, after cultivation on the poor medium. behaved an female gametes; they were larger and acquired a red pigment. Cells, grown on the rich nutrient medium had a male tendency, were smaller and possessed a green pigment.

The red gametes do not copulate with the red ones, nor the green gametes with the green ones. When mixed they begin the formation of pairs quite rapidly, the green gametes surround the red ones and conjugate with them as individuals with a sharply manifested sexual difference (according to Dogel', 1951).

Sexual potency at a relative sexuality is also noted in bacteria by some authors. According to the opinions of Lederberg, Hayes, Clark and others, cells of the coli bacillus *Bac. coli K-12* of auxotrophic variants possess different degrees of sexual tendency. As was indicated earlier, these bacteria are able to exchange some essential vital substances upon a direct contact of cells. This mutual exchange or mutual assimilation of cell metabolic products occurs only with certain combinations of organisms and is regarded as a sexual process. As in algae, variants of coli bacilli according to the authors, have a male and female sexual tendency, manifested to various degrees. In other words, bacterial cells possess different sexual valencies. According to this, the process of cell fusion expresses itself in various ways. The greatest productivity is obtained upon mixing a culture of maximal male potency with a culture of maximal female potency. Defining the degree of potency by the four-number system, Lederberg showed that productivity upon mixing $F+_4$ with F was greater than that obtained after mixing of cultures $F+_3$ with F. The productivity of mixed cultures of $F+_2$ with F is still lower. Productivity decreases gradually in combinations of cultures $F+_4 \times F+_1 > F+_4 \times F+_2 > F+_4 \times F+_3 > F+_4 \times F+_4$.

As may be seen from the above, one can assume that the sexual function in various specimens of lower organisms proceeds quite differently. In some forms the sexual process occurs between sharply differentiated male and female gametes, in others this process proceeds between organisms which are sexually homogenous; gametes of the same sex, male or female but with different sexual potency, copulate. There are organisms in which the copulation process is accomplished inside one and the same cell between separate parts of the protoplast with a different sexual potency or with different potencies of the same sexuality.

The sexual tendency and sexual potency are not constant features of the organism. Both may be altered depending on nutritional conditions, and also on environmental factors. One and the same cell during its development may become a male or female cell depending upon various growth conditions.

Consequently, sexual tendency is of a relative nature. Each cell has two elements--a male and a female one. The presence and manifestation of these elements during the physiological and morphological development of the organisms determines the latter's sexuality.

At the present stage of our knowledge, although we may experimentally succeed in affecting the direction of the sexual tendency of the organism, many basic questions in the field of sexuality remain unsolved. It should be noted that the developing theory of Hartmann on general bipolar bisexuality as a basis of the whole sexual phenomenon was preceded by statements of Bütchli and later by Schaudin. Bütchli regarded the sexual process as an act leading to the rejuvenation of the organism. Schaudin assumed that each cell is hermaphroditic or bisexual to a certain degree, and possesses elements of both the male and female sex, As a result of predominance of one or the other element, the cell becomes either male or female. Thus, the author explained cases of autogamous copulation in protozoa, algae, and bacteria.

Hartmann and his collaborators, developing the theory of relative sexuality, indicate that from the very beginning organisms on earth were of a bisexual nature and contained both male and female elements. These elements evoke a sexual tension in the cells, stimulating them to unite and to level off or stop this tension. Sexual tension has a decisive role in the whole chain of fertilization processes. It is created as a result of an irregular development of one or the other sexual element. Sexual tension may occur not only between separate cells but also between parts of one and the same cell. In bacteria and other specimens of microorganisms which are on a lower evolutionary level, sexual tension between parts of the protoplast is apparently of essential importance in the rejuvenation of the organisms. The sexual tension created in the cells also determines the nature of autogamy in microbes. It should be assumed that in nonsporeforming bacteria autogamy does not proceed as in sporeformers, that in actinomycetes it takes place in a different manner than in fungi.

In various species of sporeforming bacteria this process is morphologically manifested in a different way. The sexual process in *Bac. bütchlii* which was described by Schaudin (1902) is not found in other sporeforming bacteria. In them fusion of various cellular elements proceeds without formation of the longitudinal line of inclusions. However the product of fusion--the spore--constitutes in both cases a formation of the same order and may be regarded as a zygote. Formation of similar zygotic cells or spores in specimens of filamentons bacteria, *Oscillospira guilliermondii,* occurs after fusion of two and sometimes of three and more contiguous cells, In these organisms the sexual differentiation is on a higher level, as it involves, not parts of the protoplast, but whole cells. A more complex process of copulation occurs in organisms which are higher in evolutionary development.

Microorganisms manifest considerable variability. They are more variable and react to environmental factors more strongly and rapidly than higher plants. However, there is no basis for regarding them as highly polymorphic organisms. Bacteria, actinomycetes, fungi and other specimens of microbes undergo changes according to their species properties.

An higher plants and animals, they have their evolution and philogeny which determine the development of cultures in ontogenesis and the formation of new forms and, in general, the nature of variability of the species.

In this the microorganisms sharply differ from higher plants. Peculiarities of their structure, development, and many biochemical processes separate these creatures into independent groups or classes of organisms. Biological specificity of microorganisms is also undoubtedly reflected in the phenomenon of variability of their species. The regularities observed of variability processes in bacteria, actinomycetes and other microbes do not fit into the schemes and theories of variability of higher organisms.

Not long ago, the followers of formal genetics refuted in general any regularity in hereditary variability of microorganisms only for the reason that the nature of alterations did not correspond to the "laws" of Mendel and to other established points of view. Microbes were regarded as defective organisms, as a result of which the whole factual material on variability was disregarded.

Under the pressure of enormous material amassed in the literature, geneticists have recently also been forced to pay attention to microorganisms. Many species of bacteria, yeasts and fungi became the preferred subjects for attempts at the solution of a series of fundamental genetic problems. Many works, mono -graphs and reviews are devoted to the genetics of microbes.

Reviewing this material, many foreign authors (Dubos, 1948, Lederberg, 1954, Braun, 1953, Catcheside, 1951, Hayes, 1953, and others) treat it from positions of pure genetic conceptions. According to their opinions, all alterations of a hereditary order in microbes are conditioned by genes. Genetic conceptions, applied to higher plants and animals are applied wholly into the microbial world. The authors take into account neither the specific structure and development of microbial cells, nor the manifestations of their life activity. Even the factual material which often speaks against these conceptions is used by them in attempts to prove their points of view.

It should be noted that the changes observed in microbes were long considered as phenomena contradicting the universally recognized laws of formal genetics. However, microbiology was at that time deprived of theoretical generalization and could not offer its views against the existing genetical ones. Opinions of a conciliatory nature were put forward. The peculiarities of the variability phenomenon in microbes were regarded as a deviation from the general genetic law and other names were suggested for designation of separate manifestations of variability. Instead of mutation, the word saltation was used, etc.

Together with the development of the theory of Michurin, microbiology in our country and in others generalizes this material in an entirely different way and draws conclusions according to the basic laws of Darwin on the variability of organisms.

The large experience of microbiology shows that heredity is not a manifestation of some invariable particles of living substance, transferred from generation to generation during an infinite series of generations. Heredity cannot be regarded as a manifestation of idioplasm nor genes be regarded as carriers of hereditary features, not subjected to environmental influences,

Heredity is a property of the whole organism, the whole protoplast or living substance, and not of separate particles located in chromosomes or in chromatin corpuscles of the nucleus, although the latter is of essential importance. The transmission of traits from one generation to the other, the ability to form new properties, develops in the organism during the whole historical process and depends upon the conditions under which the given microbe exists and develops. The property of inheritance of traits is formed during a long series of generations of one or another species. Formation of new species occurs under the influence of environmental factors and strictly according to the hereditary basis. New forms are produced due to the influence of various agents on the hereditary basis. Knowing the natural properties of the organism and its adaptive features, one may obtain, and in fact does obtain, useful variants by a choice of suitable conditions of growth and development.

It is possible to create a new organism and to direct its development, varying the conditions of growth, following the history of its evolutionary development. This theory of Darwin was well confirmed by Michurin and his followers in numerous experiments in laboratories as well as in the fields.

The enormous amount of microbiological material on adaptive variability at present induces many followers of the theory of corpuscular heredity to revise and alter their views on the variability and heredity of microorganisms. The investigations by foreign scientists (Ephrussi et al., 1949, 1956, Hinshelwood et al. , 1946. 1952; Monod et al., 1942, 1952, Slonimskii, 1956 and others) show the special importance of environmental conditions in the creation and formation of new species. In many instances they showed that the environment or substrate induced the formation of new features in the organism, and stabilized and transferred them into the constructive stage of living plasma.

The spectfity of the development of microorganisms, and their unusually rapid proliferation enables us to follow a series of life activities in a large number of generations in a relatively short time and to establish regularities and particular features of species -forming processes. If, in order to establish the phenomenon of heredity in higher organisms many years are needed, in microorganisms it may be elucidated in one to two days. Thanks to the biological specificity of microbes, one succeeds in following a series of biochemical processes in them which are connected with the mechanism of variability and heredity. It is quite natural that many questions of genetics are solved on microbiological references.

Variability in microorganisms is manifested in many ways.

There is individual variability, age variability, species, adaptive, induced variability, etc.

Individual Variability or Polymorphism of Cells

When observing cultures of microorganisms under the microscope, whether they are bacteria, yeasts or others, one may always notice a greater or lesser diversity of cells. Cells differ from each other in size, form and other external features and also in their internal structure. Some of them possess a homogeneus plasma without any apparent structure, others granular, vacuolized or without vacuoles. In some cells there is much reserve food-- metachromatin, lipides or other substances; in still others--there is little or they lack it entirely (Krasil'nikov, 1943).

Individual differences of cells of one and the same culture are manifested to various degrees depending on the age, species and growth conditions. Cells of old cultures differ most sharply from each other when their involution begins. In these cases, as was indicated above, the cells may assume diverse forms and sizes.

Of great interest are very enlarged cells and cells of ultramicroscopic sizes or filterable forms.

Formation of greatly enlarged, deformed cells as a reflection of polymorphism is noted in cultures of various bacteria and actinomycetes. The enlarged cells are of various sizes and patterns. They may be spherical, bottle-shaped, spindle-shaped, filamentous, amoeboid, etc. Often the long cells form side appendages resembling branches, and in swollen cells budlike formations often appear on the surface (Figure 30).

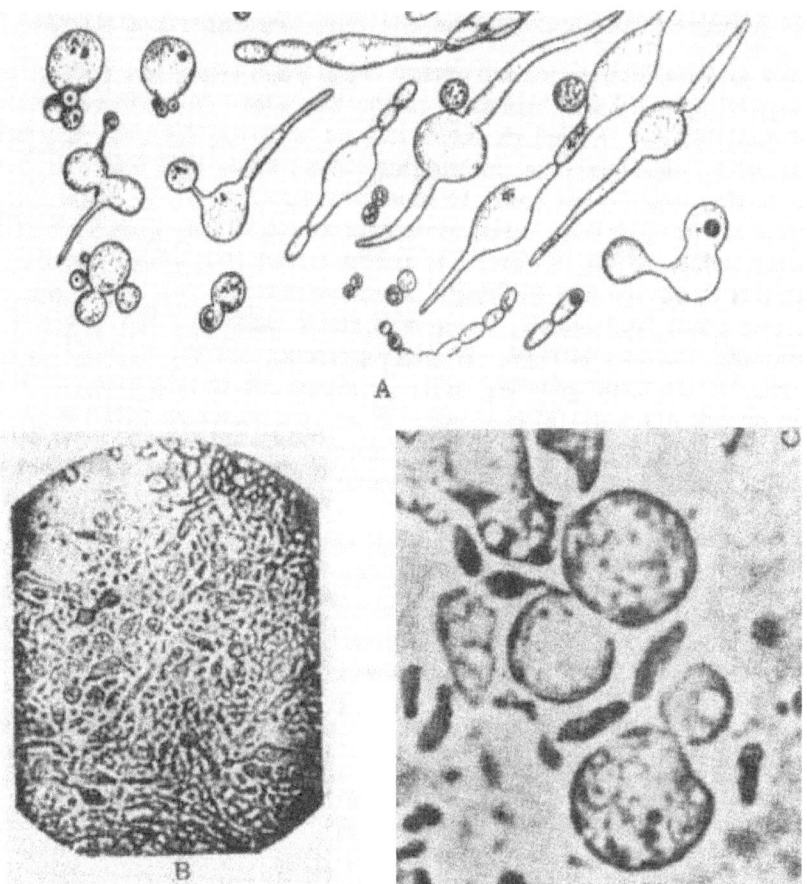

Figure 30, Involution cells of bacteria, greatly swollen and deformed: A) *Azotobacter unicapsa,* strain A--5 days on must agar (pH = 6.5) 1:1,000; B) *Azotobacter chroococcum*--1-2 days culture on must agar; C) *Achromobacter epsteinii* (Peshkov, 1955). Growth of culture at 10° C, 1:2160.

Such cells are tens and even hundreds of times larger than normal ones, often reaching gigantic dimensions. The protoplast of these greatly enlarged cells differs noticeably from the protoplast of normal ones. Their plasma is vacuolized with many inclusions of metachromatin, lipides, glycogen-like substances, etc. An accumulation of chromatin inclusions (see above) is characteristic of them.

As a rule, these cells are less viable, they do not develop and multiply; usually they perish, disintegrate and undergo autolysis. Such cells are designated as involution forms. If the process of degeneration did not go far, the cells may under certain favorable conditions regenerate and produce normal progeny.

With some bacteria, one may maintain a culture in a degenerative state for a long time, with characteristic, greatly enlarged and deformed cells. The latter multiply and give rise to organisms with degenerative features.

We have preserved, In a degenerative form, cultures of *Azotobacter,* growing on must agar, or on a medium with meat-peptone broth, and a culture of root-nodule bacteria of pea, clover, vetch and other legumes, growing on meat-peptone agar. The indicated media are hardly suitable or not at all suitable for growing these bacteria. The development of cells is weak and as a rule they are of abnormal size and form. Though they do not lose the ability to proliferate, they do it in an unusual way--by budding and fragmentation. Consequently, any degeneration or deformation of cells does not necessarily lead to death.

Strongly swollen cells are formed under the influence of physical, chemical and biological factors.

Very large, deformed cells may be obtained under the influence of chemical substances. Gamaleya obtained them in a medium with lithium chloride. They are formed under the action of increased concentrations of ordinary salts, in the presence of certain dyes, by the effect of ultraviolet rays, supraoptimal temperature, antibiotics, phages, etc. As a rule, all unfavorable agents affecting the growth of microbes may provoke the indicated forms of degenerated cells Imshenetskii, 1940; Krasil'nikov, 1954 b; Iensen, 1954; Peshkov, 1955).

The degree of morphological degeneration in various bacterial species and actinomycetes is different. In some it is strongly manifested, in others weakly, almost unnoticeable.

Morphological polymorphism is very strongly manifested in *Azotobacter*. The diversity of cells in this species amounts to several tens of forms. Lönis and Smit (1916) counted 13 different types, we (1954b) and Bachinskii (1935) have found about 200 forms. The study of a large collection of *Azotobacter* showed that such polymorphism is not manifested in all strains.

A great diversity of cellular forms of a degenerative nature is noted in some mycobacteria--*Mycob. cyaneum* and others. In actinomycetes this phenomenon is reflected mostly in *Act. violaceus* (Krasil'nikov, 1938a. 1954 c).

In sporeforming bacteria *Bac. megatherium,* deformed cells were studied by Kudryavtsev (1932). The diversity of the forms which he described was noted upon cultivation on conventional nutrient media, but more characteristic forms and better manifested forms occured on sugar media (must agar, etc).

Peshkov (1955), studying the formation of reactive forms in the microbe *Bact. epsteinii* which he isolated, regards this bacterial species as an example of an unusual polymorphism.

The formation of swollen forms is observed in various bacterial species--pathogenic and saprophytic (see Kalina, 1953, Muromtsev, 1953; Elin, 1954 and others). Numerous investigations show that bacteria react to various antibiotics--penicillin, streptomycin, aureomycin and many others--by a sharp alteration in their size and form. Deformations similar to those formed upon the action of other agents occur (Krasil'nikov, 1950, Troitskii and others, 1950).

Aside from the formation of very large cells in cultures of bacteria and actinomycetes, the formation of tiny gonidial elements is observed frequently. Under the microscope in almost every culture, particularly in old ones, one may see among the usual cells very small bodies in the form of granules, which refract light weakly. These bodies often accumulate in a great quantity, forming a fine-grained sediment.

The size of the fragmented cells is very small ranging from 0.5 *u* to a size hardly visible under good optical microscopes and even smaller --(beyond the limits of visibility).

Formation of the diminutive forms may occur in various ways, It may be observed during the process of fragmentation of cells when division is not accompanied by growth or the latter is weak. With each successive division, cells of smaller size appear. In some bacteria *(Azotobacter)* such fragmented cells are found in a mucous capsule. As a result, it looks as if a large cell, filled with gonidial elements, resembling nanocytes of algae is produced, The size of such "nanocytes" vary within the limits of 0. 1 - 0. 5 *u* in diameter (See Figure 16).

The fragmentation of cells is observed in sporeforming and nonsporeforming bacteria, in actinomycetes and mycobacteria, therefore the process is often accompanied by autolysis of the maternal cells.

As was described above, tiny cells are also formed by budding.

In separate cases inside the often deformed, and swollen cells, very small gonidia are formed. These are the so-called regenerative forms. Their formation proceeds from the protoplasm with the necessary participation of chromatin granules, as was stated above.

Owing to their small size, all those forms pass through bacterial filters and may be regarded as filterable forms. Together with the latter they evidently constitute one group of reproductive elements. These forms are also close to each other in their properties. Both are slightly viable, and do not develop or develop weakly on the usual nutrient media.

In recent years great attention has been attached to polymorphism which is connected with the formation of the highly altered, so-called L forms in some bacteria. The most detailed study was made on *Streptobacillus moniliformis,* Klieneberger-Nobel (1949). Oerskov (1942) and others; showed that in a given organism which is grown on artificial nutrient media, very different forms are manifested. Beside large, greatly swollen forms, there are extraordinarily small elements resembling the pleuropneumonia agent. The latter are so peculiar, that many investigators consider them a special branch of microbes having nothing in common with bacteria. Small forms of the pleuropneumonia type are described in many bacteria.

Cells of one and the same culture differ from each other not only morphologically but also physiologically and biochemically, therefore the differences may be quantitative and qualitative. Although there are no methods for studying the physiology of separate cells, observations on hanging drops show that this difference exists. It is manifested in the reaction of cells to environmental stimuli. Depending on individual physiological properties, metachromatin accumulates rapidly and in large quantities in some cells, while in others it accumulates in a small quantity and slowly, in still others it is altogether lacking. Accumulation of fat. glycogen, glycogen-like substances and other inclusions, occurs in some cells and not in others. Some cells multiply rapidly, others slowly. Rapidly-growing cells produce two to five generations and more, while slow-growing cells produce one to two generations during the same period, or do not start division at all. If cells which are filled with reserve nutrient substances are transferred to a poor medium, the reserves proceed to disappear in various ways: in some cells it proceeds rapidly, in others, slowly.

Bacterial cells react in a different way to the action of antibiotics--penicillin, streptomycin, aureomycin, or other preparations. Some die rapidly, others remain in a subvital condition; on a transfer to a fresh normal substrate they do not grow or grow only under special conditions of complementary nutrition; still others grow, but do not multiply or multiply slowly, and subsequently, swollen. degenerative forms appear. There are resistant cells, growing and proliferating more or less normally, without alteration of external form and size.

Hughes (1953-1955) indicated that the two daughter cells of bacteria *(Bac. coli Bac. proteus),* formed after the division of the maternal cell, differ from each other in their sensitivity to antibiotics, lack of oxygen, and some other factors.

Polymorphism of cells is also manifested in biological reactions. Cells of one and the same culture differ from each other in their resistance to various agents. Upon the action of increased temperature some cells die, while others are preserved but still others develop and multiply. This to also observed under the action of radiation, chemical reagents, etc.

The sexual process is manifested in some cells (in yeasts), in others it is lost. Some cells form spores, and others do not.

Consequently, cultures of microorganisms consist of cellular forms of great diversity; of organisms quantitatively nonhomogeneous.

These differences--morphological, physiological and biochemical--are conditioned by the biological peculiarities of microbial cells. They are not fortuitous but biologically conform with natural laws in all microbial species. Qualitative diversity is of great biological importance. Owing to the diversity of cells, the over-all microbial culture of species possesses greater possibilities of adapting itself to various conditions. If this diversity is lacking. the whole culture would die upon the action of the first unfavorable condition.

Diversity of a culture or the polymorphism of cells should be distinguished from the qualitative diversity of polymorphism of species. Polymorphism of cells in encompassed by the concept of the polymorphism of species but is not identical to it.

The qualitative diversity of microbial cultures also determines to some degree the polymorphism of the species. The species in microorganisms is characterized not only by the polymorphism of cells but also by the polymorphism of the culture. One and the same species, even one and the same culture often produce various colonies on nutrient media. It is known that *Azotobacter chroococcum* on one and the same medium of Ashby (agar medium) grows either in the form of a mucoid spreading colony, or in a compact, pastelike form; it may be black, brown or colorless; in some cases it is wrinkled or rough, in others in the form of smooth, pastelike or mucoid colonies.

In the sporeforming bacteria *Bac. mesentericus, Bac. subtilis, Bac. mycoides, Bac. brevis, Bac. cereus* and others, from five to ten types of colonies may be differentiated on agar media. One and the same culture of *Bac. mesentericus* upon inoculation into nutrient agar often forms a mixture. typical and atypical forms, rough and folded, finely-plicated; dry, spreading on the agar surface; flat, smooth, pellicle-like, glossy with a single radial fold (Figure 31).

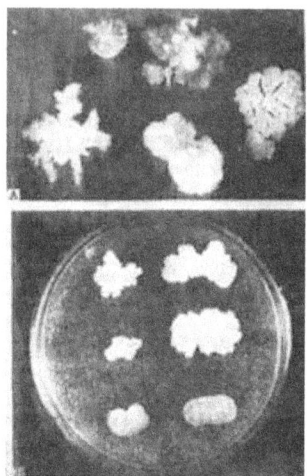

Figure 31. Polymorphism of cultures of sporeforming bacteria *Bac. mesentericus:* A) strain No 12, isolated from serozem of Central Asia; B) strain No 110, isolated from the turfy-podzol soil of Moscow Oblast'; variants of types of colonies obtained upon plating on nutrient medium (meat-peptone agar).

In *Bac. mycoides* atypical, granular, anthracoidal, folded smooth and other colors have been described (Figure 32) (Lewis, 1932; Rautenshtein, 1946;Mishustin, 1947; Afrikyan, 1954a; and others). Considerable polymorphism of cultures is observed in nonsporeforming bacteria: the root-nodule bacteria specimens of *Pseudomonas. Bacterium* in lactic acid bacteria, *Mycobacteria* and others. A great diversity of colony structure is noted in actinomycetes particularly. In these organisms polymorphism is so great, that doubts arise among investigators as to the possibility of a species differentiation. Lieske (1921), for instance, refused to divide actinomycetes into species, being of the opinion that the diversity of forms observed is a result of polymorphism and not of species difference.

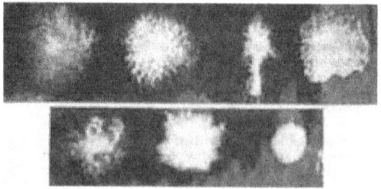

Figure 32. Polymorphism of a culture of *Bac. mycoides.* Types of colonies, obtained after plating on meat-poptone agar MPA. Strain isolated from the turfy-podsol soil of the Moscow Oblast,

According to this point of view, the sporeforming bacteria--*Bac. subtilis* and *Bac. mesentericus* are extreme variants of one and the same species.

According to our observations, *Bac. subtilis, Bac. Mesentericus, Bac. cereus, Bac. brevis* differ essentially from each other only by the form of colonies and character of growth on artificial nutrient media. Usual physiological indications, liquefaction of gelatin, curdling of milk, and decomposition of starch, are essentially the same for these organisms. More refined biochemical properties have not been revealed in them.

The homogeneity of the mentioned bacteria is also confirmed by the method of experimental variability. By the ordinary cultivation and subsequent plating on various nutrient media of a typical culture of *Bac. mesentericus,* one may obtain cultures of the type *Bac. subttlis, Bac. cereus or Bac. brevis, Bac. pumilis* and some other forms in a relatively rapid manner.

The so-called typical strains of *Bac. cereus* produce forms not differing from those of *Bac. Subtilis,* or *Bac. brevis* and *Bac. licheniformis.*

From five species of sporeforming bacteria we obtained the following monotypical variants: a) wrinkled or plicate with a glossy or fatty-glossy surface, butterlike colonies; b) thin-filmy, fine-plicate, dry, mat; c) coarse-filmy with elevated edges, saucer-like with occasional radiate, separate small folds; d) dry colonies, fine-plicate. floury-white, often merging with the agar; e) grainy-plicate colonies, moist, moist-glistening fatty with uneven, slightly diffuse edges (Table 2).

Table 2
Formation of monotypical variants by various species of sproreforming bacteria

Species	Initial strain	Variants obtained
Bac. mesentericus	wrinkled (a)	(a), (c), (e)
Bac. lichenformis	thin-filmy (b)	(a), (b), (d), (e)
Bac.cereus	thick, saucer-like (c)	(a), (b), (c), (e)
Bac. subtilis	dry, fine-plicate (d)	(a), (c), (d), (e)
Bac. brevis	smooth with hair-like edges (e)	(a), (c), (e)

We observed the formation of monotypical variants in certain species of nonsporeforming bacteria of the genera *Bacterium* and *Pseudomonas,* in actinomycetes, mycobacteria and others. These manifestations of a monotypical character of variants in the process of variability of cultures is determined by homogeneity of the living substance and occurs in organisms which are closely related. By these variants one may judge the phylogenetic closeness of the initial cultures or whether the investigated organisms belong to, one and the same species. On this is based the method of experimental variability for the establishment of species in the classification of bacteria and other microbes (see further on).

Diversity of forms and variants in cultures reflects the degree of polymorphism of species. The cellular as well as the cultural polymorphism has a defined organic connection with species variability. Among the multitude of unstable cellular elements or among unstable variants of a culture, separate organisms or colonies with hereditarily stabilized traits of the same property occur, In the yeast organisms, *Saccharomyces cerevisiae* we obtained (1934a) stable variants with properties which manifested themselves in cellular organisms as a reflection of individual variability or polymorphism. In a certain sense the species variability pre-determines individual variability. In species variability, the stable variants repeat or reproduce the traits which manifest themselves in polymorphism.

Species Variability of Microorganisms

Growth and development of the organisms and its reaction to environmental factors proceed within limits of a defined norm, separately characteristic of each species and conditioned by heredity. The alterations proceeding within the limits, of this norm do not touch the strain or species properties as long as the environment corresponds to the requirements of the organism. When the environment changes and it ceases to be suitable for the normal growth of the organism, the latter, either stops development and dies, sooner or later, or adapts itself to new conditions, altering its species properties. New variants with new hereditary features and requirements are obtained.

In laboratory practice one may often observe the production of variants under the influence of the changing environment or the action of environmental factors. In the literature there is a great accumulation of material on the species variability of microorganisms. Variants stabilized in a hereditary way have been obtained in various specimens of bacteria, actinomycetes, fungi, algae, protozoa, plant and animal viruses, bacteriophages, actinophages, etc. They have various designations--mutations, long term modifications, saltations, adaptations, sudden and discontinuous variations, etc.

All those forms of variability are of a hereditary nature, Due to this variability, the species properties of the microbe change, external morphological and cultural features as well as biochemical and biological features are thereby affected. The newly acquired properties and traits are transmitted to cells of the whole population.

New hereditary variants are produced by the alteration of a separate cell in the culture. Among the individual differences in the population, such differences appear which cover species qualities and are stabilized in the progeny. Formation of hereditary variants occurs under the influence of environmental action and, as a rule, is of an adaptive nature. Variants are often obtained in old cultures without any special external action. In this case the altered medium constitutes the activating factor. The latter changes its composition and properties with the age of the culture in an essential way, the initial food elements disappear, now ones are synthesized, various products of metabolism are accumulated, its physical and chemical features change, etc. In short, the medium becomes entirely different, sometimes obviously unfavorable. In such a medium, the cells die in large numbers. There are many half-living or subvital cells, and also cells with a shattered heredity. In these cases, upon plating on a fresh nutrient substrate, colonies with new properties develop. Variants with diverse morphological and physiological properties are produced.

With the first plating of the culture on an agar medium, as was indicated above, one may find variants with diverse cultural traits. Aside from the initial colonies, rough, wrinkled, mucoid, and hardly visible colonies which grow well appear. Under the microscope, an equivalent cell diversity in noted. The variants obtained, often differ from each other in biochemical properties. Some of them liquefy gelatin rapidly, others slowly, and still others do not liquefy it at all. The same is noted with respect to fermentation of milk, sugars and other fermentative processes.

In *Bac. coli* one may observe the formation of stable variants, which lose the ability to produce gas from sugars. Some variants of root-nodule bacteria lose the ability to form nodules on the roots of leguminous plants, other variants lose the ability to ferment sugars. In *Azotobacter* variants are formed which are unable to fix nitrogen and to develop on a nitrogen-free medium. In pathogenic and phytopathogenic bacteria avirulent cultures are obtained. In actinomycetes one may succeed in obtaining new variants of great antibacterial activity, but the formation of entirely inactive variants is also observed.

Hereditary alterations in microbes occur under the influence of various special factors--physical, chemical and biological. Resistant variants are obtained from the action of temperature, ultrasonic radiation energy, etc, Great attention in paid to X-rays, radium radiation and recently to nuclear radiations of uranium to well as to the effect of artificial isotopes. Ultraviolet rays are a very strong agent. Under the influence of radiation energy, many variants of practical importance which are of great industrial value have been obtained. For instance, very active variants of *Penicillium chrysogenum*--the producer of the antibiotic penicillin, cultures of the actinomycete *Act. streptomycini*, whose antibiotic properties exceed those of the initial strains by dozens of times. Under the influence of X rays numerous variants have been obtained in fungi and yeasts, with a complexity of features which differ from those of the initial strains. (Filippov, 1932; Rokhlina, 1930, 1954). When staphylococci are exposed to the action of these rays, the ability to produce the toxic factors dermonscrosin and hemolysin is lost. Under the influence of ß and gamma rays variants with a filamentous structure have been obtained from bacteria which never form them.

Stable alterations. with respect to heredity have been obtained through the use of ultraviolet rays in various specimens of bacteria and fungi. Variants have been described with altered cultural, morphological and biochemical properties. Great attention to given to the so-called dependent or defective variants which have lost the ability to synthesize certain growth factors or vitamins. Thirty variants have been obtained from *Bac. coli*which require the supplementary growth substances--pyrimidine, purine, threonine, proline, phenylalanine, methionine, tryptophan, arginine, cystine and others (Table 3).

Table 3

Production of *B. coli* variants which require complementary nutrient substances and other forms

Complementary substance	Formed spontaneously	Formed with x-rays	Formed with ultraviolet rays	Formed with mustard gas
Pyrimidine		1		2
Purine		1		2
Threonine	1	1		2
Proline	2	2	1	2
Phenylalaine		2		2
Methionine	2	5		2
Tryptophan	2			2
Arginine	1	1	1	
Cystine		4	1	
Leucine		2		
Lysine		1		
Glutamic acid				
Glutamine		1		
Histidine		1		
Isoleucine		2		
Tyrosine		3		2
Homocystine		2		
Thiamine	1	4		
Nicotinamide		2		
Biotin		2		
PABA		1		
Pantothenic acid		1		
Pyroxidin		1		
Sulfite-resistant	1			
Sulfide-resistant	1			
Resistant to sodium chloroacetate	2			
Resistant to lithium chloride	2			
Fermenting lactose	3			
Resistant to mustard gas, X-rays and ultraviolet rays	2			

Note. Numbers designate the number of isolations of the variants, obtained by various investigators (from the book *Microbial Genetics,* Catcheside, 1951).

A considerable number of dependent variants have been obtained in the fungus *Neurospora* (Kaplan, 1952).

Similar results are obtained upon the action of chemical agents. Great attention to being paid to the action of colchtcin and mustard gas. These substances have a particular property for inducing variability in higher and lower organisms. However, investigations show that this property in characteristic of many other chemical compounds. There is no specificity in colchicin and in other chemical substances. Demerets and his collaborators distinguish three types of chemical substances with respect to the strength of their effect on the variability of microorganisms. Strongly active are: formalin, phenol, hydrogen-peroxide, a-dinitrophenol, manganese, iron; weakly active are: boric, acetic and formic acid, trinitrophenol, acriflavin, caffeine, necrosin, etc; of very weak activity or completely inactive are: ammonia, copper sulfate, trivalent iron, divalent cobalt, divalent strontium, etc, nonactive chemicals are: sodium hydroxide, potassium hydroxide, sublimate, lactic, sulfuric, phosphoric, nitric, and hydrochloric acids, silver nitrate and many other chemicals (cited from Kaplan, 1952).

It should be noted that the production of hereditary stable and unstable variants also occurs in cultures without the special action of any agents, and essentially the same types of variants are thereby obtained as under the action of the mentioned agents. In our investigations (1933, 1934a) we compared the variants produced spontaneously and without special actions in the yeast *Sporobolomyces* and *Saccharomyces* with the variants simultaneously obtained by Nadson and Filippov (1932) by the action of radon and X-rays. In both cases monotypical groups of yeasts have been obtained (see also Nadson and Rokhlina, 1932; Rokhlina, 1954),

Kurylovich and his collaborators subjected cultures of actinomycetes and fungi which produce antibiotics to the action of ultraviolet rays. Without exerting any influence, simultaneous analysis of the variability of cultures was carried out. In both cases variants were obtained, which were identical with respect to activity (Krasil'nikov, 1955c).

The monotypical nature of the production of variants is often noted in microbiological laboratory practice and proves the great significance of the heredity of the organism.

Among biological factors which induce the formation of variants, phages and antibiotics have recently been of special interest. The agents prove to have a great effect on the variability of microbes. Upon their action diverse variants are formed, whose characteristic property in the resistance to the given stimuli.

Changes obtained in various ways in microorganisms are often of a correlative nature. When one feature is altered, other features or properties, as a rule, also become altered. This correlation may exist between the cultural, morphological and biochemical or physiological features, as well as between biochemical characteristics. For instance, in some lactic acid bacteria, the ability to ferment sorbital and mannitol in linked with a loss of the ability to synthese polysaccharide, upon which specific agglutination depends. Some strains of *Staphylococcus aureus* which are a to assimilate only normal proline, acquire the ability to assimilate both isomers, after their loss of pigment and transformation into colorless variants. A correlative connection of chemical and serological properties was noted in diverse variants of *Vibrio cholerae,* obtained under various conditions of cultivation on different media. Upon transition of smooth variants into rough forms, many bacteria lose biochemical properties, serological properties are altered, etc. Loss of the mucoid capsule in bacteria sharply alters their antigenic properties. This is also observed upon the loss of the flagellar apparatus. In microorganisms many other external properties correlate in various ways with internal biochemical processes. Alterations in the biochemical functions are accompanied by alteration of the external microscopic features and primarily of the state of the protoplasm (Meisell, 1950; Ierusalimskii, 1949, Stephenson, 1951; Dubos, 1948, Sakharov, 1952 and others).

In the practice of investigating microbes one may not always note the correlative linkage between function and form. One often obtains variants which differ only in biochemical or physiological properties; by use of present methods one may not succeed in revealing any external alterations. In capsular bacteria, variants are obtained, having properties which do not depend upon the capsule or other cytomorphological features. Often variants are formed with different physiological functions but with the same external features. From one and the same culture of yeasts, actinomycetes and bacteria variants are obtained with red, pink. and yellow pigments as well as colorless variants, and those which ferment or do not ferment sugars, decomposing starch or do not, etc.

The correlative connection is manifested in those cases when any function of the organism is related to certain particles of the protoplast, microsomes, chondriosomes or other corpuscles of living substance seen under the microscope. For instance, the capsular antigen is connected to the polysaccharide of the mucus envelope. The presence of the latter will determine the antigenic properties of the cell. It is known that some ferments are adsorbed on the surface of chondriosomes. Alteration of the latter indicates a change of a certain biochemical process. A series of properties is connected with chromatin structures. Alteration of these structures will be reflected in various ways in alteration of certain physiological functions and, in general, in biological manifestations of the organism.

Correlation maybe manifested only between physiological and biochemical properties. For instance, in some lactic acid bacteria, the ability to ferment sorbitol and mannitol is linked with a loss of the ability to synthesize polysaccharides, which in turn affects specific agglutination. Some strains of *Staphylococcus aureus,* which have the ability to assimilate only natural proline, acquire the ability to assimilate both isomers. after a loss of the pigment and transformation into colorless variants.

Correlative connection of chemical and serological properties manifests itself in diverse variants of *Vibrio cholerae,* obtained under various conditions of cultivation. There are also other manifestations of the correlative connections of physiological and biochemical processes to the variability of microorganisms. It should be assumed that variability always attacks a series of properties correlatively connected; but this connection is not always revealed by present methods of analysis.

Adaptive and Directed Variability

Laboratory experience shows that after long cultivation of microbes in a medium which is unusual for them, and has unassimilated sources of nutrients, the microbes begin to assimilate the latter and the medium becomes an ordinary one and even indispensable. For instance, if yeasts which do not ferment maltose are cultivated on a medium with this sugar as the sole source of carbon, then after some time (it may be protracted after many passages) they acquire the ability to ferment this sugar, A new enzyme is formed in the yeasts--maltase; the cells are physiologically altered, they become new variants (Kosikov, 1950, Kudryavtsev, 1954).

Such a physiological rebuilding may be induced in bacteria, actinomycetes, fungi and other microbes with respect to many sources of nutrients, carbon, nitrogen, organic and mineral sources.

The first case of such adaptive variability in bacteria was described by Wortman: then a similar phenomenon was observed by Mechnikov. Kosyakov showed that bacteria may be "trained" to antiseptics--boric acid, sublimate, borax and others. He observed that the anthrax bacillus and other bacteria might be trained to high concentrations of the indicated poisons. This training was achieved by steps, starting with small doses. Neisser and Massini observed adaptation of the colon bacillus to lactose. By protracted cultivation on a medium with this sugar, bacteria start their fermentation and assimilation. The paratyphotd, bacillus *Bac. paratyphi* "B" adapts itself to raffinose, and the typhoid bacillus--*Bac. Typhi*--to lactose, saccharose, rhamnose, dulcitol and isodulcitol. Such an adaptation to sugars to also noted in the dysenteric and other bacteria of the colon group; (see Kalina, 1953; Kudlai, 1954; Elin, 1954; Muromtsev, 1953).

A series of many sporeforming and nonsporeforming bacteria, an well as lactic acid bacteria were described as easily adaptable to rhamnose.

Adaptive changes were also observed with respect to nitrogen sources of nutrition. Strains of *Clostridium* may be trained to decompose casein and gelatin by adopting the same method of cultivation. Typhoid and paratyphoid bacteria adapt themselves relatively easily to sources of mineral nutrition and to ammonia salts.

The ability to liquefy gelatin may be induced in yeast of the *Saccharomyces* type. Some strains of *Azotobacter, Az. chroococcum,* do not grow on protein media (meat-peptone agar), but by a successive "training" they start assimilation. New forms or variants are obtained which develop well on media with organic nitrogen. If *Azotobacter* is cultivated for a long time on a medium containing nitrogen it loses the ability to fix molecular nitrogen.

Many saprophytic bacteria from the genera *Pseudomonas, Bacterium, Bacillus* and others, which do not grow on media with mineral nitrogen, acquire the ability to assimilate it during the process of adaptation.

In the literature numerous cases of quantitative and qualitative alterations of fermentative properties in bacteria are described. Under the influence of the nutritional substances of the medium, the cells of the microorganisms elaborate suitable enzymes. Production of such adaptive enzymes is noted in many bacteria, fungi, yeasts, actinomycetes and protozoa. For instance, in some variants of yeasts obtained experimentally which were deprived of the enzyme cytochromoxidase, the latter is synthesized under the influence of oxygen. Ephrussi and Slonimski showed that the mentioned enzyme is directly induced by oxygen molecules in the protoplasm. In this the novo process of formation, specific particles of the plasma which are not connected with nuclear elements take a direct part (Slonimski, 1956).

Adaptive ferments such as galactosidase, maltase and others, were found in yeasts by various investigators. Synthesis of new enzymes under the influence of specific substrates to observed in specimens of various bacterial species--*Bac. coli, Bac. typhi, Bac. typhi murium, Bac. lactis aerogenes* in sporeforming bacteria, mycobacteria, actinomycetes and other forms of microorganisms.

There are widely known microbes which produce the enzyme penicillinase in defense against penicillin. In penicillin-resistant variants experimentally obtained, this enzyme is produced in a strictly adaptive way. Some conditions were revealed under which the process of production of penicillinase is accelerated or decelerated. It was established that a dose of penicillin of 0.004 unit/ ml or 8×10^{-9} M is sufficient to induce penicillinase formation. Cells treated with penicillin at 0°C and washed afterward, produce penicillinase, on subsequent incubation in a medium without penicillin, 30 times more rapidly than untreated cells.

In separate cases of adaptation the authors attempted to elucidate the mechanism of production of enzymes. Of particular interest in this respect are the works of Monod and collaborators, Ephrussi, Hinshelwood and collaborators et al. Some stages of the consecutive synthesis and intermediary products of the formation of the enzyme ß-galactosidase in the colon bacillus, in *Bac. lactis aerogenes* and *Saccharomyces cerevisiae* were established. Separate factors affecting the process of enzyme synthesis were elucidated. Calcium, magnesium, iron and some other microelements were found to exert an essential influence on protease formation (liquefying gelatin) in *Proteus*, on phosphatase in propionic acid bacteria. It was proved that for synthesis of enzymatic systems special complementary substances coenzymes, vitamins, some amino acids and other compounds are needed. These substances enter into the composition of enzymes either as a functional part of the molecule or a binding component.

A great influence is exerted on the formation of enzymes, by environmental conditions such as temperature, pH of the medium, and various chemical and physical agents. This effect may be of a direct or an indirect nature. Organisms react to external influences in various ways. depending upon the character and nature of the acting agent (Spiegelmann and Halvorson, 1956; Pollack, 1956; Knox, 1956; Stephenson, 1951 and others).

Not all enzymes are formed with the same rapidity during the adaptation of the organism. Some of them are rapidly synthesized under the influence of the specific inducer, others slowly, and still others do not adapt themselves at all under conditions of laboratory experiments. On this basis, Karström subdivided ferments into adaptive and constitutive enzymes. The first, according to the author, are formed by the cells as a specific reaction to a corresponding substrate; the second are always in the cells as a constituent part of the living substance. They do not depend upon the substrate, their specific synthesis is not subjected to experimental investigation.

This subdivision should be regarded as highly conventional. Investigations show that there is no sufficient basis for assuming essential differences between adaptive and constitutive enzymes. Probably all enzymes may be obtained by induction with specific substrates. If one does not succeed in obtaining some enzymes it is only due to the fact that conditions of their experimental synthesis have not yet been discovered.

It was experimentally shown that one and the same enzyme may be adaptive and constitutive. In some microorganisms separately induced enzymes become constitutive and vice versa. Such transformations have been observed in variants of the colon bacillus: for arabinose by Cohen, for ß-galactosidase by Lederberg, for amylomaltase and ß-galactosidase by Cohen-Bazire and Jolly and others (Cohen and Monod, 1956).

The enzymes which take part in the biosynthesis of basic metabolites (amino acids, proteins and other essential compounds) should be included in constitutive systems. Without them the organism cannot develop. One such enzyme in the colon bacillus is N-acetylornithase which hydrolyzes N-acetylornithine with the formation of ornithine. The latter constitutes an indispensable element for the growth of the mentioned bacterium.

Some variants of the colon bacillus which are experimentally obtained, lack the indicated enzyme and do not develop on media without ornithine. In the presence of N-acetylornithine such variants synthesize N-acetylornithase. Consequently, the latter is a constitutive enzyme in the initial strain of *Bac. coli* and an adaptive enzyme in its variant. Similar phenomena were observed in experiments with other microbes. These data show the relative character of subdivision of enzymes into adaptive and constitutive.

A comparative study of adaptive ß-galactosidase and constitutive ß-galactosidase in the colon bacillus or of adaptive and constitutive penicillinases in the sporeforming *Bac. subtilis* shows that there is no essential difference between them. The affinity for the substrate, the degree of activation by ions, the coefficient of thermal inactivation, the immunochemical specificity are entirely the same in both adaptive and constitutive enzymes.

It was established that in many cases the cell reacted to the stimulation by the substrate, by the formation of the enzyme at once or almost at once. Production of some enzymes proceeds not during the process of protracted adaptation of the culture, not in successive generations of proliferating organisms but in the same cell which came into contact with the substrate. In such a cell a rebuilding of the protoplast or its parts occurs, under the influence of the specific substrate. As a result of this, the cells acquire new properties. In this way new variants appear.

Only those substances which are able to evoke a suitable reaction in the metabolizing organism may induce the cell to form enzymes. Enzymes are formed in strict correspondence to the specificity of the substrate. Galactosidase is synthesized under the influence of galactose, maltozymase under the influence of maltose, arabinose induces formation of arabinase, etc.

Substrates of the nutrient medium, and also substances synthesized in the cell may be inducers of enzymes. Compounds formed as a result of one enzymatic reaction, may serve as a substrate of other enzymatic processes or inducer for synthesis of new enzymes.

The enzymes formed anew may also be preserved in the cultures of microbes after the disappearance of the inducing substance. The duration of hereditary transmission of the ability to form enzymes to subsequent generations varies depending on the organism, enzyme and environment. The production of adaptive nitrase, cytochromoxidase (under the influence of oxygen) and ß-galactosidase by variationbacteria ceases immediately after the removal of the inducers from the medium. In yeasts, which are adapted to galactose, the enzyme galactozymase is continuously produced for a long time in a series of numerous generations, which develop in the absence of galactose. Penicillinase is synthesized by cells of some adapted strains by *Bac. subtilis* when they are grown on media without penicillin.

The longer the culture is subjected to the influence of a specific substrate, the stronger is the ability to produce a specific enzyme established. If the variant which was removed from the adaptive enzyme is grown anew on a medium with the same inducer, then the ability to form the enzyme again appears, but with greater rapidity, and is established in a more stable manner.

Adaptive enzymes are very specific with respect to the substrate which induces their synthesis. Some of them are more specific than antibodies which are obtained upon the immunization of animals. With the aid of an adaptive enzyme one may sometimes succeed in subdividing bacterial strains which cannot be differentiated by antigenic indicators. Owing to the high specificity, adaptive enzymes are used as reagents in the analysis of many organic compounds for the differentiation and recognition of separate substances.

Adaptive enzymes constitute the first manifestation of variability of the organism; with their aid, the nutritional substrate becomes an internal component of the living substance. These enzymes are like a gateway through which the milieu enters; and the nonliving becomes living. They determine the mechanism of the adaptive variability of the organism. It is possible that the study of the formation of adaptive enzymes in microorganisms will also solve some problems of species variability in general.

With suitable conditioning one may alter the nature of the microbe in relation to those properties which do not seem to be connected with its enzymatic activity. For instance one may train a culture of bacteria or actinomycetes, to increased concentrations of salts or increased temperature and vice versa, thermophilic and halophilic bacteria may be transformed into mesophilic types.

Referring to this type of adaptation, some investigators indicate that in this case also the great importance of enzymes is not excluded. The latter indirectly alter their properties or are synthesized anew under the influence of internal inducers which are formed in the cell under altered growth conditions. The presence of such inducers in cells was experimentally proved by Stainer (1956). A typical induced pyrocatecholase is formed by cells of*Pseudomonas* in the presence of tryptophan in the medium, but not in the presence of pyrocatechol.

Tryptophan is a remote precursor of pyrocatechol. Formation of such internal inducers may occur in the cell under the influence of various physical, chemical and biological agents.

Many organisms require complementary nutritional substances for their development and do not grow on special synthetic media without them. By the gradual "training" of the organisms to a medium with decreasing concentrations of these substances one may force the microorganisms to synthesize the latter and to grow on media without them. One succeeds in growing the typhoid bacillus on a medium without tryptophan, the dysentery bacillus without nicotinamide, i.e., without substances which are indispensable for the normal growth of these bacteria. Propionic acid bacteria--*Bac. pentosaceum* acquire the ability to synthesize thiamine after only several passages on media with substantial quantities of this growth factor. In this way their growth becomes independent of the presence of the given vitamin in the medium.

Adaptation of microbes to phages is a widespread phenomenon often observed in laboratory practice. As was earlier noted, many bacteria and actinomycetes are subjected to the lytic action of phages. The latter penetrate into the cells of bacteria and actinomycetes and under suitable conditions dissolve and destroy them.

Upon the interaction of microbial cells and phages, variants resistant to phages and not succumbing to their lytic effect are formed. Such variants are obtained in various specimens of bacteria (sporeforming and nonsporeforming) actinomycetes mycobacteria and micrococci. Phages, an was indicated above, are specific in their action on microbes of a given species. Those which lyse bacterial cells are designated as bacteriophages, those lysing actinomycetes--actinophages.

Variants which are resistant to phages are endowed with a well-manifested specificity and are only resistant to those phages which induced their formation. Group specificity as well as species and strain specificity are noted. The phage of the colon bacillus induces formation of resistant variants only in cultures of the given bacterium. The phage of the typhoid bacillus evokes formation of resistant strains in *Bac. typhi,* in the diphtheria bacillus resistant variants are obtained under the influence of the phage of *Mycob. diptheriae.* In the tubercle bacillus-- under the influence of the phage of *Mycob. tuberculosis,* etc.

Such a specificity is not always manifested. There are phages known which lyse cultures of *Bac. coli* and *Bac. dysentertae* equally well. Phages of staphylococci are described, which attack the diphtheria bacillus. In actinomycetes phages are often found whose action to polyvalent, they lyse not only cultures of one and the same species but also strains of various species and even of different groups. Some phages of *Act. streptomycini* also actively lyse strains of *Act. violaceus* and *Act. griseus.*

Experimentally obtained variants of actinomycetes or bacteria which are resistant to phages often acquire resistance to another phage species. However, the resistance to nonspecific phages is less strongly expressed.

The phage-resistant variants are of differing stability, depending on the species of the microbe, the individual characteristics of the phage, and on the external conditions. In some cases, strains keep their resistance to the phage for a long time, and in others the resistancy is lost very soon.

Variants which are resistant to phages differ from the initial cultures in several other properties; their antigenic characteristics, virulence and particular biochemical functions change. Often, with the appearance of resistant strains of actinomycetes, antagonists become less active or completely inactive against bacteria. Highly active strains of *Act. streptomycini* often lose the ability to synthesize streptomycin under the influence of phages.

Adaptation of phages. During the interaction of phages, and cells of actinomycetes or bacteria, mutual adaptation takes place. An inactive phage, not lysing, and not attacking a culture of actinomycetes, acquires the ability to lyse its cells after prolonged common growth.

Phages inducing the formation of adapted cultures of actinamycetes may undergo changes during the process of adaptation and adapt themselves to resistant variants. The latter are lysed under the influence of adapted phages; under certain conditions, they, in turn, form new strains which are resistant to adapted phages, etc. Thus, one may obtain a continuous series of adapted variants of microbial cultures on the one hand and phages on the other.

The adaptive nature of phage variability may sharply change, depending on the culture of the host microbe on which the given phage is inoculated. Upon maintain_ Ing an actinophage on one culture of actinomycetes one obtains very active strains with a large range of action, and on cultivating it on another actinomycetes culture, adaptive variants of phages are produced with little activity and a narrow range of action.

Adaptation of microbes to antibiotics. Adaptation of microorganisms to drugs and antibiotic substances is very strongly manifested. Adaptability of bacteria to penicillin, streptomycin, aureomycin, terramycin and many other antibiotics is widely known. Various species of sporeforming and nonsporeforming bacteria, cocci, mycobacteria, actinomycetes, fungi, yeasts, protozoa, and even insects, become adapted to antibiotics. Numerous cases of adaptation to antibiotics of various pathogenic microbes causing enteric diseases, anthrax, tuberculosis, diphtheria, pest, skin diseases, etc were described.

Observations show that it is not the whole culture but individual cells which become adapted to antibiotics. The higher the concentration of the antibiotic in the medium, the lower the number of surviving and adapted cells. Cultures adapted to small doses of antibiotic are, as a rule, resistant, to small concentrations of antibiotics in the medium. Cultures adapted to high doses, develop on media with high concentrations of the antibiotic.

Adaptation of microbes to antibiotics, as in other cases of adaptive variability, proceeds in a directed and specific manner. Variants are only resistant to that antibiotic which induced their formation. If bacteria are subjected to the action of streptomycin, variants resistant to streptomycin are obtained. Under the influence of penicillin, variants resistant to this antibiotic are formed; under the influence of aureomycin, strains resistant to aureomycin appear, etc. Such a specificity is constant and regular. It maybe used to some degree for the differentiation and identification of antibiotic substances and their producers.

In many cases, the specificity of resistance in experimentally obtained variants to not an absolute one. In acquiring a resistance to one antibiotic, bacteria often become less sensitive to some other antibiotic. However, resistance to other preparations to considerably weaker than to that antibiotic which induced the formation of the variant.

Microbes may simultaneously become adapted to two, three and more antibiotics, when exposed to suitable mixtures of the preparations. For instance, by the action of a mixture of streptomycin and pencillin on staphylococci, variants are obtained which are resistant to both antibiotics. Bacterial forms have been obtained which are simultaneously resistant to pencillin, streptomycin, and erythromycin, than to erythromycin, carbomycin, streptomycin, aureomycin, and to antibiotics in other combinations.

The acquired resistance to antibiotics is stable and is transmitted hereditarily through a long series of microbial generations. Adapted variants are preserved during many subcultures on media not containing the antibiotic. The longer the action of the antibiotic on microbial cells. the stronger the acquired properties of resistance are established. The hereditary resistance to the antibiotic to rapidly lost by the action of some specially chosen antibiotic substances or chemical reagents on the culture. For instance, chloramphenicol abolishes the resistance of staphylococci to penicillin. An opposite phenomenon in also noted: an increase of resistance to antibiotics under the influence of particular substances.

With the acquisition of resistance to antibiotics, microbial variants change some other properties. Often virulence and pathogenicity is lost, the ability to ferment various organic compounds, such as sugars, disappears, gram-positive bacteria become gra -negative, root-nodule bacteria lose their ability to form nodules on roots of leguminous plants, etc. Biochemical properties, an well as morphological properties are changed. Cultures with smooth colonies acquire a wrinkled, rough or granular structure, or become mucous.

Formation of dependent variants. As a result of the prolonged adaptation of microbes to antibiotics, variants may be formed, which require the given antibiotic for their growth. Dependent cultures are obtained, which do not grow on the medium without the antibiotic, as when formation of vitamin-dependent variants takes place.

A considerable number of dependent variants has been described in the literature. Such variants are often obtained in laboratory practice. They are very specific and require for their growth, only those antibiotics which induced their formation. Variants dependent upon antibiotics are more specific than resistant variants. This property of the given variants may be used for the differentiation of antibiotics as well as their producers.

Directed changes in microorganisms may also be obtained by the method of induction of properties. This is due to the fact that when two defined species are developing, some of their attributes and properties are transferred from one species to the other. These attributes are also transferred when the cultures are grown in a medium containing metabolic products or filtrates of the other culture. One of the organisms acquires the properties of the other.

Such changes were first observed in the nonsporeforming bacillus, *Bac. proteus* by Zilber (1928), who obtained a para-agglutinating variant of *Proteus* under the influence of the typhoid bacillus. Gracheva (1946), obtained a strain of the colon bacillus, *Bac. coli,* with the properties of another species of the same group, *Bac. breslau.* Prokhorov (1950), transformed a colon bacillus into a culture with properties of the typhimurium bacillus by the same method. Timakov (1952) and his co-workers studied induced variability in different bacteria of the colon group for a number of years. He imparted the properties of the paratyphoid (the Breslau and Schotmüller types), typhoid and dysentery bacteria to a culture of *Bac. coli,* growing it in a medium with heat-killed cultures of these species. He suggested calling the culture that acquires the new properties, the "accepting" one, and that which transfers its properties, the "directing" one. Much data have been accumulated on the transfer of antigenic properties from one culture to the other (see Kalina, 1953; Dubos, 1943; Catcheside, 1951; Braun, 1953 and others). By the method of induction, variants were obtained in various bacterial species. In addition to the bacteria of the colon group, staphylococci, many nonsporeforming and sporeforming bacteria, mycobacteria and others may be induced. Legroux and Genevray (1933), transformed a nonpigmented, avirulent strain of *Pa. pycoyanea* into a pigmented virulent culture. We have observed the transformation of nonpigmented, nonfluorescent strains into pigmented fluorescent cultures of *Pa. fluorencens,* obtained from the rhizosphere of certain plants. Nonpigmented strains were grown, or merely kept in a medium with filtrates of fluorescent bacteria. After some time, upon plating, single colonies with well-expressed properties of the inducing culture were obtained.

By use of this method Alexander and Leidy (1951) observed a change of smooth S-and rough R-forms of the pathogenic bacillus, *Bac. influenzae.*

Similar transformations are also observed in sporeforming bacteria. Manninger and Nagradi (1948) kept a virulent encapsulated nonmotile culture of *Bac. anthracis* in a nitrate of the saprophytic, nonencapsulated motile soil bacillus, *Bac. mesentericus,* and obtained variants which differed from the original culture by properties characteristic of *Bac. mesentericus.* These new variants proved to be nonencapsulated, with flagella, motile and virulent. Tomcsik (1949) did not fully confirm these data, however, he observed at the same time that nonencapsulated variants of *Bac. anthracis* possessed an antigen identical with the antigen of the directing culture--*Bac. mesentericus.*

The induction of properties by this method was observed with meningococci (Alexander and Redman, 1953) typhoid and paratyphoid bacteria, dysentery bacilli, diphtheria bacteria, micrococci and other specimens of bacteria.

We have reported data of our investigations on the induction of new properties of virulence in root-nodule bacteria, i.e., the ability to form nodules on roots of leguminous plants foreign to these bacteria (Krasil'nikov, 1945 b).

The root-nodule bacteria, as in well known, form nodules on roots of certain plants. For instance the nodule bacteria of soybean are capable of forming nodules only on roots of soybean; the bacteria of kidney beans, on roots of kidney beans, the lupine bacteria, on roots of lupine; etc. In accordance with this specificity these bacteria are subdivided into species: *Rhizobium japonicum* (sojas), *Rh. phaseoli, Rh. lupini,* etc.

Not all the rhizobia have such a strict specificity. Some of them may form nodules on roots of plants belonging to different species or even to different, closely related genera. For example, *Rh. leguminosarum* forms nodules on roots of peas vetch, lentils and *Lathyrus sativus; Rh. Meliloti*--on roots of lucerne, *Galeopsis ladanum* and *Frigonella.* Consequently, among the rhozobia there are species strictly specialized toward a single plant species and bacteria with group specificity.

By cultivating bacteria of this or that group on media with filtrates of the directing cultures, we have succeeded in changing their specificity and imparting to them properties of virulence of the directing culture. The root-nodule bacteria of peas, vetch and sweet clover acquire the ability to form nodules on roots of plants which are foreign to them: clover and lucerne; bacteria of acacia started to form nodules on roots of peas and lupine, and bacteria of broad beans on roots of kidney beans and Lucerne (Table 4).

Table 4
Changes of specificity of virulence in root-nodule bacteria by induction

Directing Cultures:	Accepting Cultures: Peas	Accepting Culture: Vetch	Accepting Culture: Lucerne	Accepting Culture: Broad Beans	Accepting Culture: Kidney beans	Accepting Culture: Acacia	Accepting Culture: Clover	Accepting Culture: Sweet Clover
Rhizobium trifolii	+	+	+	-	-	-	+	+
Rhizobium meliloti	+	+	+	+	-	+	+	+
Rhizobium phaseoli	-	-	-	+	+	-	-	-
Rhizobium leguminosarum	+	+	-	-	-	+	+	-

Note: Plus means that the induction succeeded and the accepting cultures acquired the properties of the inducing bacteria; minus means that the properties were not imparted.

Not all root-nodule bacteria can be changed in a given direction and not all directing cultures are able to transfer their specificity to other bacterial species.

Out of nine species tested, four did not acquire the specificity of the inducing cultures. Many cultures belonging to the same species did not acquire new properties. For example, out of twelve cultures of root-nodule bacteria of beans isolated in different places, seven did not change their specificity. Moreover, not all the variants obtained experimentally from the same culture lend themselves to varation. Out of twenty variants of the root-nodule bacteria of clover, *Rh. Trifolii,* more than half did not acquire the virulence of the inducing culture which was the original strain of *Rh. Trifolii.* (Krasil'nikov, 1945 b, 1955 b).

In subsequent studies we succeeded in imparting the properties of virulence to some nonroot-nodule bacteria of the genus *Pseudomonas.* Upon prolonged cultivation of these bacteria on media containing filtrates of root-nodule bacteria of clover and lucerne, strains were obtained which possessed the ability to form nodules on roots of clover or on roots of lucerne. It should be noted, that not many cultures change in this fashion. Out of 91 strains of different species and genera of bacteria, only two acquired the properties of the directing cultures, i.e., the ability to form nodules on roots of leguminous plants.

The acquired virulence can be maintained indefinitely by means of passages from plant to plant. The newly acquired attributes of virulence often do not remove the former specificity. Root-nodule bacteria of peas which acquire the ability to form nodules on the roots of clover did not lose their ability to form nodules on the roots of peas. This was also observed in relation to other bacterial variants developing on vetch, acacia, sweet clover and broad beans.

Change of properties of virulence in root-nodule bacteria have been observed by Rubenchik (1953) and Peterson (1955). Under the influence of the metabolic products of the root-nodule bacteria of clover, variants were obtained in other species of *Rhozobium* with the specificity of the directing culture.

Balassa (1956) changed the specificity of virulence of root-nodule bacteria, acting upon them with DNA of the respective cultures of *Rhizobium.* He used *Rh. meliloti* as the directing culture. With the DNA obtained from this culture he acted on nodule bacteria of soybean *(Rh. japonicum)* and lupine *(Rh. lupini).* Then the cultures were tested for virulence toward lucerne. Nodule bacteria of soybean after being acted upon by the given metabolite of *R. meliloti* acquired the ability to form nodules on the roots of lucerne.

As can be seen from the above-mentioned data, certain metabolic products exert a certain effect on the accepting species of bacteria, and lead to changes in their attributes and properties.

Of special interest are the changes obtained by "transformation", described first by Griffith in 1928. This author found that pneumococci of type III possess substances which may be transferred under certain conditions to pneumococci of type II, imparting to it the properties of type III pneumococci. It was shown later that these transformations are quite specific and may occur not only in vivo but also in vitro on appropriate media under certain conditions.

The substance causing these transformations was isolated in a chemically pure state and was thoroughly studied. It turned out to be present only in the pneumococcus type III (encapsulated, virulent type) and was acquired by the pneumococcus type II (nonpathogenic, nonencapsulated). The chemically pure substance possessed a high transforming capacity. It is sufficient to add 0.003µ g of this substance to 2 ml of the corresponding medium to transform an avirulent nonencapsulated culture of pneumococcus type II into a virulent encapsulated variant of pneumococcus type III. Chemical, enzymatic and serological analysis and data obtained by electrophoresis, ultraviolet spectroscopy, etc show that the active part of the substance contains neither protein nor free lipides or serologically active polysaccharides (Dubos, 1948; Braun, 1953; Catcheside, 1951 and others). This part consists mainly of DNA. The given acid causes the synthesis of the capsular substance. Studies have shown that the DNA of the pneumococcus S-variant is similar to the DNA obtained from other sources, for instance from cultures of the rough R-variant.

However, their effects differ. The former possesses transforming ability, transforming the R-form of pneumococcus into the S-form but the latter is devoid of this ability. Consequently, the transforming action of DNA is obviously connected with the presence of extremely small quantities of other substances, the effect of which manifests itself in the presence of the given DNA.

Not all the pneumococci are able to accept the transforming substance. Out of a great number of experimentally obtained nonencapsulated avirulent variants, only a very few show the ability to accept this substance. Consequently, the active part of the capsular substance of pneumococcus type III is effective only when the accepting culture possesses specific functionally active acceptors, determining the biochemical activity and specific nature of the pneumococcus cells.

In 1952, Austrian and MacLeod observed that in addition to the variability of the polysaccharide, the pneumococci also show changes in their proteins. According to these authors, upon transformation of pneumococci a change takes place in a special somatic protein. Together with the above-mentioned substance this protein in various combinations may produce different variants in the process of changing of the culture.

Hewitt (1956) transformed a nonencapsulated avirulent strain of the influenza bacillus into a capsulated strain. Hotchkiss (1951) obtained penicillin-resistant strains of pneumococci from a sensitive strain by growing the latter in a medium containing DNA isolated from cells of a strain adapted to penicillin.

Streptomycin-sensitive pneumococci, resistant to penicillin may be changed into streptomycin-resistant by the use of a transforming substance which is liberated upon lysis of streptomycin-resistant strains of pneumococci by the action of penicillin. This substance which is capable of imparting streptomycin-resistance to the culture and of transferring other hereditary properties from one strain to another is not a DNA, but has a different chemical composition. It is not decomposed by the enzyme desoxyribonuclease. A transforming susbstance of the same type transforms streptomycin-sensitive strains of typhi murium into streptomycin-resistant ones (Zinder and Lederberg, 1952).

The occurrence of transformation has been detected in certain nonsporeforming bacteria which possess slime capsules. By the action of corresponding substances they lose the capsules and with them some other properties are lost or new ones are acquired.

According to Zinder and Lederberg, properties of cells can be transferred from one culture to another by special tiny corpuscules or particles of the protoplast. They lyse the directing cultures by a special phage. The products of lysis, in the form of the mentioned corpuscules, pass through fine glass ultrifilters and enter the calls of the accepting culture. The latter acquire the properties of the former. The particle carriers of the transferring substance are very small, not more then 0.1μ in diameter, and visible only in the electron microscope. The authors observed this method of changing the cultures in *Bac. typhi murium.* Two cultures differing in their nutritional requirements were used in the experiments. One of them required histidine and the other required tryptophan. They were placed in one vessel divided by a glass ultrafilter, each one of the two cultures being placed in a separate hall of the vessel. As a result, nonobligatory cultures were obtained. The authors called this method of changing cultures transduction.

A similar picture of culture changes was observed by Brown and co-workers (Brown, Cherry, Moody and Gordon, 1955) in *Bac. anthracis.*

They selected a special phage (Lambda phage) which in distinction to many other phages (in all 32 phages. were examined) possessed a wide range of action. It lysed bacteria of different species and genera: 9 strains of genus *Bacillus anthrasicis,* 3 strains of genus *Bac. cereus,* 4 strains of *Bac. cereus* var. *mycoides* and others.

A motile culture of *Bac. cereus* was infected with this phage and after a twenty-four-hour incubation the culture was filtered through an ultra-bacterial filter; a nonmotile virulent strain of *Bac. anthracis* was infected by the filtrate or by the purified phage from this filtrate. After twenty-four hours motile bacilli appeared in the culture of the latter. These bacilli grew into a motile avirulent culture with certain properties of *Bac. cereus.*

The authors assume that the phage elements transduce certain particles of living matter from *Bac. cereus* to *Bac. anthracis,* and introduce certain properties with them. According to the authors, changes, as were made in the above-mentioned bacillus, appear regularly, in all repeated experiments. They think that the mode of carrying over properties from one organism to the other, which they observed, differs from that described by Zinder and Lederberg. The latter claimed that the carriers were not the phage particles but special corpuscular elements.

Spiegelman (1946) while studying induced changes in yeast, obtained a substance which evoked the ability to ferment galactose in species which were unable to ferment this sugar. He called this substance adaptin. Its nature remained obscure. By the method of transduction one may succeed in conferring on bacteria fermentative, antigenic, virulent and other properties possessed by the directing culture, and one may also obtain motile strains from nonmotile ones, to which the ability to form flagella is imparted by the phage particles (Baily, 1956).

Changes which are called "recombinations" also belong to the phenomena of induced changes. The external part of these changes is based on the fact that, upon mixing two cells from different cultures, generations with the mixed or combined properties of the initial parental organisms are formed. The cells transfer their properties and attributes upon direct contact with each other. New variants appear which are called recombinants. The formation of such recombinants was described by Tatum and Lederberg (1947) in one strain of the colon bacillus K-12. From this strain defective variants were obtained which had lost the ability to synthesize certain amino acids and vitamins essential for growth. On these variants experiments of "vegetative crossing" were performed (Lederberg and Tatum, 1954).

The colon bacillus, strain K-12, after irradiation by ultraviolet rays, forms strains which differ from the original culture by their inability to grow on simple synthetic media without additional substances. These defective variants are called auxotrophs or minus-variants. Some of them require a certain auxiliary substances these are monoauxotrophs. Others require two, three or more substances; these are polyauxotrophs.

If one grows auxotroph strains separately on a full nutrient medium, and then inoculates 10-15 hour cultures on a deficient nutrient substrate, after thorough washing in saline, then one does not observe growth of the culture and the inoculation will be sterile. Only in rare cases do single colonies grow out; the new variants, the prototrophs, are able to develop on media without additional substances. The frequency of appearance of these variants, according to Lederberg and other investigators, is not higher than one cell per 1,000,000-10,000,000 of the initial auxotrophs.

Observations show that auxotroph variants that were mixed before inoculation give a more intense growth on a noncomplete medium than that of the control inoculations.

Tatum and Lederberg an well as some other investigators experimented mainly with two variants. One of the variants (58-161) required biotin for its growth (B-)and methionine (M-), and the other variant--W-1177 required threonine (T-) leucine (L-) and vitamin B$_1$ (B$_1$-) Both variants did not grow without the indicated substances on special mineral media. Mixing these variants, strains were obtained, which differed from the initial ones in that they developed on incomplete media in the absence of biotin and methionine and also variants which did not require threonine, leucine and vitamin B$_1$; they themselves synthesized these growth substances. This can be schematically represented in the following ways strain 58-161 (B-M-) and strain W-1177 (T-L-B$_1$-) give a variant B+M+T+L+B$_1$+.

The transfer of properties upon mixing cultures may be accomplished by various strains. Some attribute or property is transferred from cells of one strain to cells of another. This property or attribute will, for the sake of simplicity, be designated by the letter F. If the attribute is there, the cells or the whole culture is designated by its name with the addition of F$^+$. If the attribute is lacking the culture is designated by F$^-$. Some attributes may be transferred from one F$^+$ strain to many other F$^-$ strains. According to Hayes (1953), a strain or, as he calls it, a mutant W-677 (-CA) (streptomycin- and azide-resistant) when present in a culture together with 58-161/F$^+$acquires the properties of the latter and becomes strain W-677 CA/F$^+$. The number of such transformations in platings on agar media reaches 75% of the inoculated and tested colonies. A nonobligatory strain or a culture of 58-161/F$^+$ transmits the factor F$^+$ to the incomplete strains W-877/CA/F$^-$, 58-161/F$^-$, W-877/F$^-$, 58-161/CA/F$^-$. Thereby new strains are obtained which are nonobligatory and retain the acquired properties quite well for a number of months on a nutrient egg Dorset agar medium at 4°C.

Factor F, determining one or another attribute is not transmitted through the filtrate. A young broth culture of 58-161/F$^+$ was filtered through a colloid filter with pores of 0.74μ A. R. D. The incomplete strain W-677/F$^-$, was kept in a filtrate after which colonies were plated on an agar medium. Out of 115 colonies sown, the acquisition of the properties of the directing culture F$^+$ was not observed in any one of them.

As was noted earlier, when the cultures are mixed with an F$^-$ strain the planting is sterile. Upon mixing the F$^+$ and F$^-$ cultures the platings are of high productivity. For instance, upon plating the variants 58-161/F$^+$ and W-677/F$^-$, 97 colonies of prototrophs were obtained out of a total number of 945 million auxotroph cells. Plating the strains 58-161/F$^+$ and W-677/F$^+$ only 4 prototroph colonies out of 477 million were observed and upon plating the strains 58-161/F$^-$ and W-677/F$^+$, 49 prototrophs appeared out of the total number of 450 million cells.

As can be seen from the data given, the largest number of prototrophs is obtained when F$^+$ variants are plated together with F$^-$ variants.

In the experiments of Hayes the incomplete variants of the colon bacillus were as follows: 58-101 required methionine and biotin (M-B-), fermented lactose, maltose, mannitol, galactose, xylose and arabinose, was sensitive to coli phage T$_1$ and resistant to T$_3$. The variants of W-677 required threonine, leucine and thiamine (T-L-B$_1$-)for growth, did not ferment lactose, maltose, mannitol, galactose, xylose and arabinose, were resistant to phage T$_1$ and sensitive to phage T$_3$.

In addition, these variants, according to Maccacaro (1955), differ from each other in certain other properties: their reaction to the acidity of the medium, their ability to absorb dyes, etc.

In experiments of other investigators with the colon bacillus K-12 analogous results were obtained. Defective strains, that have lost the ability to synthesize a certain amino acid or vitamin, reacquire the ability upon contact with cells possessing these properties. Strain U-24 which in the experiments of Clark (1953) required biotin, phenylalanine and cyotine (B-Ph-C-), does not grow on a synthetic medium lacking these substances. Strain U-10 forms biotin, phenylalanine and cystine, but does not produce threonine, leucine, and thiamine (T- L- B1-) and does not develop on a synthetic medium without these compounds. After mixing two such organisms, strains are obtained, which grow well on the indicated media and synthesize the substances necessary for growth which are lacking in the initial cultures. (Catcheside, 1951; Hayes, 1953; Braun, 1953; Clark, 1953; Lwoff, 1955, etc).

Sermonti and Spada-Sermonti (1955-56) described the formation of recombinants obtained upon crossing strains of blue actinomycetes--*A. coelicolor,* after irradiation by ultraviolet rays. One strain (5 me- hist-) required methionine and histidine and formed a blue pigment. Another requiring strain (14 pr.- glu- pigm.-)

required proline and glutamic acid and did not form pigment. In mixed platings on complete agar media three types of colonies are formed. The initial ones: a) 5 me-hist-, b) 14 pr.- glu pigm, and c) recombinants nonrequiring cultures that do not require methionine. histidine, proline or glutamic acid.

In this way Rizki (1954) observed the transmission of pigment formation ability in *Bac. prodigiosium.* Acting on a leuco-strain of this species of bacteria with a pigmented culture, the author observed that the former culture acquired the ability to form the pigment prodigiosin.

Some investigators found similar transmissions of hereditary properties in pneumococci, in typhi-murium bacteria and. others (Ephrussi, Taylor 1951; Hotchkiss, 1954; Vavali and Lederberg 1953).

Studying the process of formation of recombinants, Clark found that it was connected with the pleomorphism of the culture. As a rule, the external action that caused a degenerative development of an auxotroph culture with the formation of deformed gigantic cells and other cells, lowered the number of prototrophs formed. Haas and others noted a correlation between the formation of strongly enlarged, threadlike and other polymorphic cells and the increase of the number of prototroph variants among the auxotrophs. According to their observations, ultraviolet irradiation enhances the process of recombination in the colon bacillus with the simultaneous increase in the polymorphism of the culture. Certain chemical substances, hydrogen peroxide and others, stimulate the formation of prototrophs (Haas, Clark, Wyss, and Stone, 1952). Grigg (1952) stresses the importance of the dependence of the appearance of recombinants among auxotrophs upon the quantitative ra tios between the initial parental cells in the mixture.

If there are more F^+ cells in the mixture of the strains crossed. all the F^- cells acquire the properties of F^+. In cases where the numbers of cells of both types are equal there will be less recombinants. In case of a reversed ratio, i.e. , when there are less F^+ cells than F^- cells, the number of F^- transformed to F^+ will be the smallest (Table 5).

Table 5
The effect of quantitative ratios between components of the mixture on formation of variation with transmission of properties of the F^+ cells to the F^- cells

F^-	F^+	Ratio of $F^-:F^+$	0.5*	1*	2*
4 x 10^6 cells/ml	1 x 10^8 cells/ml	1:25	20/20	10/20	19/20
4 x 10^6 cells/ml	4 x 10^6 cells/ml	1:1	10/20	13/20	6/20
4 x 10^6 cells/ml	1 x 10^6 cells/ml	4:1	2/20	0/20	3/20

*Ratio of the number of F^- cells transformed into F^+, after being mixed together (hours)

As can be seen from this data at a ratio $F^-:F^+$ of 1:25, out of 20 cells (colonies) tested, all were variants transformed from F^- to F^+. At a 1:1 ratio of these strains, the number of transformed variants was 10 out of 20 and at a ratio of 4:1 it was 2 out of 20 cells studied. These numbers of variants are obtained half an hour after mixing; after one hour their number is smaller and after two hours it becomes even less. Ultraviolet rays increase the fertility of the F^+ cells but not that of F^- cells and enhance the maturation of prophage into phage, an a result of which the number of gene carriers increases.

In the formation of recombinant variants the following rules can be observed: variants B+M+ (not requiring biotin and methionine) are formed more often than variants B-M+, (requiring biotin). Upon planting a mixture their number reaches 14 %. B_1+L+variants (forming vitamin B1 and lecithin) are encountered more often than variants B_1+L+ or B_1+L-. Variants T-B+M+B_1+L+ (requiring threonine) are obtained more often than variants T+B+M+B_1+L+. All these data show that in the living matter of the organism there is a strong bond between various attributes.

It should be noted that such variants cannot be obtained upon mixed cross platings of all cultures. Even in the same species of the colon bacillus only the K-12 strain gives variants capable of combined variability. Another such strain of the colon bacillus (strain B) does not give such variants. Numerous attempts to obtain variants from this strain, able to give recombinants, were unsuccessful.

It was suggested that strain K-12 is a homothallic culture and strain B, a heterothallic one. The former possesses both the masculine and the feminine tendencies, whereas the other has only one of them. Therefore, the latter cannot dissociate according to this attribute and give combined variants upon mixing, as happens in K-12 strains. Strain B in monosexual or asexual, and is similar to certain imperfect fungi, for instance from the group of asporogenous yeasts: *Torulopsis, Mycotorula,* etc. According to the authors, upon contact between the minus variants of the homothallic culture a process of conjugation or exchange of substances takes place between them, similar to the sexual process in higher organisms. With these substances the carriers of hereditary properties, the genes, are also transmitted. (Lederberg and Tatum, 1954; Anderson et. al., 1957).

On the mechanism of variation. As can be seen from the above-mentioned data, the phenomenon of variation in microorganisms is quite diverse both in its appearance and in the nature of its response to external influences. The mechanism of variation and the inheritance of properties remains unexplained. There are many points of view, and many suggestions were made all of which can be reduced to two basic theories: the theory of mutation" the theory of physiological adaptation. These theories were formulated long ago. and were already known during the period of the first discoveries in the field of variation in microbes in general. However, there have not been sufficient factual data to corroborate any of these theories.

The studies performed in recent years in the field of variation are of great interest. Tatum and co-workers (1942) have shown that certain acquired fermentative properties in individual variants of the fungus *Neurospora*are regularly transmitted to the offsprings during sexual reproduction which is similar to that existing in higher plants. After these observations analogous data were obtained with yeast of the genus *Saccharomyces, Saccharomycodes,* etc. Gamete cells of the two cultures cross, unite and form one zygote with the properties of the parental organisms. In the subsequent multiplication of the zygote the properties of the original cultures segregate through the generations and combine in various groupings. As result, new variants are formed with combined properties. These variants are called combinants or recombinants.

Variations of this type take place in the evolution of microorganisms. However, it is possible only in species which multiply sexually. In bacteria, actinomycetes and also in a great number of fungi (imperfect) and yeasts the sexual process in its full expression is nonexistent. Naturally, one cannot observe variations of the type produced by sexual mixing of properties, nor conduct a genetic analysis of any sort in these organisms.

Data were given above, showing that in bacteria and other microbes hereditary changes which have a certain similarity to genetic changes of the combination and recombination type, characteristic of yeast may occur. The followers of the genic mutation theory assume that bacteria possess regular nuclei with a specific set of chromosomes and gene-like structures, which determine the rules of the process of variation. As a confirmation of this assumption the investigators use the observation of Tatum and him co-workers on the changes occurring In the K-12 variant of the colon bacillus upon crossing it with cells possessing different properties. The recombinations obtained are identified with the combinations in fungi and yeasts in sexual crosses.

The phenomenon itself really resembles combinations formed during the process of sexual reproduction; however, the mechanism here is probably different. The bacteria mentioned do not have a sexual process of the form that is observed in fungi and yeasts. Direct conjugation and mixing of living matter of two cells is not possible here. It should be assumed that upon contact between calls in the above-described cases of transduction the transforming substance is transmitted from one cell to the other across intact membranes and if there are defects, in the latter, they are not detectable.

The existence of transforming substances may now be regarded as proven. They are found in various microorganisms--bacteria, actinomycetes, yeasts and fungi. As previously noted, it was found in pneumococci, nonsporeforming bacteria, in sporeformers, etc. In certain cases the chemical structure of these substances was established or their chemical composition disclosed.

In transduction the transforming substance may be carried over from cell to cell with the aid of phages. The latter build their body from the live matter of bacterial cells (mainly of DNA). When the cells lyse the phages are liberated and if they pass to cells of another culture they evidently carry over particles of living matter of the former cells. The mixing of protoplast particles of two different cultures takes place, and with it the hereditary properties are mixed, which later seggregate out with the formation of new variants--recombinants in the progeny.

The ability of microbial cells to absorb complex organic compounds, metabolites of microorganisms, has been established by many investigators with different representatives of the lower organisms. It has been shown that compounds such as vitamins, auxins, amino acids, enzymes, antibiotics and other substances important to life are absorbed by microbial cells through an intact membrane.

It has been proved that various metabolites of microbial origin can serve as inducers of the synthesis of specific biocatalysts: vitamins, ferments and other compounds. Penicillin. absorbed by cells of sensitive bacteria, induces biosynthesis of the enzyme penicillinase, sulfonamide preparations induce the formation of para-aminobenzoic acid (PABA), etc.

Oparin and Yurkevich (1949) have shown that brewer's yeasts have the ability to absorb enzymes from the medium and use them. It was also established that absorbed enzymes stimulate the cells to form their own enzyme. Kosikov (1950) observed the formation of the induced enzyme, invertase, in yeast organisms. The cells that absorbed invertase from the nutrient medium soon began to synthesize their own invertase and used it for the splitting of saccharose. The variants thus obtained, maintained the ability to synthesize this enzyme for an indefinite length of time and transmitted this ability hereditarily through many generations.

Antibiotic substances and other microbial toxins which are absorbed by the cells, lead to the formation of antitoxins, due to which new resistant variants are formed. Various metabolites formed by microorganisms may serve as antitoxins. For instance, PABA is synthesized in order to neutralize the poisonous effect of sulfonamide preparations; penicillinase and streptomycinase--as an antidote to the corresponding antibiotics, etc. The resistance of strains to sulfonamide is determined by the degree of formation of PABA. The higher the concentration of sulfonamide in the medium, the more PABA is necessary to counteract its effect. For the neutralization of 50 μ g of sulfonamide in the medium, the hemolytic streptococcus produces 0.007 microgram of PABA; when the dose of sulfonamide is thrice increased, the streptococcus also increases its production of PABA three times. Increasing the concentration of PABA in the medium tenfold, correspondingly requires more sulfonamide in order to exert an antibacterial action (Woods, 1940).

It was observed that PABA is a specific metabolite which is antagonistic to sulfonamide preparations. It is part of the composition of the vitamin, folic acid and a vital substance for microbial cells.

Many metabolites, as is well known, are of considerable importance in the life of microbial cells and some of them are quite indispensable: their function determines various processes of life in the cells. If for some reason any of these metabolites are blocked, the cells become different and their biochemical processes are not the same as those of the initial, normal cell. If the given change is stabilized and transmitted to subsequent generations, new variants are obtained during the process of multiplication.

The blocking or change of metabolites may take place under the influence of antimetabolites or metabolites which are formed by other species of microorganisms. The interaction between metabolites and antimetabolites may manifest itself in different forms and degrees. After being absorbed by the cells from the medium, an antimetabolite may be fixed and combine with metabolites, disturbing their function. As a result of such blocking the formation of new variants of microorganisms is possible.

Many metabolites are described in the literature, which have corresponding antimetabolites inactivating them. Vitamin B_1, or thiamine, has an analogous antimetabolite--pyrithiamine and neopyrithiamine. Pyrithiamine paralyzes the action of thiamine. Analogous antimetabolites have been found for biotin, choline, folic acid, ascorbic, and nicotinic acids and for the amino acids, tryptophan, methionine, glutamic acid and other compounds (cf. Woolley, 1954; Shive, 1952).

If In all cases of induced variation--upon "induction" of properties, transformation, transduction, etc the acting substance or inducer is processed by the living protoplasm, and takes part in a series of various biochemical transformations of some metabolites, causing corresponding changes in them. The process of transmission of properties from one organism to another is evidently possible only in cases where there is a close phylogenetic relationship between them. In addition, the cells of the accepting culture must have a corresponding acceptor, capable of binding the inducing substance of the inducing culture. Loss of these acceptors deprive the cell of the ability to change in this manner.

The phenomenon of adaptation to substrate and to external agents may be caused by other reasons and may consequently have another mechanism. Some investigators connect the resistance of bacteria to antibiotics or phages with the lose by the cells of the special substances, "receptors", which capture the molecules of the antibiotic and hold them within the cells or inactivate them, without disturbing the intactness of the cell.

In a number of cases the mechanism of resistance formation to antibiotics is connected with a change in the permeability of the cell membranes or their protective barrier. Ehrlich has shown (1909), that normal trypanosomes take up acriflavin, while acriflavin-resistant strains which are adapted to it, do not absorb the drug. There are similar data concerning penicillin. Certain penicillin-resistant bacterial variants absorb less penicillin than the initial sensitive cultures. According to our data, chemically purified mycetin is absorbed by the diphtheroid bacteria of *Mycobacterium* sp. (initial culture) in the amount 32.5 units and the experimentally obtained drug-resistant variant absorbs only 16.0 units per same number of cells. According to different authors, streptomycin is absorbed by the membrane of certain bacteria and is bound within it by nucleic acids. As a result, the metabolism is disturbed and the activity of the cells differs from that of normal organisms.

The adherents of the mutation theory maintain that any culture contains a small number of mutants which are undetectable by the usual methods of microbiological analysis. According to calculations of various workers, there is one mutation per 10^7-10^9 ordinary vegetative cells, developing on nutrient media under ordinary laboratory conditions. According to others. the number of mutations may reach 5×10^2 and also 10^{10} (cf. Braun, 1953; Catcheside, 1951).

It should be noted that the authors determine the number of mutations only with respect to one characteristic. If one considers variations of different features. and they are quite numerous, then the number of mutations vastly increases. Variants which are resistant to various agents: antibiotics, phages, chemical reagents, and physical agents, like temperature, radiation energy, U.V., X-rays, etc, appear in a culture. Each type of these agents is in its term quite diverse. There exists a great number of antibiotics, phages and chemical reagents. Each acts on the cells in its particular way and against each of them ready microbial mutants ought to preexist. Furthermore, the culture should contain mutants adapted to different concentrations of each of the active factors, which are also quite numerous. The number of mutants formed for each agent is approximately equal. Therefore, if one accepts the point of view of the mutation theory, one should assume the pre-existence in each culture of mutants adapted to all possible agents and to various doses of the agents. In other words, an infinite number of mutants should exist; almost every cell of the culture is in some degree a mutant, which of course is hardly probable.

It is assumed that the mutations are spontaneous, without any effect from external agents and independent of the environment. The environment only selects the already existing mutants from the culture. Upon action of one or another agent, be it antibiotic, phage, chemical or physical, the cells resistant to the agent, that is the mutants, survive, while all the sensitive forms are removed.

The mutation theory cannot explain many phenomena of adaptive changes. It is known that microbial cells can adapt themselves to antibiotics (as well an to other agents), when they are in a resting state, in the lag phase of the culture. Under suitable conditions many adapted forms are simultaneously formed in the culture Their percentage may be increased considerably, to 1-10 per cent and more, whereas according to the mutation theory their number does not exceed 1 per 100-1,000 millions of normal cells.

The phenomenon of stepwise adaptation of bacteria to ever-increasing concentrations of an antibiotic and of other substances remains unexplained. Upon inoculation of the active agents, into media with increasing concentrations the frequency of formation of resistant variants increases sharply and this process proceeds much faster.

The appearance of forms which are simultaneously resistant to two, three or more substances is also not clear. It is even more difficult to understand the removal of acquired resistance of variants by the action of other antibiotics (or other substances). The mutation theory is also unable to explain many other aspects of variation.

The lack of basic differences between mutants and adaptations was noted above. The so-called mutants which arise as a result of changes in the genes, may become adaptive forms. Hereditary fixation of properties is observed in quite obviously adaptive variants, whereby these properties are transmitted through thousands and millions of generations. On the other hand, mutants may be unstable and revert comparatively quickly to the initial microbial forms.

The numerous observations and experiments of the recent years convince the investigators more and more that variation in microorganisms takes place by way of physiological adaptation to the environment. The living matter of the organisms specifically reacts to external agents, changing its properties to correspond to the quality of the acting substance. Under the influence of the substrate, inducer cells primarily react by changing their enzyme systems. Many investigators are of the opinion that when a substance induces the formation of an adaptive enzyme, it participates together with the specific component of the cell in organizing the synthesis of the given enzyme. Upon prolonged action of the inducer, the adaptive enzyme turns into a constitutive enzyme. The acquired ability may be carried through a long series of generations in the absence of the substance that caused its appearance.

Due to vigorously acting agents (the so-called mutagens) a relaxing of the hereditary properties of the organism takes place. This also happens in old microbial cultures. Cells with relaxed hereditary properties are much more readily subject to the action of environment. They react more readily to specific substances and show the properties of variation that have been induced by different substrates. The specificity of the latter leaves its imprint on the formation of new variants, or mutants. Mutants are created by environment, as are adaptive variants. Both are phenomena of the same order.

Variability of Soil Microorganisms

The problem of the variability of soil microbes under their natural conditions of habitation is little understood. Only a few works on this problem are available and those are of a speculative character.

In laboratory practice, as was indicated above, one often encounters the phenomenon of variability, polymorphism and species variability in microorganisms. The sporeforming bacteria--*Bac. mycoides, Bac. subtilis, Bac. mesentericus,* and other typical cultures with the characteristic structure of colonies become atypical and form variants differing from the ordinary forms.

Are similar variants indirectly formed in the soil or are they only formed under laboratory conditions?

It is known that in some places typical variants of *Bac. mycoides* are not encountered but instead, atypical forms, with fine granular, evenly edged colonies, are found.

We have found such forms in the chestnut soils of the Volga area. Typical variants of mycoid structure in their colonies were completely absent (Krasil'nikov et al., 1934b, 1936). In some soils typical forms together with atypical ones are found.

We have performed a series of experiments in order to determine the genetic link of the typical cultures with the atypical variants,

We did not succeed in obtaining typical cultures from atypical strains upon cultivation of the latter in artificial media under various growth conditions such an high or low temperatures, various pH, aerobic and anaerobic conditions. Nineteen strains, isolated from various places and various fields of the chestnut soils of the Volga area were studied. None of them produced typical mycoid variants. A certain resemblance to the mycoid outline of the colony was observed in the case of one strain. Six out of nineteen atypical strains studied were transformed into typical variants with mycold structure after a prolonged cultivation in spodsolic soil taken from fields of the Timiryazev Agricultural Academy. The remainder either did not change or gave variants of still another structure and type.

The second series of expriments was performed with typical cultures of *Bac. mycoides* isolated from podsolic soils (4 strains) and chernozems (2 strains). The cultures were placed in soils in which these bacteria are not encountered, namely in chestnut soils. They were then incubated for 3 -6 months at room temperature. The soil humidity was maintained at the level of 60% soil capacity.

Systematic platings from the soil had shown that the mycold forms soon began to disappear and with each successive plating their numbers decreased. After 1-3 months almost all the original strains disappeared. Instead, various atypical strains, characteristic of the given soil as well a strains which are not usually encountered in such soils, grew upon plating.

We have observed these transformations of *Bac. mycoides* in both sterile and nonsterile chestnut soils; in the latter the transformations proceed more rapidly.

These data show that one form of bacterium maybe converted into another. The peculiarities of the soil influence the character of the variations.

The specific effect of soils on the bacterial variability is also confirmed by experiments carried out with other cultures of soil bacteria.

We have long been interested in a group of sporeforming bacteria which, according to the classification, are named *Bac. subtilis, Bac. mesentericus, Bac. cerus, Bac. brevis,* etc.

Do they represent separate species or are they variants and forms of one and the same species? This question has been experimentally determined.

Cultures of *Bac. mesentericus* were introduced into various soils--podsolic, forest, field and garden soils; chernozems, chestnut soils and red soils. The cells of *Bac. mesentericus* died out with relative rapidity in red soils and forest podsolic soils. In weakly cultivated fields of acid sod-podsolic soils and also in chestnut soils and chernozems, the culture did not die out completely. A small number of the cells survived. After some time the survivors began to multiply. However, upon plating on nutrient media they produced colonies which differed from the original ones.

In the course of adaptation new variants were obtained in chernozems and in chestnut soils.

In podsolic soils we obtained variants of *Bac. cereus* and *Bac. subtilis* and in chestnut soils variants of *Bac. subtilis* and *Bac. licheniformis.* In chernozems, all forms given in the table were obtained, i. e. , cultures, whose colonies resembled those of *Bac. cereus, Bac. subtilis. Bac. licheniformis, Bac. brevis,* and *Bac. mesentericus.*

An can be seen from the aforesaid, one and the same culture of *Bac. mesentericus* manifests different morphological forms according to the properties of the soil. In some chernozem soils (Moldavian SSR) we have found only two types of this group of bacteria, namely, *Bac. subtilis* and *Bac. cereus.* We converted individual strains of one or another type into typical cultures of *Bac. mesentericus.*

Considering only the external appearance of the culture, the difference between the mentioned species may be thought of as a reflection of polymorphism. Apparently, the classification of this bacterial group should not be based on external differences but on biochemical and biological properties. Of the latter, antagonism and antibiotic spectrum are the most indicative (in our opinion).

Afrikyan (1954a), employing the principle of the specificity of antagonism had shown that the group of sporeforming bacteria classified as *Bac. subtilis-mesentericus* is a very diverse nonhomogenous group in its biological properties, and therefore, cannot be considered as a homogenous taxonomic entity.

The principle of the specificity of antagonistic interactions makes it possible to reveal natural ecologic strains and types which actually exist, and to differentiate them.

Some ecological pecularities of sporeforming bacteria which are in soils of the extreme north, in the Soils of the islands of the Arctic Ocean--Novaya Zemlya, Franz Joseph Land, Severnaya Zemlya and others, are devoid of spores. The cultures grow well on nutrient media under ordinary laboratory conditions and form large colonies, but spores are absent. When they are grown in many ordinary media, the cells do not form spores.

The absence of spores in such cultures may mislead the investigators. Similar cultures can be classified as nonsporeforming bacteria. However, the structure of the colonies, the size of the cells, and the structure of the protoplast show that these organisms belong to the sporeforming group. The formation of spores in such bacteria can be induced. They form the usual endospores, but, as a rule, very late, after 2-3 weeks or later. Many of them form spores at elevated temperatures, 33-37° C. Some cultures begin to form spores after prolonged cultivation on protein-rich media, and others after they are kept in soils of the more southern regions such as the podsol of Moscow Oblast', chestnut soils of the Volga area and others. This peculiarity of these bacteria is apparently caused by long habitation In unusual soil-climatic conditions. Those conditions change the growth characteristics and the whole biology of the species. Bacteria lose the capacity to form spores. In some organisms, only the initial processes of spore formation are preserved. Cells of some strains form dense sectors, in the form of a prospore, at their ends, or in the middle part of the cells.

Under the conditions of arctic climate where the soil is always frozen or thaws for a short period, spore formation does not constitute a protective reaction of species preservation and in gradually lost. The biological essence of such bacteria as a sporeforming species is nevertheless preserved.

The capacity to form spores is lost, not only in bacteria of the arctic soils, but also in bacteria which inhabit more southern regions. The formation of asporogenic forms in general is a phenomenon quite frequently encountered in specimens of the genus *Bacillus,* and it takes place in laboratory cultures as well as in nature directly in the soil. One does not always succeed in revealing the asporogenic forms in the soil since they are differentiated only with difficulty from some nonsporeforming bacterial species.

Much attention is being paid to *Azotobacter,* among the soil microflora. This microbe is widely distributed in soils.

Azotobacter grows under various climatic conditions, from the extreme north to the tropics. It can be found in various soils, virgin and cultivated; it is found in podsols, serozems, chernozems, chestnut soils, brown soils and even in red soils. Notwithstanding the diverse conditions of the habitation of *Azotobacter,* only one species is described as widespread in the literature, namely, *Azotobacter chroococcum.* Some authors think it can be subdivided into 2 to 4 species. In our manual we have given 7 species, two of them being of an entirely different group.

Laboratory experience shows that cultures of *Azotobacter* of whatever origin, are at first eight homogenous according to their culture characteristics. They all grow well on media devoid of nitrogen (Ashby agar) and form large slimy colonies. After some time the colonies become brown and eventually black. All cultures fix nitrogen, to a greater or lesser extent, and are diagnosed in the majority of cases as *Azotobacter chroococcum.*Detailed studies. however, show that the cultures of this species are far from being of one type.

Upon studying a great number of Azotobacter strains collected from different soils of the Soviet Union it was found that they differ from each other, morphologically and biologically. The size of cells, their form and structure are far from similar, as described by many authors. We have described (Krasil'nikov, 1949 a) the characteristic differences of the, individual strains of *Azotobacter,* differences which served as the basis for dividing them into separate species such as *Az. beijerinckii, v. jakutii, Az. galophilum* and others. The cells of these and other organisms differ from each other in their size and form as well as in the internal structure of the protoplast. Cultures also differ in color, the nature of structure of the colonies and other factors.

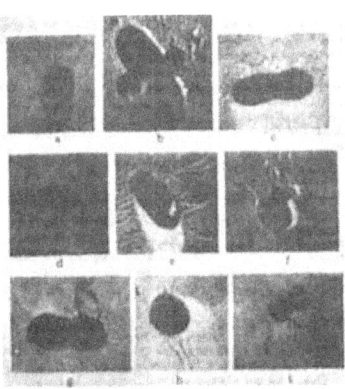

Figure 33. *Azotobacter chroococcum*. Diversity of cultures and strains, according to formation of flagella, and cell forms: a) strain 54, stock culture; b) strain 11 isolated from serozem (Central Asia); c) strain 13 from serozem of Tadzhik SSR; d) strain 14 from chestnut soil of the Volga area; e) strain 16 from chernozem of Khar'kov Oblast'; f) strain 19 from Crimean chernozem; g) strain 20 from red soils of Caucasus; h) strain 22 from podsol soil of Moscow Oblast', i) strain 25 from peat compost.

Cultures of *Azotobacter* differ sharply from one another in their form, size and location of flagella. The flagella of some organisms are long and multiple and uniformly located along the periphery of the whole cell, the flagella of other organisms are short and few and located uniformly or not uniformly along the cell. There are forms with very short flagella which are in the form of bristles (Figure 33). These characteristics are very stable and can serve as means for their differentiation (Krasil'nikov, Khudyakova and Biryuzova, 1952).

The main differentiating principle in this case as in other organisms is the specificity of antagonism phenomena. Khudyakova (1950) and then Babak (1956) showed that the cultures can be definitely divided into separate groups and subgroups according to their ability to suppress their competitors as well as according to their reaction to the action of antagonists. These groups and subgroups should be considered an independent taxonomic units--species and varieties.

An was mentioned above the cultures of *Azotobacter* are characterized by the polymorphism of their cells. They may form very small hardly discernible cells or gontdia, or very large gigantic cellular elements. The latter are frequently considered lifeless involution forms, which are nevertheless capable of forming regenerative bodies in their protoplast.

Under natural conditions of growth, *Azotobacter* is characterized by its great variability and polymorphism. This microbe changes under the influence of various soil factors, physical, chemical and biological.

A typical culture of *Az. chroococcum,* which was isolated from the garden soil of Moscow Oblast' and kept for 2-6 months in a Caucasus red soil and in a chestnut soil of the Volga area, became polymorphous. Several stable variants different from the original strain were isolated. Among them. variants were found which had very small cells and had lost the capacity to form slime capsules and to accumulate a fatlike substance in their cells. Such variants persisted and even grew in chestnut soil while the original strains perished under those conditions. In red soils, variants with obvious degenerative signs were obtained. The cells of those variants were very small and polymorphous, their protoplasm was bright, hardly discernible under the microscope, almost optically empty and without granular inlets. The colonies were small, flat, pastelike and without slime. They ceased to grow and perished upon fifth or seventh planting.

Small even filterable forms of *Azotobacter* were found in soils (Novogrudskii, 1935; Rybalkina, 1938b). *Azotobacter* forms which lost their ability to fix nitrogen are described in the literature (Rubenchik and Roisin, 1936, Stumbo and Gainey, 1938; Wyss, O., and Wyse, M., 1950).

Considering the variability of microbes in the soil one should mention the roo -nodule bacteria.

The root-nodule bacteria are more diverse and more widely distributed in the soil than *Azotobacter*. They are frequently encountered in soils where *Azotobacter* cannot grow. Each species of root-nodule bacteria to endowed with the capacity to form nodules on the roots of a given species of plants.

Apart from this symbiotic specificity, many species of root -nodule bacteria can be subdivided into strains which are markedly different from each other in many other properties.

We have studied numerous strains, of root-nodule bacteria of clover, lucerne, Onobrychis, kidney beans, vetch and peas, Strains of one and the same species differ from each other in cultural, physiological and biochemical properties and also in the extent of their activity and virulence. If their property of nodule formation were absent, they could have been classified as different species.

Under the influence of soil and climatic factors other bacteria as well as fungi, actinomycetes, yeasts and other microorganisms also change. Some of them change their species' characteristics rapidly and sharply, others are more stable and change insignificantly.

Ecological factors and not geographical locality, as some authors think, are the main reasons for such variability.

Bacterial Variability under the Influence of Plants

The vegetative cover strongly affects the variability of soil microbes. Plants,through their root excretions, change the natural properties of some species of microorganisms.

The different reaction of the various strains of *Azotobacter* upon the action of one and the same plant was frequently mentioned in the literature. Opinion was expressed as to the existence of the so-called local races of *Azotobacter* and on their specificity in regard to the root excretions of plants.

Our observations have shown that this microbe changed its properties under the influence of the root excretion of wheat. Its cells become smaller and smaller until they are hardly discernible under a microscope. With the decrease in size their metabolism also changes, and the capacity to grow on ordinary nutrient media is lost. (Krasil'nikov. 1934b).

Upon prolonged action of wheat-root excretion, the formation of new forms takes place which are markedly different from the original ones. We have obtained three such variants.

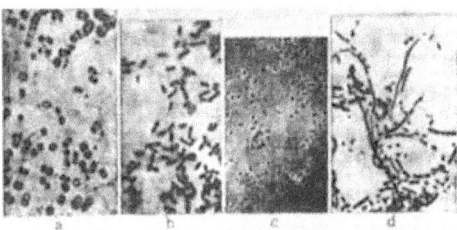

Figure 34. The variability of *Azotobacter chroococcum* under the influence of root excretion of wheat. The formation of stable variants in the rhizosphere in experiments under laboratory conditions: a) original culture, b) variant A on Ashby medium, 24 -hour culture; c) variant B on Ashby medium, 72-hour culture; d) variant C, 24-hour culture.

One of them, variant A, was of small cell size, the form of cells was bacillary, its protoplasm was homogeneous without fatlike inclusions. The presence of one or two small granules of chromophilic substance was noticed. They did not form sarcina-like packets, nor capsular resting forms (Figure 34). The cells in old cultures were shortened, often of oval or spherical form but without slime. Slime formation in this variant was either absent or weakly pronounced. The colonies were slightly convex of a pasty consistency, more rarely semislimy (Figure 35A). This variant resembles the culture of *Az. vinelandii* to a certain extent, but it does not form fluorescent pigment which is characteristic of the latter.

The second variant (B) differed from the original cells by its somewhat smaller size and weak slime formation. The cells were of oval or spherical form, and contained a fatlike substance (Figure 34 C), The colonies on Ashby agar were slimy, diffuse and transparent (Figure 35 B).

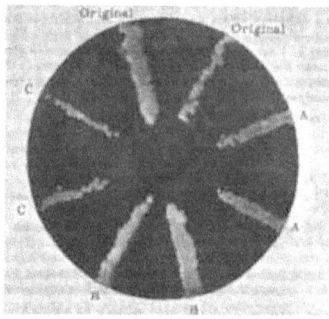

Figure 35. Colonies of Azotobacter chroococcum variants obtained under the influence of wheat-root excretions: A--variant A; B--variant B; C--variant C, 72-hour culture on Ashby agar.

The third variant (C) differed from the former two by the polymorphism of its cells. Growing cells in a young culture (2-3 days) on Ashby agar, were of diverse form; spherical, oval, rodlike and sometimes threadlike of 15 μ and more in length (Figure 34d). Their diameter never exceeded 1. 5 μ being on the average 1-1.2 μ; cells with a diameter of 2-3 μ were rarely encountered.

We found such cultures in the rhizosphere of wheat, in fields of the Experimental Station in Ershovo (Figure 36). One of them was sent to Dr. Bachinskaya, who studied it in detail. This culture was described as a new species of *Azatobacter--Az. unicapsa* (Bachinakaya and Kondratleva, 1941). This strain does not differ in morphological, cultural and physiological properties from variant A which we obtained.

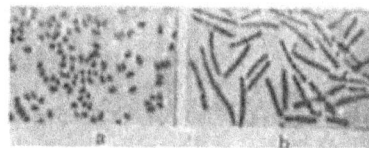

Figure 36. Culture of *Azotobacter unicapsa* Batschin, isolated from the rhizosphere of wheat in Volga area: a) 24-hour culture; b) 48-72 hour culture on Ashby agar.

The second culture corresponds to the experimental variant B and can be considered a new species or variety. The third strain resembles the experimental variant C in that it formed long threadlike cells reaching 50 μ in length and 1.5-2 μ in diameter. In its properties it differs sharply from the typical *Azotobacter--Az. chroococcum* and, according to the rules of taxonomy, can be considered as a separate variety if not as a species of the same genus.

Bukatsch and Heitzer (1952) performed detailed experiments on numerous *Azotobacter* cultures isolated from the root zone and roots of 31 species of plants--legumes, cereals, grain, fruit and industrial cultures. Cultural, morphological, and physiological properties of the isolated strains were studied. The authors have shown that the studied strains differed from each other in appearance and biochemical properties to a greater or lesser degree. They differed in the rate of respiration, fermentation activity (in reference to catalyse), their response toward the pH of the medium, and high salt concentrations. They also markedly differed in their capacity to fix nitrogen and in their reaction to root excretion of different plants; finally, they exerted a different effect on the growth and yields of plants. The authors noticed that some strains promoted the production of high yields of peas and nettle *(Lamium album* L.) but not of other plants. These active strains were isolated from the soil near the roots of the above-mentioned plants. This was the reason to think of specificity of the strains. Other strains did not show such a specificity with regard to plants.

Petrenko (1949, 1953) reached the conclusion that *Azotobacter* forms variants which are strictly specialized with regard to some plant species. Each species of plants has its *Azotobacter* strain in the same manner that each species has its own strain of root-nodule bacteria. Hence, the effectiveness of employing a given microbe is manifested only in those cases when its appropriate specific strain is employed. Analogous data were obtained by Zinovlev (1950).

Unfortunately, the authors did not confirm their assumptions by detailed laboratory experiments. The strain specificity was established only on the basis of the data of their effectiveness under field conditions without a corresponding microbiological analysis. The effectiveness of *Azotobacter,* as is well known, varies in different experiments. Even in experiments with one and the same culture of plants and in one field, the data on plant crop increments due to *Azotobacter* vary and are even contradictory.

It should be remembered that *Azotobacter,* owing to its polymorphism under various conditions, in different parts of the same field, may manifest itself differently without changing its species identity.

The sporeforming bacteria, *Bac. mycoides, Bac. megatherium, Bac. mesentericus,* and others, are subject to sharp variations when they grow in the rhizosphere of plants. Their cells lose the capacity to form spores, they become shorter, do not form threads--chains, and they decrease not only in length but also in diameter; as a result, cultures with very small, deformed elements, and pale, poorly stained plasma are obtained. The formation of asporogenic, deformed variants in the above-mentioned bacteria is described in the literature (Novogrudskii, Kononenko and Rybalkina, 1936; Clark F., 1940, 1949 and others).

Arkhipov (1954) showed that the causative agent of anthrax, *Bac. anthracis,* markedly reacts to the action of certain plants. The cells of *Bac. anthracis* lose the capacity to form spores and become asporogenic and simultaneously avirulent. Degenerative forms, characterized by unusual structure and cell growth were obtained under the influence of plants.

Data are available in the literature on the variability of root-nodule bacteria under the influence of plants. In our experiments, the root-nodule bacteria of clover and lucerne changed under the influence of the root excretion of peas, wheat, flax, corn, and cotton. The strongest influence was that of corn. After keeping the clover bacterium *Rh. trifolii* in the rhizosphere solution of the given plant for one and a half months the cells became sharply deformed. Upon plating on Petri dishes containing different media, several variants were obtained which differed from the original culture and from each other, morphologically, physiologically and culturally. Cells of some variants were swollen or spherical, spindlelike or threadlike in form, and the plasma was vacuolated with various inclusions (Figure 37). Such cells grew badly on the soy agar, they formed small colonies which were of pasty consistency and flat, without slime. They cannot be maintained for a long time in the laboratory; after several platings (5-8) they cease to grow (Krasil'nikov, 1949a, 1954b; Korenyako, 1942).

Figure 37. Rhizobium trifolii. Variation of culture under the influence of corn-root excretion:a) original one-day culture on soy agar; b) variant A; c) variant B.

Some strains acquired the ability to grow on meat-peptone medium, while the original culture cannot grow on this medium. The variants obtained also differed from each other in certain biochemical properties. They acidified milk (the original ones alkalized milk), fermented sugars, which were not attacked by the original strains, liquefied gelatin, etc (Table VI).

TABLE 6
Comparison of ability to ferment sugars displayed by variants of Rh.trifolii obtained under the influence of corn-root excretions

	External features	Glucose	Sacchrose	Maltose	Lactose	Mannitol
3	Pasty, smooth, small, rodlike cells	+	-	-	-	-
10	Slimy, normal cells	++	+++	-	+	-
12	Pasty, rough, very small and deformed cells	++++	+	-	++	+
13	Dwarf, small and deformed cells	-	-	-	-	-
	Original culture	-	-	-	-	-

Note: Plus-sugars fermented; minus-sugars unfermented

The influence of plants on the variability of microorganisms was clearly shown in investigations in which root-nodule bacteria were grafted on roots of vegetatively coalesced soy cultures. If one kind of legume is grafted to another kind, for example a stock of kidney beans to soy or vice versa, and the coalesced plants are infected with root-nodule bacteria of the stock of the graft, then bacterial cells can be found in the formed nodules which are capable of growing and forming nodules in the roots of the graft. Employing this method (mentor method), we succeeded in changing the virulence properties of root-nodule bacteria of clover, peas, kidney beans and soy, and to confer on them the property of forming nodules foreign to them. The root-nodule bacteria isolated by us from the nodules of yellow acacia grafted to *Haltmodendron halodendron* were capable of forming nodules on the root of separately grown *H. halodendron*. The original cultures form nodules on roots of yellow acacia but do not form nodules on the roots of *H. halodendron* (Krasil'nikov 1954b).

Similar results were obtained by Peterson (1953). She inoculated coalesced legumes with cultures of nodule bacteria of the stock and obtained variants with specific virulence characteristic of the graft bacteria.

Wilson and Wagner (1936) have showed that root-nodule bacteria of clover can adapt themselves to the plant host. With the change of the plant the root-nodule bacteria change their properties accordingly.

There are data in the literature on the adaptive variability of phytopathogenic bacteria and fungi. According to specialists, specific specialized forms of these microbes were formed as a result of adaptive variability (Naumov, 1940; Gorlenko, 1950).

Gorlenko and co-workers showed that some strains of the sporeforming bacterium *Bac. mesentericus* and of the nonsporeforming *Pseudomonas fluorescens* could be converted into phytopathogenic forms by appropriate culture in artificial nutrient media and in vegetative substrates (Gorlenko, Voronkevich and Chumaevskaya, 1953).

The given data provide a basis for the assumption that plants exert a definite effect on the variability of microorganisms. Unfortunately we are short of the means which are required for the detection of subtle changes which take place in many other species upon the action of root excretion or more strictly of the whole rhizosphere of this or another plant. We cannot say with certainty how this or another organism behaves when introduced under plants, whether it experiences any changes, or it remains in its original form.

The microflora is the main factor of the variability of the microorganisms. Large numbers of antagonists supressing the growth of different species inhabit the soil. For example, many sporeforming bacteria, actinomycetes, fungi, protozoa and others have a harmful effect on *Azotobacter*. Under the action of these inhibitors, *Azotobacter* either dies or adapts itself by changing its properties. Forms of *Azotobacter* which are resistant to these forms are not infrequently encountered. They differ morphologically, culturally and physiologically from the original ones. Some of them are more vigorous nitrogen fixers, others, on the contrary, become less active or lose this capacity altogether. Xonokotina (1936) obtained *Azotobacter* variants under the influence of soil amoebae. These variants were no longer edible for the amoebae, the amoebae did not engulf them. These variants differed from the original strain in their pigment, character of growth on Ashby agar, cell structure, capacity to form slime capsules, etc.

Afrikyan (1954) watched the formation of forms of *Azotobacter* resistant to microbial antagonists, bacteria, and actinomycetes.

When obtained In this way, variants prove to be genetically stable in many cases. They can be maintained for years under laboratory conditions. We are dealing here not with polymorphism of bacteria but with species variability taking place in the natural conditions of their habitation.

The root-nodule bacteria change their species properties under the influence of the soil microflora. Konokotina (1938) brings data of her observations on the variability of nodule bacteria of soy under the influence of pure culture of the sporeforming bacterium *Bac. subtilis.* The author obtained variants, among which three of them differed in their properties from the original form. Their activity was twice as high as that of the original culture.

THE PRINCIPLES OF CLASSIFICATION OF MICROORGANISMS

The classification of microorganisms is very unsatisfactory. There is no common principle of classification in microbiology. The classification of bacteria and actinomycetes is especially inadequate. This can be explained by the peculiarity of those organisms, the simplicity of their structure and growth and lack of external properties for differentiation. An immense number of species is not fully described and is not classified in an appropriate way. Besides, the classification is not an objective one, there are different directions in it, which often do not agree with each other and are often even contradictory.

The main reason for the inadequacy of classification of bacteria and actinomycetes lies in the absence of factual material on phylogeny of individual species, groups and subgroups. The natural classification should reflect the stages of evolutionary development. The principle of phylogenetic structure of classification requires a thorough knowledge of the organisms and species relationship.

Unfortunately this principle is absent in the modern classifications. The classification of bacteria is made according to separate, frequently accidental properties. One and the same homogenous group is frequently divided into separate taxonomic entities often without sufficient grounds.

The great shortcoming of every classification is the fact that it is built in only one plane, a descending or an ascending one, while in reality each taxonomic entity should represent a center of formation of new forms, species, varieties, etc.

Classification also becomes complicated by the fact that phylogenetically foreign organisms are concentrated around these centers, the morphology of which is close to that of the main specimens of the studied group. Each group of bacteria, actinomycetes or fungi comprises main and convergent forms. All these organisms may appear homogenous according to their appearance and physiology, while phylogenetically some of them comprise a uniform group and others are accidental foreign organisms. The natural classification should embrace only the former organisms which determine real affinity, while the foreign convergent forms should be excluded.

Microbiological literature contains numerous data on the morphological similarity of entirely different organisms. For example, a large group of nonsporeforming yeasts, comprising the family Torulaceae, in reality represents a mixture of different species and not only species but also genera, and generally speaking in a mixture of different unrelated groups. The bacteria of the genus *Micrococcus* are characterized by their spherical shape. Into this group organisms which in fact belong to coccoid bacteria are included and also not infrequently specimens of actinomycetes are included in the genus *Mycococcus.* To the rodlike bacteria of genus*Bacterium* very diverse microbial forms are referred. Bacteria from the genus *Bacterium* are frequently linked to *Mycobacterium* only because the former sometimes possess lateral branches which are characteristic of the genus *Mycobacterium.* Thereby the specificity of this property is not accounted for. Properties which are natural and fixed in one genus may be abnormal and unnatural in others.

The shortcomings of the bacteriological classification have their origin in our scant knowledge of the life of the organisms. In order to be able to speak of the phylogenetic relations between the organisms, it is not sufficient to know and study one randomly chosen stage of the life cycle of the microbe. A thorough knowledge of its growth, development, structure, reproduction, life cycle, polymorphism, variability, etc, is needed. In order to obtain much knowledge, the organism in question should be studied not only in laboratory conditions but also in natural surroundings.

The lack of knowledge of the life cycle of this or another microbe frequently misleads the investigator. For example for this reason mycobacteria are considered by some authors as micrococci or as rodlike bacteria.

Each property, biological, physiological or morphological, may be employed for the identification of the organism, but not each diagnostic property is of systematic value. The problems of identification and classification should be distinguished from each other although in some cases they are interwoven and overlap. Fixed properties which are transmitted by heredity, which characterize the nature of the species, and which are manifest under determined growth condittons, are taken into consideration for classification.

The Species Concept in Microorganisms

The concept of species is the most difficult problem in classification and is of the utmost importance in taxonomy of microorganisms. In microbiology this basic taxonomic entity was less studied than in any other branch of biology.

Different authors, according to their specialization, much as mycologists, bacteriologists, phytopathologists, medical bacteriologists, and applied microbiologists consider microbial species differently. Each branch of microbiology chooses different properties for the determination of species. In yeasts, for example, certain indicators are taken as a basis for species determination (sugar fermentation of *Saccharomyces, Zygosaccharomyces* and others) in bactoria, other and moreover, different properties are employed in different groups; actinomycetes have still other properties which are employed for this purpose.

Often authors consider bacterial species according to their own views and whims, which do not agree with the factual material.

The more thorough the knowledge of the organism, the more strict the term species becomes. It progresses with the development of the knowledge of the organism. It was a long way from the Linnean idea of a species as a fixed and not changing entity to Darwin's teaching and present ideas of the ever-developing species. In the determination of species, Linnaeus stressed two basic properties--the constancy of properties and their sharp demarcation in various species. Later on Korzhinskii (1892) introduced a geographical principle into the concept of species which was more thoroughly elaborated by Wetstein (1898). According to this principle, the species morphology is correlated with the place of its habitation. Approximately at the same time, a historical moment entered the concept of species; this was especially pronounced in the works of Semenov-Tyan'Shanskii (1910), who pointed out that the sum of architectonic properties is a result of interaction of a complex of physicogeographical factors which took place in a bygone geological period (according to Krasil'nikov, 1938a).

The establishment of phylogenetic relationships within the species led to the separation of the main and the subordinate taxonomic entities--varieties, forms, etc. The species was divided into main and secondary forms; binomial nomenclature became tri-and quadrinomial nomenclature.

With the development of genetics the concept of species widened according to the ideas of variability and heredity of organisms. New terms were introduced for the determination of species subdivision, such as "biotype", "pure line", "jardanon", "linneon", etc. ["Jardanon"--a simple means of classification of lower organisms. "Linneon"--the complex of "jardanons"--according to the Russian concept, the inner species variety of forms does not exceed the limits of qualitative unity of the species.]

The experimental verification is a new stage in the development of the determination of species. Of the works in this direction the most important ones are those of Touresson, (1922-1925) which showed the dependence of the external form of plants on the ecological conditions of habitation. He introduced the term "ecotype". Species is now being looked upon as a definite system. This is most sharply pronounced in works of Vavilov (1932) who showed that Linnean species are definite systems of forms, and not an accidental motley of races. Komarov in his work "Flora of the Kamtchatka Peninnsula" (1920) gives the following definition of species: species is a morphological system multiplied by geographical definiteness (Komarov, 1940).

Lysenko (1952) considers species as separate links of a single chain of the organic living world. Species, he writes, is a unique qualitative state of the living matter. The basic characteristics of animal, plant and microbial species are defined intraspecies relationshps between the individuals. These intraspecies relationships are qualitatively different from those between the individuals of different species.

This definition is based upon the biological properties of organisms and their specific properties, the knowledge of which should be the subject of a detailed study.

As can be seen from the aforesaid, the concept of species is considered differently. Every specialist has his own narrow idea about species, moreover, the factual material accumulated by others to not always taken into account. In spite of the diversity of concepts and radical biological differences, the principles of classification of different groups of organisms should be the same. Admitting the right of the different specialists to probe into the problem and to elaborate it according to their specific needs, we point out that those concerned with classification should adhere to the main outlines and taxonomic rules, established by international congresses and by conferences of specialists. Besides, there are general biological features which characterize species of higher and lower organisms alike.

Concerning the concept of species in microorganisms, certain specific properties which preclude the possibility of deploying all the principles which are being applied in the classification of higher organisms, should be mentioned.

The distinguishing feature of microorganisms is their lack of multitude of morphological characteristics. Therefore, determination of species on the basis of morphology in the overwhelming majority of cases is impossible. Morphology can be used only for the determination of higher taxonomic entities--genus, family and higher.

The geographical principle or more strictly speaking, the ecological-spatial principle, and the historical principle can not be employed for the determination of microbial species. No data are available in microbiology, which could show some definite regularity in the distribution of separate microbial species in nature. Even less data exist on the fossil microbial forms. The scant data on the presence of microbial cells in microsections of coal, shales and other deposits gave no grounds for establishment of even large taxonomic entities. Methods employed in cytology and genetics are not suitable in that case. Bacteria and actinomycetes do not possess a genuine, well-defined and isolated nucleus (only nucleoids are present), they do not possess plastids or any other organelles, characteristic of cells of higher organisms. The sexual process in those organisms is either lacking or, when present, is a unique one and cannot serve as a means for the establishment of a genetic relationship of the organisms.

The biochemical method of species determination unfortunately is not systematically elaborated. Separate attempts to divide species according to their biochemical properties are as yet of no importance.

Biochemical properties can only be employed in individual cases for determination of species. Products utilized by the microorganisms and decompositlon of some organic and inorganic compounds can serve as a basis for such determination. In other cases organisms are determined by metabolic products of a specific nature, for instance by their capacity to form antibiotics, toxins, slimes and other features.

The mode and sequence of the decomposition of organic compounds such as carbohydrates, with the formation of intermediate compounds is of great interest for classification of organisms.

Unfortunately, this direction is not sufficiently elaborated for the purpose of classification (Shaposhnikov, 1944; Knight, 1955). It should be noted that species of practical importance are the most thoroughly studied ones. In such cases species is determined by the leading biochemical or biological property. Antagonists of actinomycetes as well as of other microbes can be divided into species according to the antibiotic they produce. In higher plants species are not infrequently established according to their economic importance, or according to some property utilized by men in everyday life or industry.

Organisms of any importance in economy were more thoroughly studied and naturally are among the best known organisms. Many biochemical, physiological and biological properties of such organisms have been studied, and methods for their determination are well elaborated.

Such profound and comprehensive studies enable the revelation of the nature of the species and its relationship to other species. Often, owing to such a thorough study, the heterogeneous quality is revealed in organisms which were formerly thought to represent a homogenous species. Organisms which are assumed to belong to one species prove to be a complex group consisting of several or many species.

There are comparatively few species which are utilized by men in everyday life and in industry. The majority of organisms, however, were studied only to an extent which seemed necessary for general purposes. It is clear that the usual methods employed for the determination of bacterial species are not uniform. There are many more species in nature than we are able to determine by the relatively primitive methods at our disposal. The heterogeneity of the known species of bacteria and actinomycetes is becoming more and more apparent with the more thorough study to which they are subjected.

What are the accepted methods for the determination of species of bacteria, actinomycetes, and microorganisms in general?

The ability to grow organisms in artificial nutrient media in pure cultures and to follow their growth directly under the microscope under different growth conditions is of great advantage. The microbes proliferate at a high rate and, due to this, many problems connected with their polymorphism and variability during their life cycle can be solved in shorter time than when working with higher organisms. This is especially important in determining the variability or stability of taxonomic properties over a large number of generations.

The comparative studies of pure cultures of microbes enable one to study the species not only by itself but also with regard to the entire population during their variations and deviations which take place in the process of variability.

In spite of what was said before, that the morphological principle is in many cases unsuitable for the determination of species of bacteria and actinomycetes, it should not be neglected. In some cases it is a trustworthy method for species determination. Such organisms as *Azotobacter--Az. chroococcum, Bac. megatherium,* threadlike bacteria, many sulfur bacteria, iron bacteria and other species can be determined and identified by their external cultural and morphological properties.

Physical properties are not infrequently employed for the determination of species. This method is trustworthy in many cases. However, it is of limited importance and may lead to errors if employed without taking other properties into account. The afore-mentioned method requires further elaboration. Such indexes as microbial behavior toward oxygen, temperature, pH of the medium, fermentative capacity of the organism, its requirements of organic and inorganic nutrients may be properties of a species only in a restricted sense.

The phylogenetic principle can be employed in microbiology with success. The afore-mentioned biological properties of microbes enable one to determine their genetic relationship experimentally. The method of experimental variability reveals not only heterogeneity of cells in the culture and their polymorphism in the life cycle but also the frequency of the formation of variants and, according to the latter, the affinity to the naturally existing forms and species of microorganisms.

Employing this method, we were able to establish the affinity between specimens of actinomycetes. The transformation of mycococci into mycobacteria, mycobacteria into proactinomycetes, and the latter into actinomycetes was experimentally proven. Data were obtained on the affinity of lactic bacteria, propionic bacteria, certain micrococci and pseudobacteria to a group of organisms close to actinomycetes (Krasil'nikov, 1938a, 1947b, 1949c, 1955b).

Microbiological literature contains data on the phylogenetic affinity of some specimens of mycobacteria (Imshenetskii, 1940). Osterle and Stahl (1929) and then Rautenstein (1946) obtained variants of *Bac. mycoides* which did not differ from *Bac. effusus, Bac. olfactortus, Bac. nanus, Bac. vulgatus, Bac. cereus, Bac. brevis, Bac. panis. and Bac. mesentericus.* A multitude of variants of *Bac. mycoides* was also obtained by us (Figure 38) (1947b). Medvins'ka (1946) experimentally established that the sporeforming bacteria *Bac.*

mesentericus fuscus, Bac. mesentericus niger, Bac. mesentericus vulgatus, and *Bac. mesentericus panis*described in the literature represent one species with two varieties. Gibson (1949) brings a voluminous experimental material concerning the scope of the species *Bac. subtilis.* He identified this species with the following bacteria: *Bac. aerrimus, Bac. globigii, Bac. leptosporus, Bac. levaniformans, Bac. panis, Bac. niger, Bac. viscogenes, Bac. vulgatus* and some other varieties.

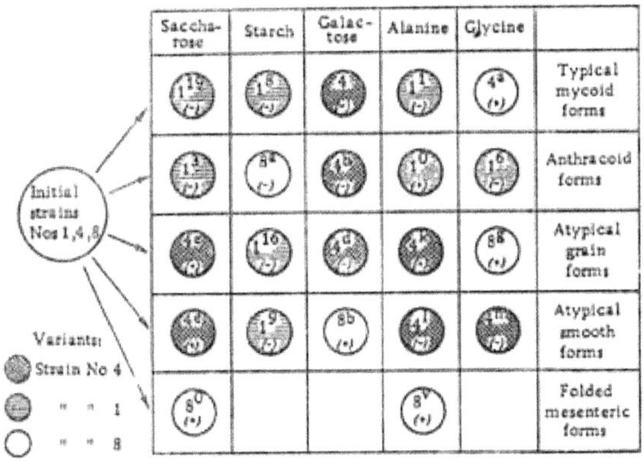

Figure 38. Variants of the culture *Bac. mycoides* obtained experimentally: (+) and (-) variants: according to the utilization of nutrients vertical rows; according to colony structure--horizontal rows.

Nakhimovskaya (1948) obtained variants of *Ps. aurantiaca* which did not differ from *Ps. fluorescens.* The typical strain of *Az. chroococcum* yielded colorless, capsuleless variants with small atypical cells which grew on protein media and did not fix nitrogen. Such variants were also isolated from the soil.

Widely known are the variants R, S, M, G and others in various specimens of bacteria. They markedly differ from the original cultures and often appear to be like independent species which exist in nature.

Even more variants are encountered which differ from the original cultures in their physiological and biochemical properties. In the majority of cases such variants are not distinguished in laboratory practice and remain unrecognized.

With the aid of the experimental variability method we have shown, the affintty between the individual species of root-nodule bacteria and also the phylogenetic closeness of the latter to certain root-nodule bacteria of the genus *Pseudomonas.* It was established that certain asporogenous species of *Pseudomonas* can be transformed into root-nodule bacteria under certain conditions (Krasil'nikov, 1954b, 1955b).

Employing this method we succeeded (Krasil'nikov 1947b) in establishing the affinity between certain species of nonsporeforming bacteria of the genus *Bacterium* and *Pseudomonas* and then between the sporeforming bacteria and other groups. During the course of the dissociation of two different cultures of the genus *Pseudomonas* similar variants (Figure 39) were obtained which were identical morphologically, physiologically and biochemically.

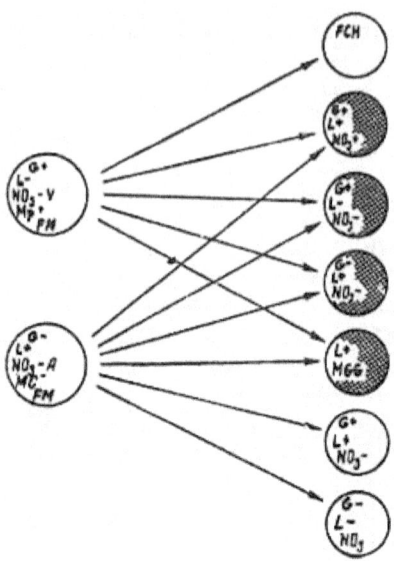

Figure 39. The scheme of formation of monotype variants by different cultures of *Pseudomonas fluorescens* (strains A and B): fermentation of: G-glucose, L-lactose; MC--milk coagulated; MP--mtlk peptonized; $N0_3$--reduction of nitrates; F--fluorescence; FM--fluorencence on meat-peptone agar or meat-peptone broth; FCh--fluorescence on Chapek medium; + reaction positive, - reaction negative (according to Krasil'nikov, 1947).

Species of sporeforming bacteria such as *Bac. mesentericus, Bac. subtilis, Bac. cereus,* and *Bac. licheniformis* yield identical variants as was mentioned above. All this indicates the similarity of hereditary and species properties of organisms.

The method of experimental variability also reveals the phylogenetic relations between yeasts. Filippov (1932) established the affinity between various specimens of nonsporeforming yeasts by this method. Cultures of the genus *Torulopsis* which were exposed to X-ray irradiation produced a number of stable variants which should be included in the genera *Torulopsis* and *Mycotorula* (Filippov 1932) according to their morphological and physiological properties, if the rules of present classification were applied.

The yeast fungus, *Sporobolomyces,* yielded through the method of dissociation many (more than twenty) variants some of which were identical with cultures of *Torula. Torulopsis,* and *Mycotorula;* others had well-developed mycelia which did not differ from the mycelia of mycelial fungi. We obtained these variants by relaxing their heredity with subsequent adaptation. These data show that expertmental variability is a valuable method for the study of a species and its relationship to other species, and therefore should be employed in the classification of microorganisms in general and of bacteria and actinomycetes in particular.

The serological method of diagnosis of pathogenic bacteria is widely used in medical bacteriological practice. This method is a very sensitive one and enables the detection of small differences between strains which are formed under the conditions of variability of these or other cultures. The sensitivity is determined by the chemical nature of the antigens and the capacity of the method to distinguish differences between the various molecules, predominantly protein molecules, which cannot be differentiated by chemical methods. With the aid of antigens the smallest differences in the protein composition can be detected and consequently subtle culture variations can be diagnosed. Nonetheless, this method is not being used in the classification of microbes to the extent that could be expected. Though it is an excellent and sensitive method for the differentiation of related strains It is at the same time unsuitable in the majority of cases for the classification and differentiation of species. Attempts to divide the root-nodule bacteria by serological methods yielded contradictory results which did not agree with the classification done by conventional methods. Entirely different species are linked to one taxonomic group and vice versa (Zipfel, 1911; Simon, 1914; Stevens, 1923; Fred et al., 1932; lzrail'skii, et al.,

1933). Similar results were also, obtained when an attempt was made to classify strains of *Azotobacter, Radiobacter,* bacteria of the genus *Pseudomonas, Bacterium, Clostridium,* streptococci, and others (Mayr and Harting, 1948; Fred et al., 1932; Frances-Shattock, 1955).

It is apparent that some points of the method are not sufficiently elaborated. Experiments showed that not all antigens of microorganisms and higher organisms could be detected (Lamana, 1940; Davis, 1951). These authors made an attempt to classify species of sporeforming bacteria of the genus *Bacillus.* They could not detect specific antigens in vegetative cells; neither the somatic antigen (0) nor the flagella antigen (H). More encouraging results are obtained with spore antigens.

Concerning the lactic bacteria of the genus *Lactobacterium* there was no success recently in classifying them by serological means. Quite recently Sharpe succeeded in obtaining precipitin, with the aid of which these bacteria are as if differentiated into types, which can be correspondingly determined by physiological-morphological methods. (Sharpe, 1955, Briggs, 1953.)

Fourteen year studies of Mez and his collaborators (1922, 1926, 1936) on plant classification by serological methods were unsuccessful.

The method of serological diagnosis is well elaborated for colic bacteria--*Bact. coli, Bact. typhi, Bact. paratyphi , Bact. dysenteriae,* and others. This method is successfully used for the classification of streptococcii, pneumococci, staphylococci, and especially of viruses (Bawden, 1955; Holmes, 1955; Andrews, 1955; Ryzhkov, 1952). It should be pointed out that groups established by this method are of a specific character and do not always correspond to the groups determined by other methods. Experience shows that the determination of species by serological methods should be carried out with precaution and the results must be carefully analyzed. It is only of secondary importance and complements other methods of microbiological analysis.

Attempts are made to identify bacterial species by means of phages. This method is based on the capacity of phages to lyse defined bacterial species. Among phages there are strains with strictly specific species properties attacking only one bacterial species. There are also group-specific phages which attack bacteria belonging to different species or even genera. It is assumed that the former phages are suitable for species differentiation and the latter for group differentiation.

Data exist on species differentiation by means of phages of bacteria of the coli group, staphylococci, nonsporeforming and sporeforming soil bacteria, and others. Katznelson and Sutton (1951) pointed out that phytopathogentc bacteria (in various substrates) can be detected by means of phages. They detected *Ps. phaseolicola* in seeds of kidney beans without isolating them in pure cultures. The seeds were washed with water, the washings were inoculated with the specific phage of *Ps. phaseolicola* and after some time the titer of the phage was determined. By this means the degree of phage multiplication was determined. The phage, according to the authors, can multiply only in the cells of the afore-mentioned species.

Smith, Gordon and Clark (1952) have shown that sporeforming bacteria can be differentiated by the phage method. Data are available on the differentiation of species and groups of the coli bacteria--*Bact. coli, Bact. typhi, Bact. paratyphi, Bact. dysenteriae.* Stocker (1955) and others think that all these organisms represent one closely related bacterial group. To this group, according to him, *Bact. pestis and Bact. pseudotuberculosis* also belong. As can be seen from this, the bacteriophages and actinophages are very sensitive, but are far from specific to the degree required in classification. The phage method as well as the serological method can be employed in separate cases and then only as auxiliary tests.

Parasitological and toxicological methods of species classification are employed in phytopathology. Certain parasites are very specific and parasitize only defined plant species, therefore this method can be employed for the differentiation of the parasite by its host.

It is known also that certain parasites form toxins, for example the diphtheria bacillus, the causative agent of gas gangrene, *Cl. botulinum* and others (Oakley, 1955).

These properties are of a narrow specific character and can he employed only for a restricted group of microbes. It may happen that when different representatives of microorganisms are studied from the point of view of formation of specific toxic and nontoxic metabolites this principle will find wide application in classification of microorganisms.

According to our data, specificity of antagonism constitutes an essential species indicator. We have shown that microbes grown together in mixtures often exhibit antagonism. One culture supresses the growth of the other. This process can be caused by specific and nonspecific substances.

Specific substances as antibiotics are of importance in species determination. Each species of an antagonist synthesizes its specific antibiotic and sometimes two and three antibiotics simultaneously. As a rule, the characteristic property of antibiotics to the fact that they do not suppress the growth of the producing organism and other organsms belonging to the same species. The action of the antibiotic and, consequently, of the organism which produces the antibiotic is directed toward the organism's competitors. This specificity is strictly constant for the antagonists and can serve as a means for species determinaion. On the basis of this principle, we have elaborated a method of species differentiation of actinomycetes and sporeforming antagonistic bacteria (Krasil'nikov, 1951c, Krasil'nikov, Korenyako, Nikitina, Skryabin, 1951, Afrikyan, 1951a). Employing this method we were able to establish a multitude of species among certain groups of organisms which were classified as one species. Similar data were obtained by Teillon (1953) and some other investigators (see below).

From the afore-mentioned it may be concluded that species is a really existing taxonomic entity, consisting of different individuals, strains or cultures (in the laboratory), forms and varieties. They all possess one property which can be manifested in different forms. This property characterizes the species an a whole, and as long as it Is preserved, the species remains unchanged. Individuals and varieties are only forms in which the species exists. Individuals and varieties may differ from each other but the species properties common for them remain unchanged.

Consequently, species as a link in the evolutionary chain of development represents, in a certain sense, a closed entity of living organisms. It is closed because all the individuals of the species have a definite relation to one another. The relations within the species differ from the relations between organisms of different species. This is clearly shown in examples of microbial antagonism.

Proximity of species is manifested in the requirements for certain life conditions, nutrition, light, temperature and other factors. All organisms of the same species, no matter how different they may be, require the same basic factors for existence, which may be different for other species. The organisms of one and the same species always assimilate the same medium regardless of the geographical zone in which the individual organisms or cultures may grow.

For example, the actinomycetes which produce streptomycin and comprise one species *Act. streptomycini* are widely distributed in nature; individual strains live in different geographical zones, in different soils, in the north and south, east and west, in the silts of rivers and lakes, and on different plant residues. Their ecological conditions of habitation are very different. Nonetheless all strains of this species are of the same heredity and of the same fundamental characteristics.

Species is an isolated link in the evolution of organisms consisting of qualitatively homogenous forms, which require the same basic life conditions, have the same origin and are characterized by definite morphological and physiological properties, and also by the degree of variability which is transmitted by heredity.

The scope of a species is determined by the extent of the variability of the organism in question, by its morphological-genetic capacity. The species manifests itself in varieties, forms and individuals (cells) which represent species variations and polymorphism under defined conditions of existence.

Species is an ever-changing and developing system of closely related organisms. Its beginnings can be seen in individual changes of cells. These changes widen and become fixed by natural selection in the course of adaptation to the environment and dominating growth.

Experiments showed that microbes are not monomorphous either under the conditions of growth in artificial nutrient media or in nature. In natural conditions microbial cells are subjected to stronger and more diverse interactions; it is natural to expect more frequent and diverse formation of variants under these conditions. This is confirmed by microbiological analyses of natural substrates. In the soil, as it will be shown later, species exist in very diverse forms, morphological and functional. Diversity of strains, forms and varieties of each species under natural conditions are determined by the properties of the latter and hereditary properties of the organism.

In classification systems species are grouped into larger taxonomic entities--genera, genera into families, etc.

SUBDIVISION OF MICROORGANISMS INTO MAIN GROUPS

In our manual for determination of species all microorganisms except protozoa are included in one series of primitive organisms of the plant kingdom--Protophyta.

This series is divided into two groups:

1) Schizophyceae--lower organisms, possessing chlorophyll, phycocyanin or phycoerythrin;

2) Schizomyceae--chlorophyll-less organisms. Here belong all bacteria, fungi and actinomycetes described in the literature. The organisms of this group are very diverse in their structure, growth, biological essence and phylogeny. They are divided into classes; each of them has its genealogy and it should be assumed, its own unique roots of origin. In other words, organisms of this group are of polyphytic origin.

We divide them into four classes:

Class I--Actinomycetes

Class II--Bacteriae

Class III--Myxobacteriae

Class IV--Spirochaetae

Actinomycetes

The class of actinomycetes is a homogenous, well-studied group of microbes. According to our data, the group of actinomycetes included various subgroups: actinomycetes, micromonospora Waksmania, proactinomycetes, mycobacteria, and mycococci; the affinity between them was established by us experimentally. Beside the above-listed organisms, the group of actinomycetes comprises to a greater or lesser extent specimens of the so-called pseudobacteria, lactic and propionic bacteria and some others (Krasil'nikov, 1938 a).

Genus Actinomyces. Actinomycetes are higher organisms than bacteria, according to their structure and growth. As it is known they form a well-developed mycelium. The mycelial threads are thin, 0.5- 1.0 μ in diameter, without septa. Branching is similar to that of fungi, from the main branch side branches of the first, second, third order, etc originate. The mycelial threads are thinner than in fungi, more fragile, are easily broken and destroyed, forming shreds and splinters.

Actinomycetes grow on dense nutrient media in the form of compact, dense, gristlelike leathery colonies. The latter have a smooth, granular, rough or plicated surface. The colonies grow into the medium with their threads, and have a flat or convex form. They are of such density that platinum wire is unsuitable for removing part of the colony, for this purpose special loops have to be used.

Actinomycetes possess a substrate and aerial mycelium. The former really represents the entire colony. The aerial mycelium consists of hyphae which originate in the threads of the substrated mycelium. It may be abundant and cover the entire colony with a fluffy, velvety or floury coating. It may also be weakly developed in the form of separate bundles of threads located on the surface of the colony.

Sporangia containing spores are formed on the threads of the aerial mycellum. The structure of sporangia varies in different species. Some actinomycetes have a spiral or more precisely spindle like sporangia, others, straight or undulant. The number of coils in the spiral sporangia can fluctuate between 0.5-1 and even 5-7 and more. The coils in some cases are densely packed, they frequently have the form of spheres, in others they are stretched (Figure 40A).

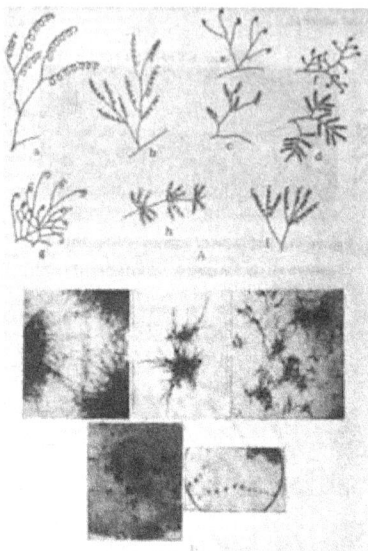

Figure 40. The structure of sporangia of actinomycetes: A) spiral sporangia with coils of different character; B) nonspiral sporangia of actinomycetes.

Nonspiral sporangia also differ from each other. In some species they are very short, straight, in the form of bristles, in others-long. straight or undulant but not spiral (Figure 40B).

Spores are often of oval, spherical forms, less frequently they are rodlike, and cylindrical forms with sharp ends (Figure 41). Spores of many actinomycetes have different forms--spherical, oval, and rodlike. In young sporangia spores are frequently rodlike, in mature sporangia they are oval and spherical. The rodlike spores gradually become round until they assume spherical form.

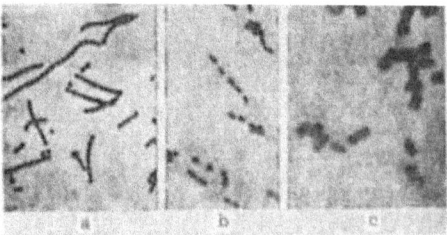

Figure 41. The form of spores of actinomycetes: a) spherical; b) elongated, c) cylindrical with cut-off ends.

The spores are formed by segmentation or fragmentation. In the first case, the sporeforming branch is fragmentated by transverse septa into separate cells which part and become mature spores (Figure 42b). In the second case, the protoplasm of the sporangium becomes fragmented into separate sectors or lumps without formation of septa. The lumps become rounded at the ends, assume rodlike, oval or spherical form, become covered with their own membrane and transform into mature spores. Afterward the membrane of the thread is destroyed and the spores are released (Figure 42a).

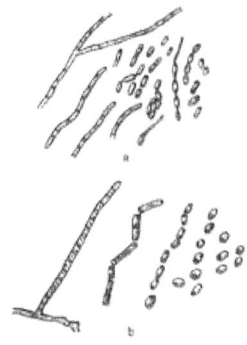

Figure 42. Spore formation by actinomycetes: a) fragmentation; b) segmentation.

Upon submerged growth in liquid media with constant shaking (on shakers) many actinomycetes have a fragmented mycelium resembling that of proactinomycetes. The threads form septa and disintegrate into rods and cocci. Frequently the disintegration of the mycelium proceeds by fragmentation (Figure 43). Cells thus formed resemble the spores described above. These cells are usually mistaken for spores, which of course is erroneous.

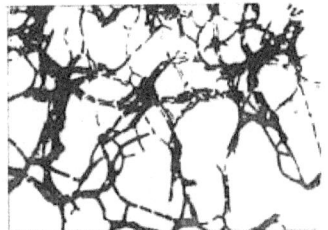

Figure 43. Fragmentation of mycelial threads of actinomycetes upon submerged growth in shakers. The threads of *Act. streptomycini* disintegrate into rodlike elements

The formation of spores ends the cycle of growth of the culture. Cells which are formed upon fragmentation of the vegetative mycelium under the condition of submerged growth are ordinary vegetative forms like those of mucor fungi.

As it is known, the mycelial threads of mucor fungi are devoid of septa and upon growth in submerged cultures transform into yeast cells called mucor yeasts through fragmentation. These cells differ from the spores formed in the sporangium of the fungus.

Many actinomycetes form various pigments on nutrient media--blue, violet, red, orange, yellow, sky-blue, green and others. The pigments are either diffuse or not. There are cultures whose pigment diffuses in the medium leaving the colony colorless.

The physiological property of actinomycetes is their capacity to utilize many organic substrates even such which are not utilized by other bacteria and fungi. In natural conditions on plant residues, actinomycetes begin to grow after all the readily assimilated substances have been decomposed and assimilated and the fungi and bacteria have stopped growing. Actinomycetes grow abundantly on semi-rotten residues. When a lump of peat or humus is inoculated with actinomycetes, it will soon be penetrated by the threads of the latter, and its surface will be covered with a white coat of the aerial mycelium.

Due to the fact that they are not fastidious, they are widespread in nature, they can be detected everywhere--in the extreme north and in the tropics, on barren rocks and in fertile chernozems, on mountaintops and in valleys. They live in water, silts, in the soil, on various plant and animal residues.

Upon growth in artificial nutrient media, actinomycetes form a number of peculiar substances. They have a characteristic smell somewhat resembling the odor of earth. Some species have a smell of camphor, fruits, etc. The chemical nature of those odors is not known.

Among the metabolic products of actinomycetes different biotic, antibiotic and toxic substances are found. Of the biotic substances the following were found: vitamins B_1, B_2, B_6, B_{12}, biotin, folic acids, auxins, pantothenic acid, nicotinic acid and others, in addition they form some amino acids which serve as complementary nutrients.

The formation of antibiotics by actinomycetes is widespread. Approximately 50-70 % of isolated actinomycetes produce antibiotics. There are species which form substances toxic for plants and species which form substances stimulating the growth of plants.

Actinomycetes not infrequently form dark-brown substances in substrates which by their chemical composition resemble the humic acids of soil (Flaig, 1952; Küster, 1952, and others).

Genus Proactinomyces. Proactinomycetes resemble actinomycetes by their structure. They form a well-developed mycelium in the early stage of growth on ordinary nutrient media. This mycelium is similar to that of actinomycetes, and soon becomes segmented into rode and cocci. The segmentation of threads is accomplished through transverse septa (Figure 44).

Figure 44. *Proactinomyces ruber.* 48 hour culture on must agar: a, b, c--disintegration of mycelium into rods and cocci.

Colonies of the typical specimens are usually bare and devoid of serial mycelium, rough, wrinkled, seldom smooth, and are less compact than the colonies of actinomycetes; sometimes they are of a doughy consistency. Hyphae near the base of the colony often grow into the agar. Some species more related to actinomycetes have colonies covered with weak coating of an aerial mycelium with straight sporangia. Spores in that case are usually rodlike. The characteristic property of proactinomycetes is the disintegration of the mycelium during their growth into rods of different length and into cocci.

There are pigmented and colorless species among the proactinomycetes. The pigments are often red or orange, more rarely yellow or brown; only the colonies are colored while the surrounding medium remains colorless. Physiologically, proactinomycetes do not differ markedly from actinomycetes. The nutritional requirements and fermentative capacity is similar in both genera. Few antagonists can be found among them. In soil proactinomycetes are rarely encountered.

Genus Mycobacterium. This genus consists of rodlike organisms,, resembling bacteria. Their cells are of irregular form (Figure 45), immobile and grampositive; they do not form genuine spores, but in some species the cell plasma becomes fragmented into 3-5 parts, in the same manner as in actinomycetes. The characteristic property of mycobacteria is their branching. In many species this is sharply pronounced. Cells possess 1-2 branches in the form of lateral appendices. In some species branching is rarely encountered. After inoculation the rodlike cells disintegrate into cocci relatively rapidly (after 1-2 days and sometimes after only 10-15 hours). In this stage of growth mycobacteria may be mistaken for micrococci. Gray and Thornton (1928) described bacteria closely resembling mycobacteria; they were curved, branched, gram-positive but they possessed flagella and were mobile. These bacteria were called *Mycoplana.* Similar bacterial forms were described by Köhler (1955), who considered them mycobacteria, which is erroneous.

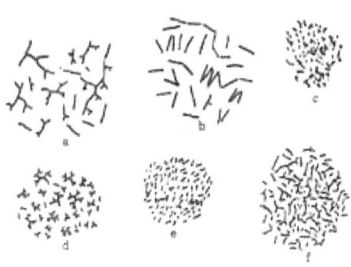

Figure 45. Mycobacteria. Different types of structure: a--*Mycob. hyalinum;* b--*Mycob. rubrum;* c--*Mycob. cyaneum;* d--*Mycob. bifiidum;* d--*Mycob. citreum;* f--*Mycob. filiforme.* Magnified about 600 times.

Their colonies resemble those of bacteria. In ordinary nutrient media they are of pasty or slimy consistency, the colonies are convex, more seldom flat, and sometimes assume gluey droplike form. The color of the colonies may be red, orange, pink, yellow, blue or brown. There are many colorless forms.

In the course of variation, mycobacteria can form variants of the structure of proactinomycetes. Mycobacteria do not possess sharply pronounced biochemical or physiological properties. This group is similar to bacteria and consists of representatives with different physiological and biochemical properties. Some species decompose proteins, carbohydrates, sugars, organic acids and alcohols. Others decompose hydrocarbons, products of oil, paraffin, tars and other substances not utilized by ordinary bacteria. Among mycobacteria, oligonitrophilic forms are encountered, which grow well on nitrogen-free medium; they apparently fix atmospheric nitrogen.

Mycobacteria do not possess antagonistic properties. They were not found to produce antibiotics.

Among mycobacteria some phytopathogenic forms are described, such as *Mycob. michiganense* and others (see Krasil'nikov, 1949 c). The well-known causative agents of tuberculosis and diphtheria belong to mycobacteria.

Mycobacteria live in soils of different geographical zones and their population there is abundant. Their numbers depend on the state of the soil and on external conditions. Mycobacteria similar to actinomycetes, grow well in soils of low humidity. Owing to this fact they are frequently the prevailing organisms in arid regions.

Genus Mycococcus. In their external appearance the mycococci resemble micrococci, or, more strictly, mycobacteria in their last phase of growth. The cells of mycococci are coccoid of 0.5-1.0 μ in diameter. They differ from micrococci by the irregular shapes of their cells. The latter are angular, pearlike, irregularly spherical (Figure 46). They multiply by fission, budding and constriction. Before fissions cells elongate slightly and assume a rodlike form. Mycococci are gram-positive, immobile, and nonsporeforming. In the course of variation of mycococci, variants are obtained of the mycobacterial type, which provides the grounds for considering them genetically related to the latter.

Figure 46. Mycococci. *Mycoc. ruber*

Physiologically and biochemically they do not differ from mycobacteria. In soil they are rarely encountered.

Genus Micromonospora. Organisms belonging to this genus resemble actinomycetes in their external appearance. They produce a well-developed mycelium in the form of thin branching threads. Their formation of spores differs from that of actinomycetes. The spores (conidia) are formed on short branches or directly on mycelial threads. The spores are single (Figure 47), of spherical or oval form (they rarely have an elongated form). Their size is similar to that of actinomycetes--about 1 μ.

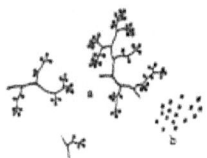

Figure 47. Micromonospora globosa: a--branches with sporangia; b--spores.

The colonies of micromonospora are compact. merge with agar, and lack the aerial mycelium; upon spore formation a weak coating appears. consisting of short aerial branches--sporangia. The colonies are red-brown, yellow-brown, and of other colors. The pigment does not diffuse into the medium.

The organisms grow slowly and poorly in ordinary media, the majority of species prefer a temperature of 40-45° C or higher. They are rarely encountered in soil.

Their physiological and biochemical properties do not differ from those of actinomycetes. Some species form antibiotics.

Bacteria

Microorganisms designated as bacteria are diverse in their structure, nature, growth, and biological properties.

According to external appearance, bacterial calls are divided into three types which are distinctly different from each other: coccoid, rodlike, and spiral forms. According to this they are divided into groups: cocci or Coccaceae, rods--Bacteriaceae and spirals--Spirillaceae. Each group is further subdivided.

Cocci--family Coccaceae. The characteristic property of this group of bacteria is the spherical shape of their cells. Their shape is regular and they possess a well-defined membrane. The diameter of the cells is 0.2-1 μ, more frequently 0.6-0.7 μ. They divide by fission, which occurs in different planes. The cells are nonmottle, and with the exception of one species *(Sarcina ureae)* do not form flagella or spores. The majority of cocci are gram-positive, only gonococci are gram-negative.

The cells are either separated from each other or united in aggregate pairs, arranged by fours into platelets of packet-shaped aggregates. According to this property, cocci are subdivided into groups: 1) micrococci--

genus*Micrococcus*, 2) diplococci--genus *Diplococcus*, 3) streptococci--genus *Streptococcus* and 4) sarcina--genus *Sarcina*.

In the genus *Micrococcus* the cells are individual and do not form complexes. Only during their division do the cells remain in pairs for some time. Sometimes the cells are glued mechanically together in small formless clusters and easily separate into individual cells.

In the genus *Diplococcus* the cells appear in pairs. Sometimes they resemble pairs of beans flattened laterally along the long axis.

The genus *Streptococus* is characterized by chain formation. Cell division takes place in one plane, perpendicular to the long axis. The length of chains varies.

In the genus *Sarcina* the cells are combined in packets in regular cubical forms. The number of cells in each packet varies from 8 to 54 and more. Cell division takes place in three reciprocally perpendicular planes. In some species the calls divide in two planes; then they form plates located in one plane. Such cell location is conditioned by growth conditions of the sarcina. Some investigators are inclined to consider them as a separate species--*Pediococcus;* however, there are not sufficient grounds for doing so.

The participation of Coccaceae in the soil, formation processes is very restricted, if it takes place at all. In soil these organisms are rarely encountered, but if so, only in small numbers.

In laboratory practice during analysis of soils, coccoid forms of mycobacteria, proactinomycetes and mycococci are mistaken for micrococci. On Kohlodny's growth plates the coccoid forms often belong to actinomycetes and sometimes to rodlike bacteria.

Rodlike bacteria--family *Bacteriaceae.* The rodlike bacteria are the most varied and widespread group in the soil. According to the structure and development of the cells, rodlike bacteria subdivide into two large subgroups: sporeforming and nonsporeforming.

Asporogenous Bacteria

Sporeforming bacteria have a rodlike form and are gram-positive. The cells are 2-10 x 0.5-1.2 μ in size. They form spores according to the character of spore formation, the shape of the calls, and the cytochemical composition. The sporeforming bacteria are divided into the genus *Bacillus* and the genus *Clostridium.*

Genus *Bacillus.* Bacteria allocated to this genus have both motile and nonmotile rodlike cells of 2-10 x O.7-1 μ. The flagella are usually peritrichous. Their plasma is homogeneous, sometimes containing granules of reserve food stuff. Granuloma is absent. They multiply by division. Many species form long chains--threads. Upon spore formation the cells do not swell or swell slightly. The spores are terminal and central or located in any part of the cell. Most of thorn are aerobes, but anerobes are also encountered among them. Physiologically and biochemically this bacterial group is very divers. Among them there are many species with sharply pronounced proteolytic activity: they do compose proteins with the formation of ammonia, H_2S and other odorous products of the putrefactive process.

There are bacteria which vigorously hydrolyze starch, sugars, alcohols, organic acids, and many other organic compounds. Among the sporeforming bacteria of the genus *Bacillus* there are autotrophs, chemotrophs and organisms which oxidize hydrogen, ammonia, methane, and other compounds. Organisms are described which are capable of fixing molecular nitrogen. Bacteria which synthesize various active organic compounds--biotics, antibiotics, toxins, and others belong to this group.

Among the sporeforming bacteria--*Bacillus,* antagonists are frequently encountered. The latter are second to actinomycetes with respect to distribution in nature.

The sporeforming bacteria are widely distributed in soils, in numbers which reach tens and hundreds of thousands and even millions.

Genus *Clostridium*. Cells of this genus of sporeforming bacteria differ from those, of the genus *Bacillus* in their structure and physiological properties.

They swell before spore formation, assuming a lemonlike, or club shape. They are of a quite large in size (7-15 x 1.5-2 μ). The cells in young cutures are somewhat smaller (5-10 x 0.8-1 μ). The cells contain granuloma which stains with I + KI in dark-blue almost black color. The spores are not fundamentally different from those of the genus *Bacillus,* in some species they are somewhat larger.

Bacteria of this genus are anerobes. They vigorously hydrolyze starch and pectinic substances and ferment sugars and other carbohydrates with the formation of butyric, proptonic and other acids. Some of them are characterized by acetone--butyrtc fermentation.

Many species fix molecular nitrogen. The best known of them is *Clostridium pastourianum* described by Vinogradskii.

The cultures of the genus *Clostridium* are widely distributed in soils, predominantly in those which contain humus and are rich in organic substances.

Sporeforming Bacteria

Asporogenous bacteria are the most diverse and most widely distributed soil organisms. Their diversity to manifest in their size, form, growth, cultural pecularities and especially in their physiological and biochemical properties.

The overwhelming majority of asporogenous bacteria do not differ from each other with respect to their external appearance--structure, size and type of colonies. They cannot be differentiated by ordinary laboratory methods according to their physiological properties. Nonetheless, the group of asporogenous bacteria comprises biologically different species, which can be established only upon thorough study of a few of them.

Asporogenous bacteria an a rule, are gram-negative and motile, although their mobility to not always apparent. The motion in due to flagella. According to the location of the flagella they can be divided into peritrichous (genus *Bacterium)* and lophotrichous (genus *Pseudomonas).* Among the peritrichous soil bacteria there to a great number of diverse groups. We shall describe here only the most important and widely distributed genera*Bacterium* and *Azotobacter.*

Genus *Bacterium*. The bacterial cells of this genus are rodlike, their size is 1.5-10 x 0.5-1 μ, more often 3-7 μ in length and 0.5-7 μ in diameter. They are gram-negative and motile. The flagella are located along the whole periphery of the cell, peritrichously. The number of the flagella vary; sometimes they are very numerous. They do not form spores. Their motility and the presence of flagella is not always apparent. In some species motility and the presence of flagella are apparent only under strictly specific conditions. The cells are flexible, their external form does not change. In many organisms the form and the size of the cells varies with the age.

Cultural, physiological and biochemical properties of the different representative of this genus are diverse. There are species which decompose organic and inorganic compounds and are capable of decomposing and synthesizing, oxidizing and reducing many substances.

There are among them pigmented and nonpigmented forms, which decompose various plant and animal residues and cause putrefaction of proteins with the formation of NH_3, H_2S, and other compounds.

This is the most numerous and varied group among the soil bacteria. Their numbers in 1g of the soil approach millions and billions.

The species diversity of the genus *Bacterium* is very great, but the number of species described to relatively small. This can be explained by the fact that the external features of these bacteria, as is the case in other bacteria, are very weakly expressed.

Genus *Azotobacter*. This genus represents a group of asporogenous bacteria with peritrichous flagella, capable of fixing molecular atmospheric nitrogen. This group will be described in detail in the chapter on nitrogen fixation. Here we shall only mention that cells of *Azotobacter* differ from those of other soil bacteria by their markedly larger size. Many cultures are rodlike in the initial stages of growth (Figure 2b, c), and afterward become spherical. The cells are often united in packetlike complexes, and are coated with a slimy, compact capsule.

The cultures grow well on synthetic nitrogen-free media (Ashby medium, Beijerink medium, and others). The majority of cultures and species do not grow on protein media nor, in general on media containing organic substances (meat-peptone agar, soybean agar, must agar).

Azotobacter to not found in all soils; the degree of its distribution to not uniform.

Great importance in soil fertility to ascribed to this group of bacteria.

Presently, the genus *Azotobacter* is represented by few specie --*Az. chroococcum, Az. azile, Az. vinelandii,* and others. There are reasons to assume that there is a much greater number of species and varieties in nature.

Bacteria with polar flagella.

Asporogenous bacteria with polar flagella, mono-and lophotrichous, are divided, according to their physiological and biochemical properties, into the following genera: *Pseudomonas, Rhizobium, Acetomonas, Azotomonas,* and others.

Genus *Peoudomonas*. Apart from bacteria of the genus *Bacterium* this group of organisms to the most widely distributed in soils. There are millions, tens and hundreds of millions and often billions per 1 g of soil, depending on the properties of the soil and climatic conditions.

Cells of this group are rodlike and 2-5 x 0. 6-0.7 μ in size. Larger as well as smaller forms are rarely encountered. At the end of the cell there is one or a few flagella (Figure 1, a-e). The spores are not formed inside the cells. They multiply like other bacteria by division.

The appearance of the culture of this group does not differ from that of other groups of asporogenous bacteria. In nutrient media they produce colonies which are smooth, shining, colorless or rarely pigmented. Many species form a fluorescent green-yellow pigment which diffuses into the medium.

Bacteria of this genus are very diverse in their physiological and biochemical properties. There are representatives among them which are similar to those of other bacterial groups, which decompose various organic and inorganic, protein and nonprotein compounds. Many species decompose organic compounds with the formation of final and intermediate decomposition compounds. There are species which fix molecular nitrogen, the so-called oligontrophils. This group includes autotrophs, heterotrophs, pathogens, saprophytes, etc.

Abundant growth of *Pseudomonas* is seen in the rhizosphere of vegetative plants as well as on rotting animal and plant residues.

Species with antagonistic properties are encountered among the representatives of *Pseudomonas* and *Bacterium*. Some species form antibiotics, others form toxins. But the majority of species of the genus *Pseudomonas* form biotic substances --vitamins (B_1, B_2, and others), auxins, various amino acids. etc.

The role of *Pseudomonas* in soil-formation processes was not studied in detail but judging from its activity, it should be assumed to be considerable.

Genus *Rhizobium*. This genus includes asporogenous bacteria with polar flagella which form nodules on the roots of leguminous plants. For this reason they are called root-nodule bacteria. They do not differ from the bacteria of the genus *Pseudomonas* in structure. The cells are rodlike, sometimes curved like mycobacteria (3-5 x 0.7 μ) and motile. In nodules and in certain artificial media they are deformed and possess bacteroid forms--

swollen, irregularly spherical, retortlike, amoeboid, etc. Sometimes they possess lateral protrusions. However. the latter do not appear during the course of branching, as they do in mycobacteria, but to a result of degenerative process. Cells with protrusions and bacteroid cells in general do not survive.

The cultures of root-nodule bacteria resemble those of ordinary asporogenous bacteria. Their colonies are colorless, smooth, slimy, convex or flat. They grow well in many synthetic media and also in media with plant organic substances (soy bean extracts and tissues of soybean plants). Some species do not grow in media with animal proteins (meat-peptone agar, meat-peptone broth, and others).

Physiologically and biochemically the root-nodule bacteria do not differ from other bacteria. Their only specific property in the capacity to form nodules on the roots of legumes. The cells penetrate the cells of root tissue and develop there symbiotically, securing the fixation of nitrogen.

The root-nodule bacteria are specific in that each culture forms nodules in the roots of strictly specific plants or groups of related species of plants. The division of these bacteria into species to made according to the plant host.

The root-nodule bacteria are widely distributed in soils. They can be found everywhere, where the corresponding legumes are found and often even when such plants are absent. The number of these bacteria in soils varies in relation to the conditions and properties of the soil itself.

Genus *Azotomonas*. Asporogenous bacteria 3-7 x 0.8 µ in size, and having polar flagella are included in the genus *Azotomonas*. They are motile, and multiply by division. Morphologically, culturally and biochemically they do not differ from other nonsporeforming bacteria. They grow well on many usual nutrient media, both protein and nonprotein, and on organic and synthetic media with a mineral source of nitrogen.

The characteristic property of these bacteria to their ability to fix molecular nitrogen. Some species form a yellow-green fluorescent pigment, which diffuses into the substratum. They resemble *Az. Vinelandii*. They are rarely encountered in soil.

Among the asporogenous bacteria, apart from the above-mentioned, there are many organisms grouped in separate genera which possess special and unique physiological functions as for example the sulfur bacteria, iron bacteria, nitrifiers, denitrifiers, etc. Different threadlike and other bacteria also belong here (see Krasil'nikov, 1949a; Bergey, 1948).

Spiral bacteria--family Spirillaceae. Bacteria of this family have spiral and corkscrew-shaped cells, which can easily be distinguished by their appearance. Some of them are long with a large number of coils, thin or thick, others are short with 1-2 coils. There are also very small forms with half a turn.

The bacteria are motile, with polar flagella, mono- and lophotrichous (Figure 1 a-f). The cells multiply by fission and constriction. They live predominantly in water but are frequently encountered in soils. Many of them participate in the sulfur cycle.

This group is subdivided into five genera.

Genus *Vibrio* or *vibrions*. The cells are small 1.5-3 µ in length and 0.5-0.7 µ in diameter. They are bent in the form of a comma, and are very motile due to their polar flagella. They are widely distributed In water reservoirs and in soil. Among them there are pathogenic species. Of the saprophytic forms, many reduce sulfates. They probably are of great importance in the soil.

Genus *Spirillum (spirilla)*. The cells are large, of various length from 3 µ to 70 µ and more, and 1-2 µ in diameter. They possess polar flagella and move actively, multiply by division and constriction, do not form pigments, and live in water. They are rarely encountered in soil.

Genus *Thiospira*. Organisms of this genus are colorless autotrophs which ozidize H_2S. Inside their cells, drops of sulfur are accumulated, which are also oxidized and serve as energy source. They live in water and are rare in soil.

Genus *Thiospirillum* and genus *Rhodospirillum*. These genera differ from the preceding one due to their red-purple pigment. They are autotrophs, and synthesize organic compounds, some at the expense of light energy (genus *Rhodospirillum)* and others at the expense of light and chemical energy (genus *Thiospirillium).* They live in water reservoirs.

Phages

Phages constitute a special group of organisms. Their size to measured in millimicrons (m μ) usually 50-100 m μ. Smaller phages are also encountered.

Phage particles have an oval or spherical shape either with a short tail or without it. They multiply only inside the microbial host and then only in a young vegetative living cell. In dead and old cells they do not reproduce.

The reproduction process is as follows: phages become attached to the cell surface with their tails and enter it through the cell membrane; after they have penetrated the cell, pronounced structural and chemical changes of the protoplast take place. Desoxyribonucleic acid is formed in large amounts and all its mass becomes fragmented into separate granules or prophages. Subsequently. the cell membrane is destroyed and prophages are liberated in the form of mature phages (Figure 48). The entire process takes 15-40 minutes. In some bacteria this process takes five or more hours.

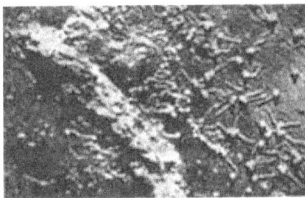

Figure 48. Phage particles, released after cell destruction. Phage of mycobacteria *Mycobacterium* sp. (according to Battagini, 1953)

The number of phage particles formed in one cell varies from 50-1,000, most often 100-200, and varies according to growth conditions, stage and bacterial species.

The characteristic property of phages is their ability to lyse bacterial cells. Phages possess defined specificity, attacking microbes of one or several species. Some phages attack only few strains of one species. There are phages which attack only bacteria--bacteriophages, and phages which attack actinomycetes-actinophages. Phages attacking other groups of microorganisms--fungi, yeasts, algae, etc, are unknown. There are descriptions of individual cases where phage particles were found in yeasts and fungi, however these data have not yet been confirmed.

The chemical composition of phages resembles that of animal viruses. They contain protein, lipides and carbohydrates, as well as deoxyribonucleic acid in large quantity. All phages possess antigenic properties, ie., when they are introduced into the body of animals, they are capable of causing the production of antibodies with specific properties. Because of this, specific antiphage sera are available.

Different phages are endowed with different resistance to the action of environmental factors--temperature, salt solutions, etc. Some phages are inactivated by a relatively low concentration of salts or low temperature, others, on the contrary, survive 75° C for one hour. Phages are more resistant to radiation than bacteria. They may be preserved for long periods of time in a dry state. They are easily inactivated by certain substances such as citric acid, phenanthrene, and others.

Phages differ from each other in their lytic activity. Some of them lyse calls quickly and completely; others weakly and not to completion. They may be differentiated according to the character of lysis of colonies. Some produce lytic zones with sharply defined borders, others with diffuse borders.

Phages are relatively easily subjected to variation. They change their specificity, and the character of the formation of "negative colonies", in other words the form of lysis plaques. Their resistance to this or other external factors can also change (Rautenstein, 1955).

Phages, as shown by studies, are widely distributed in nature. They can be found in water, animal and plant residues. They are found in large quantities in soil. The works of Rautenstein and Khavina (1954) and of others have shown that actinophages live in the soil in considerable numbers. It can be assumed, according to some data, that their role in the soil to considerable, Demolon and Dunes (1933, 1939) for instance assume that phages of root-nodule bacteria inactivate the latter.

The clover exhaustion of the soil, according to the authors, is indeed conditioned by much a phenomenon. Phages exert a great effect on the variability of bacteria and actinomycetes and favor formation of new forms and variants. In other words, phages constitute one of the powerful factors of species formation.

There are various points of view on the problem of the nature of phages which can be reduced to two essential ones. According to one-phages are ultramicroscopic organisms or microbial viruses (viruses of bacteria and actinomycetes) (Zilber, 1953; Ryzhkov, 1952; Sukhov, 1951, 1955). According to the other theory, phages are biocatalysts, i.e., active substances with biocatalytic or enzymatic properties capable of reproduction under certain conditions (Kriss, 1944). Both points of view present elements which cannot be neglected.

By 'soil' we understand (Vil'yams, 1931) a loose surface layer of earth capable of yielding plant crops. In the physical sense the soil represents a complex disperse system consisting of three phases: solid, liquid, and gaseous.

The solid phase of the soil consists of individual diverse particles, according to their chemical composition and dimensions. The size of soil particles fluctuates within a wide range.

The different properties of the soil and its fertility depend, to a considerable extent, on the composition and size of the particles. The soils are subdivided according to the size of particles as follows:

	Size of particle, diameter in mm
rocky	above 3
coarse sand	3-1
medium sand	1-0.25
fine sand	0.25-0.05
coarse dust	0.05-0.01
medium dust	0.01-0.005
fine dust	0.005-0.001
silt	less than 0.001

Every soil contains particles of all the above-mentioned dimensions but in a different quantitative relationship. The soils are subdivided according to their mechanical composition an follows (according to Kachinskii, 1956):

	percentage of particles smaller than 0.001 mm
heavy clays	80%
medium and light clays	80-50%
heavy loams	50-40%
medium loams	40-30%
light loams	30-20%
sandy loams	20-10%
compact sand	10-5%
loose sand	less than 5%

The quality of the soil as an environment for microorganisms is determined not only by the size of the particles but also by the character of their distribution, in other words, by the structure of the soil.

The soil may exist In two states: in a dusty state or in a form of aggregates. In the dusty form all properties of the soil are distributed in all directions. Such soil reacts in a uniform manner to external factors. For example, it swells uniformly in all directions upon wetting, thus, becoming inaccessible to air and water. Such soils are called structureless. As a result of their purely physical and mechanical properties they are unable to retain water for long periods, they dry up rapidly, and are poorly aerated. The conditions for plants and microorganisms are not favorable.

The structured soil consists of aggregates of different sizes, from 1 to 20 mm and more in diameter. The aggregates in the soil are more or less loose, the intervals between them are of different sizes and forms. These intervals or pores ensure the presence of water, air permeability and various foodstuffs for plants and microorganisms. In such a soil the evaporation of moisture is diminished.

The structure of soils is of the utmost importance. With each increase in the degree of orderliness of the structure, the physical properties of the soil aeration, water permeability, etc are improved. When comparing these properties of the structured and structureless soil, Remezov (1952) of the Dolgoprudnoya Experimental Station gave the following figures:

	In structureless soil, %	In structured soil, %
Porosity	50	55-60
General porosity	45-48	20-25
Capillary porosity	2-5	30-35
Noncapillary porosity	5	30-40
Air content	3-5	20-25
Water permeability (in mm/hr)	1.6	0.7

As can be seen from the above figures, the capillary porosity in the structured soil increases sharply and many other positive properties improve, thus securing high fertility and abundant proliferation of microorganisms.

The structure which contains aggregates 1-10 mm in diameter, which are water-resistant, is of the greatest importance.

In each soil aggregate the individual particles are in close contact with each other. The attraction bonds between them are quite different from those acting between the aggregates. The distribution of air, water and different elements of the soil within the aggregates is of different nature unlike those between the aggregates.

Due to the ability of the structured soil to absorb and hold water, as well as to its other properties, the structured soils are better able to provide conditions indispensable for the growth and development of plants.

The basic conditions for the formation of the structured soils is the presence of a sufficient quantity of silt-colloid particles, organic material, and other factors essential to the growth and life of microorganisms. The soil particles are glued together with the help of colloids, the products of microbial metabolism. Without these colloids the soil particles would not aggregate and would remain in an isolated state.

According to Godroits, the binding of the smallest elementary particles into microaggregates precedes the formation of aggregates; these can be destroyed only by chemical means.

In the presence of colloids small soil particles and microaggregates combine into simple aggregates and these, in turn, into complex ones. These aggregates are of different size and form. Between the simple aggregates there are intervals or pores which, as we shall see later, are of the utmost importance in the life of microorganisms. The glueing of soil particles may be stable or unstable. The strongly glued particles determine to a considerable extent the fertility of the soil.

There are various conditions and determining factors in the formation of soil structure. Humus and organic substances are of the greatest importance. The influence of manure, composts, and other organic fertilizers on structure formation is well known. P.A. Kostychev has empirically shown the influence of organic substances on the processes of soil-structure formation and noted the supreme importance of microorganisms.

The formation of soil structure is closely linked with the plant cover. It was thought until recently that only grass considerably improves the structure of the soil while woods destroy it. Recent investigations have shown that forests, too, favor the formation of soil structure.

Zonn (1954) gives the following data on the per-cent content of waterproof aggregates of soils having different plant covers. In the solonets, steppe grasses, waterproof aggregates comprise 17%; under oak forests, 58%; on chernozems plowed for 50 years, 17%; and under a forest belt, 47%.

The formation of soil structures is a subject of the investigations of many scientists. The first attempts to explain the cause of aggregates was made by biologists.

The scientific elucidation of the above process is due to the Russian scientist P. A, Kostychev. Kostychev mastered the methods of microbiological and physicochemical analysis of soils. He proved experimentally that there are the closest links between the microbiological processes of organic substance decomposition in the soil and the formation of structure. He showed that certain substances which serve as a cement are required for the soil particles to combine into aggregates. Such substances are the products of microbial metabolism and of the disintegration of animal and plant residues. According to his experimental data, the process of structure formation takes place only under conditions optimal for microbial proliferation, when the biological transformation of the organic mass, with the consequent formation of cementing substances, is possible.

When studying the decomposition of plant residues, Kostychev (1889) noticed an abundant growth of fungi and bacteria which developed until the whole mass of the decomposing substance was covered with these lower plants. He also showed that in different cases of decomposition of plant residues the course of the process varies and is caused by different microbes. Sometimes bacteria are the first to develop on decomposing material and the plant residues become covered with slime; sometimes fungi develop first and then the surface of the substance never becomes slimy.

When studying the growth of microorganisms on decomposing organic compounds, Kostychev noticed that when only bacteria develop the substrate never darkens. The substrate, according to his data, becomes colored only in the presence of fungi. On the basis of these findings, the author concluded that fungi play the most important role in the darkening of the substrate and the bacteria have no part in it. At the same time, Kostychev did not preclude the role of bacteria in this process. He pointed out that when the access of oxygen to the substrate is arrested, fungi stop their activity and their growth ceases, but the process of decomposition continues, though at a lower rate. (Kostychev--*The formation of chernozem. The soils of the chernozem belt of Russia, their origin, composition and properties,* 1889).

It should be pointed out that Kostychev (1892) not only determined the importance of microbes in the decomposition of organic compounds but was the first to notice their synthetic role in the soil. He wrote that humus does not represent a dead mass but in every way is full of life in all its diversity. Not only are processes of decomposition of complex organic compounds taking place in it, but also the synthesis of complex compounds from simple ones.

Under certain conditions the humus is destroyed with the liberation of mineral compounds utilized as nutrients of plants. The structure of the soil is then destroyed. If the soil in maintained under such conditions, an irreversible process of humus decomposition takes place, then, parallel to a certain accumulation of plant nutrients, the structure of the soil will be destroyed and, as a result, its fertility will decrease. Kostychev recommended that the structure of the soil should be continuously regenerated by growing perennial plants, pointing out that this process of regeneration takes place in the deposit (Kostychev, 1882).

The ideas of Kostychev were developed and furthered by Vil'yams. Vil'yams paid special attention to humus in the structural formation of soils, pointing out that not every type of humus is capable of forming a stable soil structure; only freshly precipitated humus of given properties may contribute to the stable soil aggregate (Vil'yams-Lectures on the soil science given at the Moscow Agricultural Insitute in 1895-1898, 1897). Vil'yams

(1897) thought that physicochemical agents and rain water play a considerable part in the destruction of soil structure. He noticed that rain water containing ammonia destroys soil aggregates liberating calcium, and, as a result, humus loses its cementing properties. According to him, the main factor in the regeneration of soil structure is the sowing of grasses. Later, Vil'yams suggested that biological processes, caused by the metabolism of microorganisms, are the most important factor in the formation of structure. The process of organic residue decomposition varies according to the degree of aeration and is caused by different types of microorganisms. Vil'yams distinguishes two entirely different processes in the decomposition of plant residues which are of great importance in the structural formation of soils and these are: aerobic (caused by fungi and bacteria) and unaerobic processes.

The aerobic process takes place in the surface layers of the soil, where plant residues are decomposed by microbes, bacteria, and fungi. The bacteria decompose the residues with the formation of humic acid, which, under conditions of full aeration, is fully mineralized. Therefore, Vil'yams does not think that humic acid is essential for the glueing together of soil particles. Only when this acid penetrates the lower layers of the soils, where unaerobic conditions prevail, it undergoes denaturation and is transformed into more stable compounds capable of coalescing the soil particles into water-resistant aggregates.

When the decomposition of plant residues is carried out by fungi, crenic and apocrenic acids are formed. These acids are highly soluble in water and are leached from the soil. They are, therefore, of no great importance in the formation of structure.

According to Vil'yams, the anaerobes are of the greatest importance in the formation of soil structure. According to him. these microbes synthesize ulmic acid through the decomposition of root residues, especially of grassy vegetation, This acid upon denaturation is transformed into a colloidal state, thus obtaining cementing properties. Upon dehydration it becomes insoluble in water and, conesquently, the particles which it glues together become water-resistant.

The ideas of Kostychev and Vil'yams, concerning the role of microorganisms in the formation and destruction of structures, were developed in the works of Soviet and some foreign investigators (Mishustin et al; 1945a, 1951, 1936; Gel'tser, 1940; McCalla, 1943, 1945; McHenry and Russel, 1943, 1944, and others).

The various investigators of the role of microbes in the process of structural formation had different approaches to the problem of the mechanism of their action. Some of them assume that microorganisms, by decomposing plant residues, form intermediate decomposition products which are responsible for glueing together the soil particles. Others assume that soil particles are coalesced by the products of microbial metabolism, while plant residues serve as a substrate for their nutrition (Gorshkov, 1940; Gusev. 1940; Rubashov, 1949, and others).

Some scientists (Kanivets, 1951, and others) suggest that fungi such as *Trichoderma lignorum, Mucor intermedius*, and *Martierella isabellina* are the most important factor in the process of structure formation. Foreign investigators tested *Trichoderma köningii, Aspergillus niger* and other fungi with positive results (Martin and Andersen, 1924; Martin and Waksman, 1940; Martin, 1945, 1946; Peels, 1940; Peels and Beale, 1944). Gel'tser (1940) suggests that lysates of fungal cultures formed by bacterial action and also colloidal protein compounds synthesized by bacteria represent the cementing substance.

Kononova (1951) assumes that the cementing substance consists of a mixture of cellulose decomposition products and protoplasm of the decomposed cellulose bacteria.

All these studies clearly demonstrate the essential, if not the exclusive, role of microorganisms in the formation of soil structure. However, the essence of this action remains unclear. The published data are not convincing and require verification.

Considerable work was carried out by Mishustin et al. He studied the action of fungi, actinomycetes, and bacteria on the decomposition of organic compounds (peptone, albumin, saccharose, starch, malic acid) artificially introduced into the soil on the formation of soil structure. The author concludes that the aggregation of soil particles is accomplished most vigorously by fungi and actinomycetes of mycelial structure. Thirty-two to fifty-one per cent of aggregates of 0.25 mm and more in size is obtained in experiments with fungi. Experiments with actinomycetes yield 25-30%. This value never exceeded 2.8% in the control soil. The various kinds of fungi have unequal powers of causing aggregation. Various sporogenous and asporogenous bacteria were studied including representatives of the genera *Azotobacter, Pseudomonas Rhizobium, Bacterium, Bacillus,* and others. Their cementing ability was very weak. The number of aggregates did not exceed 2-3%. more often 1-1.5%, and only in sporadic cases were there more than 10%. The cementing of soil particles proceeds considerably weaker under the influence of the simultaneous development of fungi and bacteria than in experiments with pure cultures of fungi. The opinion of the author is that bacteria in mixed cultures lower the aggregating action of fungi and actinomycetes. This is apparently caused by the action of bacteria in supressing the development of fungi or by their destroying the cementing compound formed by the fungi. Mishustin distinguishes two types of structure formations: biological, caused directly by microorganisms, and the so-called type of "humus structure", formed by humus compounds. The first type yields an unstable structure and is considered as an original and indispensable stage in the natural circumstances of structure formation (Mishustin and Pushkinakaya, 1942; Mishustin, 1945a). Rudakov (1951) assumes that the cementing of soil particles into aggregates in carried out by active humus, which comprises a complex of compounds of uronic acids and bacterial proteins, or products of their lysis. The uronic acids are formed, in his opinion, by pectin-destroying or protopectinase bacteria, which are found in greater or lesser quantities in the soil and on decomposing plant residues. Bacteria possessing protopectinase grow on plant roots, penetrate intracellular spaces, destroy intermediate compounds such as protopectin, and in so doing, form sugars and uronic acids. The latter, by combining with bacterial proteins, form uroprotein complexes capable of cementing soil particles. The galacturonic and other uronic acids are formed at the expense of living roots of vegetating plants as well as at the expense of dead root residues.

Not all bacteria, according to the author, possess protopectinase activity. The most intensive production of protopectinase is observed in the sporogenous bacilli, *Bac. polymyxa, Bac. macerans (Clostridium), Ps. radiobacter, Bac. mycoides, Bac. asterosporus,* and others possess weak protopectinase activity while *Bact. coli* is completely devoid of this enzyme,

Employing these bacteria together with vegetative residues of clover as a fertilizer, Rudakov obtained the following results expressed as the percentage of water resistant aggregates of 0.5-3 mm:

	%
In the control	11.48
Bac. polymyxa	20.51
Bac. mycoides	10.11
Bac. coli	4.989

According to Lazarev (1941-1945), the formation of soil-particle structure is done essentially by the ß-humatic, partially by the *a* -humatic, and partially by the complex humus formed from the products of bacterial autolysis.

Tyulin (1954) distinguishes two types of structural aggregates in forest soils of the podzol belt: unstable calcium-humate aggregates and stable iron-humate aggregates. These forms of aggregates and the differences between them can be detected by special methods devised by the author.

The nature and properties of the soil aggregates are determined by the quality of the humus compounds and by the nature of their interaction with mineral particles. The glueing ability depends on the elements contained in the humus, The presence of a sufficient quantity of calcium confers upon it certain properties; the presence of iron or aluminum confers different properties (Antipov-Karataev, Kellerman and Khan, 1948; Ponomareva, 1951, and others).

The glueing properties of humus are heavily dependent on the preponderance of humic and ulmic acids on one side, or crenic and apocrenic acids (fulvo acids) on the other. It is well known that the composition of humus varies in different soils. The humic acids of podsol soils differ from those of the chernozem or serozem soils, The fulvo acids of krasnozems and podsol soils also have a different composition (Tyurin, 1949; Kononova, 1953-1956, Ponomareva, 1956, and others).

All these variations in the humus of various soils have a definite effect on the nature of the structure of the soil particles. Ponomareva (1951) studied the structure formation in soils in relation to the development and metabolism of worms. She observed that the excrements of worms in the upper plow layer comprises, on the average, about 52 tons per one hectare which is about 1,700,000 individuals. The soil near plant roots is infested with worms to a greater degree than the soil outside the zone of roots (approximately 50%).

The excrements of worms are glued more compactly than the aggregates of ordinary soil. Their water resistance is conditioned by the cementing organic compound of worm intestines. The author showed that the degree of the soil structure due to worms varies in relationship to the vegetation. In oak forests the quantity of worms is larger than that in fields under oats and grasses, or in fir forests.

Under two-year-old grasses, 1,790,000 worms were found per hectare, in fields under oats, 560,000; in oak forests, 2,940,000, and in fir forests, 610,000,

The above data show how closely the mechanism of soil structure formation is connected with the biological and biochemical processes which take place in its depth. The plants, their composition on one hand, and microbial biocoenoses of the soil on the other, determine, under certain soil and climatic conditions, the direction and degree of the formation process of soil-particle structure.

SOIL POROSITY

Owing to the structural forms and aggregates, interspatial spaces or pores of different sizes and configurations are formed in the soil (Figure 49 A and B). The biological activity of the entire soil population is concentrated in these pores.

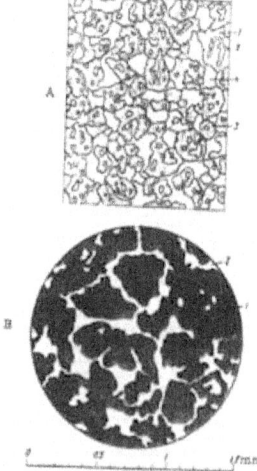

Figure 49. Soil porosity: A) porosity of cultivated structured soil (schematic): 1--thin, predominantly capillary pores in aggregates, which fill with water on wetting; 2--medium-sized pores (cells, channels), upon wetting they will fill with water for a short period and subsequently, after the resorption of water, with air; 3--capillary pores; 4--large pores, between aggregates almost always filled with air (according to Kschinakii, 1956); B), visible porosity of soil aggregate (reproduction from a microsection). Thin chernozem (southern): 1--micro-aggregates; 2--visible pores (according to Kachinskii et al., 1950).

The total volume of all the pores in 1 cm^3 of soil taken in its natural surroundings is designated as soil porosity. The porosity of soils is one of the most important factors determining their fertility. The porosity volume varies in different soils and depends on the type of soil, its state, external, seasonal, climatic influences and on the vegetation. However, the most important factors in determining soil porosity is the degree of soil structure, and the size of soil aggregates. The smaller the aggregates the less the spaces between them and the greater the total porosity. The size of the spaces between soil aggregates and particles varies from several microns to 2-3 mm and more (Kachinskii, Vadyunina and Korchagina, 1950).

The total porosity is determined according to its volume and specific gravity. It is determined by the following formula:

$$P = (1 - vol/sp) \times 100$$

where P--porosity, sp--the specific gravity of the solid phase, vol--weight by volume. The specific gravity of the solid phase (ratio of the weight of the solid phase to the weight of water) for different soils fluctuates from 2.4 (chernozem) to 2.7 (krasnozem).

The total porosity of different soils varies (Table 7). The chernozems have the highest porosity, 63-58%, and the highest total aggregation, more than 50% in the upper horizon A and somewhat less, 46%, in horizon B. Solonets has a lower porosity: the total, 50%, aggregate, 29.55% in horizon A_1. In the sod-podsolic soil the porosity is lower than that in the aforementioned ones: the total, 47-49%, aggregate, 32%, and the interaggregate, 15-16%.

Table 7
Soil porosity
(According to Kachinskii et al., 1950)

Soils	Depth of the layer, cm	Total porosity, %	Intra-aggregate total porosity, %	inter-aggregate porosity, %
Lixiviated chernozem with slght clay admixture, Kurak Oblast', steppe				
	A--0-4	63.86	40.54	23.32
	A_1--10-14	61.17	39.30	21.87
	A_2--40-44	58.75	--	--
	B_1--55-59	58.93	36.43	21.82
	B_2--80-84	57.85	36.03	21.82
Nutty-lumpy solonets with slight clay admixtures, Sverdlovak Oblast', virgin soil				
	A_1--10-14	50.00	29.55	20.45
	B_1--15.19	50.18	20.27	29.91
	C--60-64	44.40	--	--
Medium podsolic loam, Moscow Oblast'				
	A_1--0-12	49.05	31.61	16.43
	A_2--20-32	47.55	32.27	15.28
	B_1--32-35	41.70	--	--
	B_2--55-85	36.76	--	--
	B_2--85-110	34.10	23.78	10.32

Soil porosity is subdivided into intra-aggregate and interaggregate porosity. The first determines the free space between the soil particles inside the aggregate. The total size (volume) of this porosity varies considerably. The individual pores and slits inside the aggregates are, as a rule, smaller than the interaggregate pores.

The interaggregate porosity consists of the intervals between the individual aggregates. As can be seen from the data in the table, it occupies considerably smaller volume than the total intra-aggregate porosity.

Besides this porosity, there are capillary and ultracapillary porosites. The size of pores in these cases in very small. They possess peculiar physical and mechanical properties for retention of water and air.

The liquid phase of soils

Soil water as a physical body may exist in three states: solid, liquid, and gaseous. The water is subdivided according to its mode of binding with soil particles into categories, forms, and types (Rode, 1952, Dolgov, 1946). There are several classifications of soil water. The most widely used one in that of Lebodev (1930). The following forms of water are recognized (see Figure 50).

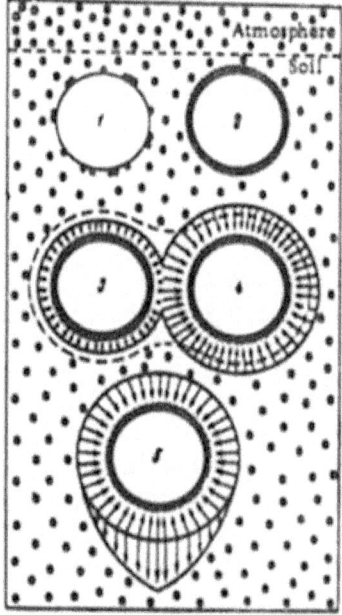

Figure 50. Schematic representation of various states of water in sand (small circles denote molecules of water in the form of vapor):

1--sand grains with incomplete hygroscopicity, 2--sand grains with maximal hygroscopicity, 3-4--sand grains with film water, water moves from the grain with a thick film (4) toward the sand grains with a thin film (3) until the thickness of the films equalizes (dotted); 5--sand grain with capillary water (according to Lebedev, 1936).

Water vapor. Water vapor saturates the soil air spaces. In the soil it moves from places of highest saturation to places of lower vapor pressure. When the soil is cooled or heated unevenly, the vapor pressure changes correspondingly. At night, for example, the upper layers of the soil cool more rapidly than the lower. Due to the difference of the vapor pressure thus formed, the latter moves from the lower layers upward where it condenses on the surface of cooled soil aggregates. During the day the reverse process is observed, owing to the warming of the upper layers of the soil.

The soil can be enriched with water at the expense of the water vapor of the air, subject to the same law. Upon cooling of the upper layers of the soil, the water vapor of the air condenses upon them. This is the diurnal distribution of water vapor in the soil.

Deeper movements of water vapor take place in accordance with seasonal temperature changes.

The water vapor is consequently of great importance in the enrichment of the soil with moisture. It should be assumed to play an especially important role in southern and regions with a distinct continental climate, where sharp differences in day and night temperatures cause considerable vapor condensation in the soil,

Hygroscopic water is physically closely bound to soil particles. It covers soil particles and is maintained there by the forces of molecular cohesion. Its density is more than 1; its specific heat is about 0. 9; it does not freeze, The forces of molecular cohesion are stronger than the earth's gravity; consequently, there in no hydrostatic pressure. When the soil is moistened to such an extent that its particles are covered with a continuous layer of water molecules, we talk of a state of maximal hygroscopicity. Similar conditions exist upon saturation of a space with water vapor.

The hygroscopic water is so strongly bound to soil particles that it can move only after it has been transformed to vapor. The removal of this water from the soil can be accomplished only by heating to 105° C.

The capacity of the soil to absorb or strongly bind water is called hygroscopicity. The latter depends upon many factors: mechanical composition of the soil. the content of organic compounds or humus, different organomineral compounds and metabolites. It increases with the content of humus and products of the metabolism of microorganisms in general.

The smaller the soil particles, the higher their adsorption capacity. The process of water adsorption by the soil particles is accompanied by the liberation of heat of wetting. Its value in various soils varies from 3 to 10 cal. For the gray forestbelt soil, solonets, and chernozem it is 5. 8, 5.92, and 6.06 cal, respectively, and for peat soil, 14.80 cal,

The film water to also physically bound to the soil, However, the bonds of its molecules to the soil particles are less strong than those of the hygroscopic water. It does not obey the law of gravitation upon movement, and there is no hydrostatic pressure either.

The molecules of film water (according to Lebodev's scheme) cover soil particles with a continuous thin layer or with a film. Its force of cohesion to the soil particles is weaker than that of hygroscopic water. The film water is bound loosely with the soil, it moves as a liquid, but differs from liquid water in certain physical properties; it has a higher viscosity and a lowered freezing point. According to some data, part of this water does not freeze at -78° C and complete freezing is noted only at -150° C (according to Vilonskii, 1954).

The amount of hygroscopic and film water in the soil changes in relation to various factors: osmotic pressure of the soil solution, soil composition, and others, In soils poor in soluble compounds such as some of the sod-podsols, the film water may exceed the maximal hygroscopicity. and in soils rich in dissolved substances it may equal or be even lower than the maximal hygroscopicity. In saline soils, loosely bound water may be completely lacking (Dolgov, 1940).

The physically bound water is assumed to be inaccessible to plants, The suction power of roots cannot overcome the force of the attachment of its molecules to the soil. It is not known if it is accessible to microbial cells, This water comprises about 1% of the soil's dry weight in sandy soils poor in humus, in loams, 3-5%, in soils rich in humus, 10%, and in peat soils, above 10%.

Apart from the aforementioned two categories of water, chemically bound water is also present in the soil. It enters into the composition of minerals and represents an integral part of compounds appearing as water of crystallization and hydration. This water can be removed only upon prolonged heating, whereby essential changes of the properties of the heated compounds take place. This water is unavailable to organisms.

Gravitational or filtration water-- free water. It moves in the soil as a liquid under the pull of gravitational and capillary forces. It filters downward through the depths of the soil, obeying the law of gravity.

Upon filtering downward it reaches subsoil water or is converted in higher levels into mobile capillary water. Its greatest quantity held by adsorption and capillary forces corresponds to the total moisture capacity of the soil. The ability of the soil to let through this water is called soil permeability. The gravitational water is accessible to plants and other organioms living in the soil. It can be converted into capillary water and, as a result, so it is assumed, it is rendered unavailable.

Capillary water-- the free part of water which fills the pores (the capillary space between soil particles) and moves in them by means of capillary forces caused by surface tension and the wetting of the surface of soil particles.

The water in capillaries rises higher, the smaller their diameter. At a pore diameter of 1μ the height of a water column is 15 m. Under natural conditions the water in soils and subsoils usually does not rise that high.

Analyses of the soil pores show that they represent a complex system. They spread out in different directions and the capillaries merge with pores of large or small dimensions and different forms.

The velocity with which the water rises in the capillaries is called the water-rising capacity. It to qualified by various conditions and also by the properties of the soil. The structure of the soil particles is of great importance. The larger the aggregates, the easier the penetration of the water through the soil. In soils without aggregates the pores are small, the water filters through these pores slowly, but it can rise quite high. It evaporates on reaching the soil surface. The less structured the soil, the more rapid is the desiccation of the surface layer. The higher the temperature of the air, the quicker the ascent of water in the capillaries and the desiccation of the soil.

Swelling of the soil strongly affects the flow and ascent of water. When mechanical composition of the soil becomes heavier, the ascent of water is slowed down and can be completely arrested, as it is in the solonets soils or illuvial horizons of the sod-podsolic soils (Kachinskii, 1956). Capillary water may flow in any direction.

Capillary water exists in three states:

1. The capillary immobile water consists of individual small droplets separated from each other. In this case, stasis of water takes place, and the conditions for the development of microorganisms are different from those in continuously flowing water. The separated water droplets in soil capillaries resemble, to some extent, the nutrient solution in glass capillaries obtained artificially in the laboratory.

2. The capillary mobile water is characterized by its discontinuous location in capillaries. It is available to the plants.

3. The easily mobile capillary water fills the pores of the soil and the subsoil over the ground water. It is easily available to plants. The capillary water represents the most important part of the soil moisture being available to all forms of life in the soil.

The classification of soil water given here is of a relative character. The quantitative content of free and bound water changes in relation to the properties of the solid and gaseous phases. The conversion of one form of water into another is also conditioned by changing external factors: temperature, pressure, various admixtures of organic and inorganic compounds, etc.

The physically bound water can, under certain conditions, be partially converted into free water and vice versa.

The soil water is in constant motion and this is of particular importance to organisms, since, together with the water, nutrients and oxygen required for respiration are also continuously supplied. Due to the difference in vapor pressure in various sections of the soil the water flows in different directions, vertical and horizontal. Only on rare occasions, when air pockets are formed in thin pores or capillaries, is the water stream interrupted and becomes static. This stasis in temporary or relative, since water in such cases still moves through conversion into the gaseous state,

It follows that the soil pores are filled with moisture in the form of vapor and with free and bound liquid water. Water is found in the intervals between the soil particles no matter how small they are, between the aggregates and within them. All forms of free and loosely bound water are available to microorganisms.

The individual intervals between aggregates and particles represent sites of concentration of individual types and groups of microbes as shown in the schematic figure (Figure 51). The latter form more or less isolated colonies in each of the intervals of the soil space. In these micro- and macrocenters the microbes live, reproduce, and carry out various biochemical processes. Products of their metabolism diffuse into the surrounding medium.

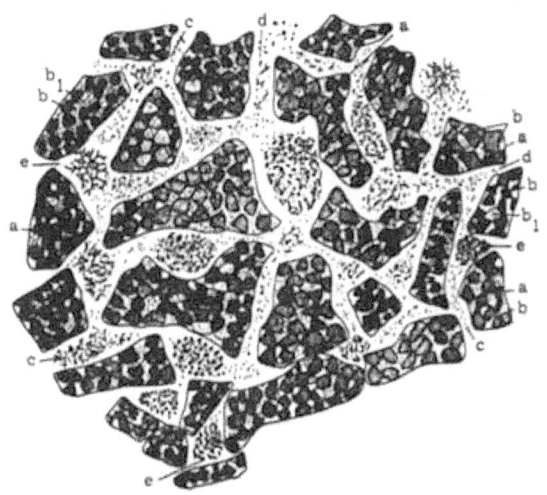

Figure 51. Schematic representation of the structure of a soil aggregate and the distribution of microorganisms: a) aggregates consisting of many microaggregates; b) pores between the microaggregates filled with bacteria (b_1); c) pores between the aggregates within which develop individual cells (d) and colonies (e) of bacteria, fungi, and actinomycetes.

The pore system determines the centers of development and the type of propagation of microorganisms. As the separate large pores are usually interconnected through smaller channels of capillary structures, the colonies of microbes which develop in large pores or centers are not separated from each other. There is an exchange of metabolites between the colonies of the individual types and groups of microbes, and an interrelationship of an antagonistic or nonantagonistic character is thus established. Investigations show that microorganisms and especially bacteria live not only in the pores and spaces between soil aggregates but also on the surface of the latter. As will be shown later, bacteria are adsorbed by soil particles and can live and grow in such a state. Different organic and inorganic substances which can be utilized by the microbial cells are also adsorbed on the surface of soil particles.

The adsorbed bacteria react with the adsorbed vitamins, antibiotics, toxins, and other compounds, according to their nature. The microbial cells grow well or badly according to the conditions created on the surface of the soil particles.

The structure of the system of pores, or porosity in general, determines, to a certain degree, the manner of distribution of microorganisms in the soil. With changing porosity the distribution of microbial cells also changes.

Diverse manifestations of life activity of soil microorganisms are also reflected in the character of the soil-porosity system.

The soil solution represents a very dynamic and indeed the most active part of the soil. Different chemical and biological processes take place in it. The composition of the soil solution is an important factor in the nutrition, growth, and reproduction of organisms. To a great extent it also determines the total productivity of the soil, G. N. Vysotskii (1902) compared the soil solution to the blood of animals.

Molecularly and colloidally dispersed compounds of mineral, organic, and organomineral composition are present in the soil solution. The following mineral compounds can be detected in the soil solution: ammonia salts, nitrites, nitrates, bicarbonates, carbonates, chlorides, sulfates, phosphates in the form of salts of calcium, magnesium, sodium and potassium; compounds of iron, manganese, aluminum, and silicon; microelements: zinc, copper, cobalt, vanadium, boron, molybdenum, radium, and others,

The amount of these compounds in the soil solution fluctuates in relation to the peculiarities of the soil and climatic conditions and also depends on the solubility of the compound (Vilenskii, 1954). The data on the solubility of the individual mineral salts present in the soil are given in Table 8. It can be seen from the table that the solubility of the salts fluctuates within a wide range. It increases with the rise of temperature. The solubility curve is different for the various salts.

Table 8
Water solubility of mineral salts present in the soil (g/l)
(according to Vilenskii, 1954)

Salt	0°C	10°C	20°C	30°C	40°C	50°C
K_2CO_3	10.53	10.83	11.05	11.37	11.69	12.13
KCl	276	310	340	370	400	426
KNO_3	133	209	316	458	639	426
K_2SO_4	74	92	111	130	148	166
Na_2CO_3	7	125	215	388	485	--
$NaHCO_3$	69	82	96	111	138	--
NaCl	357	358	360	363	366	370
$NaNO_3$	721	779	845	916	984	1,041
$NaSO_4$	50	90	194	408	--	--
$CaCO_3$	--	--	0.0145	--	--	0.0152
$Ca(HCO3)_2$	161.5	--	166	--	--	--
$CaCl_2$	595	650	745	1,020	--	--
$Ca(NO3)_2$	1,021	1,153	1,293	1,526	--	--
$Ca(H_2PO_4)_2$	--	--	153	--	170.5	--
$CaSO_4$	1.759	1.928	--	2.090	2.097	--
$MgCl_2$	528	535	545	--	575--	--
$MgSO_4$	408	423	445	454	--	504

The presence of gases in the solution strongly affects the solubility of salts. For example, carbonic acid increases the solubility of calcium carbonate, converting it into bicarbonate, the solubility of which exceeds many times that of carbonate. Sodium chloride in solution increases the solubility of gypsum; and sodium sulfate, on the contrary, lowers it. The concentration of salts in the soil solution changes relatively to the soil moisture.

When the soil dries, the solution concentration of salts increases, then the salts crystallize and precipitate, First to precipitate are the carbonates of alkali metals, then gypsum, and, finally, the easily soluble compounds.

With the increase in moisture the concentration of the solution decreases and the majority of salts redissolve.

There are also gases in the soil solution which are either absorbed from the atmosphere or formed, in the soil. Especially large amounts of carbon dioxide and oxygen are found. Their solubility changes in relation to barometric pressure, temperature, and certain other factors. The higher the temperature, the lower the gas solubility, The solubility of gases is directly proportional to the partial pressure of the gas. Since there are relatively more gases in the soil and their pressure is relatively high, it follows that their concentration in the soil solution is higher than in water in an open space, The presence of electrolytes in the soil solution decreases the solubility of gases.

Colloidal compounds in soil solutions comprise, according to Gedroits, 5-20% of the dry weight of the solution's residue. The majority are of organic origin. Silicic acid and iron and aluminum hydroxides may be found in the soil solution in the colloidal state.

The soil solution contains all the soluble organic compounds formed and released by plants, animals, and microorganisms, and also many substances synthesized directly in the soil outside the organisms by free extra-cellular enzymes. One can also find in the soil solution humus compounds, humic acids and their salts, various acids, alcohols, esters, amino acids, antibiotics, and toxins.

The soil solution represents a nutrient medium for the entire population of the soil and especially for the microorganisms. In all cases when the medium is favorable and there are no hindering factors, the amount of organisms is abundant. The more nutrients in the solution, the more intense the development and metabolism of soil microorganisms. Fertile soils and soils fertilized with large amounts of humus have a high concentration of nutrients in their soil solutions. Soils of low fertility, not containing humus, have small concentrations of nutrients in their soil solutions and growth of microbes will be slight.

We have compared the nutritional values of soil solutions of four samples of soils. One sample was taken from the garden, well fertilized; the second from fertilized fields of the sod-podsolic belt (Chashnikovo, Moscow Oblast'); the last two samples from the chernozem of the Moldavian SSR and serozem of the Uzbek SSR. The soil solutions were obtained from wetted soil (60% of total moisture capacity) under high pressure (about 100 atm.) with the aid of a press. The solution obtained in one series of experiments was sterilized in an autoclave at 110° C for 40 minutes, in the second series it was filtered through Seitz filters; the third portion remained unsterilized. All three portions of the soil solution were Inoculated with six kinds of bacteria preliminarily tested and proven to be pure cultures.

The results are given in Table 9.

Table 9
The growth of bacteria in various types of soil solutions
(millions of bacteria per 1 ml)

Soil solution	Pseudomonas No. 11	Pseudomonas No. 23	Bacterium No. 1	Bacterium No. 5	Azcroococum	Rhizob. meliloti
Solution of podsol, garden soil						
autoclaved	350	41	640	86	105	60
filtered	113	20	280	50	88	40
natural	52	20	100	20	5.0	0.1
Solution of podsol, slightly cultivated field soil						
autoclaved	1.5	0.1	3.2	0	0	0.1
filtered	0.6	0	1.5	0	0	0
natural	0.2	0.	0.6	0	0	0
Solution of						

chernozem						
autoclaved	120	900	170	1,200	450	600
filtered	50	600	100	800	200	500
natural	20.2	200	40	400	50	400
Solution of serozem						
autoclaved	100	450	50	20	100	250
filtered	80	300	40	5	60	200
natural	20	50	5	0	20	200

As can be seen from the data the garden soils and especially the chernozem soils contain more nutrients for the bacteria than soils from fields with low contents of humus. Serozems have more nutrients than sod-podsolic field soils.

Sterilization by autoclaving improves the nutritional value of the soil. Apparently it in due to the hydrolysis of certain organic compounds which become more available for assimilation by the organisms. Upon filtration of the soil solutions, some substances are hold back by the filter, reducing the amount of nutrients.

Not all bacterial genera react similarly to the nutrients of the soil solution. For example, *Bacterium* No 1 develops most abundantly in the soil solution of a garden soil of the podsol belt, and *Bacterium* No 5 in the solution of chernozem. The culture of *Bacterium* No 23 grows preferably in the solution of the serozem soil. The root-nodule bacteria of lucerne develop well in the solution of chernozem and serozem and almost do not grow at all in the solution of the field soil of the podsol belt,

The reaction of the soil solution is of great importance for life processes in the soil as well as for many physicochemical and biochemical reactions. Too acid or too alkaline a solution is of little use, or of no use at all, for the growth and development of organisms.

The reaction of the soil solution is conditioned by the dissolved salts. The acidity of the soil is caused In some cases by hydrogen ions present in the soil solution, in other cases by adsorbed ions. The first in called an active and the second , a potential, acidity. Besides these two, we also distinguish a total or titrable acidity or alkalinity which is determined by ordinary titration.

The following soils are pH distinguished according to the active acidity: Strongly acid, 3-4; Acid 4-5; Weakly acid 5-6; Neutral 6-7; Alkaline 7-8; Strongly alkaline 8-9.

Podsol soils, marshy soils, and gray forest soils have an acid reaction; serozems have a neutral or alkaline reaction; solonets soils have a strongly alkaline reaction (Vilenskii, 1954).

The presence of carbonic acid, organic acids, carbonates, and other compounds effects the reaction of the soil solution. The presence of carbonates and especially of sodium and calcium bicarbonate causes an alkaline reaction.

The oxidation-reduction potential of the soil solution. The concentration of hydrogen ions in the soil solution is of great importance to biological processes. The oxidation-reduction potential of the solution determines the direction and character of chemical and biochemical reactions and the solubility of biologically important components of the medium, as well an the products of microbial metabolism, The degree of dissociation of water ions (H and OH) has a great influence on the solubility of the different mineral salts: those of silicic acid, sesquioxides of iron, aluminum, and others. Bivalent iron (Fe^{++}) dissolves in a weakly acid solution at pH = 4-6 and precipitates at pH = 7. Trivalent iron (Fe^{+++}) dissolves in a strongly acid solution at a pH below 3, and at pH = 3 it precipitates. The same takes place with manganese and some other elements.

In the surface layer of the soil where the amount of oxygen is sufficient, oxidation processes proceed vigorously under the influence of aerobic microbes. With increasing depth, the amount of oxygen in the soil solution diminishes. The solution loses its oxidizing properties on the so-called oxidation-reduction border. Below this border, reduction processes take place. The depth of location of the oxidation-reduction border varies in different soils. It may fluctuate in one and the same soil depending an moisture, temperature, and other external factors. This border should be considered as relative. Experiments show that in the upper layers both oxidation and reduction processes may take place and aerobes, as well an anaerobes, may develop. On the other hand, In the deep layers, oxidation processes may take place an well as reducing processes. However, the former are considerably weaker than the latter.

V. R. Vil'yams assumes that aerobes in the upper layers grow in pores between the aggregates, where oxidation processes take place. Inside the aggregates, anaerobes with reducing function predominate. Anaerobic processes may be conditioned by an abundant growth of aerobes. The latter absorb oxygen and create anaerobic conditions in a closed system.

The buffering capacity of the soil solution. The buffering capacity, or the ability of the solution to resist changes of the active reaction upon acidification or alkalization, is one of its characteristic properties, It is caused by the content and composition of soil colloids and their adsorptive capacity. The higher the adsorptive capacity of the colloidal particles, the greater the buffering capacity of the solution. The buffering capacity of the soil also depends on the adsorptive capacity at the solid particles of the soil. Soil is a strong buffer. Different chemically active compounds are neutralized or inactivated by the soil: e.g. , acids, toxins, antibiotics, vitamins, and other substances of microbial and other origin.

The gaseous phase of soil

The gaseous or air phase of the soil plays an essential role in life processed. The free space between the soil aggregates is filled with air, provided it is not occupied by soil solution.

The air in the soil exists in three states: a) free, filling the free space between the soil particles and the aggregates, b) dissolved In the soil solution; c) adsorbed by the solid phase of the soil.

All these states of the air are of importance in the life of the soil. The air is adsorbed on mineral and organic colloidal particles, the largest amounts being adsorbed by organic matter in the dry state. With the increase in moisture, the adsorption of air diminishes. When the moisture somewhat exceeds the maximal hygroscopicity, the adsorption of air is completely arrested. Water molecules are adsorbed by the soil particles more tenaciously than gas molecules.

Different gases are adsorbed by soil particles with varying tenacity, according to the following series; $NH_3 > CO_2 > O_2 > N_2 > H_2S > CH_4$. The capacity of soils to adsorb and retain gases varies, and depends upon the composition of the colloidal fraction. Humus and iron hydroxide adsorb most strongly.

With increasing temperature, the capacity of the soil to retain gases decreases. The free air and the air dissolved in the soil solution in of the greatest importance for the biology of the soil. The total free air content of the soil depends on its porosity and moisture. Since water and air occupy the same sites in the soil, any increase in the volume of one of these components leads to a corresponding decrease in the volume of the other.

As already mentioned, the porosity of the soil is not a constant value. It fluctuates, for different reasons, between 25-50% and, In rare instances. rises to 60%.

Only in dry soils the soil space almost entirely filled with air. In rainy seasons the pores are filled with water which supplants the air.

The upper layer of the soil (0-10 cm) of Northern Caucasus chernozem contains the following quantities of air per 1,000 CM^3 of soil under different plants (A. A. Shmuk, 1924): Virgin soil 80 cm^3 Winter wheat 240cm^3; Oats 200cm^3; Sunflower 260cm^3; Tobacco 280cm^3; Fallow 320cm^3.

It can be seen from these data that the amount of air in one and the same soil changes in relation to Its state and degree of cultivation, the vegetative cover, and in relation to other factors. There is less air in virgin (8%) than in cultivated soil , and in soil under cereals there is less air than under intertilled crops. The largest amount of air (32%) is found in fallow soil. Similar data were obtained by studying sod-podsolic soils (experimental fields of the Agricultural Academy of Timiryazev). Their air content fluctuated from 15 to 30%; the bare fallow soil contained the greatest amount of air (Vilenskil, 1954).

The air content in the soil changes but little during the whole growth period provided that soil moisture is kept constant. The amount of air diminishes correspondingly with increasing moisture.

The composition of the soil air. The air in the soil never has the same composition as atmospheric air. It is more diverse in the qualitative and quantitative content of its constituent gases.

As in well known, the atmospheric air consists of 79.01 % nitrogen, 20.96 % oxygen, and 0.03 % carbon dioxide. In addition, negligible quantities of other gases (neon, krypton, argon, xenon, helium and others) can be detected in the earth's atmosphere.

Soil air differs from that of the atmosphere by its higher CO_2 content. The oxygen content varies within a smaller range, and that of nitrogen is almost constant. Besides atmospheric gases, there are other gases in the soil formed by the metabolism of organisms and the respiration of the soil. Various volatile organic and inorganic substances can be detected in the soil: ammonia. hydrogen sulfide, methane, organic acids, alcohols, esters, tars, and many other compounds, products of the metabolism of bacteria, plants, and animals.

The composition of the soil air is not well known, The most important component of the soil atmosphere is carbon dioxide, the final decomposition product of organic matter. The intensity of the biochemical processes taking place in the soil can be judged by the amount of carbon dioxide released.

The amount of carbonic acid in soil varies noticeably in relation to the composition and type of the soil, to the metabolic activities of the soil population, and to climatic conditions and other factors. The formation of CO_2depends to a large degree on microbial metabolism. Everything that favors growth of microorganisms increases the generation of CO_2. Lundergardh (1924) assumes that 2/3 of the total amount of CO_2 in the soil atmosphere is formed as a result of bacterial metabolism, the remaining 1/3 being formed by plant roots.

In soils rich in organic compounds of humus, the CO_2 content is, as a rule, greater than in soils poor in humus.

Vilenskii (1954) gives the following figures on the formation of CO_2 in various soils (CO_2 in kg per hectare per hour): Nonfertilized clay soil, 1.25; Nonfertilized sandy soil, 2.0; Sandy soil, high in humus, 4.0; Sandy loam, not fertilized, 4.0.

Beneath the canopy of forests the air in more saturated with CO_2 than in the field (Zonn. 1954); in autumn (10-17 September), the following amount of CO_2 was found to have been released: In an oat field between forest belts, 2.09 kg/hectare per hour; In a forest belt 60 m wide (forest 50 years of age) 17. 35 kg/hectare per hour; In a forest composed of oaks, acacias and ash trees (28 years) 10.50 kg/hectare per hour; The same, but composed of oaks and honeysuckle, 5.88 kg/hectare per hour.

The amount of carbonic acid in the soil changes sharply in relation to the composition of plant residues. According to the data of Stoklasa (1906), one gram of dry root substance released the following quantities of CO_2 on decomposition: Barley, 70. 5 mg/24 hours; Potatoes, 82.3 mg/ 24 hours; Wheat, 74.6; Beet, 130.6; Rye, 110.9; Clover, 1.46; Oats, 118.9; Lucerne,160.5.

According to the data of Makarov (1952, 1953), the liberation of this gas by the soil fluctuates in the range of 400 to 800 kg/hectare in 24 hours. In fields under crop rotation the following amount of CO_2 is released during one season (in tons/hectare): fallow, 35; under winter rye, 05; under oats, 79; under first-year grass, 98.

The formation and release of CO_2 under different vegetation varies. For example, under clover 0.550 g of CO_2 is released in unit time; under serradella, *(Ornithopus sativus),* 0.305 g; under mustard, 0.218 g; under rice, 0.285 g. All data refer to release of CO_2 per 1 m^2 of soil (Reinan, 1927).

The largest amount of CO_2 in released from soil under legumes, i.e. , clover, lucerne, etc. This can be explained by the activity of the *Rhizobium* bacteria. According to the data of Bond (1941), respiration of *Rhizobium* bacteria on soya roots was 3 times higher than the respiration of the roots per unit of dry weight. The total mass of root nodules released more CO_2 than the root mass of the entire plant.

The CO_2 in the surface layer of the air may reach 10% as a result of its release from the soil. In the deep layers of the soil the air contains more CO_2 than in upper layers.

The dynamics Of CO_2 liberation changes according to the phase of plant vegetation. The liberation of CO_2 from soil under wheat is greatest during flowering but under other grasses it is greatest before the harvest. Makarov connects the maximum CO_2 release with the greatest development of the root system in the given plant phase. According to our observations, the period of maximum CO_2 liberation coincides with the maximal growth of the microflora near the roots (Krasil'nikov, Rybalkina, and others, 1934; Krasil'nikov, Kriss, Litvinov, 1930 a).

The great effect of temperature on microbiological activity, and, consequentl y, on CO_2 liberation, should be mentioned. Experiments show that a rise In temperature from 15 to 28° C increases the formation of CO_2 in the soil twofold (Bunt and Rovina, 1955).

The dependence of soil respiration on some of these factors to shown in the following curves (Figure 52).

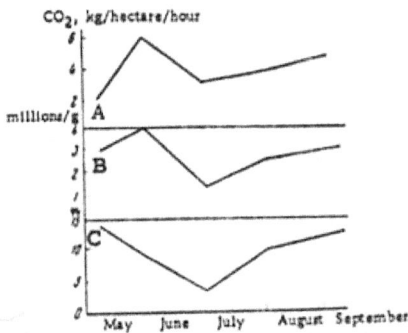

Figure 52. Dependence of soil respiration on its moisture content and the growth of microorganisms (according to Makarov, 1953): Curve A--soil respiration (release Of CO_2); curve B--total amount of microorganisms; curve C--soil humidity as of dry weight.

The quantitative fluctuations of oxygen in the soil is the reverse of that of the CO_2, In the upper layer, to a depth of 30 cm, oxygen comprises 15-20% of the total amount of gases. With increasing depth its quantity decreases sharply. In the spring the amount of oxygen, at a depth of 60-90 cm, comprises 0.3-0.8%. In the summer the amount of oxygen in deep layers of the soil rises and in July, at the same depth, reaches 15-19%; in August it comprises 11-13%, even at a depth of 180 cm. In October the amount of oxygen again diminishes. These fluctuations are caused by temperature and humidity.

It is clear that, with the varying oxygen content, all biological processes will be modified. not only quantitatively, but also qualitatively. In the presence of a sufficient influx of oxygen, oxidation processes will take place predominantly; if the amount of oxygen is inadequate, reduction processes will predominate.

The gases of the soil atmosphere can exist in a dissolved state. As already mentioned, the soil solution always contains a certain amount of air and gases present in the soil and in the atmosphere.

The solubility of gases in the soil solution depends on their characteristics, partial pressure, temperature. and on the concentration of salts in the solution. According to Henry's law, the solubility of a gas in liquid is directly

121

proportional to its pressure. If the liquid is in contact with a mixture of gases then each of the gases will dissolve, not under the influence of the total pressure, but according to its own partial pressure.

Of all the gases in the soil the most soluble are CO_2, NH_3, H_2S, and some others. Oxygen is less soluble; nitrogen dissolves with difficulty. The dependence of the solubility of gases on temperature is shown in Table 10.

Table 10
The solubility of gases at different temperatures (1 cm^3 / l)

Temperature °C	CO_2	O_2	N_2
0	17.1	0.49	0.24
10	8.8	0.31	0.15
30	6.6	0.26	--

There are always large amounts of electrolytes in the soil solution; consequently, the solubility of gases in it is lower than in pure water. The soil solution of saline soils contains less gasses than that of nonsaline soils. The adsorption of gases by soils rich in humus is higher than in soils poor in humus.

As a result of microbiological activity, ammonia, hydrogen sulfide, hydrogen, methane, and other metabolites of aerobic and anaerobic microflora. as well as such organic volatile compounds as acetic acid, butyric acid, alcohols, esters, aromatic compounds and others, can be detected in the soil. The specific scent of the earth in caused by the volatile metabolites of microbes. especially actinomycetes, whose, nature is not clear. There are many other compounds which. in the soil air, are a source of direct and supplementary nutrition, and also certain volatile compounds which suppress the growth and development of specific microbes.

By direct experimentation N.O. Cholodny (1944 a, b, c, 1951 a, b) discovered the existence of nutritional substances in the atmospheric and soil air. He showed that some bacteria and fungi, and also excised tips of plant roots, can grow satisfactorily in a drop of a medium in which soil vapors served as the only source of nutrition.

The presence of foodstuffs in the soil atmosphere was also established in our experiments in the following way: a culture of an asporogenous bacillus; *Bact. album,* when isolated from soil, is incapable of growing in the synthetic medium of Chapek. However, this bacterium placed in a drop of this medium in a soil chamber begins to grow and yields many generations. It follows that volatile substances were released from the soil and found their way into the drop of the medium, thus securing normal growth of the culture.

Meisel', et al., (1946, 1950) showed that separate components of vitamins--thiamine, nicotinic acid, para-aminobenzoic acid, present in the air, are used by microorganisms. Biotic compounds enter the air and soil atmosphere from the soil and plants. According to Cholodny (1944c), vitamins given off into the air by plants are utilized by soil bacteria and by the plants themselves. The air of forests and meadows is the richest in volatile vitamins (Grummer, 1955),

Shavlovskii (1954) detected thiamine and nicotinic acid in the soil atmosphere of gray forest soil and podsol chernozem.

In the soil atmosphere, volatile compounds, toxic to certain microbial types, can be detected, Our experiments have shown that the staphylococcus *Staph. aureus,* placed in a hanging drop in a soil chamber prepared from forest sod-podsolic soil, do not grow or grow very slowly, whereas the control cultures of *Staph. aureus* placed in a chamber with other soils (chernozem, garden soil) grow normally. The suppressing activity of soil vapors (from soil under flax) was observed, an well as the absence of such activity by the vapors of soil under clover.

Radioactive substances can be detected in the soil atmosphere, usually in the form of disintegration products of radium or other substances.

The composition of the soil air changes constantly. A continuous exchange between the atmospheric air and the soil air taken place, This exchange is of the utmost importance for the life of the soil. Without this exchange the CO_2, H_2S, methane and other gases formed would quickly fill up all the pores of the soil, the oxygen would be exhausted, and many biochemical processes would stop. The population of the soil--plants, animals, and microorganisms--would be poisoned. Without the influx of atmospheric air, without replenishment of the soil with oxygen, anaerobic conditions would be established.

The replenishment of the soil air is accomplished under the influence of many factors. The main ones are.

a) temperature fluctuations, diurnal and seasonal;

b) changes in barometric pressure;

c) diffusion of gases;

d) dynamics of life processes; utilization and formation of certain gases by the living population of the soil.

The first method of air exchange is accomplished due to the property of gases to expand upon heating and to shrink upon cooling, With the rise in soil temperature, the air in the soil increases in volume and leaks into the atmosphere. When the temperature drops the reverse process takes place; the soil air shrinks, its volume decreases, a vacuum is formed in the pores, and atmospheric (external) air is sucked in. Such fluctuations take place periodically (diurnal fluctuations). There are also seasonal fluctuations. These are less sharply pronounced and apparently are of less importance for the respiration of the soil. These periodic fluctuations of the soil temperature are responsible for a regular gas exchange between the soil and the atmosphere. It appears as if the soil respires. The characteristic feature of this respiration is, as of any other respiration, the uptake of oxygen and the release of carbon dioxide.

The respiratory activity of the soil can be increased or decreased by various factors, such as humidity, wind, and others. The water and the air are antagonists, The humidity of the soil leads to a decrease in the amount of air in the pores of the soil. During the rainy seasons the soil gets so wet that the air in almost completely supplanted. With the drying of the soil the reverse process takes place.

The CO_2 and oxygen content of the soil air varies seasonally. The largest amount of CO_2 in the upper layer of the soil is found in spring and summer. From April until September, in the temperate belt, it reaches 2-4% at the depth of 30-60 cm. In the autumn and winter the amount of CO_2 decreases considerably. Together with these findings there are observations which show that the activity of the microorganisms does not stop in winter, According to our data, in winter, at a soil temperatures of about 3-5° C above freezing point, certain forms of actinomycetes and bacteria multiply abundantly. In one gram of soil 10-15 thousand actinomycetes *Act. globisporus* were counted in the spring and summer; in winter their number reached 100-500 thousand and more. Sauerland and Groetner (1953) found that the release of CO_2 in the soil increases in winter.

The barometric pressure strongly influences the respiration of the soil. Observations have shown that with the change in pressure, the gas content in deep layers, at a depth of 2 m and more, also changes. With decreasing pressure the gas volume increases and the gases are released into the atmosphere; with rising pressure the picture is reversed.

Diffusion also plays an important role in the gas exchange of the soil.

According to some authors, diffusion alone can secure the gas exchange of the soil and maintain the composition of the soil air at a level sufficient to maintain the life processes of the soil population.

In the soil (except when frozen), biological processes of synthesis and decomposition take place continuously. In this process various organisms form CO_2, O_2, esters, acids, alcohols, ammonia, hydrogen sulfide. methane, etc. These compounds serve an nutrients for other organisms, especially microbes. The content of gases in the soil will change according to the prevalence of these or other microorganisms In a given soil and the direction of the biochemical processes.

There are indications that plant roots, not only release, but also actively absorb CO_2. The amount of CO_2 taken up, from the soil may be of the same magnitude an that coming from the atmosphere or may even exceed it. The intensity of CO_2 absorption from the soil depends on its concentration. The higher the concentration of CO_2 in the soil, the quicker it finds its way into plants via the roots (Kursanov, 1954; Samokhvalov, 1952).

The CO_2 of the soil in taken up by many microorganisms. The soil is known to be the habitat of many kinds of autotrophs which use CO_2 as a source of carbon for the synthesis of organic compounds. Besides, there are many kinds of heterotrophs in the soil which can also use CO_2.

We have cited opinions of individual workers who maintained that the CO_2 released mainly represents the product of metabolism of the soil microflora. We agree with this opinion. Experiments show that as soon as the activity of microorganisms is hindered, the release of CO_2 decreases. The reverse picture can be observed when compounds which increase the vital processes of microorganisms are introduced into the soil. Vincent and Nissen (1954) introduced into the soil small doses of antibiotics and obtained a noticeable increase in CO_2 output. In the control series the release of CO_2 amounted to 51.2-56.9 mg/40 g of soil. Upon introduction of penicillin, the amount of CO_2 released reached 112.6 mg, with chloromycetin, 85.2 mg; and with terramycin, 148.7 mg.

Numerous investigators consider the release of gases from the soil as directly linked to microbial activity, The biological processes taking place in the soil under the influence of microflora may be judged by the respiration of the soil. There can be no doubt that other organisms take part in this, too. Their role, however, is much smaller than that of microbes.

The respiration of the soil, is an indication of the biological and biochemical processes taking place in it, it can also serve as an index of soil fertility as a whole, as was maintained by Stoklasa (1905) and later on by many other investigators, Lundergardh, 1924; Makarov, 1953; Lees, 1949; Jensen, 1934; Bunt and Rovina, 1955 and others.

THERMAL REGIME OF THE SOIL

The thermal regime is of an especially great importance in the life of the soil.

The main source of heat is the solar radiation. Other sources, such as the internal heat of the planet and the heat obtained from chemical and biochemical reactions, are negligible and are not taken into account. The heat effect of radioactive reactions has not yet been studied.

As in well known, the surface of the earth absorbs heat from the sun. Air layers surrounding the earth prevent the earth from cooling and generally exert a great influence on its heat regime. The clearer the air and the less water vapors it contains, the less is the retention of the heat radiated from the earth surface.

The surface of the earth is not heated uniformly by the sun; it is most strongly heated at the equator and most weakly heated at the poles. The heat absorption is conditioned, not only by the geographical location, but also by its qualitative content, particularly by the color of the soil. Darkly colored soils absorb more heat than the lightly colored ones. The chernozems, for example, absorb 86% of the radiant energy of the sun; gray soils, 80%; and white soils, only 20%.

Soils also differ from each other in their heat capacity. This depends on various factors. Of greatest importance is humidity, since water possesses greater heat capacity than the solid particles of the soil. Dry soils warm up more rapidly than moist soils. Heat conductivity also depends on soil moisture. Dry soils conduct heat slower than moist ones.

The surface of the soil becomes warm during the day and cools in the night. This creates a diurnal fluctuation of soil temperature. The greatest amplitude of these fluctuations can be observed in the summer, especially in places with a sharp continental climate.

Heat waves are formed in the soil as a result of the alternation of warming and cooling. These waves are most sharply pronounced in the surface layers; they lessen with depth and disappear completely at one meter below the surface. In deeper layers, the temperature of the soil remains relatively constant.

Besides diurnal fluctuations, there also exist seasonal (annual) thermic fluctuations. The depth to which the soil freezes depends on regional and climatic peculiarities of the locality. There are regions where the soil does not thaw in the summer or it does so only in the upper shallow layer. This is the region of eternal freeze. The snow cover strongly influences the thermal conditions in the soil. It protects the soil against the winter freeze. In forests the soil freezes to a lesser extent than in the fields. Vegetation slows the warming up of the soil in summer and lessens the degree of freezing in winter. In the same way, it eases the diurnal temperature fluctuations in summer.

The freezing of the soil in winter exerts a definite effect on the biological processes. Microorganisms are known to be unharmed by low temperatures. Frosts of 20-30°C and more do not affect them. Many forms survive the temperature of liquid air, In our experiments *Azotobacter,* and root-nodule bacteria survived a month's storage at 180°C below the freezing point.

These are data on the increase of the metabolism of microorganisms under the influence of frost. After a three-week storage at -15 to -20° C (in frozen state) *Azotobacter,* for example, grows and multiplies more rapidly, root-nodule bacteria become more active and virulent, and yeasts are more active at fermentation of sugars, etc. This apparently is the explanation of the vigorous outburst of metabolism in the soil in spring.

The spring outburst of microbial activity is sometimes observed in laboratory conditions in bacteria grown in pure cultures. The periodicity of the change of summer and winter temperature apparently manifests itself in hereditary characteristics, which become fixed to a certain degree and are transmitted from generation to generation for some time. We have observed such an outburst of metabolic activity in some cultures of azotobacter isolated from the soil near Moscow. The influence of seasonal and meteorological conditions on the metabolic activity of bacteria was noted by some other investigators (Bortels, 1942).

Under the influence of frost a noticeable change in the chemical and physicochemical properties of the soil takes place. The concentration of the soil solution varies; a number of compounds precipitate, for example, ulminic acid into ulmin. According to our observations, the toxic substances of the soil are decomposed and inactivated. Soils exhausted by clover become less toxic after frosts. Inactivation of antibiotics produced by the soil microbes is noticed after the soil has been frozen for a prolonged period. It is assumed that many other organic and inorganic compounds in the soil are subjected to sharp changes under the influence of frost, and the soil as a whole becomes more fertile.

SOIL INSOLATION

The insolation of the soil has as yet been little studied. The soil is irradiated by the sun's rays only on the surface. The thicker the vegetation the less is the solar radiation reaching the surface of the soil.

Most rays of the spectra do not penetrate into the deep layers of the soil. There are data on penetration of infrared rays to a depth of one meter. Algologists assume that the algae found at this depth grow only because of the presence of these rays.

The importance of sun rays in the life of the soil is not clear. Undoubtedly, the effect of sun rays on the growth of microorganisms in the soil and, in particular, in the surface layer is very great. The study of microorganic metabolism on mountain summits has shown that biological processes are more vigorous there. According to our observations, the nitrogen-fixing activity of microbes on mountain summits is more vigorous than in the valleys. Some nitrogen-fixing bacteria are powerful accumulators of molecular nitrogen, in high places (Krasil'nikov, 1956b). On comparing the biochemical activity of soil bacteria in mountain soils with that of bacteria from valley soils we were able to establish an essential difference between them. The first, as a rule, possess more powerful proteolytic, amylolytic, and lipolytic activity.

Similar data were obtained by Mishustin (1947). The mountainous climate and especially insolation affect the natural characteristics of bacteria. These characteristics are not lost for some time after their transfer to valleys. These as yet isolated observations give grounds to assume that insolation strongly affects the life of soil microbes.

Summarizing, it can be said that the sites occupied by microorganisms in the soil are all the spaces between the soil particles and the aggregates. Microbes flourish in large and small pores. They inhabit microscopically small pores and capillaries. The soil solution is their nutrient medium. Its nutritional value varies, depending on the concentration of the nutrients, the presence or absence of toxic and biotic substances, the gaseous phase, intensity of air exchange in the pores, the income of atmospheric oxygen, and on the elimination Of CO_2 from the soil,

The soil solution with its nutrient properties is to a certain degree an index of the fertility of the soil and the capability of microorganisms to grow in it. It determines, not only the composition of the microbial population, but also the qualitative distribution of the individual genera and groups.

ORGANIC MATTER OF THE SOIL

Organic matter is one of the main components of the soil and conditions its fertility. According to their composition, the organic compounds of the soil are unique and complex. They are formed from plant and animal residues as a result of microbiological metabolism.

All organisms living above the earth's surface and in the soil (animals, plants, and microorganisms) find their way after death to the soil, where they are metabolized by the living cells of microbes which form various substances. These substances, in their turn, are subject to biochemical transformations, as a result of which specific, relatively stable, and complex compounds called humus are formed.

Higher plants supply the soil with organic compounds during the period of vegetation, releasing various nitrogenous, or nonnitrogenous compounds from their roots. They also shed dead fractions of roots and parts growing above the surface.

The total mass of plant residues entering the soil may reach considerable proportions. For example, in forests, the annual fall of leaves and twigs comprises 1.5-7 tons/hectare according to the type of the forest, its age, and the climate an soil conditions. Various woods yield different amounts of residue. Annual fall of leaves and twigs according to Zonn (1954), is as follows (average figures):

Deciduous forests 2.7 ton/hectare;
Oak forests 3.9 ton/hectare;
Pinewood forests 4.1 ton/hectare;
Fir tree forests 6.0 ton/hectare.

Thus, fir-tree forests rank first according to the amount of leaves and twigs shed, with pinewood, oak trees, and deciduous forests following.

The amount of the forest litter formed varies. The largest amount is to be found in fir-tree forests (50 tons/hectare and more).

According to the data given, the amount of organic compounds in soils of various kinds of forests varies. The amount of organic compounds shed in fir-tree forests reaches 5.85 tons/hectare; in pinewood forests, 3.96 tons/hectare; and in oak forests, 3.5 tons/hectare (Zonn, 1954),

As should be expected, the "sheds" of different forests differ qualitatively, too. According to the data of Zonn, the fall of fir trees is more acid than that of pine or oak trees. According to our observations, the leaves of birch and lime trees in June are decomposed in the soil more rapidly than the leaves of oak, aspen, or the needles of pine trees.

Meadow vegetation yields dry mass (from the parts growing above the surface) amounting to about 2-6 tons/hectare and roots, 7-11 tons/hectare. In the chernozem meadows of the steppes about 7 tons/hectare of dry mass (parts grown above the surface) were found and 25 tons of roots. In steppes on solonets soils 5 tons/hectare of the dry mass of the parts growing above the surface and 13 tons of roots were found (Savvinov and Pankova, 1942). In the desert steppes on serozem, about 1 ton/hectare of the mass growing above the surface and 15 tons of roots were found (Kul'tiasov, 1925). According to Kononova (1951), grass yields about 21 tons/hectare of root mass and, according to Belyakova (1953), the weight of roots of lucerne reaches 40 tons/hectare. Annual grasses yield less root mass than perennials (Vilenskii, 1954).

Plant tissues are composed of various carbon and nitrogen compounds. They contain sugars, dextrins, starch, pectic and tannic substances, organic acids, fats, waxes, tars, and many other compounds.

The main component of the plant material is cellulose $(C_6H_{10}O_5)n$. It constitutes the cell wall. Cellulose comprises 85-90% of the total weight of cottonseed fibers, and about 50% of bark.

The cellulose is decomposed by special cellulose microbes--bacteria, myxobacteria, actinomycetes, and fungi. Various intermediate compounds are formed in the decomposition process: organic acids, alcohols, sugars, and others.

Hemicellulose, in addition to cellulose, also appears in plant cells. Hemicellulose is easily hydrolyzed by acids and alkalis with the formation of sugars, uronic acids, and other compounds.

In wood, cellulose is impregnated with lignin, the content of which reaches 34%. Lignin differs from cellulose by its higher content of carbon (62-69%, in cellulose only 49.4 %) and lower content of oxygen. Upon oxidation it yields aromatic compounds. The chemical structure of lignin has not been ascertained. Lignin in the soil is decomposed by microbes with the formation of final decomposition products, CO_2 and water, or intermediate products.

Proteins are the most common nitrogenous compounds present in the cells of plants, animals, and microbes. They are present in protoplasm, nuclei, and in various protein reserve substances (metachromatin, protein crystals, aleuron grains, etc). Complex proteins are known--proteids and proteins proper such as globulins which are insoluble in water but soluble in dilute salt solutions; water-soluble albumins; prolamins--proteins of the gluten of the wheat grain (gliadin) which are soluble in 80% alcohol; glutelins--plant proteins, soluble in dilute alkaline solutions; sclero-proteins--insoluble proteins of horny tissues such as collagen, keratin, and others.

Many complex protein compounds are known, such as phosphoproteids containing phosphorus, nucleoproteids--proteins of the cellular nuclei and nuclear inclusions (which upon hydrolysis are decomposed into simple proteins and nucleic acids containing phosphorus), chromoproteids--proteins containing pigments (e.g., blood hemoglobin and some antibiotics formed by microbes), and glucoproteids (mucoproteins), which are proteins containing carbohydrates.

Other complex proteins which are present in plant, animal and microbial cells are: albumoses and peptones, which are protein compounds forming colloidal solutions and giving biuret reaction, and amino acids, which are colorless watersoluble compounds containing amino groups ($-NH_2$) and carboxyl groups (OH-C=O).

Plant residues as well as the dead cells of microbes and animal organisms find their way into the soil, where they are subject to physicochemical and biological processes.

The main transformations of plant residues are carried out under the influence of biotic factors. The dead parts of plants in the soil begin to decompose immediately, at first under the action of their own enzymes and then quite rapidly (perhaps simultaneously) under the action of microbial enzymes.

The first to be decomposed are the easily assimilated organic compounds: sugars, organic acids, and alcohols; then follow proteins, amino acids, fats, pectins, gums, hemicellulose, and lastly cellulose and lignin. The soil microbes also decompose waxes, tars, and many other stable compounds. It can be said that no organic compound exists which cannot be decomposed by microorganisms. Some of them are decomposed rapidly (carbohydrates, proteins, etc.), and others, slowly (tars, waxes, etc.).

The decomposition of organic compounds may be carried out to the final products, CO_2 and water, or may stop with the formation of intermediate compounds. The latter may be in the form of organic acids, alcohols, amino acids, etc.

Simultaneously with the decomposition of organic compounds, synthetic processes are taking place in the soil. The so-called autotrophs are known to synthesize organic compounds by assimilation of CO_2. The first of these are the photoautotrophic algae, which may be present in considerable numbers. Many colorless chemitrophs and pigmented bacteria possess the same capability. They assimilate carbon dioxide and synthesize organic compounds at the expense of chemical or light energy. Among these are the nitrate, sulfur, iron, hydrogen, and methane-oxidizing bacteria. The sole source of carbon for these organisms is CO_2. Their energy requirements are satisfied by the following simple compounds: ammonia, nitrates, sulfurous and ferrous compounds, hydrogen, methane, and others. Many heterotrophic microorganisms are capable of assimilation of CO_2 and of synthesizing organic compounds. This capability was detected in the representative of the genera *Pseudomonas* and *Azotobacter*, in sporogenous and asporogenous bacteria, in yeasts, fungi, and in actinomycetes. The synthesis of organic compounds may reach considerable dimensions, 5% and more of the CO_2 supplied during the experiment (Liener and Buchanan, 1951). Cells dividing in the logarithmic phase of growth assimilate ten times more CO_2 than in other stages of growth (Mac Lean et al. 1951; Shaposhnikov, 1952; Rabotnova, 1950; Linsh and Calvin, 1952; Citterman and Knight, 1952, and others).

According to Werkman and Wilson (1954), all microorganisms, autotrophs and heterotrophs, are endowed with the ability to assimilate CO_2, but to a different extent according to the type and conditions of culture growth.

The synthesized compounds and decomposition products of plant residues, as well as other organic compounds, find their way into the soil solution and are utilized as nutrients by microbes and plants.

Shmuk (1930) noted the presence of the following compounds in the soil: nitrogenous compounds (methylamine, choline, histidine, arginine, lysine, cytosine, xanthine), fats, organic acids (oxalic, succinic, crotonic, acrylic, benzoic, etc. esters (glycerides of caprylic and oleic acids), carbohydrates (pentoses, pentosans, hexoses, cellulose and its decomposition products), alcohols, aldehydes, tars, paraffins, and other compounds.

Davidson, Sowden, and Atkinson (1951), employing the method of paper chromatography, detected about 30 compounds in the organic fraction of the soil: such as arginine, histidine, lysine, alanine, leucine, proline, isoleucine, valine, aminovaleric acid, aspartic acid, tyrosine, threonine, glutamic acid, and others.

According to Kejima (1947), the following acids can be detected in the soil: 6-7% aspartic acid, 5% glutamic acid, and 18% of other amino acids, totaling 31.9% total nitrogen. According to the author, 66-75% of soil nitrogen is not in the humus but in microbial proteins.

Such compounds as polyuronic acids which comprise either components of plant tissue (hemicellulose) or products of microbial synthesis (slimy compounds constituting bacterial capsules) can also be detected In the soil.

Schreiner and Reed (1907) isolated various organic nitrogen and carbon products from fertile soils. Creatine, xanthine, hypoxanthine, adenine, and cysteine were among the first detected.

Rudakov and Birkel' (1949) found uronic acids among the metabolites of plant roots. The liberation of these acids takes place with the participation of bacteria possessing protopectinase.

Shori isolated allantoin from the soil and Enders obtained methyl glyoxalate, a compound which, according to Neiberg, is an intermediate in hexose fermentation and, according to Gebert, a primary structural element of protolignin. It is assumed that methyl glyoxalate is an intermediate compound, "a bridge" which links the lignin and cellulose theories of the origin of humic acids (Kononova, 1951). There are various biologically active compounds in the soil: vitamins (B_1, B_2, B_6, B_{12}), auxins, pantothenic, nicotinic, folic, and para-aminobenzoic acids, biotin, and other compounds activating the growth of plants and microbes, There are also inhibitors that may suppress the growth of plants (toxins) and microbes (antibiotics), and free enzymes--catalase, peroxidase, invertase, amylase, tyrosinase, and others.

Investigations show that enzymes exist in the soil in the active state. Their quantity varies in accordance with the soil composition, season, and climatic conditions. In fertile cultivated soils there are more enzymes than in poor nonfertile soils. The more organic compounds in the soil, the more active is the growth of microbes and the greater the enzymatic activity of the soil (Hoffmann, 1952). The upper layers contain more enzymes than the deeper ones.

The liberation of CO_2 has been observed to depend on the enzymatic activity of the soil (Seegerer, 1953; Ukhtomskaya, 1952). According to the data of Ukhtomskaya, the amount of enzymes in the soil increases proportionally to the amount of organic compounds introduced (Table 11). The enzymatic activity of the soil is more pronounced in May than in October, when the microbiological processes diminish.

Table 11
The enzyme content of the soil

The amount present in 100g of soil, expressed as decomposed substrate: catalase and peroxidase in ml of 0.1 N $KMnO_4$ solution; protease, in mg of nitrogen; amylase, in mg of maltose; invertase according to inversion, in mg of glucose (Ukhtomskaya, 1952)

Enzymes	May	May	May	May	October	October	October	October
	Control	500 tons*/ hectare	1,000 tons/ hectare	2,000 tons/ hectare	Control	500 tons/ hectare	1,000 tons/ hectare	2,000 tons/ hectare
Amylase	29	1,132	3,568	4,320	71.0	596	1,606	1,870
Invertase	29.69	428.2	2,012	2,173	87.5	333.7	712	1,182
Protease	48.0	62.68	61.25	84.0	42.5	53.2	54.2	68.48
Catalyse	279	601	671	723	260	741	470	980

*The organic compounds were introduced with sewage waters.

Kuprevich (1949) detected the presence of the following enzymes in the soil: catalase, tyrosinase, phenolase, asparaginase, urease, invertase, amylase, and protease, noting that their accumulation depends on soil cultivation. The quantitative figures for catalase, invertase, and urease present in soils, according to his data, are given in Table 12.

Table 12
The activity of extracellular soil enzymes
Catalase in cm^3 of O_2 produced in three minutes at 18i C; invertase in mg of inverted saccharose; urease in mg of decomposed urea (Kuprevich, 1949)

Soils	Catalyise	Invertase	Urease
The soil of the garden of the Botanical Institute of the USSR Academy of Sciences in Leningrad	6.0	167	34
The soil of a pine forest	6.4	202	41
The soil of an orchard	7.9	220	70
Washed river sand under barley	0.4	0.0	15(?)

129

Sorensen noted greater activity of xylanase in cultivated soils than in noncultivated soils. The enzymatic activity increased six times and more when straw or xylan were applied to the soil (Sorensen, 1955).

Scheffer and others (1953) and Seegerer (1953) pointed out the increased activity of invertase and urease after the application of organic fertilizers, especially manure.

The amount of enzymes in the soil also depends on the vegetative cover. When a green crop of serradella was plowed in, the amount of catalase and invertase was greater than after the plowing in of green lupine (Table 13).

Table 13
The catalase and invertase content in 2 g of soil
Catalase, in cm^3 O_2 from 5 ml 3% H_2O_2. Invertase, in mg of inverted saccharose

Soil	Catalase August	Catalase Sept.	Invertase August	Invertase Sept.
Fallow (control)	4.5	4.6	17.43	6.02
After introduction of lupine	5.6	6.7	27.63	32.41
After introduction of serradella	5.3	7.3	29.34	39.67

As can be seen from the given data, the enzymatic activity is closely correlated with the activity of microorganisms. Any increase In the amount of the latter leads to enhancement of enzymatic soil processes. Hoffmann (1951) considers that the enzymatic activity of the soil is an index of its fertility.

There are data in the literature indicating that plant roots excrete various enzymes into the soil, such as catalase, tyrosinase, amylase, protease, lipase and others.

All these organic compounds comprise only 10-15% (approximately) of the total organic mass of the soil. However, owing to their great activity, they are of considerable importance. Many of these organic compounds (vitamins, auxins, certain amino acids) are catalysts of biological and biochemical processes in the soil.

The part played by free extracellular enzymes is not yet clear, but they may be assumed to be important in transformations of many types of organic compounds and, in particular, in the synthesis of humus compounds.

We should note the considerable role of antibiotics in the life of the soil. These substances influence the composition of the microbial populations and this affects many soil properties.

Humic substances of the soil. Humus comprises the bulk of the organic soil compounds and is responsible for the dark coloration.

Humus is a mixture of various and very complex natural compounds. The uniqueness of these compounds does not allow for their classification into any of the groups of compounds known to organic chemistry. These substances are synthesized in the soil, apparently exogenically, by the action of extracellular enzymes. The composition of humus is more complex than that of many compounds of plant and microbial organisms. Humus comprises 85-90% of the total organic matter of the soil.

The chemical composition and origin of humus is not as yet clear. Characterization and subdivision of humic soil substances is based on external features.. color, and its relation to solvents. The main components of humus are assumed to be the three acids: ulmic, humic, and crenic.

According to Vil'yams, ulmic acid is formed during the anaerobic decomposition of organic compounds by anaerobic microbes. It is easily soluble in water imparting a dark brown color. It forms water-soluble salts with monovalent cations (potassium and sodium) and insoluble salts with bi- and trivalent cations. Under the influence of external factors, such as low temperature (freezing) or drying, ulmic acid is converted to water-insoluble ulmin.

Humic acid is formed under aerobic conditions and is considered to be a product of bacterial and fungal metabolic activity. Its properties are close to those of ulmic acid. It is less soluble in water than ulmic acid and gives the soil a black color. It is also denatured and converted into an insoluble compound, humin. It forms water-soluble salts with monovalent cations and insoluble salts with biand trivalent cations.

Humic acid has been studied in more detail. The following organic groups were detected: carboxyl (COOH), hydroxyl (OH), carbonyl (CO), and methoxyl (CH_2O), (Kononova, 1951).

Humus contains from 10-40% humic acids. The largest amount can be found in chernozems.

Humic acid contains 3.5-5% nitrogen. After acid hydrolysis about 50-60% of the nitrogen goes into solution in the form of amides and mono-and diamino acids. The molecular structure of humic acids has not been determined. According to the available data, more than one humic acid exists. Dragunov (1948) found that two samples of humic acid, one obtained from peat and the other from chernozem, differed from each other in their chemical composition, in the amount and structural type of their functional groups, as well as in the structure of their nuclei,

Bremmer (1955) subjected samples of humic acids, obtained by him from nine different soils, to chemical analyses. Each sample of the acid was analyzed for total nitrogen, ammonia-nitrogen, amino-nitrogen, and a-amino-acid nitrogen. The solutions obtained after hydrolysis were analyzed by paper chromatography for amino acids.

It was found that the samples of humic acids studied differed from each other in the composition of their nitrogen compounds and amino acids. Alkali extracts contain much of the nitrogen in the form of acid-soluble nitrogen compounds. About 20-60% of the nitrogen does not dissolve after acid hydrolysis. From 3-10% of the nitrogen is in the form of amino sugars. Nineteen amino acids were identified by means of paper chromatography: phenylalanine, leucine, threonine, isoleucine, valine, alanine, serine, aspartic acid, glutamic acid, lysine, arginine, histidine, proline, hydroxyproline, a- amino-butyric acid and others.

Humin and humic acids are decomposed by bacteria and fungi, especially by actinomycetes. Many actinomycetes grow well, bear fruit, and form antibiotic compounds on media containing humic acids as a sole source of carbon and nitrogen. Many forms of bacteria also grow on humic acid substrates.

Crenic acid was first found in spring water. According to Vil'yams, it is formed by fungi under aerobic conditions during the decomposition of forest vegetation and forest litter. Its properties differ sharply from other humic acids. It is colorless, highly soluble in water and acids, is not subject to denaturation and forms salts which can be crystallized.

Crenic acid possesses sharply pronounced acidic properties. According to Vil'yams, it can raise the soil acidity to such an extent that the activity and growth of many microorganisms is arrested.

It is difficult to accept this assumption. Organic acids as such are by themselves nutrients for many forms of microorganisms. It is quite clear, therefore, that their accumulation in the soil will be accompanied by an increase in the number of microbes.

Owing to its solubility, crenic acid penetrates deep layers of soil and there, combining with bases, forms crenates. They are harmless to microorganisms and are utilized by them as nutrients. Crenates are highly soluble in water, are easily leached from the soil, and may find their way either into ground or surface water. Thus, due to the high solubility of crenic acid and its salts, their accumulation in large concentrations is prevented.

Crenic acid may be reduced by nascent hydrogen with the formation of apocrenates. The reduction is carried out with the participation of anaerobic bacteria. Apocrenates are the salts of apocrenic acids. They have not been obtained in pure form. The salts of monobasic cations are highly soluble in water. Calcium apocrenate is slightly soluble in water and apocrenates of trivalent metals--iron, mangane se, and aluminum--are completely insoluble. These compounds are deposited in the soil in the form of voluminous amorphous sediments.

Crenic and apocrenic acids (fulvo acids) are widely distributed in soils. Their properties vary according to the soil. Kononova (1953) found that the acids from podsols differ from those of krasnozems.

The diversity of the natural conditions of soil formation both of a geographical and an ecological character, influence humus formation as a whole and, in particular, the composition of its individual components: humic, ulmic, and fulvo acids and other organic and organomineral compounds.

V. V. Dokuchaev was the first to point out the regular nature of the formation and transformation of humus under the varying conditions of different soils. climates. and zones. P. A. Kostychev and V. R. Vil'yams conceived the idea of the regularity of humus-compound formation in soils, in relation to the vegetative cover arid biochemical activity of the microflora.

Later investigations proceeding from chernozems to podsol soils, proved the regularity in the formation of the individual components of humus. Tyurin. (1949). developing the thesis of Dokuchaev on the basis of data from the literature and the results of his own investigations, showed that the geographical regularity of humus formation manifests itself not only quantitatively but also qualitatively. As a rule, the humus of the coniferous forests of the northern and central belt of the USSR and, in general, of the podsol soils is of a bright color, it contains few stable humic and ulmic acids but many compounds highly soluble in water which are easily leached from the soil, e. g. , crenic acid and apocrenates. Their concentration in podsol soils is 2-3 times higher than that of humic and ulmic acids (Kachinskii, 1956). In southern steppe regions having a grass vegetation, the soils contain humus with a different ratio of humic and fulvo acids.

The composition of humus in various soils also differs. Chernozem-type soils contain humic acids of different properties from those of podsol soils. Kononova (1956) showed the regularity in the variations of humic acids in the main soil types of the USSR. She found variations in the elementary composition of the acids, their optical density, and their distribution. The humic acids of chernozem soils are the most highly condensed, they are followed by the humic acids of the dark-gray forest soils, chestnut soils, and the bright-gray soils of the serozem; the humic acids of podsol soils and krasnozems are but weakly condensed, By applying the methods of X-ray structural analysis, the author determined the main structural outlines of humic and fulvo acids, which varied in relation to the type of the soil. These investigations disclosed the unity of the soil-forming process. While studying the genesis of humus and its components in various soils, Ponomareva (1956) reached analogous conclusions.

Investigators express three different points of view on the mechanism of humic-acid synthesis (see Kononova, 1951). The majority of workers consider that the formation of these compounds is outside the activity of microorganisms. This was criticized by Kostychev and then by Vil'yams.

At present, microorganisms are considered to play an increasingly important role in the process of humus formation. Reistric et al., (1938,1941), by means of molds, detected the formation of compounds of the aromatic quinone series from sugars.

These investigations stimulated the study of products of microbial metabolism; products which could serve as building material for the synthesis of humic acid. At present, many foreign (especially German) and Soviet investigators are busy studying microorganisms, their metabolic products, and the synthesis of humus compounds.

Great attention has been drawn to molds, actinomycetes, and heterotrophic bacteria as producers of humus-like compounds. While studying the products of bacterial metabolism, Martin, J. (1945) found that about 30% of the humus is synthesized at the expense of bacterial polysaccharides of the uronic type. The most stable of them, "levan", is formed by sporogenous bacteria *Bac. mesentericus* and *Bac. subtilis*.

Flaig (1952) isolated 42 cultures of actinomycetes from the soil which, under certain conditions, form a dark-brown or almost black humus-like compound. Kuster (1950-1952) concentrated his attention on fungi which produce compounds similar in color and certain chemical properties to humin substances. Laatsch. Hoops, and Bieneck (1952) found that the fungus *Spicaria* and certain actinomycetes, when grown on artificial protein media, are capable of forming a compound closely related to humin. Scheffer and Twaditmann (1953), Plotho (1950), Laatsch and others succeeded in finding a medium in which, under given conditions, fungi or actinomycetes formed substances of the phenol type. These investigators assumed that the oxidation-reduction systems--quinones < = > polyphenols--are in a state of continuous activity in the living cell, being oxidized by polyphenol oxidases and reduced by dehydrases. With the cessation of respiration the quinones are released from the cell, being irreversibly oxidized; they then combine with organic nitrogen compounds (protein decomposition products) to form humic acids. Consequently, according to the above-mentioned authors, the reaction of quinones with microbial nitrogen compounds is the basis of humus formation.

Wilts (1952) noticed that humic substances are formed from various organic compounds. The building blocks of the humus particles may be products of decomposition of lignin and of tannic compounds--aromatic compounds of the phenylpropans series, easily hydrolyzed carbohydrates (cellulose and others), and proteins which are subject to complex transformations as a result of bacterial metabolism.

The works of Soviet investigators Mishustin, Gel'tser, Rudakov, Kononova, and others should be mentioned. Mishustin (1938) studied the formation of humus substances upon self-heating of grain; Gel'tser (1940), upon the decomposition of fungi; Rudakov (1949) ascribes the main role in humus formation to pectin compounds. Troitskii (1943) assumes that humic acids are formed by microbes from decomposition products of vegetative residues. Tepper (1949, 1952) has shown that humin substances are formed at the expense of pigments formed by fungi and actinomycetes (see Rudakov, 1949 and 1951).

Kononova (1951), in her monograph, proposes that various plant residues and products of resynthesis, as well as the microbial protoplasm participating in the process of humus formation, may serve as sources of humus. According to her. the primary molecule of humic acid emerges as a result of the condensation of aromatic compounds with an amino acid or polypeptide. This process takes place with the participation of microorganisms under the conditions of biocatalysis maintained by the oxidative bacterial enzymes. As a result, nitrogen-containing compounds of a cyclic structure are formed.

RADIOACTIVE SUBSTANCES OF THE SOIL

Among the mineral elements of the soil a special place is occupied by radioactive substances: radium, uranium, thorium, and others. According to Baranov and Tseitlin (1941) their content (weight %) in different soils in as follows:

	Ra (x 10 $^{-11}$)	U (x 10 $^{-5}$)	Th (x 10 $^{-4}$)
Krasnozem, Batumi	6.71	20.13	9.19
Desert serozem	2.96	8.8	2.61
Bright-chestnut	8.22	24.66	5.63
Medium podsol loams, Moscow Oblast'	8.88	24.66	5.63
Dark forest	7.45	22.35	5.99
Podsol, Leningrad Oblast'	9.46	28.38	4.79
Loamy chernozem	9.08	29.24	5.14
Mountainous tundra, Khibiny	7.46	22.58	4.10
Marshy tundra, peat	1.94	58.3	9.5

The biological significance of natural radioactive elements remains unknown. It should be assumed that it is of considerable importance for the plant, animal, and microbial population of the soil. Existing data show that these substances in small concentrations activate biological processes, increase metabolism, and exert a positive influence on the growth of plants. The natural radioactive substances of soil find their way into plants, may concentrate there, and cause definite effects (Drobkov, 1951; Vlasyuk, 1955; Popov, 1956, and others).

The biological action of radioactive substances (radium, uranium, radium emanations and others) has been studied for a long time by microbiologists. Nadson et al. (1920, 1932), Filippov (1932). and Rokhlina (1930, 1954) studied in detail some of the biological processes of yeasts, fungi, and bacteria caused by radium, radium emanations. X-rays, etc. These authors were the first to establish the effect of radium and other sources of radiation energy in promoting genetic mutations.

We (Krasil'nikov, 1938) have shown that various types of luminescent actinomycetes react differently to the radiation of radon. Certain species were more sensitive than others. Radon rays have a stimulating or suppressing effect on the growth of mycobacteria, actinomycetes, and proactinomycetes. According to our observations, pigmented cultures are more sensitive to radon than nonpigmented ones.

In recent years we have studied soil bacteria--*Azotobacter,* root-nodule, and some others and their relation to certain radioactive substances, such as radium, thorium, and uranium. It was found that the bacteria absorb these substances from the soil and accumulate them in their cells in considerable quantities which many times exceed the concentrations of these substances in the soil.

Attention should be drawn to the fact that the degree of accumulation of radioactive substances in cells varies in different kinds of bacteria. Some kinds of bacteria, especially *Azotobacter,* accumulate radium in large quantities, others, in small quantities, or are completely devoid of this capacity. Even in the same genus different strains accumulate natural-radioactive substances to a varying extent.

The radioactive substances inside the bacterial cell stimulate growth and metabolism. Nitrogen fixation by *Azotobacter* is enlarged under the influence of radium and thorium. The ability of root-nodule bacteria to penetrate the roots of legumes and to form nodules is also increased (Krasil'nikov, Drobkov, Shirokov, and Shevyakova, 1955).

As a rule, the activating doses of the substances studied by us cannot be detected by ordinary electronic counters (radiomer B-2 and others). The microorganisms are sensitive to irradiation by radioactive substances in doses which cannot be detected by modern instruments.

The microbial population of the soil as well as plants are adjusted to small concentrations of radioactive elements. High doses given to them artificially under experimental conditions are harmful. Minimal concentrations of radium, uranium or thorium which can be detected by electronic counters damage even the least sensitive species of bacteria. Under the influence of such doses the cells undergo degeneration, increase in size and deform, their protoplasm becomes coarsely-granular, vacuoles appear, and their reproduction slows down and eventually stops altogether (Figure 53). Similar changes were observed by Filippov, Shtern, Rokhlina, and other collaborators of Nadson, in yeasts, fungi, and certain plants when irradiated by X-rays, radon, or ultraviolet rays. The same picture of degeneration in yeasts, under the action of large doses of radium and other sources of radioactive irradiation, was noted by Meisel' (1955). His investigations led to the emergence of a scheme of consecutive damage to the structure and function of cells.

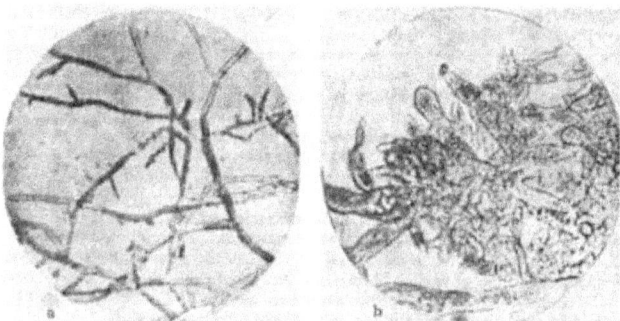

Figure 53. The effect of radioactive compounds (U) applied in the minimum doses detectable by the electronic counter (B-2) on the culture of *Aspergillus niger:* a) control; growth on a medium without radioactive substances; b) growth on a medium containing uranium. Swollen hyphae of the mycelium with degenerative coarsely-granular protoplasm.

As noted by Vernadskii (1926, 1929), radioactive substances possess free energy and continuously carry out considerable chemical activity in the soil. The energy of radioactive elements affects chemical and biochemical processes of microbes and organisms. Vernadskii stresses the fact that life in the biosphere originates from two energy sources: solar radiation and atomic radioactive energy. According to his calculations, only three radioactive elements, uranium, thorium, and radium supply the earth with heat, the quantity of which exceeds a thousand times that received by the earth's surface.

The biosphere of the earth accumulates dispersed radioactive elements and concentrates them on the surface, thus essentially changing the energetics of the whole population. It should be assumed that plants, animals, and microbes have, during their long evolution, acquired the ability to utilize these powerful energy sources. Analyses show that radioactive substances are present, to a larger or smaller extent, in all organisms and almost always in concentrations exceeding those in the surrounding environment. In many instances, plants contain ten times or hundreds of times higher concentrations of radioactive substances than the surrounding substrate (Vinogradov, 1932; Baranov and Tseitlin, 1941; Drobkov, 1951, and others).

The problem as to whether the organisms require the radioactive substances remains experimentally unsolved. Opinions are held according to which these substances, in small doses, do not play any role in the life of organisms and, in large doses, are harmful. Recently, data have accumulated which prove the reverse: small doses of radium, uranium or thorium stimulate the growth and increase the dry mass yield. Studies on the importance of natural-radioactive substances in soil fertility and in the life of plants and microbes are still inadequate. There are a number of observations which give reason to believe that these substances play an essential role in nitrogen fixation. A question arises as to the energy source needed to fix 100 to 150 kg and more of molecular nitrogen per one hectare of soil in one season. To fix such amounts of nitrogen and they are actually of this magnitude, *Azotobacter,* the most powerful nitrogen-fixing organism, requires 5-10 tons of glucose. Such quantities of this energy-yielding material are hardly to be found even in the most fertile soils.

Perhaps in this case the radioactive soil substances constitute the energy source which is indispensable for nitrogen fixation, as well an for many other processes taking place in the natural environment.

'The natural-radioactive substances deserve the most painstaking studies as biocatalysts on the earth's surface. When they enter into the chemical composition of living organisms, it should be assumed that they are not destroyers but creators, participating in many transformations and stimulating various enzymatic processes.

THE ADSORPTION CAPACITY OF SOILS

Soil is known to adsorb various substances. There are a number of forms of soil adsorption: mechanical, physical, chemical, biological, and physicochemical.

Mechanical adsorption. The soil, as any other porous body, retains particles present in the filterable liquid. In other words the soil acts as a filter. Ordinarily, the size of retained particles exceeds the size of the soil pores but even smaller particles can be retained.

Physical adsorption is linked to the phenomenon of surface tension and is manifested by the fact that increase or decrease in the molecular concentration of compounds in the solution takes place an the surface of particles.

Physicochemical or exchange adsorption--consists in cation exchange. The cations from the solid phase of the soil are being exchanged for an equivalent quantity of cations present in the surrounding soil solution.

Chemical adsoerption expresses itself in adsorption of certain ions from the soil solution, which form insoluble salts in soil. Consequently, a precipitate is formed which enters into the solid phase. Thus, for example, the ion of phosphoric acid precipitates in the presence of calcium salt (carbonic, hydrofluoric. or sulfuric acid). An insoluble salt of tricalcium phosphate is obtained. The latter precipitates and enters into the composition of the solid phase of the soil.

Biological adsorption according to Gedrolts is characterized by the adsorption of compounds from the soil solution, by microbial cells and green plants.

The adsorption of microbial cells by the soil also belongs here.

According to Gedroits, the physicochemical adsorption is of the greatest importance. In his opinion it is conditioned by the soil-adsorbing complex, consisting of chemical substances capable of exchange reactions. These substances may be organic and inorganic compounds or colloids undissolved in the soil solution. The latter represent the smallest particles, in size less than 1μ more often from 1-100m μ which do not precipitate in water and pass through fine filters. They are only visible in the ultramicroscope.

Organic compounds of the soil such as the humic acids, organomineral and inorganic compounds such as aluminum-silicates, iron hydroxide, argillaceous minerals and others may exist in the colloidal state. They represent a finely dispersed system, the particles of which possess high surface-reaction capacity for adsorbing substances present in solution. The soil colloids can be divided into hydrophiles and hydrophobes. The first adsorb water molecules and hydrated ions of the soil solution on their surface. The latter do not absorb molecules of the liquid phase of the solution.

The soil colloids adsorb cations. This adsorption is an exchange process, since with the adsorption of these cations, other cations are being released in equivalent quantity.

The sum of all adsorbed or exchanged cations which can be eliminated from the soil is a constant value for a given soil (Gedroits). It varies only with the acquisition of new properties by the soil and with the change of its essential nature.

The sum of adsorbed bases comprises the volume capacity of adsorption. It in expressed in milliequivalents per 100g of soil. The adsorption capacity varies in different soils. It is smallest in podsol soils and largest in chernozems. The former is conditioned essentially by a mineral adsorbing complex and the latter by the organomineral part of the soil.

Soils saturated with bases (chernozem, etc), contain magnesium and calcium in the adsorbing complex. Saline soils. besides these two elements also contain sodium. There are soils which are not saturated with bases. To these belong the podsol soils which contain hydrogen.

The adsorption capacity is conditioned by the composition, properties and degree of dispersion of the soil. The greater the number of small particles in the soil, the higher the specific adsorption surface. Soils having a large percentage of highly dispersed organic humus compounds possess higher adsorption capacity than soils poor in organic compounds. The adsorption capacity of humus is 150-250 milliequivalents and humic acid, 300 milliequivalents per 100g.

Soils possess exchange capacity not only in regard to cations but also anions. The adsorption of anions takes place in the soil in the presence of iron and aluminum hydroxides.

The adsorption of microbial cells by soil particles is also of great importance. This phenomenon has been inadequately studied and much of the data requires experimental verification; certain data are contradictory. Nevertheless, the little data available are of great interest.

Adsorption of bacteria by soil

It has long been noted, in the laboratory practice, that bacterial cells are adsorbed by various materials in powder form. Kruger, (1889) demonstrated the adsorption of bacterial cells from their aqueous solutions by coke, clay, brickflour, magnesium oxide and other substances. Later Eisenberg (1918) showed that bacterial cells can be adsorbed by animal charcoal. Michaelis (1909) noticed that different bacterial genera are adsorbed to a varying degree.

Bacterial adsorption by soil particles was demonstrated by Chudiakov, N. N. (1926) and his collaborators Dianova, Voroshilova (1925), Karpinskaya (1925) and others. These investigators have shown that the soil adsorbs considerable quantities of bacterial cells. According to Dianova and Voroshilova (1925), between 252 and 4,350 million bacterial cells par hectare are adsorbed depending on the kind of soil and generic peculiarities of the bacteria (Table 14).

Table 14
The adsorption of *Bact. prodigiosum* by different soils
(number of cells in millions, per 5 g of soil)

	Cells introduced	Total adsorbed	% adsorbed
Podsol exp. fields of Agr. Acad. im Timiryazev			
	58,600	4,470	58.8
	5,860	4,988	85.9
	58.6	52	90.4
Chernozem of the Voronnezh Oblast'			
	32,800	28,000	87.5
	16,400	16,200	98.8
	3,280	3,230	98.5
	328	327	99.7

The loam soils of the experimental fields of the Agricultural Academy im. Timiryazev adsorb *Bac. mycoides* 95.5%; *Bac. ellenbachensis* 57.5%; *Bac. mesentericus,* 40.5%; *Ps. fluorescens liquefaciens* 79.7 %; *Staph. pyogenes,* 80%; *Bact. prodigiosum,*98%; *Bact.coli,* 10-20%.

Novogrudskii (1936c) studied the adsorption of podsol soils of the experimental fields of the Agricultural Academy im. Timiryazev, the soils of the Moscow Botanical Garden and chernozem of the Voronezh Oblast'. We have compiled the result of these studies in the tables. Table 15 shows the data on bacterial adsorption and Table 16 the adsorption of fungi and actinomycetes.

Table 15
Adsorption of bacteria by different soils (in %)

Soils	% Bac.Mycoides	% Bac. Mesentericus	% Bac. megatherium	% Bac. chroococcum	% Bac. fluorescens	% Bac. denitrificans	% Bac. leguminosarum
The timiryazev Agricult. St. (podsol)	71	10	61	64	8	36	44
Botanical Garden	82	76	62	44	20	20	45
Voronnezh Oblast'	99	99	93	95	50	82	88

Table 16
Adsorption of spores of fungi and actinomycetes by different soils (in %)

Soils	% Asperg. niger	% Penic. glaucum	% Mucor mucedo	% Fus. sp.	% Botrytis cinerea	% Act. 154	% Act 105	% Act. 110
Podsol	14	43	57	99	93	8	10	13
Botanical Garden	20	43	27	97	97	15	31	28
Voronezh chernozem	97	94	97	99	99	99	75	94

According to our data the Moldavian chernozem (medium loams, carbonate) adsorbs two to three times more cells of *Azotobacter* than the upper layer of the podsol soil which was previously under forest (Experimental Station of the Moscow State University, Chashnikovo, Moscow Oblast'). In the former soil, 4,000 million *Az. chroococcum* and 600 million *Az. vinelandii* were adsorbed per gram and in the latter soils 2,400 million and 200 million per gram respectively.

The adsorption capacity of soil varies according to the depth. The upper layers of the soil are characterized by higher adsorption capacity than the lower, The poorly cultivated soil of the Experimental Station of Chasnikov adsorbed *Az. chroococcum,* according to the different horizons, as follows:

a) Layer A-A_1; 80 % in May and 92 % in August
b) Layer A2 (10-20 cm); 50% in May and 53% in August
c) Layer B_1 (30-40 cm); 35 % in May and 25% in August
d) Layer B_2 (50-70 cm); 60% in May and 75% in August

Under the same conditions soil of the same type. but well cultivated, adsorbed cells of *Azotobacter* as follows:

In the layer 0- 10 cm; 85 % in May and 93 % in August
In the layer 10-20 cm; 80% in May and 83% in August
In the layer 30-40 cm; 65% in May and 65% in August
In the layer 50-70 cm; 78% in May and 87% in August

The adsorption capacity of soils is closely connected with their mechanical composition. Sand contains particles 1.0 to 0.25 mm in diameter and sand dust with particles 0.25-0.05 mm in diameter, adsorb bacterial cells weakly. Dust containing particles 0.05-0.01 mm 0.01-0.005 mm and 0.005-0.0015 mm in diameter adsorbs microbial cells most actively. The slimes of rivers and lakes having particles of 0.0015 mm and less in diameter are devoid of adsorbing capacity. since the size of the particles of slimes (1-1. 5 μ and less) does not exceed that of the ordinary bacterial cell. In such an environment bacterial cells are themselves adsorbants. The greater the adsorption of bacteria, the fewer the cells found afterward in the suspension.

The character and degree of adsorption of microbial cells by the soil is conditioned to a large extent by the qualitative features of the organisms proper, and their generic properties. The degree of adsorption depends on their metabolic state and their vital potential. Some bacterial genera are adsorbed more vigorously and in a larger quantity than others. According to some authors, many nonsporiferous bacteria are adsorbed considerably weaker, by the same adsorbent, than the sporiferous bacteria or micrococci.

For example, podsol soil adsorbs bacterial cells as follows:
Bac. mycoides 71%
Bac. megatherium 61%
Az. chroococcum 64%
Ps. fluorescens 18%
Bact. coli 10%
Bact. denitrificans 36%
Rhizob. leguminosarum 44%

Gram-positive bacteria are adsorbed by the soil in larger quantitites than the gram -negative bacteria. Bogopol'skii (1933) gives the following data. Peat of medium decomposition adsorbs 74% of *Bac. mycoides,* 81% of*Urobact. pasteurianum,* 21-22% of *Bact. coli* and *Ps. fluorescens.* The same results were obtained by Eisenberg (1918) while studying the adsorption of bacteria by charcoal and other adsorbants. According to his data, the adsorption of gram-positive bacteria, *Micr. pyogenes, Micr. candicans Sarcina lutea* and others was 500 times larger than that of the gram-negative bacteria, *Bact. coli, Bact. typhi, Ps. pyocyanea, Vibrio cholera* and others.

The degree of adsorption of microbial cells by the same soils depends upon the pH of the suspension from which the cells are being adsorbed. The spores of *Bac. mycoides* were maximally adsorbed at PH 4.5, with the increase of pH to 5.8-6.7 the percentage of spore adsorption decreases. When the pH of the medium is raised to the neutral (pH 7.0), and higher into the alkali zone (pH 7.8), the degree of bacterial adsorption remains on the same level or even decreases slightly (Table 17).

Table 17
Bacterial adsorption by podsol soil at different pH
(number of cells grown in the Petri dish)
(according to Eisenberg, 1918)

	pH of the suspension	suspension	soil	adsorbed %
Bac. mycoides				
	4.6	82	8	90
	5.8	66	33	50
	6.7	67	59	12
	7.4	63	46.5	27
	7.8	66	45.5	28
Bact. coli				
	4.4	384	311	19
	5.6	361	384	0
	6.3	384	178	35
	6.9	360	199	45
	7.5	358	294	20

The adsorption of *Bact. coli* under the same conditions is different. The highest adsorption takes place at a pH of 6.3-7.5; in an acid or strongly alkaline medium the cells are adsorbed much more weakly.

The adsorption capacity of the soil varies in relation to its moisture. Very wet soil adsorbs less bacteria. Upon rinsing with water a considerable quantity of adsorbed bacterial cells is desorbed and washed out, A single washing of podsol soils, according to our experiments, releases about 11% of *Az. chroococcum*. The larger the volume of water, and the more prolonged the elution of the soil, the more bacteria are washed out. Of the 2,900 million adsorbed cells of *Azotobacter* the following quantities were eluted:

First elute, after one minute, 330 million (11.4%)
Second elute, after two minutes, 54.0 million (0.18%)
Third elute, after two minutes, 35.2 million (0.12%)
Fourth elute, after two minutes, 0.2 million (0.007%)

The eluted bacteria comprised 14% of the total. One hundred ml of water were used for each washing per 5g of soil.

The elution of the bacteria is apparently, limited. Above this limit the bacteria are not desorbed even after prolonged washing. The quantity of desorbed cells in different soils varies. The desorption of bacteria by water from their natural surroundings is observed after rain or irrigation. This is closely connected with the seasons.

The adsorption capacity of soils varies during the vegetation cycle. According to the data of Novogrudskii (1937), less bacteria are being adsorbed in spring and autumn than in summer,

In April and September, in the podsol soils of the fields of the Agricultural Academy im. Timiryazev, 40-66% of bacteria were adsorbed and in the summer 60-90%. The seasonal variations of the adsorption capacity of the soil are conditioned not only by moisture but also by temperature. The less moisture in the soil, and the higher its temperature, the stronger the adsorbing capacity for microbial cells.

Samples of soils at 25% moisture per dry weight maintained at 0° and at 25°, adsorbed 57% and 68% *Bac. mycoides* respectively.

The adsorption of bacterial cells by the soil is a reversible process. Upon change of pH, temperature, moisture and other factors. the bacteria are desorbed.

An exchange adsorption is observed, similar to that of mineral substances. if the soil is saturated with one type of bacteria and is then saturated with cells of another more adsorbable kind, then an exchange of bacterial cells will take place. The formerly adsorbed cells will be released or displaced and will reappear in the suspension.

When two or more kinds of bacteria are simultaneously adsorbed, the more adsorbable ones will be preferentially adsorbed (Novogrudskii, 1936).

According to Chudyakov and collaborators, the adsorbed cells preserve their viability but their metabolism slows down or stops altogether.

Dianova and Voroshilova (1925) determined the biological activity of bacteria in strongly adsorbing soils and in sand. The substrates were wetted with nutrient mediums (such as peptone, glucose and others), were sterilized and inoculated with *Bac. mycoides, Bact. prodigiosum* and *Sarcina flava.* Their biological activity was determined by the CO_2 released.

Under all experimental conditions the authors noticed a strongly diminished liberation of CO_2 from the soil.

For example, in experiments with *Bac. mycoides* the following amounts of CO_2 were released: in sand, 106.3-126.8 mg, in soil, 0-7.8 mg; with *Bact. prodigiosum* 24-52 mg of CO_2 was released in sand and 0 in soil;*Bact. megatherium* released 66 mg of CO_2 in sand and 2.6 mg in soil; *Bact. coli* released 41-69 mg of CO_2 in sand and 10-23 mg in soil.

The stronger the adsorption of bacteria, the less their activity. The activity of *Bact. coli* in sand is 2-4 times higher than in soil, while that of *Bact. mycoides* and *Sarcina flava* is 10-30 times higher than in soil.

The activity of adsorbed bacteria increases with the rise of soil moisture. At 60% of total moisture capacity, 5.2 mg of CO_2 was released from the soil, at 100% of moisture capacity, 28.6 mg (experiments with *Bac. mesentericus).*

According to Lipman (1912), the nitrification process by bacteria in a clay soil proceeds slower than in sand. *Bact. proteus* releases 40% more ammonia nitrogen in sand than in clay; *Sarcina lutea* 80% and *Bac. mycoides* 87%.

According to our observations, cells in the adsorbed state reproduce quite actively. Thus, for example, after careful washing of the soil *Azotobacter* remained in the adsorbed state at a level of 55 millions per gram of a total 100 million per gram introduced. We washed the soil samples daily for a month. About 300 million cells per gram were washed out. However, after the last elution, a considerable quantity of cells remained in the adsorbed state. Thus, they increased daily by 10 million bacterial cells for each gram of soil.

Apparently, the process of adsorption of microbial cells by soil particles is also of a biological, and not on ly a physiochemical character, and it must not be considered only from the point of view of physical or chemical forces. Krishnamurti and Soman (1951), analyzing the literature data and their own investigations, reached the conclusion that the phenomenon of adsorption of bacterial cells is of a specific character. The percentage cell adsorption is conditioned by the adsorbent properties as well as by the generic properties of the microbe. The adsorption coefficient is strictly constant under given conditions. The authors even recommend the differentiation of bacterial species on this basis.

There are almost no data in the literature on the adsorption by the soil of products of microbial metabolism although this problem is of considerable theoretical and practical interest. Microbes, as was pointed out above, grow abundantly in soils, proliferate, display high biological activity and synthesize and release various metabolic products into the milieu. Among these products there are many biologically active substances: enzymes, vitamins, auxins, amino acids and other biotic substances. Antibiotic metabolites, toxins, etc can also be found, Once these substances are excreted from the cell into the soil, part of them undergo decomposition and inactivation, another part is adsorbed by soil elements. The degree of adsorption of such active metabolites is unknown.

It should be noted that the literature contains very little data on the adsorption of organic compounds by the soil. The adsorption capacity of the soils, as was stressed above, was studied mainly in reference to mineral elements: cations and anions. No attention was paid to organic compounds. However, these processes of interaction between the soil and organic compounds should provide an explanation for the formation of organomineral compounds which determine the essence of soil fertility or the formation of soils as such.

The available data on adsorption of organic compounds by the soil mainly refer to the problem of humus formation.

Kravkov (1937) introduced aqueous extracts of grasses and straw into the soil and observed their fixation. According to his data, the water-soluble plant compounds are adsorbed by soil particles to a varying extent which depends on the type and properties of the soil. A different adsorption capacity was recorded for each soil. The organomineral compounds so formed are considered by the author to be the humus of the soil.

Persin (1944) introduced aqueous extracts of fresh straw and hay of various grasses as well as extracts of straw and hay, after they had been subjected to decomposition by microorganisms. It was found that the water-soluble extracts of fresh straw and hay are not adsorbed by the soil and that extracts of decomposed straw and hay are adsorbed to a varying degree, depending on the stage of decomposition. The greatest adsorption of water-soluble compost substances was observed after 75 days of decomposition at the optimal temperature for microbial activity.

According to the observations of the author, chernozems adsorb organic compounds in greater quantities than podsol soils. The adsorption capacity of soils for organic substances is conditioned by their mechanical composition. The larger the clay fraction in the soil, the greater its adsorption capacity, and, consequently, it retains the adsorbed substances more tenaciously.

It should be noted that Kravkov, Persin and some other authors (Chizhevskii and Makarov, 1939) carried out their experiments in nonsterile soil. Consequently, a great part of the introduced organic compounds (if not all of them) was decomposed by microorganisms and was lost to the investigators. The real magnitude of adsorption in these experiments cannot be precisely determined.

It should be noted that in the investigations of Persin, the soil adsorbed only those water-soluble organic substances which are obtained from decomposed plant residues, i. e., substances formed by bacterial activity.

Simakov (1938) carried out experiments with tannin and xylan. These substances were differentially adsorbed by the soil, xylan less than tannin. The author in his work during 1944 carried out experiments on adsorption of amino acids and sugars by lignin, which represented one of the components of the soil complex. The experments have shown that the afore-mentioned substances were strongly adsorbed by lignin and equally strongly retained. During this process their properties changed; they became more stable.

According to Simakov (1944), the amino acids asparagine and glycine adsorbed by lignin are decomposed slowly by microorganisms.

A considerable number of papers have been devoted to the adsorption of humic substances by the soil (see Zyrin, 1945; Khan, 1950-51; Aleksandrova, 1944; Tyurin, Gutkina, 1940, and others). These investigations have shown that humic substances form stable organomineral compounds with the mineral parts of the soil, the bond between the mineral particles and organic substances of the humus may be of a physical or a chemical nature.

We (Krasil'nikov, 1954c) have tested antibiotics of actinomycetes, bacteria and fungi.

Antibiotics, due to their specific antibacterial action are easily detectable, and can be found in various natural substrates, as for example in the soil. They are, therefore, convenient objects for the determination of the adsorption capacity of soil particles. Antibiotics were introduced into various soils and their adsorption determined. We tested penicillin, streptomycin, globisporin, aureomycin, terramycin, subtilin, gramicidin, and other antibiotics, and have shown that they are adsorbed in considerable quantities. For example, we introduced streptomycin at a concentration of 2,000 units/ g; after some time 1,120 units/g were adsorbed by the chernozem, 1,800 units/g by podsol, 1,080 units/g by the serozern and 1,540 units/g by krasnozem. Similar quantities of globisporin were adsorbed by the different soils. Penicillin was adsorbed as follows: 380 units/g by chernozem, 280 units/g by podsol soils, 380 units/g by serozem, and 200 units/g by krasnozem. Aureomycin and terramycin as well as antibiotics of bacterial origin such as subtilin and gramicidin were adsorbed by the afore-mentioned soils in varying quantities. The antibiotic 1609 was only adsorbed by the podsol soils, and then only in a very small quantity, 20-30 units/g. This antibiotic was not held by any other soil.

Thus, the various soils adsorb different quantities of antibiotics, however, the nature of their adsorption is different from that of mineral compounds. Soils poor in humus (podsols, krasnozems) adsorbed antibiotics in considerably larger quantities than soils rich in humus. Different layers of the same soil possess different adsorption capacities. We have studied the streptomycin adsorption capacity of the podsol soils (Experimental Station Chashnikovo, Moscow Oblast') of cultivated and noncultivated soils. The results are given in Table 18.

Table 18
Adsorption of streptomycin by the podsol soils at various layers (units/ g)

Soil	A_0 Layer	A_2 layer	B_1-B_2 layers
Forest mixed	1,300	700	5,400
Meadow	1,800	1,200	7,400
Roima* of the river Klyaz'ma	3,000	2,400	3,100
Glade	1,500	300	3,000

*[Russian term designating a very low and broad river terace,which may be flooded at the highest water mark. Flood basin but very extensive.]

The B_1-B_2 layer has the strongest adsorption capacity and the layer A_0-A_2 the smallest. The degree of adsorption of antibiotics by the soil does not depend solely on the soil properties but also upon the properties of the antibiotics themselves. In one and the same soil, for example in chernozem, we have observed the following adsorption:

	Units/g	μ g
Streptomycin	1,120	2.2
Globiosporin	1,080	1.8
Terramycin	900	1.0
Pennicillin	380	0.1
Preparation 1609	0	--

The antibiotics in the adsorbed state preserve their antibacterial activity for some time. The period for which a given antibiotic preserves its activity depends on the soil and the properties of the antibiotic itself. For example, in same soils penicillin remains active for 20-30 hours, in others 2-3 hours; terramycin remains active for 3-5 days in podsol and 1-2 days in chernozem. Some antibiotics (preparation 1600) lose their activity immediately.

Organic compounds adsorbed by the soil undergo various transformations; they are decomposed, inactivated and disappear, They are replaced by other compounds.

The adsorbed fraction of the antibiotics retains its antibacterial properties for a more prolonged period than the antibiotics in the free state present in the soil solution.

For example, free streptomycin disappears from the podsol soil after 10-12 hours, while adsorbed streptomycin is preserved for more than 30 hours. In the serozem, the nonadsorbed streptomycin is inactivated after 20-25 hours; its absorbed fraction can be detected even after two days. A still greater difference was observed in the experiment with aureomycin. In podsol it can be detected in the free state after 20 hours, in the adsorbed state after 5 days In the serozem the free antibiotic is preserved for no more than two days, while in the adsorbed state it retains its activity for more than seven days.

The antibiotics in the soil are partially inactivated by the soil solution and by microorganisms.

Not only antibiotics but also other microbial metabolites, as well as intermediate decomposition products of plant residues, and various compounds of humus, are adsorbed by the soil.

The biologically active metabolites present on the surface of soil particles exert a great influence on their physicochemical state. The soil particles holding the substance on their surface gain new properties.

The presence of living microbial cells adsorbed on the soil particles should be regarded as a complex system of biotic-mineral complex. Each soil particle carries elements of living organisms, the study of which is of the utmost importance to soil biologists.

THE MICROFLORA OF THE SOIL

Microorganisms are an integral part of the soil. If the soil should lose these organisms it would lose its main property--fertility--and it would turn into a dead, barren, geological body.

The soils are inhabited by numerous representatives of the microflora--bacteria, actinomycetes, yeasts, fungi, algae, protozoa, insects, worms, and others. Besides, there are in the soils various ultramicroscopic organisms: bacteriophages and actinophages.

No accurate data on the numbers of microorganisms in the soil are available. Methods for the detection of the entire soil population are not available. The existing methods give only a relative idea of the density of the microbial population.

Two essentially different methods are employed for the quantitative estimation of microbes in the soil: a) determination by means of soil inoculation of artificial media-liquid and solid, b) direct count of cells.

These two methods give different data on the quantitative aspect of the microbial population of the soil.

In practice, the inoculation method is more extensively used. There are various methods of inoculation and media.

The amounts of micoorganisms detectable in the soil vary, depending on whether they are inoculated into a solid or liquid medium, or inoculated by sowing the surface of agar medium, or dispersed in aqueous suspension by serial dilutions. Inoculation with soil is often carried out by placing small soil particles on agar medium.

In all instances the number of bacteria grown on agar media is smaller than upon growth in liquid media inoculated by the method of serial dilutions.

Data from the literature on the amount of bacteria, actinomycetes and fungi in the soil are obtained, in the majority of cases, from growth on agar media. According to these data, the number of bacteria per gram fluctuates within the range of from several tens or hundred thousands to many millions depending upon the soil composition and the medium (Starkey, 1929, 1931, 1955; Gray and Thornton, 1928; Clark, 1940; Timonin, 1940-41, Waksman, 1952; Jensen, 1934-36; Mishustin, 1956, and others).

Thom (1938), summing up data from the literature and the results of his own investigations on the quantitative determination of the bacterial population of the soil, considers that the total number of bacteria in one gram of soil reaches 50 millions. Since the greatest number of bacteria is concentrated in the plant rhizosphere many authors give data on microbial composition of this zone. Starkey showed, by the method of counting on agar media, the presence of 199 to 3,470 million bacterial cells per gram, depending upon the species of plant.

Humfeld and Smith (1932), counted 5-8 billion bacteria in one gram of soil with green fertilizer. Clark (1949) found 5 billion bacteria per gram of well-fertilized soil and also in soils under a mixture of grasses. Rippel (1939), analyzing the soils of Germany, and Feher (1933) analyzing the soils of Hungary and Austria, counted from one hundred thousand to 5 billion bacteria per gram of soil, depending upon soil composition and climatic conditions.

Nonfertilized soils have a smaller microbial population; ranging between hundred thousands and millions, but on the average 3-7 million per gram. Bunt and Rovina (1955) counted from 400,000 to 15 million bacteria per gram in the subartic soils of Iceland.

We have obtained similar figures. We have counted from several hundred thousands up to 15 million bacteria per gram, in the soils of the Kola Peninsula, the Islands of the Arctic Ocean and in mountainous soils of Pamir and Caucasus. The podsols of noncultivated or poorly cultivated soils contain, according to our investigations, 300 thousand to 10 million cells per gram of soil. Chernozems rich in humus contain 10-1,000 million cells per gram. Similar data were obtained by many other investigators studying various soils.

Higher counts (ten and hundred times higher) are obtained by the method of inocculation into liquid media and by the serial dilution method. For example, soils poor in organic compounds (podsols) gave from one to 100 million cells per gram, and fertile soils (chernozem) from one hundred to 1, 000 million bacteria per gram estimated by the solid-media method, meat-peptone agar (MPA). The method of inoculations on liquid media, meat-pe ptone broth (MPB) revealed to 10-500 million and 1,000-10,000 million bacteria per gram of soil respectively.

While studying the rhizosphere soil of lucerne in Central Asia (serozems) we have found (by the method of serial dilutions 50-100 billion bacteria per gram, and Raznitsina (1947) and Korenyako (1942) obtained even higher numbers. Such high figures are constantly obtained during the investigation of plant rhizosphere under given conditions.

Such high indices of microbial population throws doubt on the accuracy of the methods employed. Experiments especially designed to check this method were carried out. We have assumed that the high figures obtained by this method can be explained by the adsorption of bacterial cells on the walls of pipettes and with their subsequent elution (desorption). Experiments have shown that adsorpton of cells does indeed take place. The number of bacteria is 2-5 times, and sometimes even 10 times less if the pipettes are changed upon each dilution. The numbers obtained, when the pipettes are changed. are of an order of 1-10 billion per gram of soil, upon dilution of the soil with one pipette the numbers increase to 5-100 billion per gram of soil.

We checked the trustworthiness of this method by using pure cultures of *Bact. prodigiosum, Ps. fluorescens, Mycob. rubrum, Az. vinelandii* and *Bac. subtilis*. Aqueous suspensions of these bacteria were diluted with and without change of pipettes.

The number of bacteria in billions per 1 m/ obtained in such an experiment are as follows:

	With change of pipettes	Without change
Bact. prodigiosum	35.5	38.1
Ps. flourescens	22.8	100

Mycob. rubrum	1.0	4.5
Az. vinelandii	2.1	2.5
Bac. subtilis	3.9	7.3

In this experiment the change of pipettes lowered the number of bacteria 1.5-4 times depending upon the bacterial species. Similar lowering of bacterial numbers was observed on studying soil samples. The difference in the numbers is more pronounced if the number of bacteria in the soil is large. Hundred billions and more of bacteria were detected in the rhizosphere of lucerne grown in the Vakhsh valley when the soil suspension was diluted with one pipette; the number was five times less. when the pipettes were changed. The control soil contained 100-500 millions per gram upon dilution with one pipette, 50-150 millions were counted upon dilution, when the pipettes were changed, i.e., 30-50% less.

According to Vinogradskii, the method of direct counting of the soil bacteria also gives higher numbers than the method of growth on agar media, i.e., approximately the same are obtained upon serial dilution, or even higher (Table 19).

Table 19
The number of bacteria detected by various methods in the plow layer of soils under perennial grasses
(in thousands/ g)

Soil	Direct count method	Liquid inoculation	Solid inoculation
Podsol soil, field	560,000	500,000	7,500
Turf-podsol soil, garden, Moscow Oblast'	6,800,000	5,600,000	15,600
Chernozem, Moldavia	8,700,000	7,200,000	25,000
Chestnut soil, Trans-Volga region	3,500,000	1,000,000	9,500
Serozem, Central Asia	9,300,000	7,500,000	90,000

The difficulty of the method of direct counting is that the smears contain living cells and dead particles of the same soil. and they cannot be differentiated with certainty. The soil always contains a large amount of small particles which can be stained and thus become indistinguishable from the bacteria themselves. It is especially difficult to tell the coccoid cells from the small globular bodies and granules.

In recent years some investigators attempted to use fluorescent dyes for the differentiation of bacteria from the dead soil particles. Burrichter (1953). employed acridine-orange for the staining of soil smears and studied them under a fluorescent microscope in ultraviolet light. In soils rich in humus (9. 98 %)the author counted 9,453 million bacteria per gram of soil and the total number of microbes was 18,331 million bacteria per gram of soil; in soil poor in humus (1.80%) 1,230 million bacteria per gram were found. The number of colonies of slimy bacteria in the first soil amount ed to 157 million per gram. Soil, fertilized with compost. contained a total number of microbial cells of about 16,132 million per gram, and soils poor in organic substances 25 million per gram. Strugger (1948, 1949) found from 1,038 to 8,640 million bacterial cells per gram of soil employing the same method of fluorochrome staining.

It should be noted that the method of fluorescent staining also has shortcom -ings. The green color of living cells or the red color of dead cells and other particles

may often depend not on the cell viability but upon many other factors, such as the concentration of the dye, pH of the medium, temperature and others (Krasil'nikov and Bekhtereva, 1956). It is sometimes difficult to say what are the green or especially the red-stained bodies; are they living bacterial cells or dead, or soil particles.

The method of direct microscopy of soil smears (Kubiena, 1932) is also of little use for the quantitative estimation of cells.

As can be seen from the given data, the existing methods of microscopic analysis are inadequate for quantitative determination of microbial numbers in the soil and for the determination of their forms. Therefore, when comparing data obtained by employing one of the aforementioned methods, the investigators limit themselves to relative figures.

The bacterial numbers vary in different soils, according to their fertility and nutritional qualities. The more fertile the soil, the richer it is in humus, the denser its microbial population. The podsol soils (Moscow Oblast) contain, in well-cultivated fields, 3-10 millions per gram, and the chernozem soil of the Kuban, contains (similar method of counting) 15-50 million bacterial cells per gram of soil.

One and the same type of soil also varies in the amount of microbes it contains. The podsol soils, not well cultivated and poor in humus, contain 500,000 to 1.5 million cells per gram and in some cases only a few thousand per gram (the soils of the Kola Peninsula). Well-cultivated, systematically fertilized soils contain 3-25 million cells per gram. Garden soils, as a rule, are richer in microbes than the soil of fields.

Virgin soils contain less microbes than cultivated soils. (Table 20).

Table 20
The number of bacteria and actinomycetes in various soils
(in thousands / g on meat-peptone agar, in Petri dishes)

Soils	Bacteria	Actinomycetes
Podsol containing iron, Kola Peninsula	10-30	5-25
Podsol of Moscow Oblast' from under forest	100-300	70-100
Podsol of Moscow Oblast', garden	1,000-10,000	500-1,000
Chernozem, Kuban', under wheat	5,000-15,000	400-800
Serozem, Central Asia, virgin soil	850-1,500	600-1,000
Serozem, Central Asia, under lucerne	3,000-10,000	500-800
Chestnut soils, Trans-Volga region, virgin soil	400-1,500	450-860
Chestnut soils, Trans Volga region, under lucerne	5,000-15,000	500-1,000

The upper layer of the soil s richer in microbes than the deeper layers. For example, we have found the following amount of bacterial cells in podsol soils of the experimental fields of the Academy of Agriculture im. Timiryazev:

in the layer 0-20 cm deep, 5.7 million/g
in the layer 20-35 cm deep, 2.4 million/g
in the layer 40-60 cm deep, 0.5 million/g
in the layer 80-100 cm deep, 0.001 million/g

In the root zone of the vegetating plants, in other words, in the rhizosphere, the soil is saturated with bacteria to a greater extent than in the zone outside the roots. The vegetative cover, as will be shown later, exerts a strong influence on the concentration of microbes in the soil.

The number of microorganisms in the soil varies with the season. According to the literature and our own data, their total number in winter is smaller than in summer. This is especially noticeable in the soils of the north.

The analysis of soils of Severnaya Zemlya and other islands of the Northern Ocean showed that in May, when the soil was still in the frozen state, it contained tens of thousands of organisms per gram and in August many millions of bacteria per gram (Table 21).

Table 21
Seasonal variations of microbial numbers in the soil of the Severnaya Zemlya
(in thousands/g, counted on meat -peptone agar MPA)

Soil sample	May	August

Sector I, loam	23	1,340
Sector II, loam	40	4,380
Sector III, loose calcareous soil	91	16,600
Sector IV, loam	14	3,600
Sector V, loose calcareous soil	112	6,600

The number of microbes in the soil of the temperate zone is greatest in spring, smaller in summer, it increases somewhat in autumn.

The data on numbers of bacteria in winter are few and contradictory. The majority of investigators think that life in the soil stops altogether during the winter. A considerable number of microbes die from cold and their total number decreases.

According to our observations, the microbial activity does not always cease in winter. Under a deep snow cover the earth is not always frozen and in such a soil microbiological processes take place. This can be found by studying the growth dynamics of individual species of actinomycetes. Korenyako has shown that during the winter months of 1952-1954 certain species of actinomycetes *(A. globisporus)* grew more abundantly, in Moscow Oblast' soils, than during the summer and autumn.

Besides, certain biochemical processes, leading to detoxification of the soil take place in winter (Krasil'nikov, Korenyako and Mirchink, 1955).

The vigorous growth of microbes in spring is, according to our opinion, not only caused by the warm temperature and by moisture, but also by other factors, First, the toxins are inactivated or decomposed in winter due to low temperature. Second, low temperatures, as was noted above, stimulate the growth and activity of microbes. in addition, many soil nutrients under the action of low temperature, change and become more available to microbes.

It was pointed out above that microbial growth in the soil depends on the presence of organic substances of humus. This is not always true. The amount of organic substances in the soil may be very high (peats, marshy soils) while the growth of microorganisms is rare. Not infrequently a reverse picture may be observed. Certain primitive soils of mountainous regions are poor in organic substances and at the same time rich in bacteria.

The concentration of microorganisms in the soil depends mainly on the presence of such organic substances as can be easily utilized by bacteria. There are fresh plant and animal residues and products of their primary decomposition which have not yet been transformed into humus, as well as a number of products of synthesis, etc.

Of great importance for the life of microbes are organic growth factors; vitamins, auxins, various biotic elements and substances which suppress their growth and multiplication.

Small doses of these substances markedly enhance the growth and multiplication of microbial cells as well as that of plants, by promoting various biochemical and physiological processes.

This part of the organic compounds, or soil humus, is, in our opinion, of the greatest importance, and a correlation should be found between their quantity and the total number of the microbial population. Unfortunately, such a correlation is very difficult to study and has not as yet been methodically worked out.

Adsorption should be taken into account in the determination of microbial numbers in the soil. The data of observations and experiments given above showed the degree of bacterial adsorption by soil particles. Bacteria in the adsorbed state can be found in tens, hundreds, millions and billions in one gram of soil. The method employed by us (inoculation on media) in most cases accounts only for microbes in the free state as well as for a fraction of those adsorbed. The majority of adsorbed cells remains unaccounted for; the number differing from case to case.

The majority of investigators give data obtained by analysis of dry soil samples. Naturally, these data are far from the real figures. It is known that the number of microbes is decreasing in dry soil. During prolonged storage a large number of microbial cells die. Sometimes, upon drying, the total number of bacteria decrease by a factor of 2-3 and not infrequently 5-10 times (Table 22).

Table 22
The decrease in bacterial cells during dry storage in the laboratory
(in thousands per gram)

Soils	Fresh samples	Samples after 10 days storage
Chernozem of the Kuibyshev Oblast'	500,000	50,000
Serozem of the Uzbek SSR	150,000	45,000
Podsol of the Moscow Oblast'	3,500	1,500
Chestnut, Trans-Volga region	60,000	10,000
Severnaya Zemlya*	9,300	1,300

* Samples of the soil of Severnaya Zemlya (taken in August) were analyzed the same day and then after one month.

Upon storage of samples in the dry state the qualitative composition of the bacteria also changes. Some bacterial genera disappear almost completely, others remain in small quantities, still others do not decrease in numbers at all.

Actinomycetes and then mycobacteria are the most stable in this respect. The highest percentage of destruction is noted among the bacteria (Table 23).

Table 23
The survival of various groups of microbes upon storage in dry soil
(in thousands/g)

Soils	Sporiferous bacteria	Nonsporiferous bacteria	Mycobacteria	Actinomycetes
Chestnuts of the Trans-Volga region				
Fresh	1,500	56,000	1,000	1,500
Dry	450	5,000	900	1,000
Podsol of Moscow Oblast'				
Fresh	650	5,500	850	1,250
Dry	325	1,400	600	980
Chernozem of Kuibyshev Oblast'				
Fresh	2,500	400,000	25,000	25,000
Dry	280	46,000	16,000	26,000

In those cases where the soil dries up slowly, an increase in number of actinomycetes and certain species of mycobacteria is observed. These organisms can grow in soil of minimal moisture, when the growth of other microorganisms ceases (Krasil'nikov, 1940c).

Differences in survival capacity have been observed not only in different groups but also in different species, and even different strains of the same microbe show different ability to survive. According to our observation, cultures of *Bact. herbicola, Az. vinelandii,* nodule bacteria of soya and *Ps. aurantiaca,* die out rapidly in the dry soils of podsol (Moscow Oblast') and in virgin serozern soils. Of some hundred million cells only a few (10-100 cells/g) remained viable after two weeks storage. Mycobacteria such as *Mycob. rubrum* and some other species remain viable in considerable quantities (100,000 cells/g and more.)

Not all strains of *Azotobacter,* in dry samples of soil, die out at the same rate. Of twenty cultures of *Az. chroococcum* studied, eight strains of *Azotobacter* survived in considerable numbers--up to 10% and more of the cells. Of 10 strains of root nodule bacteria of lucerne only about 10% of 4 strains survived in dry samples of serozem soils; in the podsol soil only one strain survived and then only in a negligible amount (0.5% and less).

The sporiferous bacteria show the same diversity as far as their survival capacity is concerned. About 80-90% of *Bac. megatherium* dies out in dry podsol soils of the Moscow Oblast', and 30-40% of *Bac. subtilis* and *Bac. mesentericus.* Only 5 strains out of 20 of the latter, when isolated from various podsol soils, were resistant to storage in dry soil samples.

No complete drying out of bacteria in dry soils was observed. Even in the cultures most sensitive to drying, there are single cells which are stable and survive for long periods in the dry state. Thanks to such cells the species does not die out under conditions of prolonged drought.

Great variations in the composition of the microflora also take place when the soil samples are kept moist.

It is clear that the soil as a whole, and the separate soil aggregates possess different physicochemical conditions for the life of microbes than those present in isolated soil samples. Some bacterial species grow quicker in natural surroundings, others, slower.

Table 24 shows data from an analysis of samples of a frozen soil, taken from the archipelago of Severnaya Zemlya in May.

Table 24
Variations in the microflora in various samples of soil during moist storage
(in thousands/g; counted on meat-peptone agar)

Soils	Bacteria, fresh	Bacteria, after 10 days	Myco-bacteria fresh	Mycob. after 10 days	Actino-mycetes fresh	Actino-mycetes after 10 days
Sector 1, loam	23	40	0.5	8	0	0.5
Sector II, loam	40	56	0.8	22	0	0
Sector III, loose calcareous soil	91	150	1.5	65	0	1.0
Sector IV, loam	13	62	1.0	37	0	0.8
Sector V, loose calcareous soil	112	346	2.5	120	0	1.5

The total number of bacteria In the samples after 10 days increased 2 to 4 times, and the number of mycobacteria 16 to 50 times. Actinomycetes in fresh samples, were almost nonexistent, and after 10 days their numbers reached 500-1,500 per 1 gram of soil.

Similar changes in the composition of the microflora is noted in other soils during their storage in the moist state. In samples of podsol soil no *Azotobacter* could be detected by us after 2-3 days. The reason for this is the abundant growth of its antagonists--*Bac. subtilis* and *Bac. mesentericum.* In samples of chestnut soil mucolytic bacteria grew abundantly, and fungi of the genus *Fusarium* disappeared almost entirely.

For the determination of the microflora of the soil one has to take into account the composition of the medium into which the microorganisms are being inoculated. Experiments show that organisms from many soils grow better on synthetic media of Chapek, CPI, etc, than on media containing protein. On starvation media (water agar) and semistarvation media (Ashby agar) the number of bacteria is often 2-5 times greater than on rich nutrient media (Table 25).

Table 25
The number of soil bacteria after inoculation of various media
(in thousands/g)

Medium	Garden soil	Primitive soil, mountainous, 3,800 m	Primitive soil, Severnaya Zemlya
Meat-peptone agar (MPA)	3,500	54	23
Synthetic medium of Chapek	3,800	270	154
Synthetic medium CPI	4,400	850	--
Ashby medium	4,200	680	187
Water agar	3,800	800	--

It should be pointed out that it is easier to isolate bacteria from primitive soils ,or mountainous soils (mountain summits, islands of the Arctic Ocean, etc) on starvation, semistarvation or synthetic media which do not contain protein. The bacterial colonies on such media are very small and can often be seen only with the aid of a a magnifying glass, or even only under a microscope. Such microcolonies usually consist of a few cells only.

The microflora detected after inoculation on starvation and semistarvation media differs, from that found in peptone media.

The predominant organisms capable of growing on the synthetic medium of Chapek, CPI, etc are the auxotrophs (prototrophs), which do not require any growth factors or organic nitrogen. They can synthesize all the necessary biotic substances such as vitamins, auxins, etc.

On water agar and on the Ashby medium microorganisms usually grow at the expense of their food reserves. Among these organisms auxoautotrophs and auxoheterotrophe can be detected likewise. Often the so-called oligonitrophils grow on the nitrogen-less medium of Ashby. These are unique forms of bacteria and mycobacteria which are capable of nitrogen fixation in small quantities, satisfying their growth requirements (Mishustina, 1953).

On peptone media, and generally on media rich in organic substances, predominantly auxoheterotrophs (metatrophs) grow. Auxoautotrophs also grow on these media. The quantitative ratio of prototrophs to metatrophs varies from soil to soil. In soils rich in humus, and well-fertilized with organic fertilizers, the amount of the former and the latter is approximately the same.

Organic, protein-containing media, are toxic for many soil microorganisms.

Apart from the afore-mentioned microorganisms, there are great numbers of organisms in the soil possessing specific functions. Such organisms can be detected on special, so-called selective media. To such bacteria belong the nitrifiers, sulfur bacteria, iron bacteria, Azotobacter, cellulose-decomposing bacteria and others. Special media are required for the cultivation of such bacteria.

The principle underlying the use of selective media (Vinogradskii, 1952) is as follows: in the selective medium, favorable conditions exist for the detection of a given function.

It should be noticed, that the selective media are of relative importance. Investigations have shown, that many if not all prototroph bacteria have the capacity of growing on complex nonselective media. For example,Azotobacter can grow on nitrogen-less media due to the capacity of nitrogen fixation, but it can also grow on media containing inorganic and organic nitrogen.

Experiments have shown, that selective media are not strictly selective. No matter what the composition of the selective medium and how carefully it is prepared, other bacterial forms grow on it in addition to the desired bacteria.

On the nitrogen-less medium of Ashby or Beijerinck, apart from *Azotobacter,* many oligonitrophils and metatrophs grow well. On the medium of Vinogradskii, used for the nitrifiers, bacterial satellites also grow well. On media containing cellulose, not only bacteria capable of decomposing cellulose grow, but also other forms of bacteria.

Selective media are not optimal for the growth of bacteria. In. many cases, bacteria grow better upon addition to these media of ready sources of nutrition. Nitrogen compounds may be added for the growth of*Azotobacter,* sugar and other organic compounds for the growth of the cellulose-decomposing organisms, protein and nonprotein substances for others, etc (Rotmistrov, 1950).

According to Kalinenko (1953a, b) iron bacteria and nitrifiers grow well on ordinary organic and even protein-containing media.

It is evident that there are no strictly selective media. Universal media do not exist either.

The data presented in this chapter provide the basis for the assumption that the data on bacterial numbers in the soil are rather lower than in reality. Knowing their numbers, their total mass can be determined, or in other words, the soil productivity.

Cocci are 0.7 μ in diameter, their volume is 0.18 μ^3, and their weight 7×10^{-10} mg. About 5×10^9 cells are present in 1 ml. The size of nonsporiferous bacteria is on the average $3 \mu \times 0.7 \mu$, the cell volume 1.15 μ^3, and the weight 10^{-9} mg. About 900-1,000 million cells may be present in 1 ml. Cells of larger size (5 $\mu \times 1$ μ) have a volume of 3. 9 μ^3, their weight is 10-8 mg. In 1 ml there are about 350 million cells.

According to the data of Tanson (1948), in 1 ml there are 1,000 cocci of 1 μ in diameter; 330 million sporiferous bacteria of the size of 3 $\mu \times 1$ μ, and 1 million spores of fungi, 10 μ in diameter. According to Van Niel (1936),there can be 1,400 million cells of *Bact. coli* per 1 ml; according to Butkevich (1938), 10^9 cells weigh 0.5 mg; Jensen (1940) found that 10^9 cells of *Azotobacter* weigh about 5 mg. Similar data are given by Kendall (1928), Strugger (1948) and some other authors.

We have obtained the following data on the total microflora of the rhizosphere of vegetative plants. There are 2-2.5 kg of cells in a soil under lucerne in Central Asia, per 120 kg of soil; i.e., 6,000-7,000 kg of cells per hectare. Outside the root zone there are, according to our calculations, 1,500-2,000 kg bacterial cells per hectare of the upper (plow) layer. Consequently, there are about 7-9 tons of bacterial mass per hectare (Krasil'nikov, 1944).

In soils of medium fertility the total mass is considerably smaller. For example, in podsol soils under two-year clover and frequently fertilized we have found 1,000-3,000 millions of organisms per gram of soil in the rhizosphere and in the zone outside the roots, 300-800 million organisms per gram of soil. The total bacterial mass in the root zone amounted to 1,200-3,000 kg and outside the root zone about 350-1,000 kg. The total bacterial mass per hectare was 1,500-4,000 kg.

In the same soil under wheat, there were 800-1,200 million organisms per kg in the rhizosphere, and 100-200 million outside the roots. The total mass of bacteria was 1,100 kg per hectare.

In a poor. lightly cultivated soil (podsol) we have found under wheat, only 100-150 kg of bacterial mass per hectare in the upper (plow) layer. Eighty per cent of this mass was found in the rhizosphere.

Strugger (1948). on the basis of his investigations and those of Kendall, calculated that the total bacterial mass comprises 0.03-0.28% of the weight of the soil. Clark (1949) has shown that the bacteria constitutes 300-3,000 parts per million by weight of the soil. These data agree with our own.

Similar numbers are given by Khudyakov (1953c), Mishustin (1954), Berezova (1953) and others.

It should be recalled that our calculation takes into account only bacteria, whereas other organisms living in the soil, such an actinomycetes, fungi, algae, and protozoa are unaccounted for. They comprise a considerable mass of living substance.

The total number of fungi and actinomycetes per gram of soil runs to tens and hundred thousands, and not infrequently millions of organisms per gram of soil. The number of algae reaches thousands and hundred thousands and the diatomaceous algae 100 million per gram of soil (Brendemuhl 1949). The total mass of these organisms cannot be calculated owing to the peculiarities of their structure and growth. Nevertheless. according to the investigators, it is only slightly less than the total bacterial mass.

The total mass of protozoa and insects per hectare is 2-3 tons (Gilyarov, 1949, 1953).

The total mass of the living organisms does not represent merely a static reserve of organic substances, but a living active mass with a large potential, This mass is in constant growth. The individual cells of this mass grow, reproduce, grow old, and die. A constant change and regeneration of the whole living mass takes place.

Under natural soil conditions bacteria give on the average no less than two generations per month during the whole vegetation period, which lasts 7-9 months in the south and 3-5 months in the moderate belt. Consequently, the entire bacterial mass undergoes regeneration 14-18 times during the summer (in the southern belt), and 6-10 times in the moderate belt. The total bacterial production in the upper (plow) layer reaches tens of tons of living mass for one vegetative period.

The intensity of bacterial growth in the soil was determined by the time required for the doubling of their numbers. Three organisms were used in the experiment. *Az. chroococcum, Ps. aurantiaca* and *Bact. prodigiosum.* A sample of a garden soil was placed in an asbestos bag inoculated with the above-listed organisms , carefully mixed and immediately subjected to a microbiological analysis. The soil in the bag was washed with water, until all the desorbed bacteria were removed, and then the bag was buried in the same garden soil from which the samples were taken. After 1-2 days the bags were taken out and the soil was subjected to the same procedure as before. This was repeated for a month. The experiments were carried out in May, July-August and September-October, three series in all. In each series 100 million organiams,were introduced into the soil. In the first analysis (immediately after mixing) 26 million *Azotobacter* cells were washed out. 18 million cells of *Ps. aurantiaca* and 34 million cells of *Bact. prodigiosum* (all these numbers are per gram of soil). The rest of the cells were in the adsorbed state, but they did not lose their capacity to reproduce.

Upon subsequent analyses the following amount of cells was washed out (the May experiment), in millions/ g:

Analysis	Az. chroococcum	Ps. aurantiaca	Bact. prodigiosum
Second	16	22	14
Third	23	26	10
Fourth	20	28	8
Fifth	24	34	10

As can be seen from the above data, the doubling of *Azotobacter* cells took place every 5 days, *Ps. aurantiaca* every 4 days. and *Bact. prodigiosum* every 10 days. In other words the number of generations of the first was 6, of the second, 7 and the third 3. In July-August the number of generations was 4, 4 and 2 respectively, and in September-October, 4, 3, 1 respectively.

The results of these experiments served as a basis for the calculation of the speed of growth of the bacterial maps in the soil.

DISTRIBUTION OF MICROORGANISMS IN SOIL

The problem of microbial distribution in the soil is scarcely dealt with in the literature. We know almost nothing about the localization of microorganism in the soil. It is generally assumed that microbial cells are uniformly distributed in the soil penetrating all pores by diffusion. Therefore, upon quantitative calculation of seperate groups and bacterial species one usually limits himself to one or two soil samples.

Such an assumption does not conform to reality, and the data so obtained lead to erroneous conclusions as to the distribution of the individual microbial species in the soil,

Our studies show that the distribution of microbial cells in soil is not diffuse but focal. In each focus large or small cells of one species or several nonantagonistic species, grow and concentrate. Microbes, especially bacteria and mycobacteria inhabit soils in colonies (Krasil'nikov, 1936).

The focal character of microbial distribution in the soil is being confirmed by the daily practice of microbiological soil studies. It to known that in one and the same field, *Azotobacter*, for example, may be found in one sample and not in another. If one takes 100-200 samples from 1 hectare of a given soil, cells of *Azotobacter* will not be found in all samples taken, depending upon the numbers of *Azotobacter* in this soil. The latter is determined by the state of the soil. In fertile, well-cultivated soils, rich in humus, *Azotobacter* will be found in every sample. In poor, nonfertile or virgin soils *Azotobacter* is rarely encountered in all the samples.

Azotobacter was detected by us In all 200 samples taken from 1 hectare of a cultivated serozem soil under 3-year lucerne in Central Asia. In virgin soils we have found this microbe only in 3 samples (out of 2 00) and in poorly-cultivated soils in 45-85 samples. Similar results were obtained when studying the soils of the Volga area (chestnut soil and others) and the podsol soils. The soils of the Moscow Oblast' which were under forest until 5-10 years previously and now under various plants did not contain *Azotobacter*. We did not detect *Azotobacter* cells in any of the 1,250 samples studied. In garden soil this microbe was found in almost all samples (400 samples studied). In field soils (well -cultivated) *Azotobacter* was found in 60% of samples (860 samples were studied).

For a more accurate determination of the focal distribution of *Azotobacter* in the soil we have carried out the following experiment (in the experimental stations near Moscow). One-hectare plots in three fields containing different numbers of *Azotobacter* were studied. On these hectare plots, 3 one-meter sectors, along the diagonal, were thoroughly analyzed for the presence of *Azotobacter*. To this end these sectors were divided into 100 small squares and 15-20 grams of soil were taken from each square. A total of 300 samples were analyzed from each hectare plot. The results of these analyses are given in Table 26, and the plan of *Azotobacter* distribution in the one-square-meter sectors is shown in Figure 54.

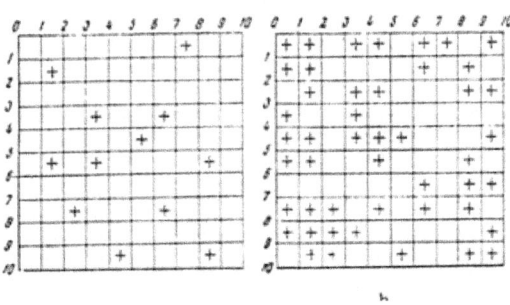

Figure 54. Schematic representation of the character of distribution of *Azotobacter* in soil. The sign "+" shows the presence of *Azotobacter* in 1 square meter soil sectors

a) poorly cultivated soil; b) well-cultivated soil (podsol, Moscow Oblast'). 1-10-numeration of squares.

Table 26
Distribution of *Azotobacter* in the soil
(number of samples containing *Azotobacter* in %)

Sectors in fields	Sector I	Sector II	Sector III	Average

Garden soil	92	90	96	92.7
Field, well-cultivated, under 3-year clover	53	60	34	49
Field, poorly cultivated, under 3-year clover	16	8	12	12

These data show that even in garden soil well-cultivated and systematically fertilized with mineral and organic fertilizers, *Azotobacter* cannot be detected in every sample tested. Of 100 samples taken from the 1-square-meter sectors, it was found in 90-96 samples. In field soils, well-cultivated and fertilized the number of samples containing *Azotobacter* was 34-60 and in poorly-cultivated soils, as well as in soils only recently under cultivation (5-10 years of cultivation, previously under forest) it was found in 8-16 samples out of 100.

More detailed examination of the soil sample reveals small foci in which the *Azotobacter* cells are located. It is known from laboratory practice that when the soil sample is placed in small lumps on Ashby agar or gel plates, not each lump gives colonies of *Azotobacter*. The percentage of growth from each lump varies from 0%-100, depending on the soil (Figure 55). The method of placing soil lumps for the detection of *Azotobacter* and other species of bacteria (nitrifiers, cellulose decomposers, etc) is extensively used in microbiological practice.

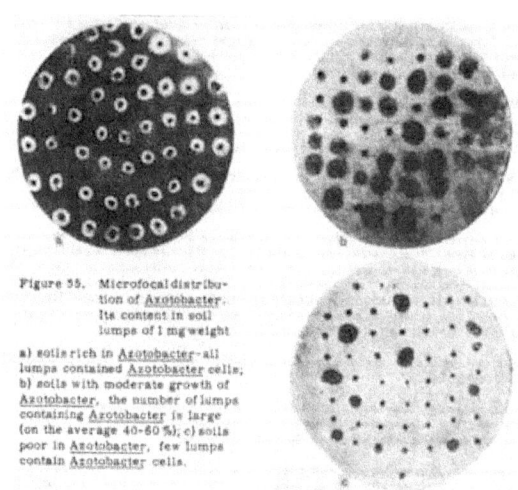

Figure 55. Microfocal distribution of *Azotobacter*. Its content in soil lumps of 1 mg weight: a) soils rich in *Azotobacter*--all lumps contained Azotobacter cells; b) soils with moderate growth of *Azotobacter*, the number of lumps containing *Azotobacter* is large (on the average 40-60%); c) soils poor in *Azotobacter*, few lumps contain *Azotobacter* cells.

We have analyzed samples of the well-cultivated soil of the same field as was the object of our previous experiments. Samples were taken from square-meter sectors, in the form of monoliths.

Each sample after thorough mixing was divided into 5 samples of 0.1 g weight. Each such sample was divided into 100 small lumps, which were placed on the surface of Ashby agar (in Petri dishes). Five samples were taken from each of the sectors studied. The results are given in Table 27.

Table 27
Azotobacter distribution in soil samples
(growth from lumps, %)

No .of sample	Sector I	Sector II	Sector III
1	23	70	28
2	99	39	15

3	97	55	31
4	88	86	8
5	56	42	21
Average	72.6	58.4	20.6

The data presented in Table 27, show that the distribution of *Azotobacter* even in small lumps of soil is focal. The number of microfoci containing *Azotobacter* in such a lump is determined by the total number of *Azotobacter* in such soil and by other factors. In the samples studied by us, the lumps of 0.1 g contained in some cases from 21-99 and in others from 0 to 3-5 microfoci where *Azotobacter* could be detected (Figure 55).

It should be noted that these foci are so small that they are not destroyed during ordinary mechanical crushing of the soil samples. In our experiments, we have carefully mlxed the soil samples and in spite of this, uniform distribution of *Azotobacter* was not achieved. The microfoci were thereby not destroyed or only a small fraction of them was destroyed. The percentage of lumps containing *Azotobacter* was almost identical to that of samples not subjected to mechanical crushing. In crushed samples 50-60 microfoci were found and in the intact noncrushed 30-50. Only when the lumps were ground into dust were the foci destroyed.

The focal distribution and *Azotobacter* cell concentration described is also characteristic of other bacteria. It is well-defined in nitrifiers, cellulose decomposers, root nodule-bacteria, mycobacteria and others. It is less well-defined in fungi, and actinomycetes as a resuit of their biological pecularities. Rippel-Baldes (1952) noted focal development of *Aspergillus niger* in the soil. In square-meter sectors he found this fungus only in 14 squares out of a total of 100.

The focal concentration of microbial cells in soils can directly be seen under the microscope, employing the method of Rossi-Cholodny, To this end cover glasses are immersed in the soil and after some period their surface is overgrown with microbes, bacteria, actinomycetes, fungi, yeasts and others.

We have found colonies consisting of several cells of a size not greater than 10 µ, Clusters of microorganisms can be encountered which occupy an area of a 100 µ cross section. More frequently, colonies of moderate size (20-70 µ in diameter) and consisting of several tens of cells are encountered (Figure 56).

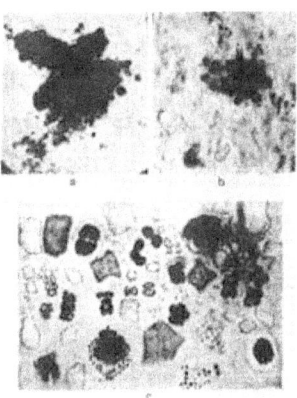

Figure 56. Distribution of microorganisms in soil: a) large colonies of *Azotobacter;* b) small colonies; c) general view on the distribution of microbes in the preparation (according to Vinogradskii, 1952).

In our experiments we have noted the formation of colonies of *Azotobbacter,* sporiferous and nonsporiferous bacteria. Frequently, formation of colonies of mycobacteria, proactinomycetes and actinomycetes could be observed (Figure 57). Actinomycetes thereby form conidia with well-developed spores. Not infrequently actinomycetal hyphae develop into rodlike and coccoid cells, in the same way as can be observed on artificial nutrient media.

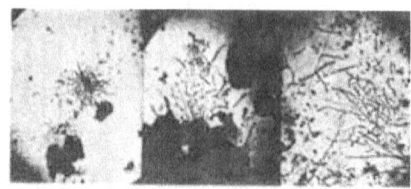

Figure 57. Colonies of actinomycetes in the soil: Branches with sporeforming cells are seen (spiral and straight). Many threads disintegrate into rods and cocci as in proactinomycetes (microphotos from glass imprints according to the method of Cholodny).

The frequency of colony formation on cover glasses depends upon the soil properties. Especially great numbers of colonies are formed in the rhizosphere of plants. Bacteria and mycobacteria grow around thin root tips and around root hairs either in confluent layers as was noted earlier, or in separate foci, in formless clusters or in colonies,

The formation of colonies in the soil has also been described by Cholodny (1934), Kubiena (1932), Rossi (1936), Vinogradskii (1952) and others. The microphotos given by us show clearly enough the concentrations of bacteria, fungi and actinomycetes.

It is self-evident that together with the colonies on the cover glass, individual cells can be seen. They result from the destruction of the integrity of the colonies. The picture of microbial distribution in the soil is clearly seen under the microscope in ultraviolet light, especially after staining with acridine- orange. The individual cells and colonies of bacteria, actinomycetes and fungi standout sharply, glowing with a green color. Only a few cells glow with a red color. We have studied different soils according to this method, and always obtained positive results.

In those cases when pores contain air, colonies of fungi and actinomycetes form fruit-bearing hyphae.

In large pores, under favorable conditions, microbes proliferate and are concentrated in larger numbers than in the small pores. Bacteria, fungi and actinomycetes can penetrate adjacent pores through capillaries.

Some investigators assume that microbes do not penetrate the very small pores of the soil. It was shown above that there exist organisms of ultramicroscopic dimensions lying on the border or beyond the border of visibility in optical microscopes (0.05-0.1 µ). To these belong phages (bacteriophages, actinophages), filterable bacterial forms, then certain cellular elements the so called L-forms, special regenerative bodies, etc of the ordinary species of bacteria and actinomycetes, and finally, small cells of individual organisms.

Investigations showed that not only ultramicroscopic organisms can penetrate through the pores and small capillaries but also many bacteria and actinomycetes of normal size, of a diameter of 0.3-0.7 µ and more.

It is known that bacteria and actinomycetes do not pass through filter candles upon ordinary filtration under pressure. Sterilization by filters is based on this observation., In laboratory practice the most widely used filters are the Berkefield filter (L_3 L_5), Chamberlain filter (N); Seitz filters or membrane filters with pores of a definite size.

If such filters are filled with liquid bacterial suspension and immersed in a nutrient solution (meat-peptone broth), then after an incubation period at 25-37° C, cells will pass through their walls and start growing outside them. In our experiments the following microorganisms passed through such filters: *Bact. coli, Bact. prodigiosum,.Ps. aurantiaca*, larger sporiferous species, *Bac. mycoides, Bac. mesentericus, Bac. megatherium*,actinomycetes. *A.violaceus, A. coelicolor*, and *A. globisporus*.

Besides bacteria, actinomycetes and fungi in the course of growth pass comparatively easily through small-pore clays. We studied the clays of natural deposits underlying the soils of the Moscow district fields. Clay samples were placed in Koch dishes, wetted with water uniformly mixed and distributed in a layer of 1.5-2.0 cm. Bacteria were introduced into the center well. After incubation at 25° or 37° C samples, from different

156

distances from the center well, were taken at various time intervals and analyzed. The experiments were carried out in sterile and nonsterile conditions. In nonsterile conditions easily detectable microbial species were employed: *Bact. prodigiosum , Ps. aurantiaca, Bact. coli, Az. chroococcum, Bac. Mycoides, A. violaceus*. In sterile experiments, besides the afore-mentioned *Bac. mesentericus, Ps. fluorescens* was also employed. The results of these experiments are given in Table 28.

Table 28
Overgrowth of clay by bacteria and actinomycetes
(in cm per 24 hours at 25i C)

Microorganisms	Sterile clay samples from Experimental Station at Chashinkovo	Nonsterile clay samples from Experimental Station at Chashinkovo	Sterile clay samples from Lenin Hills	Nonsterile clay samples from Lenin Hills	Sterile clay samples from Demitrov region	Nonsterile clay samples from Demitrov region
Bact. prodigiosum	1.0	2.5	1.5	2.0	1.0	2.0
Bact. coli	1.0	1.5	1.5	1.0	1.0	1.5
Ps. aurantiaca	1.5	2.0	1.5	2.5	1.5	2.0
Ps. flourescens	2.0	--	1.0	--	1.0	--
Az. chroococcum	1.0	1.0	--	--	1.2	1.0
Bac. subtilis	1.0	--	1.5	--	--	--
Bac. mesentericus	2.5	--	2.5	--	2.0	--
Bac. mycoides	0.5	0.5	0	0	0.5	0.5
A. violaceus	1.8	2.0	--	--	2.5	4.0
A. coelicolor	1.5	2.5	1.5	2.0	--	--

As can be seen from the data, the speed of growth and bacterial mobility is not the same in the different bacteria and actinomycetes and varies in relation to the properties of the clay. The threads of actinomycetal mycelium move with the greatest velocity. Some nonsporiferous bacteria grow rapidly. Cells of *Bact. coli* grow slowly and *Bac. mycoides* is the slowest of them all.

The microbial cells do not move through small pores of the natural and artificial substrates mechanically, or under the action of external pressure, as they do under filtration, but they move as a result of overgrowth. Dead cells do not pass through filters. There is no movement of cells, or only very weakly, when filters with living bacteria are immersed in pure water.

The more intense the growth of microbes, the more rapidly they pass through small pores and capillaries. The optimal temperature for the growth and multiplication of bacteria Is, at the same time, the most favorable for their passage through small pores. At a temperature of 5-7° C *Bact. coli* and *Bact. prodigiosum* multiply slowly and pass through a Chamberlain candle in 60-80 hours. At a temperature of 25° C this period is shortened to 20-30 hours.

The motility of microbes in clay is increased upon introduction of organic substances into the clay as meat-peptone broth, or saccharose. To do this, a well is made in the clay which is prepared in the same way as in the preceding experiment, and, at some distance from it, an elongated groove is dug. Bacteria are placed in the central well (aqueous suspension), and nutrient substances (those listed above) in the groove. It was noted that microbial cultures of *Bac. mycoides, Bact. prodigiosum, Ps. aurantiaca*, and *A. coelicolor* moved in the direction of the nutrient broth quicker than in the control experiments. In one day *Bact. prodigiosum* moved, in the control experiment, 1.5-2.0 cm, in the presence of the meat-peptone broth 3.0-3.5 cm, in the presence of saccharose 2.5-3.0 cm. *Bac. mycoides* in the control moved 0.7 cm and in MPB 1.5 cm; *Ps. aurantiaca* moved 2.0 cm in the control as compared with 3.0-4.0 cm in the experiment. The movement of actinomycetes in the presence of organic substances was by 1.0-1.5 cm greater than in the control. It should also be pointed out that the degree of permeability of small porous substrates also depends on a great number of external factors, such as the pH of the medium, aeration, and the composition of the soil solution. Anything that favors the growth and proliferation of organisms also enhances the overgrowth of the substrate.

Thus, the data obtained established the possibility of penetration of microbial cells through the smallest soil pores, under natural conditions.

Apparently no soil pores exist which cannot be penetrated by microbes.

The morphology of microbes in soil

In the chapters devoted to structure and growth of bacteria, it was shown that microorganisms may exist under laboratory conditions in a polymorphous state. Along with normal or ordinary cells there exist many forms of bacteria and actinomycetes which deviate from the norm in size and in form.

In which form do these microbes exist in the soil? What is their cellular form, and is the polymorphism of cells in the soil as characteristic as in the artificial cultures?

These problems are very little dealt with in the literature. If our knowledge on the quantitative and qualitative composition of soil microbes is slight, then the problem of the forms of the soil microbes is even less known.

It should be recalled that during direct counting of soil smears according to Vinogradskii, or overgrown glass according to Cholodny, not infrequently, the usual forms of bacteria, actinomycetes, fungi, protozoa and others were seen.

Consequently, the cells of microorganisms in the soil are of the same form and size as those grown on artificial media.

But there are instances when the smears enriched in soil microbes, as well as the overgrown glass do not contain the usual microbial cells, or they contain them in small numbers only.

We purposely enriched the soil with microbes before the analyses. To this end in one series of experiments we have introduced into the soil such nutrients as sugars and meat-peptone broth. Subsequent inoculation on agar media revealed hundreds of millions of bacterial cells in the analyzed sample. An enormous number of bacteria was found by using the method of serial dilutions (10-30 millions per gram of soil). With such a large number of bacteria in the smears on Vinogradskii's glass, we should have been able to detect them in the microscopic field (lens 90, ocular 10) in hundreds and thousands, while in reality their number did not exceed 30-50 and more often there were only 10-20 cells*. (* The determination of the number of the bacteria was carried out as follows: a 4 cm^2 square was drawn on a slide, One drop of soil suspension (0.05 ml) was uniformly spread an the square. The smear was air-dried and stained with fluorochrome dyes. It was studied under the microscope with lens 90, oc. 10. The number of fields was calculated by the following simple calculation. The radius of a microscopic field = 85 μ, its area--3.14 x 85 μ = 22,686.5 μ^2. Since there are 100,000,000 μ^2 in 1 cm, then in the given area there are 4,402 fields.)

Two to three days after the introduction of MPB into the soil, a large number of cells in the form of rods (150-200 in a microscopic field) could be seen in smears and in the glasses of overgrowth. These cells are mostly very small and apparently belong to the group of nonsporiferous bacteria of the genus *Bacterium* and *Pseudomonas*. Not infrequently, cells of the mycobacterium type are encountered, larger cells of the sporiferous bacteria are very rarely found. As a rule, the latter are without spores.

Three to five days after the introduction of the afore-mentioned broth, the bacteria seem to vanish and are no longer detectable in the smears or on the glasses of overgrowth. At the same time hundreds of millions of bacteria can be detected upon inoculation on nutrient media. Consequently, no less than 100 cells ought to have been seen in the microscopic field upon the examination of such slides.

In the other series of our experiments pure cultures of *Ps. fluorescens, Bact. prodigiosum and Az. chroococcum,* grown on artificial nutrient media, were introduced into the soil. The amount of the bacteria introduced was 100-1,000 million cells per gram of soil. The analyses were made by the method of direct counting. Already after one or two days no cells of the first two genera could be detected. At the same time millions of them could be detected upon inoculation of nutrient media. *Azotobacter* was detectable for 3-6 days upon inoculation in artificial media; in smears the cells of *Azotobacter* were found in larger numbers, but they were all dead. They did not grow in nutrient agar and did not change their form for long periods of time, as If they were in a state of fixation. Their protoplasm became less dense and the disappearance of the reserve food was sometimes observed.

The number of denitrifying bacteria in the rhizosphere of lucerne under the conditions of the Vakhsh valley reaches 10^9-10^{11} cells per gram, whereas we have found only individual cells in the smears.

We have studied the rhizosphere of pea and corn, grown in special containers filled with soil or quartz sand. When the plants grew in the sand all the cellular elements of bacteria, fungi and actinomycetes growing around the plant roots were clearly seen in the imprints on the glass. Bacterial cells present near the roots and also on the roots and on the root hairs were of the same form and size as those seen after growth in artificial media. Cells of nonsporiferous and sporiferous bacteria, mycobacteria, proactinomycetes and actinomycetes were clearly seen (Figure 58). The imprints. clearly revealed cells of *Bac. megatherium, Azotobacter,* and other bacteria of characteristic cell structure.

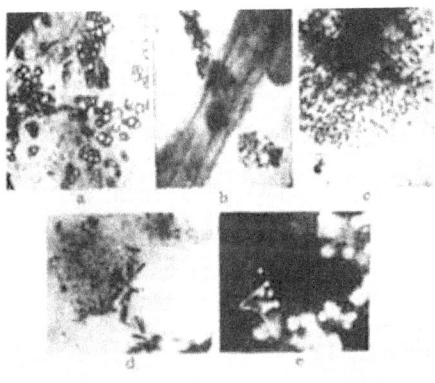

Figure 58. The growth of microorganisms in sand, in the root zone of corn: a) *Azotobacter,* introduced from the outside; the cells are dispersed around a small sector of the root; b) *Azotobacter,* small cells growing in colonies; c) nonsporiferous bacteria; d) sporiferous bacteria; e) colonies of fungi with fruiting branches.

While analyzing the imprints of the same plants, but grown in soil, we almost failed to detect normal bacterial cells. Individual clusters, or small colonies of cells of coccoid form were infrequently observed. Frequently threads of actinomycetes and fungi were seen. Inoculation of this rhizosphere soil by the method of sterile dilution revealed hundreds of millions of cells. Analogous analysis of soil in smears should have revealed not less than 200-300 cells per field; however, almost no cells were seen on using this method.

The disappearance of the bacterial cells that were introduced into the soil was observed by Dianova and Voroshilova (1925). Chudiakov (1926) explained this phenomenon by the adsorption of bacterial cells by soil particles. No doubt, there is adsorption of bacterial cells in the soil but not in such proportions, furthermore this process is of quite a different character.

Novogrudskii et al. (1936) gave much attention to the problem of the state of bacterial cells in the soil. They introduced slides containing bacterial cells into the soil and studied them under the microscope after various periods of time. In this way they found that the bacterial cells undergo deformation, degeneration, autolyzation and disappear.

Vinogradskii (1952) and many other investigators noted the presence of a large number of coccoid cells in the slides of soil smears, taking them for micrococci. According to our observations, the number of cocci in the soil is small and the coccoid cells are coccoid forms of other microbes.

By employing different methods of investigation, including the method of overgrowth, we were able to establish that these cocci are most frequently the cells of mycobacteria, proactinomycetes, and mycococci; nonsporiferous and sporiferous bacteria may also often exist in such a form.

Minute corpuscles in the form of debris can always be detected in smaller or larger numbers during the examination of soil smears. These corpuscles strongly adsorb dyes. We assume that this granular mass consists mainly of soil colloids and partly of decomposition products of microorganisms. It has been proven experimentally that they contain cell gonidia. So, if the smear or the imprint on the glass as well as the glasses of overgrowth, are covered with a thin layer of nutrient agar (or better with Chapek agar or Ashby agar) and placed in a moist chamber at 18-20° C, then, after some time, minute colonies of germinated gonidia can be detected.

The formation of numerous colonies was observed in the absence of well-defined cellular forms in the smears. At the same time the majority of forms, thought by us to be cells, did not grow.

Our investigations as well as the investigations of Novogrudskii, show that the ,cells introduced Into the soil are subject to various deformations. The majority of the cells degenerate. In so doing they swell slightly, their protoplasm becomes granulated, lightens and after the dissolution of the membrane, disappears altogether. Small granules and cell debris remain. Other cells begin to divide but do not elongate and the daughter cells in their turn divide without corresponding growth. As a result, minute cell elements are formed in the form of small granules or corpuscles.

Such cell division was observed in the sporiferous bacteria, *Bac. megatherium, Bac. mesentericus* and *Bac. subtilis,* in mycobacteria, in certain strains of nonsporiferous bacteria of the genus *Pseudomonas* and in*Bacterium,* in root-nodule bacteria *(Rhizob. trifolii)* and in *Azotobacter.* In the latter, the decrease in size of the cell due to cell division without concomitant cell growth is quite often observed (Figure 59). Great changes in the cellular elements are observed in the nonsporiferous bacteria (Figure 60).

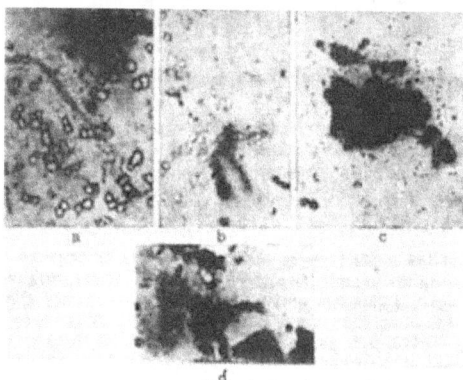

Figure 59. The development and transformation of the bacteria cells in soil (podsolic) in the rootzone. *Azotobacter* introduced from the outside: a) few cells remained unchanged (dead cells); b) most of the cells are of much diminished sizes (dimensions) with weakly refracting plasma; c) on some slides the *Azotobacter* cells are of small sizes with dense plasma; d) a colony of transformed *Azotobacter* cells (marked by x).

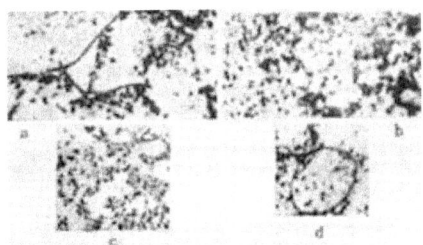

Figure 60. Transfor mation of bacterial cells in the soil into minute forms. The cells transform into the state of small granular elements--conidia--and proliferate in this state: a) *Bac. megatherium;* b) *Bac. mesentericus;* c) *Pseudomonas sp.,* normal cells on a medium containing soil extract; d) cells introduced into the soil, small and deformed; small granular conidia of *Pseudomonas sp.,* are inside the circle; among them are large oval-coccoid cells of other microbes.

These small cells preserve their viability to a certain degree. In favorable conditions, on artificial media, they grow, giving normal offspring.

We observed transformations of bacterial cells during the decomposition of plant residues (roots or parts growing on the surface). At first the bacteria grow together with other microbes in the usual rodlike forms. After a certain period of time, when the vegetative residues have to a certain extent decomposed, the bacteria degenerate and disintegrate with the formation of a mass composed of small granules. This mass contains many gonidial forms and small cellular elements.

In such a manner it can be shown that, under natural conditions, bacteria have different forms of existence. In addition to those forms, which, from our point of view, are the ordinary forms, they also most frequently exist in the form of minute cellular elements, as coccoid cells of decreased size or In the form of small gonidial corpuscles. These forms lead an independent existence for an unlimited period of time. Only under special conditions (excessive nutrition, etc) do they assume a rodlike form and a size characteristic of each species.

Many if not all actinomycetes also exist in the soil in cellular forms differing from those in the artificial nutrient media. In nutrient medium they form a well-developed nonfragmented mycelium, whereas in the soil their mycelium is frequently fragmented. Transverse partitions are formed in its threads, with a resultIng fragmentation into rodlike elements (Figure 54).

On the glass of overgrowth one can observe the successive changes in one and the same colony of actinomycetes and the formation of threadlike, rodlike and coccoid cells. This process is similar to that observed in proactinomycetes. (Krasil'nikov, 1938a). Hyphae or their branches frequently become fragmented with subsequent rounding and the formation of oval or spherical cells.

Actinomycetes in the soil grow either as actinomycetes, proactinomycetes or even as mycobacteria. Their colonies resemble the colonies of proactinomycetes or mycobacteria. However, by the use of the method of covering the glass of overgrowth with a layer of nutrient agar, such colonies grew and gave the characteristic culture of actinomycetes *(Act. globisporus).*

The process of fragmentation of threads into rodlike and coccoid cells can also be induced in the actinomycetes on nutrient media, if they are grown with constant shaking (on shakers) or during submerged growth (in fermenters).

Analogous transformations were not detected by us in fungi. Apparently, they preserve their mycelial structure in the soil. However, the possibility of sharp changes and transformations of individual species of this group of microorganisms cannot be precluded.

When considering the problem of the distribution of microorganisms in soils, the climatic or geographical conditions, in other words the ecological and geographical factors, should be taken into account, It is well known that there is no place on earth where microorganisms could not be found. They can be found in the extreme points of the Arctic and Antarctic. In places where the earth thaws even for short periods, it is populated to a greater or lesser extent with microbes. They exist in dry, hot steppes and deserts, in naked sands and rocky terrain, in valleys and on mountain summits.

Microorganisms also grow on the surface of eternal snow, often covering it with a thick layer of bright colors.

The soil-climatic conditions of existence cannot but reflect on the qualitative composition of the microflora. Microbial species change their natural properties in the process of their adaptation to the external conditions.

The available data on the regularity of distribution and formation of microbial coenoses in the soil are scarce and they deal with only a few genera : *Azotobacter,* root-nodule bacteria, *Bac. mycoides,* and some other well-described bacteria.

A great number of works are devoted to the distribution of *Azotobacter* in the soil. Attempts were made to determine some regularities of the ecological order involving this genus (Sushkina, 1949).

This problem was given attention in the works of Mishustin (1947). From a study on the distribution of the sporiferous bacillus *Bac. mycoides* in soils of the USSR, the author gives numerous data collected by him during many years. According to his data, the microorganisms are distributed in the same way as the higher plants in strict relation with the geographical location. Certain species exist in the North and others in the South. According to the author, microbes change their biological properties according to the geographical conditions.

Mishustin's scheme of the zonal distribution of microbial species in soils, is only a first attempt to establish regularity in the distribution of microorganisms and therefore it is of great interest. It needs, however, essential corrections and a number of statements require verification by factual data.

Before one can speak of regularity of microbial distribution, one must possess accurate data on distribution of individual species, or in other words a chart of microbial distribution. For the compilation of such a chart numerous analyses of soils of every region and every geographical zone are necessary. Such analyses are, not at present available.

This problem becomes complicated by the fact that the analysis can be made only in regard of a few well-known species. The majority of microbes, especially among the bacteria cannot be distinguished from each other. The inability to distinguish accurately, by morphology, one species from another, narrows the scope of microbiological studies of an ecological and geographical character. One has to limit oneself to a few species.

Another difficulty in the compilation of a geographical chart of microorganisms lies in the fact that individual bacterial species are distributed in the soil, not diffusely, but in separate colonies and foci and, in addition, they often grow and can be detected only in certain seasons.

The number and the size of the foci or individual concentrations of microbes also depend on the kind and state of the soil, as well as upon the season and other conditions. Sometimes, analyses have to be carried out in many dozens or even hundreds of samples, in order to be able to detect the presence of this or another microbe, and the degree of its growth. This explains the variability of the data on the distribution of *Azotobacter* or other bacterial species in one and the same soil.

It should be noted that it is much more difficult to establish the habitation areas of microorganisms than that of plants.

It is well known that the distribution of plants is characterized by more or less sharply pronounced localization of species. The study-of plant localization is the basis of plant geography (Alekhin, 1950).

Among higher plants there are species widely distributed over the earth. These are the cosmopolites. Other species are present in restricted areas. These are the "stenochore" species. Among them there are plants which can be found only in a few localities. They are called the endemics.

When considering the microorganisms we cannot say whether there are among them endemics and stenochores--adapted for growth in certain restricted geographical zones. Microbes are known which live in hot springs, the temperature of which exceeds 60-70° C. Microbes maybe encountered in a milieu saturated with H_2S, CH_4, and other substances. These microorganisms are endemics, restricted to the ecological zone. No data are available in microbiology on the existence of a strict differentiation of the microflora of tropic, subtropic and arctic soils. The well-known species of microbes can be detected in all the geographical zones, tropical and polar, For example, the *Azotobacter, Az. chroococcum* thrives in soils of the extreme North (Igarka, Yakutiya, Arkhangel'sk Oblast', Kola Peninsula, Severnaya Zemlya, etc), In the tropics and subtropics (Trans Caucasus, India, Australia, Egypt, etc). *Bac. mycoides, Bac. megatherium* and *Bac. mesentericus* can be encountered in all geographical zones. The root-nodule bacteria of astragals and clover are found in the soils of Central Asia, Caucasus, Crimea, the moderate belt of the RSFSR and also in the soil of the Kola Peninsula, the islands of the Arctic Ocean, Severnaya Zemlya, Franz Joseph Land and others. (Kriss, Korenyako and Migulina, 1941; Kazanskii, 1930). The root-nodule bacteria of southern plants--soya, Onobrychis and lucerne can be found in the soil of the Leningrad Oblast', Kola Peninsula and Moscow Oblast',

For a long time the yeast *Schizosaccharomyces octosporus* was considered in the mycological literature to represent the typical population of the southern microflora. We have, however, found them In many localities around Leningrad, in the sap of the oak.

Of the actinomycetes, such well -known species as *A. globisporus* and *A. streptomycini* are found in the soil of the Kola Peninsula, in the Moscow Oblast' and in the Caucasus, in the Crimea and in Central Asia. The violet actinomycete *A. violaceus* and the blue *A. coelicolor* live in soils of the Vakhsh valley, Crimea, Caucasus, Moscow Oblast', Igarka, Astrakhan' and Leningrad. According to data in the literature, the same species live in soils of Latin America, Australia, Japan, Italy, and other southern countries.

All these data support the hypothesis that the known microbial species are cosmopolites.

Certain representatives of physiological groups of microorganisms are even more widely distributed. Thermophiles and psychrophiles can be detected in considerable numbers everywhere, In the Far North, in the tropics, on mountain summits and in valleys (Egorova, 1938, Mishustin, 1945b, Kosmachev, 1955).

Hydrophiles, xerophiles, nitrifters, denitrifiers, cellulose decomposers, root nodule bacteria and others live in soil. An impression is created that all these organisms are cosmopolites. However, a more detailed study shows this to be wrong. The afore-mentioned organisms do not represent individual species but heterogenous groups. In each group there are species sharply differing from each other. Some of them are widely distributed, others restricted to certain localities.

In medical microbiology pathogenic bacteria are known, whose distribution is restricted to some countries. Among the saprophytes there are forms, some of which are adapted to a warmer climate, others to a lose warm or even to a cold climate.

Is their distribution caused by geographical or ecological factors?

In our opinion, the geographical locality by itself is not a factor. Every area or space populated by microbes is characterized by many external conditions. Temperature, light, humidity, aeration, winds, soil composition, etc. All these conditions are, to a certain degree, determined by the geographical locality; only through these conditions, may the locality affect the biology of the living organisms. Consequently,, the regularity of microbial distribution is determined by ecological factors and not by geographical locality as such.

Different external factors will predominate in each place of bacterial habitation. When the localities are considered from the point of view of latitudes, then the dominant factor will be, in all probability, the temperature. The same factor will also predominate with the altitude.

Assuming the temperature to be of the utmost importance, many authors attempted to establish the regularity of microbial adaptation in nature according to this factor. Mishustin brings forward material on the correlation of microbial distribution to the temperature of the zones in which they are found. According to his data, *Bac. mycoides* of northern soils grows at a lower temperature than the strains of the same species isolated from southern soils. The optimal temperatures for growth become lower with the latitude. For example, the optimal temperature for growth of bacteria in the Crimean soils is 36- 38° C. The maximal temperature at which their growth is I arrested Is 44-45° C. The optimal temperature for the growth of bacteria in the Batumi soils is 38° C and the maximum is 46° C; In the soils of

Kursk Oblast'; Optimal ° C 33-35, Maximum ° C 43-45
Moscow; Optimal ° C 29-31, Maximum ° C 42
Leningrad; Optimal ° C 29-31, Maximum ° C 40
Arkhangel'sk; Optimal ° C 27-30, Maximum ° C 36-39

The temperature as an ecological factor effects the total number of psychrophiles and thermophiles. Investigations show, that thermophiles are encountered in arctic and subarctic soils but in a lesser amount than in the southern soils. For example, in the islands of the Arctic Ocean, Severnaya Zemlya and Franz Joseph Land, thermophiles growing at 55° C contain about 1-2 cells per 100,000 mesophiles; in the soils of Crimea (southern shore under vine) and in the soils of the Caucasus (krasnozems under tea plantations in Anaseuli) the number of thermophiles increases 10-50 times.

It should be noted that the bulk of the microflora of northern and southern soils consists of mesophiles, growing within the temperature range of 3-5 to 35° C. We have carried out special experiments on the comparison of the microflora of soils of the northern, southern and moderate belts. The soils were inoculated into MPA, Chapek and Ashby agar media. The inoculation was at 3-4, 10, 20, 25, 30, 37 and 42° C. The results are given in Table 29.

Table 29
The number of microorganisms, growing at various temperatures
(in thousands/ gram)

The soil taken from	4° C	10° C	20° C	25° C	30° C	37° C	42°C
Severnaya Zemlya	3,100	3,300	29,800	3,000	2,200	60	0.5
Franz Joseph Land	2,900	2,500	30,000	2,400	2,100	30	0.1
Kola Peninsula, cultivated	1,900	1,70	1,300	1,800	1,800	50	0.1
Moscow Oblast', cultivated	2,400	2,800	3,100	2,800	3,400	100	2.0
The southern shore of Crimea, under vines	3,800	4,600	6,700	5,400	5,800	250	10.0
Araseuli, Caucasus, under tea bushes	1,800	1,900	2,500	2,000	2,200	250	15.0

Most bacteria have a different minimal temperature of growth, depending on the place of isolation. Cultures isolated from the soil of Franz Joseph Land have a minimum at 3° C and their growth starts after 10-12 days; those from the soil of Moscow district start to grow after 15-17 days and bacteria from the Crimea and Caucasus start to grow after 18-20 days at the same temperature. The maximum temperature of bacterial growth is even more sharply pronounced. Most bacteria from the soils of Franz Joseph Land and Severnaya Zemlya cease to grow at 30-32° C. At a higher temperature only individual organisms grow, from 100 to 1,000 cells per gram of soil. Among the bacteria from Moscow soils there are about 100,000 cells per gram of soil which can grow at 32° C. There are many forms that can grow at 42° C. In the southern soils of Crimea and Caucasus the number of thermophiles is even higher.

The given data show that in all geographical localities the mesophiles prevail. They are the forms which are of the greatest importance in the processes taking part in the soil.

The behavior of the soil microflora toward higher temperature is even more sharply differentiated. The further south is the soil, the more thermophlles and actinomycetes can be found. Apparently this part of the microflora can serve as an index in geographical distribution and formation of biocoenoses. However, this problem needs to be further studied.

The essential factor in the ecology of microorganisms is the humidity of the climate and the soil. The characteristic and general feature of marshy soils of any geographical zone is their supersaturation with moisture. The latter creates conditions of anaerobiosis, which lead to the formation of the corresponding microbial biocoenoses. In such soils anaerobes are the characteristic and prevailing form.

In dry soils of the southern deserts and steppes the lack of moisture creates special conditions for the growth and formation of xerophytes. These forms are frequently found among mycobacteria and actinomycetes. According to our observations in the Volga steppes, in the Kara Kum desert and in other. dry regions, mycobacteria prevail in periods of low humidity.

Saline soils are inhabited by halophytes, organisms which can tolerate salt, including many groups of bacteria, mycobacteria and actinomycetes. Sushkina (1949) described a halophyte form of *Azotobacter.* The halophils can be encountered among the sporiferous and nonsporiferous bacteria. They can be isolated from normal nonsaline soils, but their number in the latter is smaller than that in saline soils. In normal soils only single cells can be detected, whereas in the saline soils they reach tens and hundreds per gram.

Under other soil and climatic conditions other factors exist which influence the biology and composition of the microbial population.

It is self-evident that in each case not one single factor is operative but an intricate complex of factors, determining the formation of the microbial population. The predominant factor is accompanied by others less important, but characteristic for each locality.

One of the main reasons for our ignorance of the ecological geographical distribution of microorganisms is the difficulty in the determination of the individual species. This is made more difficult by the variation and polymorphism of the bacteria, 'The degree of polymorphism, even of the better known bacteria, is unknown. For example, *Bac. mycoides, Bac. mesentericus* and others may give 7-10 different variations differing from each other to a greater or lesser extent.

As noted above, one and the same culture of *Bac. subtilis, Bac. mesentericus, Bac. cereus* or other microorganisms can be detected in the soil in various forms.

The colonies of these bacteria are either of the typical mesenteroid form or undulate-smoothly or undulate-granularly with edges, even, wavy, or otherwise. Some of them resemble *Bac. cereus,* others *Bac. brevis,* still others *Bac. vulgaris,* etc.

A granular form of *Bac. mycoides* is not infrequently encountered. In many soils (podsols, serozems and others) together with typical mucoid and actinomycetal forms (Crimea soils). In the chestnut soils of the Trans-Volga region only the granular variant in encountered (Krasil'nikov, Rybalkina, Gabrielyan and Kondrat'eva, 1934).

Not knowing the origin of the variants, one may think them to be independent species, as indeed frequently happens.

A reverse phenomenon takes place in some other species. Different variants may have the same external appearance and the differences in their physiological and biochemical properties are not easy to determine. It is very difficult to distinguish such variants. Such forms are encountered in *Az. chroococcum* and in many species of the genera *Bacterium* and *Pseudomonas.*

It was shown in the chapter on variations that the species *Az. chroococcum* is a group of organisms consisting of sufficiently diverse cultures representing separate taxonomic entities, strains, forms, varieties or even species ("sufficient" to justify further taxonomic division).

The genera *Bacterium, Pseudomonas* and others are even more diverse. The inadequacy of the methods of differentiation of these organisms does not enable us to detect this diversity. Bacteria taxonomically belonging to a single species in reality represent whole groups, which in turn should be subdivided into separate taxonomic entities.

Differentiation of such species or groups in actinomycetes can be accomplished. By employing the method based on the specificity of antagonism, we were able to disclose the complexity of some species of monolithic taxonomic entities which were designated as species. For example, the former species *A. globisporus* is now divided into approximately 10 species and *A. coeliccolor* into 7-8 sufficiently easily differentiated species, etc.

All this supports the hypothesis that there is much greater diversity in nature than can be revealed by our methods. This diversity is caused by species variability and adaptation to different environmental conditions.

In what manner are all these forms distributed in the soil? Are they distributed according to geographical zones, or according to ecological conditions? Investigations show that the different forms, variants and even species can be frequently detected in the same soil and in the same sector. For example, root-nodule bacteria, of clover in the podsol soil near Moscow have many diverse forms and variants which differ from one another by their cultural, physiological and biological properties. Among the 150 cultures isolated from the soils of the experimental fields of the Academy of Agriculture im. Timiryazev and from the soil of the Experimental Station of the Moscow State University in Chashnikovo, more than 20 variants were detected.

Such variants were observed among the root-nodule bacteria of lucerne, isolate from the serozem soil of the Vakhsh valley (Tadzhik SSR) Two hundred strains of root-nodule bacteria of lucerne were isolated from the soil of two regions of Azerbaijan and studied. Among them more than 50 variants were found. These variants differed sharply from one another. Strains which differed from each other even more, were detected among the root-nodule bacteria of Onobrychis isolated from the very same sector. Different forms and variants of *Az. chroococcum* can be found in one field. In Moldavian soils 10 forms each differing from the others were found (Babak 1956). No loss a diversity of strains may be seen in the soil of Latvia (Pavlovich, 1053), In the Kola Peninsula (Ezrukh, 1956), in (Soviet) Central Asia In the Crimea and In the Caucasus (author's observations). Petrenko (1953) detected a large number of different forms of *Azotobacter,* in podsolic soils near Moscow.

What is the cause of such diversity? In our opinion it is the action of microecological factors or microzonality in the soil. No soil is uniform in its properties. It consists of microfoci or foci which differ from each other in the content of nutrients, humidity, temperature, composition of the microflora, vegetation, etc. The influence of the microorganisms and especially of antagonists, on the growth and development of now forms and individual species, is very great. Man's activity and the vegetative coverage are of great importance in the formation of microbial biocoenosis.

The cultivation of the soil changes its microbial composition. The draining of marshes favors the growth and concentration of aerophiles and xerophiles. Irrigation of waterless deserts increases the number of hydrophiles. The cultivation of soils of northern regions or mountain summits transforms the microflora of those soils.

It is known that many virgin soils do not contain *Azotobacter;* but it is sufficient to plow and cultivate them in order to enable this microorganism to grow. In chestnut soils of the Trans-Volga region *Azotobacter* cannot be detected, but it appears there immediately after the soils have been irrigated. Fertilization of soils, especially with organic fertilizers, sharply increases the growth of this organism in the podsol soils.

As already pointed out, the better the soil is cultivated, the more microorganisms it contains and the higher its fertility. With cultivation, the number of microorganisms increases. The increase in number involves such genera as *Pseudomonas , Bacterium, Azotobacter,* sporiferous bacteria, actinomycetes, fungi and others. The thermophiles, mesophiles, aerobes, anaerobes, antagonists, activators, and many others also increase in number. It is difficult to say which of those microorganisms is the best indicator of a cultivated soil.

Moshustin (1945a, 1947) suggested that the thermophilic bacteria be used as an indicator of the degree to which the soil in being cultivated. The author* thinks that these organisms are introduced into the soil together with organic fertilizers. * [This is ambiguous in the Russian text but probably refers to Mishustin.]

According to our investigations the thermophilic organisms, bacteria as well as actinomycetes, are natural inhabitants of the soil. Their numbers increase with the degree of cultivation of the soil. The other organisms proliferate at the same time. In some instances the thermophiles increase under more intense cultivation of the soil and sometimes other groups of bacteria and actinomycetes predominate.

Thermophiles can serve as an indicator of soil cultivation to the same extent as many other organisms or groups.

To our mind *Azotobacter* is the best indicator of the intensity of soil cultivation.

Thus, in order to draw up a chart of the distribution of microorganisms in soils, detailed information on the growth of individual species in each region, or even in each field during different seasons of the year is needed. Such data are not available at present and those which are, cover restricted areas only. An immense and painstaking work is called for, a work which can only be performed by a large team of microbiologists, working in different institutions but according to one plan and using the same methods.

Part III – BIOLOGICAL FACTORS OF SOIL FERTILITY

The diversity and abundance of the living population of the soil was discussed in the preceding chapter. The living mass of microbes: bacteria, fungi, actinomycetes, and algae, by itself comprises more than 10 tons in the plow layer of 1 hectare of a well-cultivated, fertile soil. Besides, a great number of protozoa, insects, worms, and other representatives of the living world exist in the soil.

All this soil population performs an immense work, processing considerable amounts of different mineral and organic substances, decomposing plant and animal residues, participating in the transformation of their decomposition products.

Microbes synthesize and excrete different metabolic products, which, upon entering the soil, confer upon it fertility.

Thanks to the microbial population the dead, loose, geological formation assumes life and becomes a productive body. Conditions necessary for the nutrition and growth of higher plants are thereby created. Microorganisms constitute the most important and indispensable link in the nutrition of plants.

Microorganisms of the soil, not only create the conditions necessary to the growth of higher plants, but also have a direct effect on them through their metabolic products. This chapter deals with the effect of microorganisms as agents of mineralization of organic compounds of plant and animal residues and as a biological factor necessary for the normal nutrition and growth of plants.

PLANT NUTRITION

To obtain high yield, fertilizers have been used in agriculture from ancient times. Men used fertilizers long before science established and solved the main problems of soil science, agriculture, agrochemistry, and plant physiology. The rules of fertilization were worked out empirically but, as Pryanishnikov pointed out, many of these rules attained high accuracy.

The Romans, for example, knew of the valuable fertilizer properties of animal excrements and of some mineral substances such as ash, gypsum, calcium and marl. Moreover, they knew that the fertilizing value of excrements of different animals varied. Of highest esteem was the excrement of birds.

The Romans also knew about green fertilizers. Thus, they plowed in green manure on the slopes of Vesuvius in order to increase the fertility of the soil.

The theory of plant nutrition had not yet been elaborated. Vague assumptions on the presence of some "fat" of the earth (terrae adeps) were made, This "fat" was responsible for the fertility of the soil. In order to -increase the amount of this "fat", animal excrements were necessary.

Those assumptions contained the nucleus of the humus theory, which subsequently became widespread. This theory assumes that the organic compounds are of utmost importance in the nutrition of plants. "Echoes of these ideas can be heard in the language of different nations. Let us recall that our word 'TUK' (fertilizer) once had two meanings, one of them being 'fat'." (Pryanishnikov, D. N., Plant Nutrition, Coll. Papers, Vol 1, 1952, p 58).

The humus theory of plant nutrition stated that plants feed on organic substances present in the excrements of animals and in general in humus. This theory, widespread in the 18th century, was favored by the plant breeders and agrotechnicians of that time, since it was well confirmed by agricultural practice.

The explanations of the favorable effect which organic substances had on the growth and fertility of plants differed. The most widespread conception was the assumption that the plant derives (from the organic substances) carbon which it subsequently incorporates into its body (Davy, 1813).

Other authors thought that the excrements contained certain special substances. Thus, for example, Prof. Vallerius in 1766 assumed that only the organic substances or "fat" substances of humus of soil play any role in plant nutrition, other components of the soil serve merely as "fat" solvents.

Some scientists of the 16th--18th centuries and in particular Olivier de Serres, (1600), assumed that the cause of the action of fertilizer is in its heat.

Bernard Palissy in his paper (1563) put forward the idea that the salt content of the fertilizer plays a role in the growth of plants.

First experimental attempts to elucidate the problem of plant nutrition were made by Van-Helmont. In his paper (1629) he presented results of five years of experiments on growing ivy branches in soil which was given only rain water. In 5 years the branch grew and attained a final weight which was 33 times higher than the initial weight. The soil thereby did not lose weight. Since the composition of the air was not known in those times, Helmont concluded that the plant utilized only water and this was sufficient for the plant to build its body.

Analogous conclusions were reached by the French investigator Duhamel. He grew plants in Seine water. He did not have distilled water at his disposal for control experiments, and positive results obtained by him are therefore not trustworthy, since the water from the Seine contained various organic and inorganic substances.

Woodword showed that the point of view of Van-Helmont was erroneous. In his experiments he demonstrated that mint grows better in river water than in rain water and it grows even better and gives greater increase if soil is added to the water. He presents the following data: the plant-weight gain in grains* during 77 days, was as follows: in rain water, 17; in the sewage water of Hyde Park, 139; the same with addition of soil, 284. [*"Grains" is a unit of weight.]

Woodword concluded that not only water but also something else which is present in the soil is necessary for the growth of plants.

The agricultural practice of those times did not possess any scientific basis for the elucidation of the observed link between the growth of plants, its yield and the soil. Foundations of sound scientific knowledge were layed down when first M. V. Lomonosov and then Lavoisier discovered the law of the conservation of matter.

Lavoisier discovered the composition of air and the essence of the processes of oxidation, burning and respiration. Not long before his death (1794) he wrote, concerning the nutrition and growth of plants, that plants get materials necessary for their organization from the surrounding air, from water, and generally from the mineral kingdom (according to Pryanishnikov, 1952).

New methods of chemical investigations elaborated by Lavoisier were employed by Priestley, Senebe and Sossur, in order to study metabolism in plants. These scientists showed that plants obtain carbon from air, and that apart from carbon plants require salts.

While noting the importance of atmospheric carbon dioxide as a source of carbon, Sossur also thought that the humus of the soil is of great importance. Humus contains a certain substance indispensable for plants; the daily experience of farmers pointed to a close link between the fertility of the soil and the presence of organic substances in it.

The English scientist Davy (1813) drew attention to various sugary, glueing, oily, slimy, extractable substances, and to the carbon dioxide of the soil. These elements, according to him, contain all the elements needed for the life of plants.

The humus theory became very popular after the well-known works of Teer (1752-1828), the founder of the first agricultural school and the propagator of crop rotation instead of the three-field system. He wrote that fertility of the soil depends entirely on humus. since apart from water, humus is the only substance of the soil which can serve as a nutrient for the plants (according to Pryanishnikov,1952).

These ideas were the basis of a theory of the exhaustion and regeneration of the soil's fertility. It was assumed that the more the plants absorb humic substances, the better they grow, and the greater the crop. Different plants absorb different amounts of organic substances. For example, wheat requires more humus than rye does.

Teer and his adherents considered humus to be an important product of plants and a substance of the utmost importance for the life of plants. Humus is formed at the expense of plants; some plants excrete more humus than they absorb. In other words, some plants enrich the soil in humus and others exhaust its reserves. To the former category belong clover and other leguminous plants.

The propounders of the humus theory did not take into account mineral substances. The latter, according to them, enhance the decomposition of humus and transform it into a form which can be assimilated by plants.

The humus theory was very popular even in the thirties of the last century in different countries such as Germany, England, France and Russia.

Of the Russian investigators of the late 18th and the beginning of 19th centuries, who drew attention to the importance of organic fertilizers, A. Bolotov, I. Komov and A. Poshman should be mentioned. The works of these scientists showed that fertilizers have an immense influence on the crops, Komov assumed that organic substances are necessary for plants to the same extent as they are necessary for animals, and he recommended to use them in practice. According to him, the organic fertilizers cannot be replaced by salts.

At the beginning of the 19th century the role of nitrogenous compounds in the nutrition of plants began to be better understood.

Liebig assumed that plants obtain nitrogen at the expense of atmospheric ammonia and that its presence in the atmosphere is sufficient for them. However, experiments disproved this. It was found, on the contrary, that this substance is most insufficient for plants.

The well-known studies of Bussengo revealed the sources of the nitrogenous nutrition of plants, In 1837-38 he developed his theory on nitrogenous fertilizers and recommended the use of fertilizers rich in nitrogen. He connected soil exhaustion with the depletion of sources of nitrogenous nutrition. In this process, he ascribed varying importance to different plants. Some plants absorb more nitrogen from the soil, others less. He ascribed an active role in the enhancement of soil fertility to certain plants, e. g., clover. "One should think", said he "that cultures which improve the soil do not limit themselves to its enrichment with C, H, and O_2 only, but also enrich it with nitrogen" (according to Pryanishnikov, 1945).

Hellriegel (1889), discovered the reason for the peculiar action of leguminous plants on the fertility of the soil. By his experiments he found that leguminous plants assimilate nitrogen from the air. Voronin (1886) studied the root nodules of leguminous plants and found microorganisms in them, which in his words are: "The culprits of the formation of the nodules." Later, Hellriegel by thorough experimentation showed that these symbiotic organisms are the cause of the nitrogen fixation by leguminous plants.

Thus, the conclusions of the old investigators, for example Teer, have been confirmed by the later experiments of Bussengo, Hellriegel, and others, who found that plants not only exhaust but also enrich the soil with nutrient organic substances.

Thus, the humus theory of plant nutrition, of the 18th and first half of the 19th centuries, was very popular. Fertilization with organic substances was considered to be an essential measure, not only for the increase of yields, but on the whole for the increase of soil fertility.

The attitude toward the theory of humus or organic nutrition of plants, and in general toward fertilizers changed abruptly after Liebig put forward his theory of the mineral nutrition of plants. He severely criticized the humus theory of nutrition and considered it basically wrong. He considered all the studies, performed before him by physiologists and agronomists, to be inconsistent and meaningless for the solution of the problem of plant nutrition. He criticized the humus theory of nutrition: "Taking it", in Pryanishnikov's words,"in its extreme expression, and bringing it ad absurdum."

Liebig completely rejected even the possibility of assimilation of the organic substances by plants. In his opinion only inorganic compounds can serve as sources of nutrition for plants. He considered humus as a source of CO_2 which enhances the process of the erosion of silicates and prepares the mineral food for the plant.

Liebig did not recognize the role of plants in the enrichment of soil. He severely negated and criticized the notion of "enrichment of soil in sources of nutrition." Plants, according to him, deplete soil, carrying off elements of mineral nutrition with the crops. But the depletion of the soil is carried out by various plants at different rates and in different directions. Some of them take out from the soil mainly calcium (for instance, peas), others--potassium and silicon (wheat grains). Therefore, the crop rotation, recommended by previous investigators, only slows down the depletion of the soil.

In accordance with his theory, Liebig recommended the introduction into the soil of mineral fertilizers. The amount of these fertilizers applied should take into account the utilization by plants.

Owing to the authority of its author--Liebig, the theory of mineral nutrition of plants was accepted by his contemporaries with hardly any criticism at all. The authority of Liebig's chemical school supressed all the previous ideas and theories of organic nutrition of plants. Liebig, says Ressel (1933), gave the final blow to the theory of humus. Only the most daring would still venture to maintain that plants take the necessary carbon not from CO_2 but from another source. Although. one should admit that we have no proof that plants obtain all the necessary CO_2 in this way.

Liebig's theory was developed and subjected to changes corresponding to the newly acquired factual data. In our country the theory of Liebig was subjected to a thorough revision by the Academician Pryanishnikov and his followers. Pryanishnikov originated his own direction in agrochemistry; he and his students elaborated a series of valuable postulates laying a basis for practical measures in agriculture. To him goes the honour of solving the problem of basic improvements of the nitrogen balance in the agriculture of the USSR. In counterbalance to the theory of Liebig, he ascribed a major importance to the biological processes of the soil and especially of nitrogen accumulation. Pryanishnikov did not deny the possibility of the assimilation of organic substances by plants and he himself showed it experimentally.

Notwithstanding this, the theory of Liebig is still reflected in the studies of many specialists. In the theory of plant nutrition one observes an obvious underestimation of the role of the organic substances of the soil, and its importance of the nutrition of plants is often denied altogether. As a rule, the significance of the organic substances of the soil is basically formulated in two postulates:

1. Humus substances are reserves of plant-nutrient elements which become available only after they are mineralized.

2. Humus improves the physicochemical properties of the soil, increases its absorptive capacity, and therefore, promotes also the accumulation of nutrient substances--it strengthens the structure of soil particles and with it improves many other soil properties. Organic fertilizers--manure, compost, etc, prepare the soil for the acceptance of mineral fertilizers, increase its buffer capacity, etc.

All this is quite true and is confirmed by age-old practice and by many experiments. However, to reduce the importance of these fertilizers only to the given postulates will, in our opinion, be inadequate.

The fact of positive action of humus and organic fertilizers on the growth of plants cannot be explained by the action of the mineral elements of nutrition present in them. Russel (1933), summarizing the experimental data of 60 years work of the Rothamstead Station, said that, although plants can grow satisfactorily and reach full development on inorganic nutritious substances only, under natural conditions, however, their nutrition takes place in the presence of organic substances. The question whether these substances play any active role in this process has been very much discussed. The experimental data are not very conclusive. In the Rothamstead field experiments none of the combinations of the artificial fertilizers is as effective as manure in maintaining crops from year to year.

The extensive field experiments of the German investigators Gerlach, Hansen, Schultz, Wagner, Schneidewind, and others, lead us to the conclusion that whatever the amount of the mineral fertilizer may be, if using mineral fertilizers only, one cannot reach as high yields as by the simultaneous introduction of manure (Schneidewind, 1933). Summarizing experiments of many years in the Lauchstadt Station, Schneidewind gives us as an example the following data from sugar-beet yields in centners per hectare: with only mineral fertilization--roots, 487.6; sugar content, 77.66; tops, 291.7; under mixed fertilization mineral and organic (manure)--roots, 533.6; sugar content, 88.11 and tops, 366.6.

Similar data are given by Schneidewind for potato crops. By using artificial fertilizers only he could not get more than 240 centners of potatoes per hectare, while by the introduction of mineral fertilizers and manure the yield rose to 306-312 centners per hectare. This effect of manure repeated itself year after year and was independent of the amount of precipitation whether it was a dry, wet, or average year.

Russel (1933) and his co-workers have shown that the organic substances of the soil stimulate the growth of plants and increase crops. He concludes that no mixture of artificial fertilizers can be as effective as manure in maintaining steady high crops year after year.

Academician V. I. Palladin (1924), concerning the problem of plant nutrition, wrote that green plants can nourish themselves on ready-made organic compounds. Such nutrition may go on parallel to the assimilation of carbon from the atmosphere. According to the author's observations, green isolated leaves of plants grown with artificial, sugar-containing nourishment always accumulate more organic substances in their tissues and have a higher turgor than plants nourished with mineral elements. The carbohydrate increment on sugar nutrition reaches 5 grams per $1 M^2$ of leaf area.

Lebedyantsev (1936) in an annotation to his translation of Liebig's book (Chemistry as applied to Agriculture and Physiology) mentioning the work of Sossur on nutrition of plants, has written: "Sossur considered the CO_2 of the air to be the main source of carbon nutrition, not denying, however, the possibility of utilizing carbon of organic compounds of the soil, since he was not in possession of facts enabling him to deny this. We shall note that today we do not possess such facts and the question of the assimilation of organic compounds from the soil still remains to a large extent an open one, although, undoubtedly, the main source of carbon for green plants is after all CO_2." (page 396)

As can be seen from the above, the question of plant nutrition, notwithstanding the numerous studies which have been carried out, remains unresolved in may respects.

It is hard to agree with the concept according to which during all the history of their evolution, plants, although having been in contact with organic substances of the soil, did not acquire the ability to assimilate them in one form or another.

There is no basis for denying the well-known facts and experimental data showing that plants utilize mineral compounds for their nutrition. Numerous observations speak in favor of this method of plant nutrition being the most important one under natural conditions. However, is such a nutrition sufficient to obtain high yields and fully viable seeds year after year? This question seems to us not to have been sufficiently solved.

Not so long ago, the assimilation by roots of CO_2 from the medium was said not to take place, but now this may be considered an established fact. Plants absorb CO_2 not only from the atmosphere but also from the soil (Samokhvalov, 1952; Kursanov, 1953, 1954).

It should be assumed that plants that experience lack of CO_2 in the atmosphere gladly assimilate it from the soil. Under various unfavorable conditions the photosynthetic activity of plants may decrease considerably. Thus, for example, during drought the stomata are closed, and the influx of CO_2 stops or weakens. Respiration of plants, however, under these conditions does not cease and may even increase. Starving ensues, to a larger or a smaller degree. Weakening of photosynthesis may also be caused by other factors. In all such cases, evidently, plants can switch to a heterotrophic nutrition, assimilating organic compounds from the soil supplementing their nutrition.

The ability to assimilate ready-formed organic compounds and use them as nourishment is observed in many representatives of the plant kingdom; from the lower forms such as blue-green and green algae, to the higher flowering plants.

It is known that many (if not all) algae, may under certain conditions, grow and develop on synthetic media with mineral sources of nutrition, as well as on organic media containing different nitrogenous and carbonaceous complex compounds. It was shown by special experiments that these organisms, and especially the green unicellular organisms, which usually grow on pure mineral nutrious media, grow much better upon the addition of organic substances to the solution. They can assimilate carbonaceous and nitrogenous substances (Gollerbach and Polyanskii, 1951).

Blue-green algae of the Cyanophyceae--*Oscillaria, Nostoc Anabaena, Lyngbya,* and others, also green algae of the groups *Chroococcum, Spirogyra,* etc, cannot be cultivated indefinitely on artificial organic media without a noticeable weakening of their viability. In our laboratory we have been maintaining a pure culture of the green alga *Chlorella* for already more than 25 years. We do not observe any essential lowering of life functions. This alga assimilates nitrogen from peptone and amino acids and grows quite well on must agar and even on meat-peptone agar (MPA) and gelatin.

According to the observations of many investigators, green algae *Chroococcum* and *Spirogyra,* blue-green algae *Phormidium, Nostoc, Anabaena, Gloeocapsa,* and others, readily assimilated sugars, amino acids, alcohols, organic acids, urea, casein, and other organic substances. Studies with pure cultures of algae performed first by Pringsheim (1913) and later by many others (Harder and Humfeld, 1917; Cataldi, 1941; Gerloff, 1950) have shown that these organisms grow on organic media in the dark as well as in the light. *Chlorella* assimilates sodium and potassium oxalates in the light, and malic, tartaric and some other acids in the dark.

Allen (1952) studied 26 cultures of blue-green and some green algae belonging to 11 genera--*Anabaena, Nostoc, Oscillatoria, Lyngbya. Phormidium. Gloeocapsa. Aphanocapsa, Plectonema Cladothrix, Chroococcum.* and *Synechococcus.* These cultures were grown in mineral media with different sources of organic nutrients, Yeast extract, various sugars (glucose, saccharose, lactose, maltose, etc) mannitol, glycerol, ethanol, organic acids, amino acids, casein, urea, casein hydrolysate, etc were added to the media. In the presence of these substances many algae grew better than in media with mineral sources of nutrition only. Assimilation of organic substances takes place in many species at the same intensity in the dark as in light. Upon prolonged cultivation on organic media some algae lose their pigment and continue to live as typical saprophytes (Fogg and Wolff, 1955).

Authors who studied algae concluded that they can live and grow as heterotrophs, utilizing organic compounds, such as nitrogenous and carbonaceous substances, In the same way as the ordinary saphrophytes. This is understandable because the two kinds of organisms live in a milieu rich in organic substances--in soils and water reservoirs, where animal and plant residues serve as sources of organic nutrients after their decomposition.

Green moss of the genera *Splachnum* and *Getrapladon,* settle and grow on excrements of animals, utilizing organic substances for their nutrition.

Among higher flowering plants a group is known as the so-called "humus plants"--various representatives of which can be observed to be in different stages of degradation of the chlorophyll apparatus. The humus plants are so called because they grow on substrates rich in humus and decomposing organic animal and plant residues.

The character of nutrition and the conditions of the existence of these organisms have left a certain imprint on their structure and appearance. Some of them lose the green coloration, the leaves are reduced (but only in the lower part of the stem) and lose the capacity to assimilate carbon dioxide from the atmosphere. To such plants belong the species of the following genera: *Monotropa Dum., Zymodorum, Neottia* Sw., *Corallorhiza* Rale,*Epipogon* S. G. Gmel., *Voyria* Aubl, *Fletia, Pogonia* Less, *Voyriella* Miq., and others (see below).

In other representatives of humus plants the green leaves are preserved and they remain autotrophs by assimilating carbon dioxide of the atmosphere but, nevertheless, they can utilize ready organic substances. To such plants species belong the species: *Dentaria* L., *Pyrola* L., *Goodyera* R. Br., *Cephalanthera* Rich., *Epipactis* Adans,*Platanthera* Rich., and others. All these plants, to a greater or lesser degree, live and nourish like heterotrophs, assimilating organic compounds together with mineral sources of nutrition.

The group of insectivorous plants is well known. In our latitudes such flowering plants are encountered an *Drosera rotundifolia* L., and *Utricularia vulgaris* L. To this group belong such southern plants as *Cephalotus follicularis* Labill, many species of *Nepenthe* L., *Dionaea muscipula* Ellis, *Drosophillum* Link, *Aldrovanda* Monti, *Sarracenia* L., and others.

About 500 species of such insectivorous plants are described in the literature. These plants have a complex apparatus for catching insects and special glands for their digestion.

Small animals falling into the trap of these plants are decomposed by enzymes to the soluble forms of organic compounds containing nitrogen and nonnitrogenous compounds which are then absorbed by the plants and utilized by them for nutrition.

It is characteristic that those insectivorous clearly heterotrophic plants do not lose their capacity to assimilate atmospheric CO_2. Many species which obtain ready-made organic nourishment from the living body of other plant species also belong to the heterotrophic plants. These are the parasitic plants. They are divided into facultative and obligatory parasites.

Among the obligatory parasite plants, the climbing plants of the genus *Cassytha* L. (fam. Lauraceae L.) and *Cuscuta* L. (fam. Convolvulaceae L.) are the well-known species.

The biology of *Cuscuta* L. is known in great detail (up to 50 species are recognized). They are parasites of the cereals, bushes and trees. They penetrate the tissues of the host with their haustoria and such nutrient substances out of the host. Leaves of this plant are weakly developed and can hardly be seen, the stems lack chlorophyll.

Orobanche L. are also etiolated plants with weakly developed leaves (in the form of scales) lacking chlorophyll. They are parasitic on the roots of sunflower, flax, cabbage, and others. About 180 parasitic plants of this genus are known. The shoots of *Orobanche* are so coalesced with the roots of the plant host that it becomes impossible to distinguish between the cells of the host and the parasite.

Lathraea squaniaria also belongs to the group of parasitic plants. In contrast to the aforesaid this plant does not climb; it consists of a thick, colorless, watery stem. The leaves of this plant are weakly developed and are in the form of colorless scales. The root system of this plant is close to the root of the host, from which the parasitic plant sucks the sap of the host with the aid of haustoria. The plant at first grows under the earth's surface, then its shoots pierce through the surface. The aerial parts are of a violet-red color.

The Balanophoraceae are tropical plants--parasites living on roots of various trees. Forty species of these parasites are known (14 genera). These plants, like the *Orobanche* grow in such intimacy with the roots of the host that no visible border between them is discernible.

Parasitic plants related to Balanophoraceae are the Rafflesiaceae Dum., which parasitize trees of tropic and subtropic regions. These plants penetrate the bark of the tree host, embrace its stem and roots and suck the, sap of the latter.

Parasitic plants from the family Loranthaceae D. Dow include more than 300 species. They live exclusively on stems and branches of different trees. A typical representative of this group of plants is our ordinary *Viscum album* L., which lives on deciduous and coniferous trees. This plant penetrates the branches of the host with its roots and haustoria. The stem of *Viscum album* which has the form of a dichotomically branched bush carries elongated leathery leaves of a yellowish-green color.

The group of plants of the mistletoe family is characterized by the fact that its representatives--arboreal plants-- did not completely lose the capacity of photosynthesis and preserved the normal appearance in their stems. They are the intermediate group placed between the nonchlorophyll, flowering, obligatory parasites, which nourish on ready-made organic food, and the facultative parasites which preserve their normal structure in all their parts, as well as the capacity of assimilating atmospheric CO_2.

To the facultative parasites belong about 100 species of the family Santalaceae R. Br., and more than 500 species of the family Scrophulariaceae R. Br. In our flora, species of the genus *Thesium* L. (family Santalaceae) and species of the
genus *Euphrasia* L., *Alectorolophus* Boehm., *Melampyrum* L., *Pedicularis* Boehm., *Odontites* Pers and others (family Scrophulariaceae) are encountered.

They are all grasses growing in meadows or in forests. They possess green leaves and a well-defined photosynthetic capacity. In the initial stage they develop as typical autotrophs without the slightest inclination to become parasites. However, when their roots reach a definite size (1- 2 cm) haustoria appear, with the aid of which they become attached to the root of another plant host, encountered during their growth. From this moment on the semiparasite plants begin to supplement their nourishment at the expense of the plant host. They can, however, grow without the host in the event of their not having encountered it in the course of growth. They can also parasitize each other.

There are in nature a great many epiphytic plants developing and growing on plants as on substrate. It is assumed that they nourish independently, at the expense of mineral substances present on the plant barks. However, the possibility of their using organic substances such as dead tissues, or their decomposition products, is not excluded. This problem is still inadequately studied.

Many plants assimilate organic substances, metabolic products of microbes which live on and inside the roots. The best known plants are those in the tissues of which there are bacterial microbe-symbionts, actinomycetes, and fungi.

The best known symbiosis is that of plants with mycorhiza fungi. According to the data in the literature, there are mycorhiza fungi on the roots of almost all plants which are to a greater or lesser degree symbionts. The symbiosis is such, that some plant species cannot grow without the fungi. Such plants are apparently obligatory parasites, heterotrophs whose nutrition depends on the metabolic products of the fungus. Some orchids serve as an example of such parasitism. They require fungi for their normal growth and development. They grow badly without them, and certain species such as *Neottia nidusavis* Rich., encountered in our oak and pine forests cannot grow at all without the symbiotic fungi. This orchid grows and develops under the soil surface and appears above the surface only for flowering and fruit bearing. It has no green color and its pale, thick stems carry weakly developed, slightly yellowish sprouts with scaly leaves. *Neottia nidusavis* Rich. Has completely lost the capacity of independent nourishment on mineral compounds. It is almost completely devoid of chlorophyll. Its nutrition is at the expense of decomposition products of fungi which grow inside its stem and roots.

There is a theory that orchids of this species can assimilate organic compounds directly from the soil, and that the fungi serve merely for the processing of these compounds into the form in which they are more easily assimilated. The cytological picture of the growth of fungi inside the orchid tissues shows that the mycelium grows to a certain size followed by coiling and lysis. The products of lysis are assimilated by the plant.

Other species of orchids which have lost their green color and are entirely dependent on fungi can be encountered in our forests. To such orchids belong: *Epipogon aphyllus Sw.* and *Corallorhiza trifida* Chat., which grow in forests of the moderate belt. Such orchid parasites are also encountered in tropical forests (Burgeff, 1932). In the literature, orchids are described in which parasitism becomes manifest only at a certain growth phase. The species of the genus *Myrmechis (Myrm. glabra* Bl., *Myrm. gracilis* Bl.) when young grow in the form of a root obtaining their nourishment entirely at the expense of fungi; afterward they form aerial organs, flowering sprouts with green leaves and start to nourish themselves. When the sprouts fade away they become parasites again.

According to the data of Bernard, the seeds of orchids cannot germinate without the mycorhiza fungi. Bernard has shown that the orchids cannot germinate without fungi and he elaborated methods for the artificial infection of these plants with the mycorhiza.

Parasitic plants which have lost their green coloration are encountered among the other representatives of green plants. Such for example is *Monotropa hypopithys* L., of the family Pirolaceae Drude. They are devoid of green coloration and their leaves have been transformed into brownish scales. This and other plants of the genus *Monotropa* L., receive their nourishment entirely at the expense of fungi growing inside their tissues. Gordyagin (1922) described a fern *Ophioglossum simplex* the nonsexual sporiferous generation of which is nourished at the expense of fungi, absorbing organic products of their metabolism and decomposition.

According to the observations of that author, ferns and other green plants are encountered in the forests of the Tatar Autonomous Republic, which assimilate carbon from the atmospheric CO_2 but cannot exist without symbiosis with fungi. Such plants are widespread in nature. Higher plants of such a type preserve the green coloration and the capacity of assimilation of atmospheric CO_2, but cannot normally develop and reach the stage of flowering and fruit bearing in the absence of the fungus.

The majority of plants have fungi on their roots. They are useful though not indispensable. Fungi, in such cases, promote their growth and nutrition. The plants, as experience shows, grow better and adapt themselves to new places and give a higher mass increment (Lobanov, 1953, Reiner and Nelson-Jones, 1949).

Keller (1948) and Lyubimenko (1923). when comparing the degree of symbiosis of different plants with fungi, and the degree of parasitism on the fungi, considered these phenomena as subsequent stages of evolution, as stages of transition from autotrophism to saprophitism and parasitism, i.e., to heterotrophy. In his book "The Fundamentals of Plant Evolution" Keller writes: "In the course of evolution the individual higher green plants of different families pass from this stage, and, under the pressure of appropriate natural conditions, change to nutrition at the expense of fungi. Thereby, the leaves lose their significance as organs of assimilation and become undeveloped, remaining only in the form of scales devoid of green coloration."

Lyubimenko (1923) stressed that there is no sharp difference between autotrophic and heterotrophic plants. Saprophitism, according to him, is the natural consequence of the synthesis of organic compounds: "Organisms capable of synthesizing organic compounds from mineral ones, preferably use ready organic compounds and are capable of receiving their nutrition in the same way as saprophites. Saprophitism is not per se, a special specific property of a certain group of organisms; on the contrary, it is characteristic of all organisms with a possible exception of nitrifying bacteria, and therefore, it appears in nature in different degrees. On one hand we find plants which only accidentally assimilate organic compounds from the environment; these are the facultative saprophytes, for them the saprophytic nutrition is not indispensable." "On the other hand we find plants which are true saprophytes, which have completely lost the capacity of synthesizing organic from inorganic compounds. Between these two categories there is a gradual transition of forms lessening the division between them and in reality cancelling this division, (Lyublmenko, 1923, page 186).

As can be seen from the above citation, Lyubimenko considers that all plants are capable of organic nutrition to a greater or lesser extent. The only exceptions are, according to him, the nitrifiers. We can add that, recently, material is being accumulated showing that even these organisms may utilize organic nutrients and grow on organic nutritious media In the same way as many bacteria-saprophytes.

It to known that plants may be in close symbiosis, not only with fungi, but also with bacteria. The root-nodule bacteria which penetrate the roots of leguminous plants, with the formation of colonies in the form of nodules are well known. The nodules represent swollen roots filled with bacterial cells. These bacteria grow abundantly, multiply in the cells, and produce nitrogenous substances which are utilized by the plants.

Analogous nodules are formed on roots of some other plants (nonleguminous). For example, nodules have been described on roots of *Alnus* Gaertn., *Myrica* L., *Podocarpus* L'Her., representatives of the family Elaeagnaceae Lindl.; *E. multiflora* Thunb., *E. angustifolius* L., on *Shepherdia canadensis* Nutt., on various species of *Ceanothus* L., on *Coraria japonica* Gray, and others. The nodules on these plants are formed by different representatives of the microbes such as bacteria, mycobacteria, actinomycetes and, according to our investigations, also by proactinomycetes.

These organisms have a definite positive effect on plants much the same as the root-nodule bacteria have on leguminous plants. Under conditions where the bacteria and, consequently, the nodules are absent. the plants grow weakly and often they do not reach the stage of flowering and fruiting (Fred, Baldwin and McCoy, 1932; Jongh, 1938).

There are many microbial symbionts which live on plant leaves, forming unique swellings--knots in the tissues. Such knots are formed by bacteria and mycobacteria. These organisms upon multiplying fill the cells of the knots in the same way as the root-nodule bacteria fill the nodules in the root tissues. These microorganisms, according to our investigations, exert a positive effect on the growth of plants. Without them the plants grow weakly and some species do not reach the stage of flowering and fruiting, as can be observed in *Ardisia* Thunb. The plants remain dwarfs, but after they are infected with the corresponding bacteria, knots appear on their leaves, the plants improve, they assume normal appearance, begin to flower and bear fruit (Jongh, 1938).

About 400 species of plants with the above-described knots are described in the literature, including 30 species of the genus *Ardisia* Thunb., 244 species of the genus *Pavetta* L., 42 species of the genus *Psychotri* L., 5 species of the genera *Amblyanthopsis* and *Amblyanthus* D. C., 1 species of the genus *Dioscorea* L., and others. The bacteria forming the knots pass from plant to plant through the seeds,

According to the data of many investigators, all these microbe-symbionts which live in the tissues of roots or in leaves of green plants affect the plants through their metabolic products. Some of them form and excrete biotic compounds which activate the growth of the plants, others, in addition, fix atmospheric nitrogen and pass it on to the plant in the fixed state.

Having ascribed such a great importance to the above-listed bacteria and mycorhiza fungi which provide organic nutrition for the green plants, we must draw attention to other species of microorganisms which are closely linked with the root system of plants, although this link is of a different character. We have in mind the rhizosphere microflora.

An immense number of microorganisms live around the plant roots. The most numerous among them are the bacteria. Have they any effect on the nutrition of the plants? What effect do their metabolic products have on the plant? Are they assimilated by plants and what role do they play in the growth of the latter? If plants utilize the metabolic products of microbial symbionts with such readiness why then can they not use analogous substances formed by the free-living organisms?

Apparently, the difference lies in the fact that in the first case, symbiosis between the plant and the microbe, the metabolic products of the microbes remain inside the tissue, while, in the latter case, these compounds are being formed and accumulated outside the plant in the surrounding milieu, and, in order to become available to the plant, they must find their way inside the plant via the roots or some other way.

Many authors think that the permeability of the root system for mineral substances differs from that for organic compounds. The suction power of the roots is different in these two cases. The suction power of the heterotroph plant parasite *(Lathraea squamaria)* is 22.7 atm, and the suction power of the root of the plant host *(Prunus padus* L.) is only 3.7 atm (Kostychev, 1933).

Many typical autotrophs, whose suction power is no more than average, readily assimilate organic compounds, Hutchinson and Miller (1912), studying nitrogen absorption by the pea seedling, found a phenomenon which, from the point of view of the prevalent opinion that they absorbed nitrogenous organic compounds more readily than nitrate nitrogen, was paradoxical. They tested many organic nitrogenous compounds, and in all cases the results were the same.

These data were confirmed by numerous investigators (see later).

As can be seen from the above, the subdivision of plants, according to their nutrition, to autotrophs and heterotrophs, should be considered as relative. Each of the so-called autotrophic plants is capable of assimilating organic compounds to a greater or lesser extent.

Indeed the meaning of "nutrition" should be distinguished from the term "synthesis of organic substances." Nutrition in the real sense of the word, should be called all the processes which are directly involved in assimilation of substances by the living parts of the organism from the moment when the organic compound in prepared. (Lyubimenko, 1923, page 180). This process proceeds similarly in all organisms belonging either to the plant or animal kingdoms. They all feed on organic compounds. To build a living body, all cells, whether they are microbial, animal or plant, should possess ready-made elementary organic compounds or their building blocks--nitrogenous and nonnitrogenous compounds, which participate in the general process.

The difference between plant and animal nutrition lies only in that the latter obtain ready-made synthesized organic compounds from the outside and the plants synthesize them for themselves.

The process of plant nutrition consists therefore of two elements: a) synthesis of organic substance from inorganic elements; this process Lyubimenko considers as the preparative one, dispensable, from the point of view of nutrition in the narrow sense of the word; and b) nutrition proper, i.e., uptake or assimilation of elementary organic particles for the formation of a living organism; this process proceeds in the same way in plants and animals.

The nutrition of plants and microorganisms may proceed according to the first or the second process, but each process may be manifested to a various degree in relation to the conditions of existence and to generic differences.

Vinogradskii, in his brilliant studies on the chemotrophy of microorganisms, drew a sharp line between these two types of nutrition by dividing all the microorganisms into autotrophs and heterotrophs. Lebedev (1921). on the contrary, assumed that these two types of microorganisms differ from each other, not by their manner of nutrition, but by the way they obtain their energy. Lebedev thinks that both the autotrophs and heterotrophs assimilate CO_2, but while the former obtain their energy at the expense of the oxidation of inorganic substances the latter obtain their energy at the expense of the oxidation of organic compounds. This scientist was the first who showed experimentally that the heterotrophs utilize CO_2. His data were confirmed by many other investigators.

At present, it has been confirmed that many microorganisms, even heterotrophs can assimilate CO_2, and Werkman even assumes that all the organisms in general assimilate CO_2 and that this is an important physiological function, insuring the synthesis of indispensable intermediate metabolic products (Werkman and Wilson, 1954).

The effect of humus on the growth of plants

Numerous observations and the results of accurate laboratory experiments have proved the positive effect of humus on the growth and metabolism of plants.

Bottomley and his co-workers (1914) grew duckweed *Lemna minor* L., In the nutrient solution of Knopp supplemented with small doses of aqueous extracts of well-composed peat. He obtained the following results: after 35 days 30 individuals grew from the 10 originally in the solution, in the absence of humus; in the presence of humus 132 individuals grew in the same time. Those which grew in the presence of organic substances were much bigger and their green color was more intense. Their dry weight was 50.5 mg per 100 individuals, while the dry weight of the same number of control plants was only 29.4 mg. In his later experiments Bottomley (1920) showed that the peat extract has a positive effect also on some other plants such as *Lemna minor* L., *Salvinia natans* All., *Limnobium stolonifer* L. The yield of dry mass, when they were grown in the presence of peat extract, was higher than in the controls.

Peat, after processing with microorganisms, stimulated the growth of plants more markedly than did the uncomposted natural peat. On this basis, Bottomley prepared a special fertilizer--humogen. This preparation gave positive results under laboratory conditions as well as under field conditions. It was patented and recommended for use in agriculture for various plants. According to the prescription in the patent, peptone was added to the peat to enhance the growth and metabolism of microorganisms. For a more rapid decomposition of the organic part of peat, *Azotobacter, Rhizobium,* and other bacteria were added. The inoculated peat was kept at a temperature of 25° C for 3-4 weeks. Afterward the preparation was ready for use.

Mockeridge (1917, 1924) repeated the experiments of Bottomley and fully confirmed his data, She added extracts of composted and noncomposted peat to the nutrient solution and grew the duckweed. After 9 weeks, out of 10 individuals, 249 plants grew in the control, 3,134 in the presence of the processed peat, and 1,080 in the presence of nonprocessed peat. The dry weight of the duckweed was 6.5 mg in the control and 19.5 mg in the presence of peat humus.

In our experiments we have employed well-processed composts prepared from various plant residues such as straw of Euagropyrum, wheat, oats, lucerne hay, and also from well-processed manure. The concentrated aqueous extracts from these composts and from manure were added to the nutrient mineral solution in quantities of 5-10 drops per 100 ml of the medium, which amounts to 0.005-0.01 mg of the dry substance. This nutrient solution was employed by us as a substrate for the growth of *Lemna minor* as well as for the fertilization of the sandy substrate in which we grew cereals and legumes.

Lemna minor L. was grown under sterile conditions in glass vessels placed in the window of the laboratory. In the control vessels the medium did not contain the above-mentioned extracts. The crop was counted after 85 days. The number of individuals grown was counted and their dry weight was determined. Plants were chosen which were similar in their appearance. The results of these experiments are given in Table 30.

Table 30
The effect of plant composts on the growth of
(after 85 days)

Composts	Number of specimens in the vessel	Dry weight (mg) total	Dry weight (mg) of 100 plants
Control (without humus)	180	29	16
Compost of Euagropyrum	2,500	800	32
Compost of wheat	1,200	324	27
Compost of oats	1,800	540	30
Compost of clover	2,200	726	33
Dungwash	1,800	540	30

As can be seen from these data, plants *(Lemna minor)* grow much better in containers with the organic substance. The plants were bigger, more intensely colored and their root system better developed.

Different composts have different effect on the growth of *Lemna minor*. The greatest effect was obtained by the use of the compost of Euagropyrum and dungwash. The plants grown on them were the most developed. The extract of oatstraw compost favored the growth of the plants in numbers similar to those obtained with the extract of dungwash, but their size was smaller. The least effect was obtained when the compost of wheat straw was employed.

Voelcker (1915) tested the effect of the Bottomley's preparation in vegetation containers and under field conditions, on peas, oats and on buckwheat. The greatest crop increment was obtained in the case of buckwheat, it amounted to 407%; the effect on oat was weaker, the crop increment of the latter was 131%. In the experiments with peas, increment in the green mass only was observed; this amounted to 231%. The yield of grain, on the contrary, was less (87%) than in the control (100%). Similar data were obtained by Clark and Roller (1924). This plant grew much less in a nutrient medium under sterile conditions than in the presence of humus. The most abundant growth of duckweed was observed in vessels containing manure and composted peat. The extract of lucerne compost had the smallest effect. These authors had found (1931) that composts are more effective when contaminated with bacteria than when sterile.

Aschby (1929) points out that organic substances of the processed plant residues, though not indispensable for the growth of plants *(Lemna minor* L.), none the less have a considerable effect on their growth. Saeger (1925) introduced a mineral solution of alkaline extract of humus into the nutrient medium of growing duckweeds. The yield was twice as high as in the control. He found that the action of humus and yeast extract on the growth of plants was equal.

Hillitzer (1932) on the grounds of his experiments and the analysis of the available data concluded that humus of the soil and composts stimulate the growth of plants. The components of humus act specifically on the root system enhancing its growth.

Olsen (1930) performed a whole series of experiments on the growth of duckweed and sunflower in the presence and absence of composts in the nutrient medium. By his experiments he confirmed the results of Bottomley and Voelker. In the presence of small amounts of humus (aqueous extracts of processed peat) the duckweed as well as the sunflower grew considerably better than in their absence. The nutrient medium employed by him was the solution of Dettmer. The dry weight of the duckweed grown in the presence of humus was 482 mg and in the control vessel 189 mg; the weight of the sunflower in the presence of humus--1.33 g and in the control vessel--0.80 g. In the presence of iron citrate the positive action of humus was smaller or nonexistent (according to the author). On the basis of this observation he assumes that the acting principle in humus and in composts are some forms of iron compounds.

Our experiments in raising agricultural crops with the above-mentioned composts yielded similar positive results. We have grown wheat, rye, rye grams and clover on pure quartz sand, wetted with the mineral nutrient solution in the presence and absence of aqueous extracts from composts. The amounts of compost added were the same as in the experiments with duckweed,

The experiments were carried out under sterile conditions. The plants were grown for 20-40 days, plucked out together with their roots and analyzed. Their total weight, height and general appearance were noted. The most spectacular results were obtained with wheat and rye grass (Table 31).

Table 31
The effect of humus an the growth of plants
(dry weight in g)

Composts	Wheat: height of plant in cm	Wheat: weight of plant in gm	Rye grass: height of plant in cm	Rye grass: weight of plant in gm.
Control (without organic compounds)	14.9	0.95	10.2	0.64
Euagropyrum compost	18.5	1.63	12.6	0.98
Manure	17.1	1.52	12.1	0.71

As can be seen from the table the positive effect of Eumgropyrum compost and of manure an the growth of rye grams and wheat was of the same magnitude. All the plants have greater mass when grown in the presence of organic substances than in control vessels. The greatest mass increment was obtained in cereals, clover reacted less to the introduction of organic compounds,

In the second series of experiments we tested the effect of well-processed manure and fresh straw. One and a half g of the manure and 20 g of straw together with the nutrient solution of Knopp were applied per kg of sand substratum, The experiments were performed under sterile conditions. Wheat, oats and peas were grown. The results obtained in the presence of manure were similar to those in the previous experiment. In the presence of humus the yield of oats was 35% higher, wheat 25% higher and peas (fresh weight) 42% higher. In the presence of extracts of fresh straw the crop was similar to that in the control vessels.

In one series of experiments we compared the effect of aqueous extracts of fresh straw of Euagropyrum and an extract of composted straw of the same plant on the growth of rye grass. The dry weight of the plants in the control vessels (without the organic fertilizer) was 0.5 g and in the presence of the extract from fresh straw 0.7 g, and in the presence of composted Euagropyrum 1.1 g.

In experiments with pine-tree saplings it was found that small amounts of Euagropyrum compost markedly stimulated the growth of the experimental plants. The latter grow higher than the controls, their stems were thicker, their needles longer and of a brighter color (Figure 61) Krasil'nikov and Raznitsyna, 1946, Raznitsyna, 1942).

Chester and Street (1948) grew lettuce in sand in a full nutrient solution with and without the addition of organic substances, Extracts of soil humus, aqueous extracts of casein, and yeast hydrolysates were tested. All solutions contained 6.05 mg N per 10 m3 of the medium. The plant yields were as follows:

Control (in the absence of organic compounds); dry weight (in gm) 0.157
With the admixture of soil humus; dry weight (in gm) 0.199
With the admixture of casein hydrolsate; dry weight (in gm) 0.158
With the admixture of yeast extract; dry weight (in gm) 0.172

Figure 61. The stimulating effect of Euagropyrum compost on the growth of pine-tree seedlings

Swaby (1942) did not obtain in his experiments an increment in the yield of legumes grown in the presence of organic compounds, but he noted the stimulating effect of microbes when the experiments were carried out under nonsterile conditions.

According to Schaffnit and Neumann (1953), composted peat had a stimulating effect on the growth of potatoes and one the germination of lucerne seeds. They have ascribed the stimulating effect to the action of microorganism which grew abundantly in those composts.

Andreyuk (1954) studied the effect of specially prepared composts of peat on the growth of grain cultures. The composts were prepared from peat, 25-50 % (by weight) manure and 1.5-2.0% phosphate flour. Then they were incubated under laboratory conditions or in the field for 4-7 months and longer. In some experiments *Azotobacter,* cellulose bacteria and other bacteria were introduced. The yield of winter rye in the presence of these composts was 2-2.5 times higher, and the crop of oats was 1.7-3 centners per hectare greater than in the control (5.3 centners).

Many other authors have noted the positive effect of small doses of humus (Nikishkina, 1948; Logvinova, 1939; Street, 1950, and others). The detailed studies of Khristeva (1948) had shown that solutions of humic acids exert a direct effect on higher plants. In negligibly small concentrations (0.001 % and 0.0001%) they enhanced growth and increased the yield of wheat, oats, barley, sugar beet, tomatoes and other plants. Of the tested plants, potatoes, tomatoes, and sugar beet gave the best reaction to the application of humus. Good reaction of wheat, barley, oats, millet, corn, rice, buckwheat, Euagropyrum, and lucerne was noted. Humus had a small effect on peas, mung bean,, beans, lentiles, peanut, cotton, and sesame and hardly any effect on sunflower, castor-oil plant, pumpkin, hibiscus, etc. The greatest increment in plant crops under the influence of humus substances exceeds that obtained by application of equivalent amounts of mineral fertilizers by about 10-50 %. The action of humic fertilizers was tested by the author in different soils such as podsols, serozems, chernozems, chestnut soils, and others, In all cases the effect was positive.

According to Khristeva, humic substances find their way, in small amounts, into plants, there stimulate the phenol-oxidase system, and participate in the general metabolism of the plant. The physiological function of humic substances, is in the promotion of plant respiration. As a result. an increased influx of nutrients, activation of synthetic processes, and better growth of the root and aerial parts takes place.

The strongest reaction to the humic substances is observed in young plants. The root weight and, consequently, the growth of the whole plant is increased.

Kock (1955), stressing the positive effect of humus substances on plants, explains it by the action of iron which is present in humus, This, however, was not confirmed by the studies of other investigators.

Tovarnitskii and Rivking (1937), Tovarnitskii and Statkovskaya (1938), Thiman, Lane and others (1933, 1939) employed urine and yeast extracts for presowing processing of oats, wheat, and other plants in order to stimulate their growth.

In southern Italy urine of cattle is widely used for presowing processing of the seeds of cereals (Zeding, 1955). Virtanen and Hausen (1933-1934) added yeast extract to aqueous and sandy cultures of peas. Flowering occurred 5-10 days earlier and the crop was 50 % higher than in the control plants. These authors also observed that such effect could not be obtained by growing the plants in soil rich in humus. This could be explained by the fact that in soils rich in humus there is a large amount of biotic substances.

A large amount of material on bacterial fertilizers should be added to the data given in this chapter. The practice of employing bacteria-containing fertilizers gives, in many cases, certain positive results. The addition of peat "azotogen" brings about a 10-25 % increment in crops, this increment may in some cases be even higher.

The bacterial preparations used as fertilizers under the name of "azotogen" are extensively employed in our country. They are prepared from cultures of *Azotobacter* on crumbs of peat. The selected peat must be of an appropriate quality. It must not be acidic and should be well processed. During the production of the preparation, easily assimilated sources of nutrition are added. These are sugars, alcohol, beet juice, etc. The *Azotobacter*--inoculated peat is incubated at optimal temperature and humidity; it is frequently mixed to ensure better aeration.

The incubation lasts 10- 20 days. During this time the peat is well composted. Not only *Azotobacter,* but also many other bacteria, fungi and actinomycetes, grow abundantly in the peat mass.

The number of nonsporeforming bacteria at the time of their maximal growth in a good preparation reaches several billion cells per gram. The number of *Azotobacter* amounts to 100-200 million per gram (Table 32).

Table 32
The quantitative composition of the microflora in peat azotogen
(number of cells, in thousands per g, in the period of maximal accumulation)

Preparation	Bacteria, sporeformers	Bacteria, non-sporeformers	Myco- bacteria	Actino-mycetes	Fungi
Noncomposted peat	100	40,000	1,000	1,500	750
Peat composted without *Azotobacter*	1,500	5,500,000	1,000,000	3,500,000	150,000
Peat composted with *Azotobacter*	1,700	7,500,000	1,500,000	3,200,000	250,000

As can be seen from the table, peat composted whether in the presence or in the absence of Azotobacter contains the same number of microbes. Their total number considerably exceeds that of the initial peat.

The group composition of the compost microflora varies with the increase of the maturity of the compost. In the first days of incubation the nonsporeforming bacteria of the genera *Bacterium* and *Pseudomonas,* and fungi, grow abundantly. The fungi grow only on the surface. Mycobacteria can be detected in large numbers in the composted peat. At the end of the incubation (maturation) of the preparation, the bacteria and fungi decrease in number, their place being taken by actinomycetes. The latter attain such vast numbers that the peat lumps are covered with them. This covering is of a white flour color and is visible with the naked eye. The maturing of the compost can be judged by the intensity of growth of the actinomycetes.

Analyses show that this consecutive development and change of microflora is observed in approximately the same quantitative relations in peat composted without *Azotobacter.* The introduction of the latter, however, may cause a change in the composition of the species of the nonsporeforming bacteria, but it can be marked only in certain species. The general group composition is not changed.

Trials of peat (noncomposted, or composted in the presence or in the absence of bacteria) were carried out by us in pots and in fields during a number of years. Different plants--grains and cultivated crops were involved. The general character of the efficacy of these preparations under field conditions is given in Table 33.

Table 33
Comparison of the action of fertilizing preparations on plant yields

Bacterial inocculation of the preparation	Corn: centner per hectare	Corn: %	Potatoes: center per hectare	Potatoes: %	Beets: centner per hectare	Beets: %	Wheat: centner per hectare	Wheat: %
Control	32.0	100.0	230	100	559	100	--	--
Azotobacter strain 54	33.9	106.0	255	110.8	610	109	--	--
Peat azotogen	37.1	116.0	261	113.7	649	116	--	--
Composted peat	36.7	114.7	263	114.3	639	114	--	--
Noncomposted peat	34.1	108.7	254	110.4	615	110	--	--

As can be seen from the table the pure culture of *Azotobacter* (collection strain. No 54) is less effective than the peat azotogen. Peat composted without *Azotobacter* produces approximately the same yield increment as the azotogen prepared from peat. An already mentioned, noncomposted peat is less effective than composted peat.

We obtained similar data in experiments carried out under various conditions in podsol soils of fields near Moscow and in the serozems of Kirgiz SSR and Tadzhik SSR (Vakhsh valley). In the majority of cases azotogen prepared on peat, and well-composted peat with *Azotobacter* gave similar plant-yield increments.

In a number of cases bacterialized composts were, more effective than those prepared without bacteria (Table 34).

Table 34
The growth of duckweed in water culture in the presence of small amounts of peat compost inoculated with bacteria

Conditions of experiments	Total samples	Dry weight: 100 samples in mg
Control (without compost)	106	26
Compost not inoculated by bacteria	450	37
Compost inoculated by bacteria:		
Az. chroococcum	510	39
Ps. flourescens No. 14	580	38
Ps. flourescens No. 15	85	27

Many investigators assume that the favorable effect of humus, manure and composts in caused by their ash content, of which nitrogen is the constituent of greatest importance. The latter, according to these investigators determines the efficacy of compost mass and humus of the soil. The higher the nitrogen content in the compost the more it is effective. According to the adherents of this point of view, after the mineralization of the organic compounds the nitrogenous substances are transformed into inorganic substances and so become available to plants.

According to our observations and the available data in literature, the active principles of humus and composts are not the mineral nutrients present in them but the organic substances and the biologically active metabolites of microbes. Mineral substances applied in amounts equivalent to the composts do not have an effect comparable to the latter.

Mineral elements obtained by burning manure, composted peat or soil humus do not produce an effect comparable to that obtained by the application of organic substances. We have carried out experiments with duckweed in nutrient solutions, supplemented with small quantities of extracts prepared from manure, compost and humus. In other experiments ash was added to the nutrient solution. The ash was obtained by burning equivalent amounts of these substances. The results are given in Table 35.

Table 35
Effect of humus ash on growth of *Lemna minor* L.
(on the 50th day of growth)

Experiment	The number of specimens in the vessel	Dry weight of 100 specimens, mg
Control	750	27
Manaure	1,800	39
Ash of the manure	850	33
Peat composted with bacteria	1,150	40
Ash	900	31
Soil humus	1,100	33
Burned soil	700	29

Lochhead and Thexton (1952) obtained similar results. According to their data, the ash obtained from composts have a smaller effect on the growth of microbes than the composts themselves.

All this speaks in favor of the assumption that the observed stimulation of plant growth by humus is caused, not by its mineral fraction, but by other substances.

The effect of humus on the vitamin content of plants

Vegetative composts, manure, and humic substances not only activate plant growth and increase their yields but also improve their nutrient value which in a matter of great importance. Plants grown in fields fertilized with manure are richer in vitamins and other valuable substances than those grown in nonfertilized fields. McCarrison (1926) found that seeds of millet and wheat, harvested from fertilized fields, contain more vitamins than seeds from fields under mineral fertilizers. Animals, fed on fodder from fields fertilized with manure, were more resistent to infections. and their appearance was healthier than those fed on fodder from nonfertilized fields.

Roulands and Wilkenson (1930) determined vitamins of the B-complex in the hay of clover grown in fertilized and nonfertilized fields. In the former case the clover contained more vitamins. Rats fed on clover from fertilized fields gained weight more rapidly than those fed on clover from nonfertilized fields. In the 30 days of the experiment the first gained 110 g and the latter 60 g.

Clark (1935) found more vitamin B and C in cultures of duckweed grown in nutrient solution supplemented with humus, than in a similar culture grown without humus.

Nath and co-workers (1927, 1932) have shown that organic substances of manure and composts, as well as cod-liver oil and some other substances, enhance the growth of plants and increase their vitamin content. Cereal seeds grown on fertilized fields have a greater viability and the percentage germination is higher than in seeds from fields fertilized with mineral fertilizers.

According to Graff (1928) timothy grass, fescue and meadow grass *(Phleum pratense* L., *Festuca rubra* L., *F. pratensis* Huds. and *Poa. pratensis* L.) contained more vitamins of the B-group when grown in soil fertilized with urea than when grown in soil to which mineral fertilizers had been applied.

Antoniani and Monzini (1950) analyzed plants which grew in fields irrigated with sewage waters, and plants which grew in fields irrigated with water supplemented with mineral nutrient salts. The vitamin B_1 content was greater in the former plants.

Leong (1939) compared the effect of manure and mineral fertilizers on the vitamin B_1 content of wheat and barley tissues. The vitamin B_1 content of wheat was similar in both cases; barley, however, contained twice as much vitamins when grown in the presence of manure than in the presence of full mineral fertilizer (Table 36).

Table 36
The vitamin B$_1$ content in plants in relation to fertilization
(µg per 1g of plant mass)

Fertilizer	Wheat	Barley
Without fertilizer (control)	1.0	1.1
Manure	1.2	2.0
Full mineral fertilizer P, K, Mg, NO$_3$	1.2	1.1

Hurni (1944, 1945) grew plants in sand in the presence of a full mineral fertilizer and found that in these conditions the formation of thiamine was less than during growth in the presence of humus substrate.

In Lebedev's experiments (1953), lucerne grown in fields fertilized with manure or composts, contained 81 mg/kg carotene during the period of bud formation, while plants from nonfertilized fields contained only 27 mg/kg.

Ott (1937) noticed an increase in vitamin content of plants grown in fields fertilized with a mixture of organic and mineral fertilizers.

Hammer and Maynard (1942), summarizing the literature, noted that the vitamin content of plants varies in accordance with the soil and climatic conditions, season, age of plants, etc. They ascribed the greatest importance to the nutrient value of the soil and especially to the presence of humus and fertilizers in general.

At the present time the majority of investigators concentrate their attention on mineral sources of nutrition as the factors affecting the vitamin content of plants. Vitamins A, C, B$_1$ and B$_2$ were determined in the tissues of many plants (leguminous and cereal) grown in fields fertilized with various salts of potassium, phosphorus, and nitrogen (Scheunert and Wagner, 1938; Scharer and Preissner, 1954, and others).

After the application to the soil of a full mineral fertilizer, black currant gave a berry yield of 5,158 kg/hectare (Stepanova, 1950). Their total vitamin C content was 1,827 mg/kg. The berry yield without fertilizers was 3,199 kg/hectare and the vitamin content was 1,599 mg/kg. According to Dyakova (1945), application of one portion of nitrogen increased the carotene content of oats from 10 to 28 mg and an application of triple nitrogen increased the carotene content from 10 to 53 mg (in the stalk-formation state). The application of calcium to the soil also increased the carotene content of plants. Lack of calcium in the soil decreases the synthesis of thiamine in tobacco plants (Ovcharov, 1955).

Some authors studied the variation of vitamin content in plants in connection with the application of microelements to the soil. Mc Harque (1924) noted a positive effect of manganese on the accumulation of vitamin B1 in the seeds of wheat, rice, tomatoes, and citrus fruits. Hammer (1945) gives data on the effect of microelements on the formation and accumulation of ascorbic acid and carotene in plants.

Scharer and Preissner (1954), on the basis of their experiments, reached the conclusion that the more complete the mixture of fertilizer employed, the higher the vitamin content of the plants. The amount of vitamins does not always correspond to the weight index of the crop yield. It is not infrequently observed that the vitamin content is high at relatively low crop yields. For example in an average barley crop of 4.3 g* the vitamin B$_1$ was 440 µ g and in a crop of 11.93g-39 0 µ g per 100 of the grain. *[The unit used is not clear in the Russian; it probably refers to an arbitrary yield index.]

Considerable increase in the vitamin B$_1$ content of plants was observed in the experiments in which granulated phosphorus was applied. In a weakly acidic loamy soil the following results were obtained: control--grain yield 7.12 g, vitamin content--558.7 µ g /100 g; after the application of N, K and superphosphate the crop yield was 24.5 g and the vitamin content was 798 µ g /100 g of grain,

The greatest accumulation of vitamins was noted in leguminous plants, This could be explained by the presence of root-nodule bacteria. The latter growing in the root tissue, form vitamins which find their way from the nodules into the plant.

Schounert and Wagner (10939, 1940) studied the vitamin B_1 and B_2 content in the seeds of barley and rye, grown in fields which were fertilized for many years as well as in fields with no fertilizer at all, Comparison of the analyses did not show any marked difference in the vitamin content of plants grown in fertilized and nonfertilized fields.

The lack of any effect of mineral or even organic fertilizers was also noted by several other investigators (Hornemann, 1925; Harris, 1934), These authors did not give an explanation. It should be assumed that their experiments were carried out under conditions unfavorable for the synthesis of vitamins.

Data exist which show variations in amino-acid composition of plant tissues in the presence of different fertilizers. According to Sheldon, Blue and Albrecht (1948), lucerne and some other plants grown on fertilized fields, have a different quantitative and qualitative amino-acid composition than those plants grown in nonfertilized fields. The highest amino-acid content was found in plants grown with manure,

It should be noted that the majority of investigators studying the effect of mineral fertilizers on the plant vitamin content did not take into account the microflora, and especially that which inhabits the rhizosphere.

Mineral and organic fertilizers are known to have a great effect on the life of microbes in the soil. The latter grow and synthesize various biologically active substances including vitamins. For example, in many soils, phosphates and nitrates are known to considerably increase the growth and accumulation of bacteria, fungi and actinomycetes. *Azotobacter,* root-nodule bacteria, certain species of the genus *Pseudomonas* and other forms of microbes grow well around lumps of superphosphate.

The material presented by us shows that animal and plant residues and organic substances in general have a favorable effect only after their decomposition and processing by microorganisms, i.e., after they have been transformed into humus. The active substances of humus are not the animal or plant residues but the products of microbial metabolism, the products of secondary synthesis.

We assume that the active principles of composts, manure, and of humus in general are not the mineral elements of nutrition or more strictly, not so much they, as some special compounds of an organic character. These compounds vary in their nature and comprise a special group of the so-called biotic substances.

BIOTIC SUBSTANCES OF THE SOIL

It can be seen from the above data that small doses of organic substances, present in processed compost or manure, are needed for the improvement of the growth of plants.

The question arises which of these organic substances are the activators of plant growth.

It was said above that the organic part of the soil consists of a multitude of complex compounds with specific and nonspecific proportion. The majority of these compounds are related to humic acids.

Beside the humic acids many other substances were found in the organic part of the soil. They include substances with biocatalytic properties, Among these are enzymes, vitamins, auxins, certain amino acids and other biotic substances.

Bottomley (1914-1920), Mockeridge (1924), Voelcker (1915), Clark (1935), Saeger (1925), and others, have assumed that the activating effect of humus fertilizers, processing manure and different composts with and without bacteria, in conditioned by special substances present in them, the so-called "auximones."

Their action is similar to that of the yeast bios. The analysis of peat infected with bacteria showed the presence of purine bases. The latter were also found in the cells of *Azotobacter.* This was the reason for Mockeridge's conclusion that the activating compound of composts and *Azotobacter* to one and the same substance.

It wax shown in later studies that the action of humus is caused by special substances produced by microbes (Nath, 1932, McCarrison, 1924, And others).

Anstead (1935) proposed that these substances be called phytamins. According to him the phytamins are formed by microorganisms only. Upon entering the plants they are transformed into vitamins. The latter entering with food into the body of animals and human beings are subjected to certain transformations and are converted into hormones. Hormones and vitamins are excreted from the animal body with urine or excrements, find their way into the soil or manure, And are there transformed by microorganisms into phytamins.

In such a way, according to Anstead, phytamins, vitamins and hormones are products of one and the same substance.

Analogous views were hold by Daineko (1939). He assumed that there is no difference between the animal hormones and vitamins, and that they are interlinked.

The link between plant and animal biocatalysts is stressed by many inventigators. However, there are no grounds for speaking of any cycle of these substances in nature.

Lochhead and Chase (1943) attempted to elucidate the nature of the activating substance of humus. They prepared extracts from soil humus and subjected it to various procedures, such as extraction with organic solvents, absorption with carbon, and other absorbents followed by elution. The experiments showed that the ash of humus and composts had no effect whatsoever on the growth of plants and microorganism, Of 63 microbial cultures, only one grew in the presence of ash obtained from composts. The extracts obtained from composts and humus of the soil, however, had a positive effect on their growth. The acetone extract was the one most effective, In its presence 26 cultures grew well, 26 cultures adequately, and only 11 cultures did not grow. Alcohol and ether extracts also gave good results, Filtrates and eluates from charcoal, each gave a small effect, their effect was larger if they were combined.

Allison and Hoover (1936) noted a positive effect of humic acids on the growth and activity of root-nodule bacteria. They have discovered the presence in humin of a special substance which they called "Factor R" or "Coenzyme."

Robinson and Endington (1946) found a biotic substance in soils "fluorin." This substance, according to them, is absorbed in considerable amounts by plants.

Parker-Rodes (1940) found auxins in the soil. The quantity of auxins varied according to the soil properties. In manure-fertilized soils the amount of auxins was 0. 200 μ g/ kg and in poor, nonfertilized soils only 0.06 g/ kg of soil. These substances are found in lesser amounts (0.09-0.106 μ g/ kg) in soils under mineral fertilizers. The soil is enriched in auxins in the course of composting. Sterilized soil contained 0.042 μ g auxins per kg of soil and after 6 days incubation in a glass house, under favorable temperature and humidity conditions, they increased to 0.146 μ g/ kg of soil.

Williams, Stewart, Kejes and Anderson (1942) found larger quantities of auxins in soils than did Parker-Rodes. According to their data, the A-horizon of fertile soils contains up to 175 μ g auxins per kg of soil, Nonfertile soils contain 40-60 μ g auxins per kg of soil. In horizon B auxins were not detected.

Hamense (1946) by the use of improved methods of analysis found 100 times more auxins in soils than did previous workers. According to his data the auxin content was 160-450 μ g per kg of soil (Schmidt, 1951, and Hamense, 1946).

Stewart and Anderson (1942) had shown that the amount of auxins in fertile soils is entirely sufficient for the stimulation of plant growth. Hamense (1946) found in different soils 0.16-0.045 μ g equivalents of ß-indoleacetic acid in 1 g of dry soil. The auxin content increases after the application of organic fertilizers and then falls below the initial level and afterward rises again to the normal level, characteristic of the given soil. According to the author's observations, chemically pure preparations of auxins are not preserved in the soil for long periods.

According to Matskov (1954), chemical preparations of 2. 4-dichlorophenoxyacetic acid are preserved in the soil for the duration of a whole winter (5-6 months). Another growth stimulant--heteroauxin (ß-indoleacetic acid) is not preserved in the soil that long.

Auxins were also detected in the soil by Roberts I. and Roberts E. (1939) and some other investigators.

Beside auxins, soils also contain many other biotic substances such as vitamins, biotin, nicotinic acid, pantothenic acid, folic acid, amino acids, various growth factors R, Z, X, and others, These substances have been found in the following quantities:

Thiamine	0.29-1.93 μ g	Roulte and Schopfer, 1950; Schopfer, 1943
Riboflaven	9.0-980 μ g	Schmidt and Starkey, 1951; Carpenter, 1943
Biotin	23.0-62.0 μ g	Roulet and Schopfer, 1950
Vitamin B_6	amounts not indicated	
Vitamin B_{12}	0.2-1.5 μ g	Robbins, Hervey and Stebbing, 1951, 1952
Inositol	amounts not indicated	
Nicotinic acid	amounts not indicated	Roulet, 1948
Para-aminobenzoic acid	amounts not indicated	Roulet, 1948
Pantothenic acid	amounts not indicated	Roulet, 1948
Folic acid	amounts not indicated	Roulet, 1948
Factor X	amounts not indicated	Lochhead and Texton, 1950
Unknown factor	amounts not indicated	Lochhead and Texton, 1950

Apart from complete vitamin molecules, molecule fractions are encountered in nature which have the same effect on some organisms as the whole molecules, for example, the pyrimidine, thiazole and others (Lilly and Leonian, 1939).

We have found vitamins in different soils; as a rule, they were in larger quantities in places where the microbiological processes were more intense. Thus, they appear in chernozems in larger amounts than in podsol soils (Table 37).

Table 37
Vitamin and bacterial content of various soils

Soils	Riboflavin μ g/100 gm	Thiamine μ g/100 gm	Biotin μ g/100 gm	Bacteria millions/gram
Chernozem (Moldavian SSR)	98.0	4.5	45.0	1,500
Podsol (Moscow Oblast')	5.0	1.2	25.0	0.5

Cultivated soils contain more vitamins than virgin soils.

The vitamin content of the soil is qualified not only by the extent of cultivation but also by the nature of the vegetative cover. Plants which favor abundant growth of microorganisms, as a rule, assimilate more biotic substances in the soil. In the serozems of Central Asia the highest concentrations of biotic substances were found under 2-3-year-old lucerne, their concentration was lower under cotton and very little was found in virgin soils (Table 38).

Table 38
Vitamin and microbe content of the soils of Central Asia, under various plants

Soil	Thiamine μ g/100 gm	Biotin μ g/100 gm	Microorganisms millions/gram
Virgin soil, the valley of the river Vakhsh	1.5	10.0	0.5
Virgin soil	0	+	0.1
Cultivated soil (2 year-old lucerne)	6.5	38.0	4,500
Cultivated soil (cotton long under cultivation)	3.0	18.0	1,500

According to Shavlovskii (1954, 1955), soil under potatoes contains 0.5 μ g/kg biotin, and soil under clover-- 1.3 μ g/kg. In the Lvov Oblast' in serozem forest soils and podsol chernozem the author obtained the data shown in Table 39.

Table 39
Vitamin content of soils under various plants, in μ g/kg

Soils	Biotin	Riboflavin	Nicotinic acid
Serozem forest, under wheat	0.3	4.0	100.0
The same, under grasses	0.7	7.0	230.0
Podsol chernozem, under wheat	0.8	10.0	280.0
The same, under grasses	1.5	14.0	350.6

In soil adjoining the root system there are more biotic substances than outside the rhizosphere. In one kg of soil from the rhizosphere of wheat grown in fertilized fields of the Experimental Station of Dolgoprudnoe (Moscow Oblast') we found per 100 g soil 10 μ g of thiamine, 150 μ g riboflavin, 35 μ g biotin; outside the rhizosphere, 1.2 μ g thiamine, 25 μ g riboflavin, 3 μ g biotin; in the rhizosphere of tobacco, 10-15 μ g thiamine, and outside the root zone 1.5-4.0 μ g per 100 g of soil.

According to Shavlovskii, the amount of vitamins in the rhizosphere of buckwheat is twice as high as that outside the root zone. On the 20th day of growth he found:

.	in the rhizosphere, μ g / kg	outside the rootzone μ g / kg
Nicotinic acid	600	260
Biotin	2	0.5
Vitamin B$_6$	8	--

Higher concentrations of vitamins in the rhizosphere were noted by Roulet (1954). The amount of biotic substances in the upper layer of the soil is higher than in the lower layers. The highest concentrations are found in the upper layer (0-20 cm) (Table 40).

Table 40
The distribution of vitamins in soil layers (μ g/kg)

Soil	Layer, cm	Thiamine	Biotin
Podsol of Moscow Oblast under 2 year clover	0-20	2.3	14
	20-30	0	3.0
	50-70	1.9	8.0

	80-100	0	0
Serozem of Vakhsh Valley under 3 year lucerne	0-20	5.6	42
.	30-40	1.8	14
	45-60	2.4	21
	70-90	1.1	4.2

In some soils a small increase in the concentration of biotic substances is noted at a depth of 60-70 cm.

Some other investigators have noted the decrease in concentration of vitamins in deep layers of the soil. Roulet and Schopfer (1950) give the following data on the distribution of vitamins according to, layers (per 100 g of soil):

layer, cm	thiamine, μ g	biotin, μ g
10	1.93	62
20	0.86	39
30	0.62	27
50	0.29	23

These authors found thiamine and biotin even at a depth of 2.0 to 8.5 m.

Lilly and Leonian (1939) found considerable quantities of thiamine and biotin, as well as vitamin B_1 and its components thiazole and pyrimidine, in the soil. They were concentrated in the upper layer of the soil. At a depth of 60 cm they were not detected. West and Wilson (1938, 1939) found thiamine and biotin in the root zone of tobacco and flax.

According to Schmidt and Starkey (1951), riboflavin can be found in the soil in varying amounts, according to soil fertility, vegetative cover, etc. In soils under forests the authors found 500 μ g of riboflavin and in soils under plow and in fertile soils about 10 μ g per 100 g of soil.

After the application to the soil of organic fertilizers such as straw, grass or sugar, the amount of riboflavin increased. The more organic substances applied, the higher was the concentration of riboflavin in the soil. For example, after application of grass 15%*, about200 μ g riboflavin were found; if the amount of grass added was 10% 120 μ g riboflavin were found, in the presence of grass of 5% only 60 μ g riboflavin was found. The amounts of riboflavin are given per 100 g of soil (Figure 62). *[The 15% refers to the grass, It is not clear what the percentage refers to.]

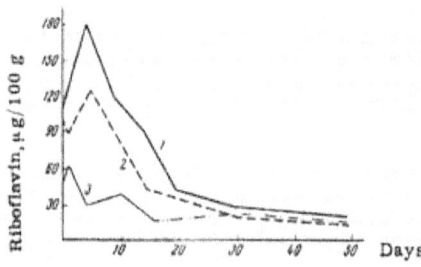

Figure 62. The formation of riboflavin in the soil, in the presence of various quantities of organic substances (lucerne grass): 1--15%; 3--5%

190

By the end of the growth period, in August-October, more vitamins are found in the soil than in the spring. The amount of riboflavin found in spring was 80-300 μ g, in autumn it reached 600-980 μ g per 100 g of soil.

Roulet (1954) found that the amount of biotin in the upper layer of a botanical garden increases in the autumn and decreases in winter and in spring.

According to this author, forest and marshy soils contain even more vitamins than garden or meadow soils.

The amount of biotic substances in the soil constantly changes, according to external conditions such as temperature, humidity, season, etc.

The vitamins are preserved in the soil for various periods of time, the length of which also depends upon soil and climatic conditions. According to Stewart and Anderson (1942), growth-stimulating substances can persist in dry soil for 3-4 years, According to Schmidt and Starkey, 50% riboflavin can be detected in fresh soil after 3 days, pantothenic acid is completely decomposed in one day.

Some amino acids can be listed among the growth factors of lower and higher plants. Plants synthesize these amino acids in the same way they synthesize vitamins, but, nevertheless, the addition of small doses of amino acids has a positive effect on the growth of plants.

Nielsen and Hartelius (1938) tested amino compounds, Six of them--ß-alanine, asparagine, aspartic acid, glutamic acid, lysine and arginine markedly enhanced the growth of lower organisms; ß--alanine had the greatest effect. Five μ g of ß-alanine per 50 ml of medium was sufficient to enhance the growth and increase the yield by 66% dry weight. The maximal stimulation was found at the concentration of 1:100,000. Arginine acts at a dilution of 1:20,000 and lysine at a dilution of 1:4,000. Glutamine exerts a positive effect on the growth of organisms in a dose of 1:1,000. Substances in such doses can be considered as sources of nutrition. Other amino acids such as asparagine and others act at even higher concentrations and cannot be regarded as growth factors but as nutrients.

The amount of amino acids in soils varies; it depends upon the properties of the latter and upon climatic conditions. The more fertile the soil, the more amino acids it contains. The concentration of amino acids is determined by the rate of their influx into the soil and by the length of time of their preservation.

This list does not include all the biotic substances of the soil. It should be assumed that the soil, the same as other natural substrates, contains many other substances unknown to us, which act as biocatalysts, enhancing the metabolism of organisms.

Lochhead and Texton (1940, 1950, 1952) detected activating substances in the soil which were of an unknown nature and could not be replaced by any of the known growth factors.

The origin of the biotic substances of soil

All the diverse biotic substances have their origin in the metabolic activity of plants and microbes.

Formation of biotic substances by the plants. It was mentioned above that, during their lifetime, the roots of many plants excrete substances which stimulate the growth of organisms, For example, the seeds of broomrape germinate, only in the vicinity of roots of sunflower, flax, corn, soy and some other plants, The root excrements of these plants activate the growth of broomrape sprouts. The activating factor is thermostable and is, not decomposed on boiling and prolonged drying (Bartsinakii, 1935; Beilin, 1941).

According to Golubinskii (1950), morning glory stimulates the germination of melon seeds. Pollen grains of angiosperms mutually stimulate each other on germination by excreting activating substances (Golubinskii, 1946).

Timonin (1941) found thiamine and biotin in the root secretions of flax. According to him, these vitamins secreted by the roots, considerably promote the growth and proliferation of microorganisms in the rhizosphere.

West (1939) found thiamine and bios substances in root secretions of flax and tobacco.

Meshkov (1952) found biotin and thiamine in root secretions of corn and peas. According to him, the more of these substances are secreted, the more intense is the growth of the plants. Corn secretes more biotin, and peas more thiamine. The root excretions contain vitamins in the following amounts per g dry weight of the plants: corn--0.5402 μ g of thiamine and 0.2308 μ g of biotin; peas--0.6634 μ g of thiamine and 0.2658 μ g biotin.

According to our observations, peas, wheat, and corn secrete more biotic substances in their early growth period than in the period of fruiting (Table 41).

Table 41
The presence of vitamins in root secretions of various plants
(μ g per ml of nutrient solution)

Plants	After 10 days growth: Thiamine	After 10 days growth: Biotin	After 45 days growth: Thiamine	After 45 days growth: Biotin
Wheat	0.1	0.6	0	0.1
Corn	0.2	0.5	0	0
Peas	0.5	1.5	0.1	0.7

Plants grown in aerated solutions secrete more biotic substances than those grown in poorly aerated solutions. For example, wheat grown with good aeration secreted 0.21 μ g thiamine, and wheat grown in the presence of small amounts of oxygen secreted 0.11 μ g thiamine. The respective amounts of biotin secreted were 0.8 and 0.6 μ g (per cm^3 of medium).

There are indications in the literature that germinating seeds of various plants secrete small amounts of vitamins into the medium (Meisel, 1950; Schopfer, 1943).

Biotic compounds can find their way into the soil together with decomposing residues. It is known that plants contain considerable amounts of various compounds which can stimulate the growth and development of organisms. For example, according to Burcholder and others (1944), the following amounts of thiamine are present in the tissues of: soya $47-61 \times 10^{-7}$; barley $28-51.8 \times 10^{-7}$; corn $17.6-31.2 \times 10^{-7}$ (in mols per kg dry weight). The aerial parts contain more thiamine than the roots: in the leaves of corn were observed $17.6-30.2 \times 10^{-7}$; and in the roots--$4.06-9.78 \times 10^{-7}$ mols per 1 kg of dry mass.

Auxins, vitamins of the B-group, bios, vitamins, D, K, C, H and P, pantothenic acid, paraaminobenzoic acid, nicotinic acid, purine derivatives and various hormones, etc, have been found in plants (Zeding, 1955; Ovcharov, 1955, and others.

Many biotic substances are excreted in the feces of animals and human beings. For example, human beings excrete daily in urine 60 μ g of thiamine, 600 μ g riboflavin, 626 μ g inositol and large quantities of pantothenic acid, as well as other compounds (Meisel, 1950). Vitamins of the B-group, pantothenic acid, nicotinic acid, folic acid, and other substances are found in the faces of human beings and animals. The investigations show that these substances are synthesized by the microflora of the intestines (Perets, Gryazno and Agibalova, 1948; Nepomnyashchaya, 1950; Najjar and others, 1943-1950; Perets, 1955), The vitamins and other substances enter the soil with manure. According to Bonner and others (1938), 1kg of manure contains 130 μ g thiamine. Riboflavin, pantothenic acid, nicotinic acid, and other activating substances are present in manure. Sauerland (1948) found considerable amounts of substances of the bios type in manure, feces and urine of cattle, The quantities of these substances fluctuate with the seasons. The largest amounts being found in summer, less in winter. The amount of growth factors in manure and animal feces also varies according to the quality of the fodder.

All these substances accidentally find their way into the soil, and on the whole they constitute only a small traction of the total amount of biotic substances in the soil. The microorganisms are the main factor in the enrichment of the soil with these substances.

The capacity of microorganisms to synthesize biotic substances has been known for a long time,

Wildiers (1901) showed the presence of activating substances in yeast cultures, These substances were called by the author bios substances, Fifteen to twenty years later the attention of various specialists was drawn to these substances, They were found in various organic substrates, in plant tissues and in cultures of many microbes.

Investigations showed that biotic substances are synthesized by various microorganisms--bacteria, fungi, yeasts, actinomycetes and others (Meisel, 1950, Ierusaimskii, 1940 Kudryashov, 1948; Bukin, 1940, Stephenson. 1951, Schopfer, 1943).

The organisms are divided according to their capacity to synthesize biotic substances into auxoautotrophs and auxoheterotrophs. The former synthesize all the compounds necessary for their growth and can therefore grow on synthetic, vitaminless media; the latter do not synthesize, or more precisely, they do not synthesize all the necessary biotic substances and therefore they cannot grow on vitaminless media,

The capacity to synthesize various growth factors is present in many if not in all species of soil bacteria.

Chemosynthetic bacteria are the most active in synthesizing biotic substances, They can grow in purely mineral, completely vitaminless media, being capable of synthesizing organic substances from atmospheric CO_2. For example, thiamine, riboflavin, pantothenic acid, nicotinic acid, vitamin Be and some other compounds were found in the culture of *Thiobact. thioxidans* (O'Kane, 1943). These bacteria and other similar chemosynthetic bacteria--the nitrifters and hydrogen-and methane-oxidizing bacteria, synthesize their full quota of biotic substances. Without this ability they could not have developed in mineral media,

Bacteria incapable of CO_2 assimilation, but growing well in synthetic vitaminless media with organic carbon sources, are also able to synthesize biotic substances, To such bacteria belong the bulk of the soil microflora*Azotobacter*, root-nodule bacteria, representatives of the genus *Peoudomonas, Bacterium*, oligonitrophils, mycobacteria, and others.

Boysen-Jonsen (1931) found that hetroauxin is synthesized by 16 bacterial species among which were *Ps. radiobacter, Bact. denitrificans, Bac. mycoides, Bac. subtilis, Bacterium* sp., and others. Raznitsyna (1938) employing the coleoptile method has detected the formation of this compound by various representatives of bacteria and mycobacteria. She divided microorganisms into three groups, according to their ability to synthesize auxins: a) organisms which do not synthesize them (or synthesize them in small quantities) *Mycobacterium rubrum, Az. vinelandii, Bac. mycoides*, and others, b) bacteria of medium activity -- *Az. agile, Az. chroococcum*strains 31, 35, *Bact. coli, Myob. luteum, Ps. flourescens*, strains F. 24 and others, c) bacteria synthesizing large amounts of auxins--*Bact. proteus, Ps. fluorescens* strain 21, *Az. chroococcum* strain 54, *Mycob. album,*and others.

Roberts I, and Roberts E. (1939) studied the capacity of bacteria, fungi and actinomycetes to synthesize heteroauxins. This compound was synthesized by 99 cultures out of a total 150 studied, According to the authors, the most active producers of heteroauxins were bacteria and actinomycetes. Heteroauxin was found in different species of nonsporeforming bacteria--*Az. chroococcum, Pseudomonas, Bacterium, Vibrio, Mycobacterium*, and many other bacteria,

Beside auxins, a number of other biotic compounds such as thiamine, riboflavin, para-aminobonzoic acid, nicotinic acid, pantothenic acid, folio acid, vitamin C, vitamin K, vitamin B_3, vitamin B_{12}, inositol, biotin, provitamin D_2, alpha- and ßcarotenes, factor R, factor Z, and others can be detected in bacterial cultures (Detinova, 1937, Leo and Burris, 1943, Jones and Grooves, 1943; Burton and Lochhead, 1951, Lochhead, 1952; Lochhead and Burton, 1955), The ability of the root-nodule bacteria to synthesize thiamine, riboflavin, pantothenic acid, vitamin B_{12} and others was discovered by Burton and Lochhead (1951), Went and Wilson (1938, 1939). These and other vitamins were found in cultures of *Az. chroococcum*, in various species and strains of the genus *Pseudomonas* and in mycobacteria. In recent years vitamin B_{12} was found in many bacteria and especially in actinomycetes, Some of these organisms such as *Mycob. propionicum, A. rimosus, A. aureofaciens*, and others, are employed in industry for the production of this vitamin. According to our observations, 90-95% of the actinomycetes isolated from soil synthesiss vitamin B_{12} (Krasil'nikov, 1954 c), According to Darken, 64-66% of soil actinomycetes synthesize this vitamin, (Darken, 1953).

Yamagutschi and Usami (1930) found vitamin B_2 in 15 bacterial cultures *(Bac. subtilis, Bac. Mesentericus, Bac. mycoides, Bact. prodigiosum,* various micrococci, and others).

Landy, Larkum and Oswald (1943) found paraminobenzoic acid in cultures of 35 species of bacteria and actinomycetes. Among these were: *Bact. proteus , Ps. pyogenes, Bac. aerogenes, Bac. subtilis, Bac. megatherium, Mycob. diptheriae, Mycob. stereosis,* and others.

Herrick and Alexopoulus (1943) found thiamine in cultures of 22 species of bacteria and actinomycetes.

According to Schmidt and Starkey (1951), 99 species of bacteria and actinomycetes, out of a total of 150 species studied, synthesized heteroauxin. Twenty-two bacterial cultures out of a total of 75 cultures studied, synthesized vitamins on synthetic media.

We have studied 192 bacterial cultures isolated from different soils of the USSR. These cultures were tested for their capacity to synthesize vitamin B_1 and heteroauxin. The results are given in Table 42. It can be seen from the table that more than 50% of the bacteria studied synthesize vitamin B_1 and almost 40% synthesize heteroauxin.

Table 42
Biotic compounds in bacteria

Bacteria	No of cultures tested	Number of cultures synthesizing vitamin B_1	Number of cultures synthesizing vitamin B_2
Bac. subtilis	18	10	10
Bac. mesentericus	15	12	5
Bac. sp	13	3	1
Ps. flourescens	8	8	8
Ps. denitrificans	12	10	12
Ps. mycolytica	3	3	3
Bact. coli	2	0	0
Bact. proteus	2	0	0
Bact. liquefaciens	8	5	3
Bact. sp.	13	6	8
Rhizobium trifolii	15	10	0
Rhizobium phaseoli	15	12	0
Rhizobium leguminosarum	8	6	8
Az. chroococcum	16	16	10
Az. vinelandii	4	4	2
Mycob. album	12	7	6
Mycob. citreum	10	10	8
Mycob. rubrum	3	0	0
Microc. albus	3	0	0
Microc. flavus	5	1	1
Microc. aureus	6	0	0

Shavlovskii (1954, 1955) studied cultures of Ps. *fluorescens, Ps. aurantiaca, Bact. herbicola and Ps. radiobacter* from the rhizosphere of plants. These bacteria were grown in synthetic media of a defined composition and after 8 days incubation the vitamin contents of the bacterial cells and of the culture filtrate were determined. The results are given in Table 43,

Table 43
The amount of vitamins synthesized by various bacteria, in μ g,
(after 8 days of growth)

Bacteria	Thiamine in 10 ml of medium	Thiamine in cells 1 g dry weight	Nicotinic acid in 10 ml of medium	Nicotinic acid in cells 1 g dry weight	Riboflavin in 10 ml of medium	Riboflavin in cells 1 g dry weight	Biotin in 10 ml of medium	Biotin in cells 1 g dry weight
Ps. aurantiaca	4.0	203	7.0	355	1.8	91	3.2	162
Ps. flourescens	0.2	23.3	4.4	511	0.14	162	0.18	20.9
Ps. radiobacter II	0.4	6.2	5.2	80.2	2.8	43	3.0	46
Ps. radiobacter III	0.3	13.2	4.0	176	3.6	158	0.6	26
Bact. herbicola	0.05	14.7	1.7	470	0.04	11.7	0.03	8.8

Many fungi produce biotic substances. Thus, for example, thiamine was detected in cultures of *Aspergillus niger, A. oryzae, Penicillium glaucum, Mucor mucedo, Mucor racemosus* in some species of *Phytophithora, Rhizopus, Fusarium* and others; and also in cultures of yeasts such as *Torula utilis, Sacchar. Cereviseae, Sacchar. logos, Endomyces vernalis, Willia anomala* (Bilai, 1955; Goinman, 1954; Bukin, 1940, and others),

Fungi have been found in the soil which synthesize vitamin B_2. Some of them synthesize this vitamin in such quantities that they are used in industry. These are *Candida (Oidium), Guilliermondella, Eremothecium ashbyi*(Dikanskaya, 1951). Biotin, pantothenic acid, nicotinic acid, paraaminobenzoic acid, vitamins C and K, and many others have been found in many fungi (Meisel, 1950).

Rogosa (1943) studied 114 various yeast cultures such as *Torula sphaerica* (26 strains) *T. cremoris* (20 strains), *Sacch. fragilis, Monilia pseudotropialis, Mycotorula lactis, Sacch. anamensis, Torulaopsis kefyr, Torula lactosa, Zygosaccharomyces lactis* and others. In all the cultures they found vitamin B_2 in amounts ranging from 0.6 to 0.11 μ g per ml of synthetic medium.

Biotic substances were also found in the mycorhiza fungi. According to Schaffstein (1938), mycorhiza fungi of orchids synthesize growth factors required for the normal growth of the host plant.

Biotic compounds are also synthesized by algae in the soil. It was mentioned above that these organisms are widely distributed in soil. Frequently algae grow on the earth's surface, forming a blue-green coating, visible to the naked eye.

It is known that algae secrete various organic substances--metabolic products (Goryunova, 1950). Biotic substances are among these products.

Ondratschek (1940) found ascorbic acid, (vitamin C) in the secretions of such algae as *Hormidium borlowi, H. flaccidum, H. nitens* and *H. stoechidium*. The green alga *Chlorella* synthesizes heteroauxin (Lilly and Leonian, 1941).

The majority of soil microorganisms are partially auxoautotrophs, i. e., they require only some biotic substances. For example, *Clostridium butyricum* requires only biotin and synthesizes its other requisites. Some bacterial species require only thiamine or pantothenic acid. According to Burcholder, McWeigh and Mayer (1944) of 163 yeast cultures 87% required biotin, 35% thiamine and pantothenic acid and 12 % required only inositol.

There are many bacteria which can synthesize only a fraction of a vitamin's molecule. For example, some organisms synthesize only part of the thiamine molecule, either thiazole or pyrimidine components of vitamin B_1. Consequently, the former would require pyrimidine and the latter--thiazole.

There are many soil microbes which require ß-alanine, desthiobiotin. pimelic acid and some other compound s-- components of this or other molecule of a biotic substance (Meisel, 1950; lerumalimskii, 1949; Stephenson, 1951, and others).

Thompson (1942) has shown that bacteria synthesize vitamins in amounts exceeding those required for their metabolism. The excess of vitamins is released into the environment. The enrichment of the substrate with vitamins takes place, not only at the expense of decomposing cells, but also through the secretion of vitamins by living organisms. The possibility in not excluded that some biotic substances are waste products of growing cells.

According to Thompson, about 50% of the synthesized thiamine remains in the cells of bacteria in a bound state. This fraction finds its way into the soil only after the death and decomposition of the cells.

An idea on the quantitative aspect of the synthesis of biotic substances by microorganisms can be obtained from the data in Table 44. These data have been compiled from various sources.

Table 44
The synthesis of vitamins by microorganisms grown on nutrient media
(1 per μ g of, dry weight of cells)

Microorganisms	B_1	B_2	Nicotinic acid	Panto- thenic acid	B_6	Biotin	Ino- sitol	Folic acid
Bact. aerogenes	19.9	154	630	780	26.8	47.9	1,400	105
Ps. flourescens	74	377	560	311	75.7	68.1	1,700	74.8
Bact. proteus	23	95	330	130	16.4	21.4	1,000	42
Clostr. butyricum	39.3	235	1,930	318	23.2	0	870	18.8
Az. vinelandii	96	351	593	184	--	4.2	--	--
Penicill chrysogenm	2.6	47	212	212	23.0	1.5	--	14.6
Sacchr. cerevisiae	360	42	1,000	100	100	1.2	5,000	31.2
Torula utiis	52.8	62	535	180	1.9	35	3,500	31.2

According to West and Wilson (1938, 1939) there are 19.6 μ g thiamine and 0.37 μ g riboflavin per 1 g of dry weight of the root-nodule bacteria of clover, grown on a synthetic medium. Clostridium (Clostr. butyricum)synthesizes 0.9 μ g/g riboflavin and Micro. ochraceus, Micr. citreus, Ps. pyocyanea, about 10-15 μ g.

Yamagutschi and Usami (1939) found about 1.5 μ g thiamine per g of dryweight of cells in cultures of Ps. fluorescens, Ps. alba, and Bact. prodigiosum. In cultures of Bact. proteus they found 9-14 μ g thiamine per 1 g dry weight of cells.

Considerable quantities of biotic substances are synthesized by many mycobacterial species. For example, Mycob. smegmatis synthesizes about 135 μ g Oof vitamin B_2 per ml of synthetic medium and 36 μ g per g of the dry weight of cells (Mayer and Rodbart, 1946). Forty to eighty μ g per ml of medium. of the active substance mycobactin was found in cultures of Mycob. phlei (Francis at al., 1953).

Ostrowsky and others (1954) studied vitamins in various representative microorganisms. Their results are given in Table 45.

Table 45
Vitamin content of bacterial cells
(μ g per g of dry weight of cells)

Microorganisms	B_1	B_2	Nicotinic acid	Biotin	Peteroyl-glutamic acid	Pantothenic acid
Thiobact. thioparus	21	31	92	0.77	0.46	0.75
Thobact thiooxydans	23	60	15	0.64	1.89	57.0
Ps. pyocyanea	15	43	240	2.4	1.0	140.0
Ps. chroococcum	96	--	590	--	--	--
Propionibact. pentosaceum	6.4	--	--	--	--	93.0
Clostr. butyricum	9.3	55	250	1.7	0.5	92.0
Bact. proteus	21.0	--	250	3.4	4.2	100.0
Ps. flourescens	26.0	68	210	7.1	1.8	90.0
Bact. prodigiosum	27.0	35	240	4.1	3.2	120.0

The above-given quantitative data are not strictly constant. They may vary, depending on the growth phase and culture conditions of the individual species of the bacteria, fungi or actinomycetes. In some cases old cells contain less vitamins than young cells, whereas in some species the reverse picture is observed: the old cells contain more vitamins than the young cells. For example, some cultures of Ps. radiobacter contain more vitamin B_1 after 8 days growth than after 2 days growth.

In some media the microbes synthesize many growth factors, in others only a few or none.

The intensity of the formation of vitamins by soil organisms is greatly influenced by the symbiotic microbes. Some of them suppress vitamin synthesis and other stimulate this process. According to Smalii (1954),Azotobacter (Az. chroococcum) in pure culture synthesize 173 μ g of heteroauxin (per cell mass in 1 Petri dish, on Ashby agar) and in the presence of the following microorganisms synthesizes this substance in the noted amounts:

Bact. mycoides, 220 μ g
Bact. denitrificans, 196 μ g
Ps. radiobacter, 243 μ g
Torula rosea, 234 μ g
Act. Coelicolor, 188 μ g
Penicill. Nigricans, 149 μ g

The capacity to synthesize auxins or vitamins does not characterize a species Different strains of one and the same species differ markedly from each other. For example, of more than 100 strains of Az. chroococcumwhich we isolated in different soils and places in the USSR, some of them synthesized large quantities of heteroauxins and others synthesized only small quantities, or none at all. Formation of heteroauxin by the different representatives of these cultures in shown in the coleoptile photograms (Figure 63).

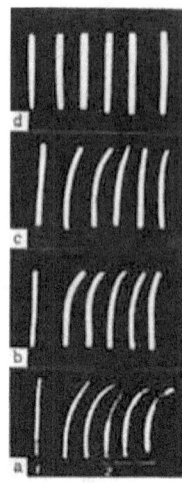

Figure 63. The formation of heteroauxin by different cultures of *Azotobacter chroococcum*. The coleoptile curvature after immersion in the culture, expressed in degrees: a) museum strain 54, angle of deviation--32°; b) strain isolated from garden soil in Moscow vicinity, angle of deviation10°; c) strain isolated from the soils of Kara Kum, angle of deviation-8°; d) strain isolated from cultivated podsol soil (Experimental Station Chashnikovo, Moscow Oblast, angle of deviation--0°; 1--control coleoptile; 2--experiment immersed in the bacterial culture.

Similar data were obtained when studying other bacterial species and not only for heteroauxin but also for biotin, thiamine, riboflavin, and other biotic substances.

Although there is no strict species specificity as far as the synthesis of biotic substances is concerned, nevertheless mass analysis does show group differences in this respect. More strains of *Azotobacter* synthesize vitamins and auxins than bacteria of the genus *Bacterium*.

Only a few species of root-nodule bacteria are capable of synthesizing heteroauxin and even these are weak forms. We have investigated 12 species of root-nodule bacteria of clover, lucerne, kidney beans, vetch,*Lathyrus vermus,* lungwort, peas, Onobrychis, soya, lupine, acacia and astragalus. All these species either did not synthesize heteroauxin at all, or synthesized it in small amounts only (Figure 64).

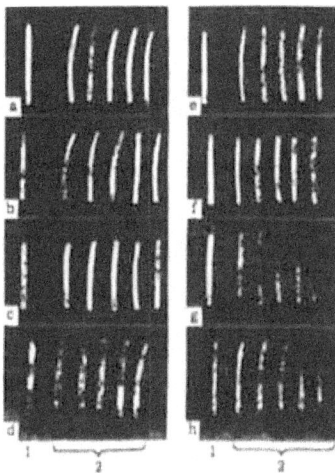

Figure 64, The formation of heteroauxins by various species of root-nodule bacteria. The magnitude of the curvature on immersion in cultures of: a) red clover, the angle of curvature-6°; b) soya, angle of curvature- 6°; c) broad beans, angle of curvature- 3°; d) peas, angle of curvature-4°; e) vetch, 4°• angle; f) sweet clover, 2° angle; g) beans, 2° angle, h) proactinomycetes from the nodules of alder tree, 3° angle; 1--control coleoptile; 2-- experiment immersed in bacteria.

Out of 60 strains of root-nodule bacteria of lucerne, only 9 strains synthesized this compound in amounts able to give a barely perceptible coleoptile curvature. Fifteen strains of *Rhizobium trifolii*, 8 strains of *Rhizobium leguminosarum* and 15 strains of *Rh. phaseoli* were examined. In all cases the picture was the same.

We have not detected any synthesis of heteroauxins by proactinomycetes which form nodules on the roots of alder tree; actinomycetes and proactinomycetes of the soil do synthesize heteroauxin to a greater or lesser extent.

According to Starkey (1944), the nicotinic-acid content of plant residues ranges from 2.4 to 85 μ g per gram of dry weight, in the majority of cases it is lower than 30 μ g/g. The same substance in microbial cells amounts to 150-1,920 μ g/g, i. e., approximately 25-60 times more.

It should be noted that the studies of biotic substances were carried out on relatively few species of soil microorganisms. The choice of organisms was taken at random, and the studies were confined to a few vitamins only, in the majority of cases to thiamine and riboflavin.

It should be assumed that in reality many and possibly all soil microorganisms synthesize these or other biotic substances which play an essential role in the life and metabolism of lower and higher organisms.

It is obvious that under natural conditions (life in the soil) the microbial metabolism and the synthesis of biotic substances would differ from that under laboratory conditions (on artificial nutrient media).

Schmidt and Starkey (1951) have shown that if plant residues which do not contain vitamins are introduced into soil which also does not contain vitamins, the latter appear and accumulate in greater or lesser amounts due to the decomposition of the residues by microbes. The increase in the riboflavin content of the soil is concomitant with the intensification of microbial metabolism (Figure 65). The more plant residues introduced into the soil, the more intense the microbial growth and the formation of riboflavin (Table 46).

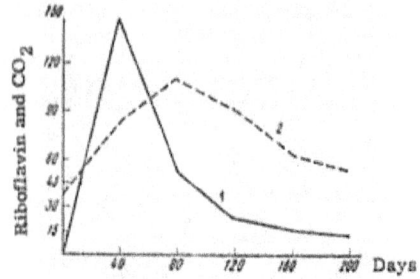

Figure 65. The formation of riboflavin in the soil as a product of the metabolism of microorganisms, the activity of which is determined by the evolution of CO_2 in mg per 100 g of soil: 1--riboflavin, in μ g/100 g; 2--CO_2 in mg/100 g.

Table 46
The formation of riboflavin in the soil during the decomposition of oat straw
(μg per 100 g of soil)

Accumulation of riboflavin in days:	0 days	1 day	3days	4 days	7 days	56 days
1.25 grams of straw applied	11	19	26	27	26	13
2.5 grams of straw applied	20	22	60	55	38	19

Similar results are obtained if glucose or saccharose are introduced into the soil instead of straw; the bacteria inoculated into the soil lacking the vitamins begin to grow at the expense of the sugars; and riboflavin, biotin, heteroauxins, etc accumulate in the soil.

According to Meisel's calculations (1950), about 400 g of vitamin B_1, 300 g of vitamin B_6 and 1 kg of nicotinic acid are synthesized by microbes in the surface layer of one hectare of the fertile soils of the southern regions, during one season (9 months).

Biotic substances are preserved in the soil for varying periods of time. A pure preparation of a vitamin introduced into the soil can be detected for several days. According to Schmidt and Starkey (1951), riboflavin and pantothenic acid persist in the soil from 3 to 20 days or longer (Table 47).

Table 47
The preservation of riboflavin and pantothenic acid in soil
(μg/100 g soil)

Vitamin	Amount introduced into the soil	Soil	Present after 0 days	Present after 1 day	Present after 2 days	Present after 3 days	Present after 6 days	Present after 21 days
Riboflavin								
	40	Sterile	38	33	--	40	34	34
	40	Nonsterile	36	36	--	43	16	12
	80	Sterile	69	68	--	81	67	64
	80	Nonsterile	65	68	--	81	49	13

Pantothenic Acid								
	50	Sterile	34	34	35	35		
	50	Nonsterile	32	34	10	10		
	100	Sterile	72	77	80	73		
	100	Nonsterile	68	64	18	10		

Riboflavin persists in soil longer than pantothenic acid. Both compounds last longer in sterile than in contaminated soil, since biotic substances, like all other compounds, are subjected to microbial decomposition.

If the microbial metabolism is artificially arrested, vitamins introduced into contaminated soil persist for the same periods as in sterile soil. It was found that biotic compounds (vitamins and heteroauxin) persist in samples of dry soil taken from cultivated and fertilized fields, from 3 to 4 months to 4 years depending on the kind of soil and its properties and also on the properties of the vitamins themselves (Stewart and Anderson, 1942).

Vitamins and other biotic substances entering the soil by one or another route are decomposed and synthesized de novo by microorganisms, Some vitamins disappear others appear. There is a continuous turnover of these substances in the soil. Biotic substances can be found in the soil during the entire vegetative period an long as the microbes live, reproduce and exhibit metabolic activity. The amount of the biotic substances is determined by the rate of their synthesis and introduction into the soil and also by the rate of their destruction, or their stability,

The effect of biotic substances on plants

It was noted above that green plants synthesize for themselves the necessary biotic substances or phytohormones. Under conditions favorable for their growth this synthesis meets all their requirements for normal growth. In certain, not infrequent, circumstances, apparently under some unfavorable conditions, the plant synthesizes inadequate amounts of these substances. Then specific avitaminoses develop which are expressed to a greater or lesser degree in the form of certain physiological disturbances and diseases.

Different plants react variously to the addition to the substrate of growth factors and vitamins. Some respond by enhanced growth or by changes in the course of biochemical processes, others react weakly and still others do not react at all. This permits us to assume that the first produce only minimal amounts of the active substances which are insufficient for their normal metabolism, the second synthesize them quite actively but in amounts still insufficient to satisfy all their needs, and the third synthesize them in adequate quantities.

Investigations show that even the last group of plants by no means always synthesize adequate amounts of biotic substances. The vitamin content of plants varies within a wide range, depending on external conditions of growth. It varies according to the soil and climate conditions (Murry, 1948; Rakitin, 1953). Fertilizers have a great effect on the quantity of vitamins present in plants.

In all cases of avitaminosis the vitamins from the substrate are absorbed by the plant. Even under normal conditions of growth, plants utilize ready-made biotic substances, if available.

The utilization of vitamins, auxins and other compounds from the soil has been confirmed in many experiments. Many plants and biotic substances were studied under laboratory and field conditions, in sterile and nonsterile experiments.

The plants' requirements for vitamins and auxins has been thoroughly studied in experiments with isolated organs and tissues, and especially with isolated roots.

It is known that excised roots of many plants will not grow in synthetic media in the absence of biotic substances and a carbon source. If a root 2-3 mm long is excised from a plant which grew under sterile conditions and placed in a synthetic artificial medium (Bonner's medium or other) it will grow in length to reach considerable dimensions and will form lateral roots, etc, only if the necessary biotic substances are present in the medium. In the absence of the latter, or if their concentration is insufficient. the roots will not grow at all or the growth will be weak.

Investigations show that roots of different plants demand different growth factors. For example, roots of flax require vitamin B_1, roots of peas, horse-radish, lucerne, clover and cotton require vitamins B_1 and B_6; roots of tomatoes. thorn apple, and sunflower require vitamins B_1, B_6 and pantothenic acid (Bonner et al., 1937; Robbins and Bartley, 1922-1938; Robbins and Schmidt, 1939, 1945). Excised roots of many plants, growing on synthetic media, synthesize all the required growth factors. Some of them synthesize them in amounts sufficient for their normal growth, others form too little. The first grow well in artificial media, the latter require the addition of the missing factors (Bonner, 1942). Bonner and Bonner (1948) give the following data on the vitamin requirements of isolated roots (Table 48).

Table 48
Vitamin requirements of isolated roots of various plants
(according to Bonner T. and Bonner H., 1948)

Plants	Vitamin B_1 requirement	Nicotinic acid requirment	Vitamin B_6 requirement
Linum usitatissimum Boenn	stimulates	-	-
Raphanus sativus L.	+	+	-
Medicago sativa L.	+	+	-
Trifolium repens L.	stimulates	+	-
Gossypium hirsutum L.	+	+	-
Crepis rubra L.	+	+	-
Cosmos sulfureus	+	+	-
Pisum sativum Gov.	+	+	-
Daucus carota L.	+	-	+
Lycopersicum esculentum Mill	+	-	+
Lycopersicum esculentum pimpinellifolium	+	stimulates	+
Dun	+	stimulates	+
Helianthus annuus L.	+	stimulates	+
Acacia melanoxylon R. Br.	+	stimulates	+
Datura stramonium L.	+	+	+

The following data show the effect of vitamins, on the growth of isolated roots. The roots of flax in the presence of vitamin B_1 elongated by 185 mm. and in its absence by 31 mm in one week. The roots of flax are calculated to synthesize vitamin B_1 at a rate of 0.02 µ g per week. Their vitamin B_1 requirement for normal growth is 2 µ g, i.e., 100 times more than they synthesize.

The roots of white clover grow well in a medium containing vitamins B_1 and PP*. *[The correct designation of this vitamin is unclear.] They increase in length with each successive transfer into a fresh nutrient medium. In the first 5 weeks the roots elongate by 84 mm, the increment in the next 5 weeks amounts to 109 mm, in the third five-week period the increment amounts to 129 mm, in the fourth five-week period--136 mm, and in the following 5 weeks 151 mm. The increment becomes uniform upon subsequent transfer amounting to about 22 mm per week.

The roots of sunflower in the absence of the vitamin complex or in the presence of only one of the vitamins PP or B_6 cease to grow after 7 consecutive resowings. Roots which were supplied with all three vitamins, PP, $B_1$and B_6 grew well for along period of time allowing for many transfers into fresh media. In the first five weeks the increment was 74 mm, in the next 5 weeks it amounted to 96 mm, in a further 5 weeks--120 mm, and in the following fortnight--150 mm.

The roots of plants belonging to diverse varieties of one and the same species react differently to vitamins. For example, one variety of tomatoes requires vitamin B_6 and does not respond to vitamin PP and on the contrary another variety requires vitamin PP and does not react to vitamin B_6 (Bonner and Bonner, 1948).

Ovcharov (1955) introduced vitamin PP into a medium in which he grew cotton plants whose leaves were cut off. He observed enhanced formation of new roots on the old roots.

Went, Bonner and Warner (1938) had shown that thiamine stimulates the growth of roots of peas, lemons and camellia. The results were more markedly pronounced when a mixture of thiamine and heteroauxin was employed. Positive results were also obtained in these cases with a mixture of vitamin B_1 and indoleacetic acid (Grebenskii and Kaplan, 1948) and also with vitamin K, biotin and pantothenic acid with biotin (Scheurmann, 1952).

Psarev and Veselovskaya (1947) noted the stimulating effect of thiamine on the formation and growth of wheat roots.

In some cases roots of certain plants required only parts of the vitamin molecule, for example only thiazole or pyrimidine (components of vitamin B_1).

Some roots require unknown biotic compounds and cannot, therefore, be grown in vitro.

Robbins (1951) in his review, brings a list of plant species the roots of which can grow on nutrient media. There are 22 such species. Roots of 27 species could not be grown in isolation despite the addition of various vitamins, auxins, amino acids and other biotic substances.

It should be noted that even the roots which can grow in vitro do not grow in the same manner as when attached to the plant. They grow in length and branch, but do not get thicker or if they do thicken, then only very slightly. The activity of the cambium is completely or almost completely suppressed. Consequently, no entirely adequate medium has as yet been found for isolated roots.

The requirement for biotic substances is well pronounced in seedlings. The need embryos of some plants develop better and quicker in the presence of certain vitamins added to the substrate. For example, the growth of pea seedlings separated from the cotyledons considerably increases in the presence of thiamine and biotin (Kögl and Haagen-Smit, 1936). Pantothenic and ascorbic acids also act favorably on pea embryos (Bonner T. and Bonner H. , 1948).

Plant embryos do not synthesize biotic substances, they utilize the food reserves present in the seeds. Even the green sprouts of many plants in the early growth period synthesize vitamins weakly or not at all (Bonner et al., 1939). Ripe embryos of thorn apple are easily grown on artificial media without vitamins while the nonripe embryos require vitamins PP, B_1, B_6, C and others.

Pantothenic acid also has a favorable effect on lucerne sprouts. Treating pea seeds with vitamin C enhances their growth by 213% as compared to the controls. Sprouts of meadow grass react positively to the addition of vitamins B_1, PP, H and pantothenic acid to the medium. The grape seeds germinate quicker in the presence of an 0.01 % solution of vitamin PP and, moreover, the formation of roots and the growth of aerial parts is more intense (Flerov and Kovalenko, 1947).

The presowing treatment of cotton seeds with vitamins B_1 and PP considerably enhances their germination and the subsequent growth of the sprouts, Seventy-five per cent of the seeds germinated after treatment with the vitamin as compared to 45 % in the control. The length of the sprouts in the former case was on the average 1.35 cm and 1.65 in the latter Zakharyants, Gorbacheva and Zglinskaya, 1950).

An increase in the growth and subsequent yield of bean seeds after treatment with vitamins B_1 and PP was observed. The height of the plants (from the treated seeds) was greater by 18%, the increment of the vegetative mass was greater by 39%, and the yield of seeds was 28% higher then that of the control plants (Dagis, 1954).

Bonner et al. reached the conclusion that the lower the vitamin content of plant leaves, the stronger they react to the addition of these substances. According to them, peas and tomatoes, contain 13-18 μ g of vitamin B_1 per kg of dry leaves and do not react to its addition. Cabbage, cosmos, Japanese camellia and others contain small amounts of vitamins in their leaves and react positively to the addition of these substances. However, this does not hold for all plants. There are species or even varieties of one and the same species which contain a small amount of vitamins in their leaves and react less to their addition than plants with higher vitamin content.

The addition of vitamins has a favorable effect even on mature plants. Tung trees after the addition of 0.5 mg of vitamin B_1 grew in 70 days twice as much as the controls. The application of low concentrations of vitamin B_1 to poppies increased the weight of their bolls as well as the crop in general. Application of vitamin B_1 together with water had a favorable effect on the growth of spinach. The weight increment during the 63 days of the experiment exceeded many times that of the control plants (Table 49).

Table 49
Dry weight of spinach leaves (in mg) at the end of 63 days of the experiment, after 13 irrigations with a solution of vitamin B_1

Treatment	1*	2*	3*	4*	5*	6*	7*	8*	9*
Control	32.4	58.1	56.2	35.8	9.6	--	--	--	--
Vitamin B_1	28.5	103.9	255.6	390.0	504.1	534.9	375.9	205.0	45.7

*This is unexplained in Russian text. Probably refers to position of leaves on plants.

Denisov introduced vitamin B_2 into a substrate where egg plants were grown and obtained a marked increase in yields. After 77 days growth the control plants had stems 7.1 cm long, the weight of the tops was 48 5 g and the weight of the roots 12.5 g. The corresponding figures for plants grown in the presence of the vitamin B_2 were 12.2 cm 121 g and 22.9 g (Ovcharov, 1955).

The growth of vine grafts, soya and other plants in increased under the influence of vitamin PP. Lemon seedlings react markedly to the addition of this vitamin (Table 50).

Table 50
The growth rate of lemon seedlings under the influence of vitamin PP
(according to Kocherzhenko and Snegirev, 1946)

Treatment	Average height of plants on 15/ VIII	Average height of plants on 15/ IX	Average height of plants on 15/ X	Average height of plants on 15/ XI
Control	27.2	29.2	35.5	38.3
Vitamin PP	23.8	34.7	44.5	47.7

Analogous data were obtained by Matveev and Ovcharov (1940) in their experiments with a Bukhara almond. The plants were sprayed with an aqueous solution of vitamin PP and adenine. Earlier opening of buds and more rapid development of leaves were observed. The number of leaves was 4 times greater than in the control plants.

Vitamins play a considerable role in the development of orchids. These plants, as already mentioned, grow badly, or do not grow at all without the micorhizal fungi. It was found that the seeds of orchids contain only small amounts of vitamin PP, which are not sufficient for normal germination. This shortage is remedied thanks to the mycorhizal fungi. Treating the seeds with vitamin PP secures their normal germination in the absence of the fungi. The dry weight increment was 3 times higher than that in control plants. It was shown that orchids of the group Vanda grow well in the presence of substances obtained from the mycorhizal fungi. These substances resemble, in their action, bios II (according to Kelly, 1952). According to Noggle and Wynd (1943), some orchids grow well in the presence of nicotinic acid. Henrikson (1951) noted the positive effect of thiamine, vitamin B_6 and nicotinic acid on the germination and subsequent growth of *Thunia marschaliana* Rchb. f. (Table 51).

Table 51
The effect of vitamins on the growth of the orchid Thunia marschaliana Rchb. f.
(according to Henrikson, 1951)

Vitamins	Height of plants, mm	Number of leaves	Length of roots, mm	Dry weight, mg
Control	48.0	5	60.5	30.8
B_1	83.5	6	92.5	59.1
B_6	48.0	6	58.5	59.1
C	55.5	6	55.5	30.1
PP	101.0	8	176.5	86.6

Rakitin and Ovcharov (1948) employed vitamin PP and adenine for increasing the growth of cotton plants in their early growth stages. Thereby, not only growth, but also fruiting was increased. The number of bolls was increased, and the cotton yield (raw material) was considerably higher than that of the controls. Similar data were obtained by Zakhar'yants, Gorbacheva and Zglinskaya (1950). They sprayed cotton plants with solutions of vitamin PP and thiamine, The cotton crop increased by 34.1% as compared to the control.

The fat-soluble vitamins, A, E, K, in contrast to vitamins of the B-group suppress the growth and diminish the crop of plants. Carotene suppresses the growth of safranin which is itself rich in carotene. Vitamin K suppresses the growth of fungi, some bacteria and the roots of higher plants. Vitamin PP antagonizes the action of vitamin K. Vitamin E, according to Schopfer (1950), arrests the growth of certain plants. The height of plants in the control was 35.75 cm and in the presence of vitamin E--6.14 cm; the number of flowers in the former was 80.4 and in the latter, 10.

Ovcharov (1955) immersed the seeds of plants in a solution of thia ine and yeast extract, Such procedure markedly stimulated the growth and increased the crops, the seeds too were larger.

Söding, Bömke and Funke (1949) obtained 30% higher yields of carrots after treating their seeds with nicotinic acid, vitamin B_1, vitamin C and other substances.

Experiments with vitamins under sterile conditions are worthy of mention. McBorney, Bollen and Williams (1935) tested the action of pantothenic acid on the growth of lucerne under sterile conditions in sand cultures, in a medium which did not contain nitrogen. Pantothenic acid was added in high concentrations. Plants under these conditions grew in the presence of pantothenic acid (in high concentrations of pantothenic acid) much better and the yield was higher.

Magrau and Mariatt, (1950) showed that a number of vitamins, such as thiamine, nicotinic acid, biotin and pantothenic acid had a positive effect on the growth of Poa annua L. under sterile conditions. Swaby (1942) tested the effect of certain organic substances including some containing vitamins, on the growth of cereal and leguminous plants, in the presence and absence of microorganisms. The experiments showed that in the presence of microorganisms organic substances rich in vitamins have a favorable effect on the growth of plants.

Shavlovskii (1954) tested the effect of pantothenic acid, vitamin B_1, nicotinic acid and vitamin B_6 on the growth of lucerne. The latter was grown on agar medium under sterile conditions for 30 days. The results are given in Table 52.

Table 52
The effect of vitamins on the growth of lucerne
(vitamin concentration in the medium = 0. 1 μ g/ml)

Vitamins	Dry-mass weight of 20 plants in mg: Tops	Dry-mass weight of 20 plants in mg: Roots	Dry-mass weight of 20 plants in mg: Total
Control (without vitamins)	38.6	8.0	46.6
Pantothenic acid	36.4	11.6	48.0
Vitamin mixture	37.2	12.2	49.4

Analogous experiments were carried out by Shavlovskii with buckwheat. The plants were grown in sand wetted with the nutrient solution of Hellrigel. containing 1 μ g of the vitamin per ml. Other containers were supplemented with yeast extract and vitaminless casein hydrolysate. In one series of experiments the bacterial culture of *Ps. aurantiaca*--vitamin producers were introduced. Plants were grown for 2 days and then analyzed. The results are given in Table 53.

Table 53
The effect of biotic substances on the growth of buckwheat

Biotic compound	Dry mass weight of 10 plants in mg: Cotyledons	Dry mass weight of 10 plants in mg: Stems	Dry mass weight of 10 plants in mg: Roots	Dry mass weight of 10 plants in mg: Whole plant
Control without vitamins	73.0	64.0	32.0	168.0
Bacteria *Ps. aurantiaca*	76.0	64.0	41.0	181.0
Vitamin B1	82.0	63.0	39.0	184.0
Vitamin B12	79.0	64.0	32.5	175.5
Vitamin mixture	80.0	65.0	36.0	181.0
Yeast extract 0.01%	80.0	66.0	40.0	186.5
Yeast extract 0.1%	90.0	65.0	40.0	195.0
Casein hydrolyste 0.1%	80.0	66.0	43.0	189.0

It can be seen from the given data that the substances tested, markedly increase the increment of the roots and aerial parts of the plants.

The biological role of vitamins has been little studied, but, according to the available data, it is important. It is well known that many of them are components of various enzymatic systems. The so-called coenzymes which enter into chemical interaction with the substrate include many vitamins, It has been found experimentally that vitamin B_1 in a compound together with phosphoric acid is the coenzyme of carboxylase-cocarboxylase.

Carboxylase is an enzyme participating in transformations of carbohydrates. It is widely distributed in plants, animals and microbes. Without it the various transformations of carbohydrate compounds, including pyruvic acid, are not feasible. The latter is the key intermediate in the metabolism of living cells, which links the metabolism of carbohydrates, proteins and fats.

In the absence or shortage of vitamin B_1 the synthesis of cocarboxylase is slowed down or arrested and, consequently, carbohydrate metabolism is slowed The latter is frequently arrested at the stage of pyruvic said, which leads to the accumulation of pyruvic acid in the cell and the complete cessation of metabolism.

Vitamin B_1 participates not only in decarboxylation of pyruvic acid but also in the reverse reaction--the fixation of CO_2 in pyruvic acid. The role of vitamins in the fixation of CO_2, as the investigations of recent years have shown, is very great.

Vitamins also play a considerable part in the formation and transformation of proteins. It has been shown that vitamins B_2, B_6, B_{12}, PP and H participate in the formation of amino acids and their transaminations. The shortage of vitamin B_6 leads to a decrease in the formation of amino acids from organic acids and ammonia. Vitamin B_6 takes part in the formation of amino acids from organic acids and ammonia. Transamination, i.e., transfer of an amino group (NH_2) from one acid to another, takes place in the presence of vitamin B_6.

In fat synthesis from sugars, vitamins B_1, B_2, PP and pantothenic acid participate, and the transformation proteins into fats also requires vitamin B_6.

Vitamins play an immense role in respiration. It was shown that enzymes participating in respiration consist of proteins and a coenzyme. The latter consists of vitamin B_2 and phosphoric acid. Vitamin B_2 in enzymatic systems plays a role in oxidation-reduction processes. Folio acid is of great importance in respiration. The germination of seeds and the respiration of sprouts increases under the action of this acid (Stephenson, 1951, Schopfer, 1943, Zeding, 1955).

Growth stimulators--auxins and heteroauxins--have a positive effect on the colloidal and chemical properties of protoplasm. According to some authors, they participate in the general metabolism of the cell as separate components. By increasing the metabolism they influence the growth of the cells of the aerial parts and especially of the roots. Under the influence of heteroauxin the influx of plastic substances increases which leads to the formation of now roots in greater quantities. During the rooting of grafts, hydrolysis of starch and fats increases in the cells of the latter. The activity of peroxidase is also increased and the tissues are better hydrated (Maksimov, 1940, Turetskaya, 1955, Zeding, 1955, and others). The action of these substances does not affect the turgor of cells only, as previously assumed, but affects the general metabolism of the plant. In this respect they resemble other biotic substances. Data exist which show that heteroauxin stimulates the formation of auxins, (Zeding, 1955).

Kuhn (1941) has shown that carotene and carotenoids have a great effect on the formation of sexual cells and on conjugation of a *Chlamydomonas* alga. According to him, there exist carotenoids with specific properties of male and female hormones. He found a carotenoid--safranol--with properties of a male hormone and a carotenoid--picrocrocin--with the properties of a female hormone.

Vitamins have a favorable effect on the fertilization of plants. It was found that the sexual organs are rich in vitamins especially the pollen. For example, the pollen of the pea tree contains 2,300 mg of carotene and that of sunflower, 1,460 mg per kg. Pollen of some plants do not contain large amounts of carotene. Vitamins decompose under the action of light and pollen decolorizes and loses its activity. Processing of such pollens with carotene increases its capacity to germinate. Thus, according to Lebedev (1952), without the addition of carotene the percentage of germinated hemp pollen was 39%, the length of the pollen tubes was on the average 100 μ ; in the presence of carotene the percentage of germinated pollen was 53.5 and the length of the pollen tubes 312 μ . The lower the vitamin content, the sharper the reaction to the addition of carotene. Pollens rich in this vitamin stimulate the germination of pollen which contains small amounts of the provitamin if they are left to germinate together.

Other vitamins (C_1, B_6, B_1, B_2, PP) also have an effect on the germination of pollen. Pollens of different species and also of different varieties of the same species do not give the same reaction to the addition of vitamins. For example, pollens of one variety of tobacco require 0.0002 mg of vitamin B_1, and pollens of another variety of the same plant require 0.005 mg per liter of the solution. Thirty one per cent of pine pollens germinated in a medium vitamin PP and in the presence of this vitamin 54% of the pollens germinated. About 10-12% of the pollen grains of one variety germinate in the presence of vitamin B_1 and in another variety--52% germinate (Polyakov, 1949).

The assimilation of biotic and other substances by plants

The problem of plant absorption of biotic substances-vitamins. auxins, and other organic substances, has interested investigators for a long time.

Assimilation of vitamins

Many investigators have studied the uptake of vitamins and auxins through the root system or leaf surface. Carpenter (1943) introduced riboflavin by spraying the crowns of decapitated plants such as tomatoes, tobacco, fuchsia and carrots, which were subsequently kept in a dark room. The analysis of their sap showed the presence of riboflavin in much higher concentrations than that in the control plants, sprayed only with water. Plants sprayed with a thiamine solution contained thiamine in higher concentrations than the control plants (Hurni, 1944; Schopfer, 1943).

Bonner et al., (1939) analyzed plant tissues grown in a solution containing vitamin B_1. The results of these experiments are given in Table 54.

Table 54

Thiamine concentration in leaves of plants after its artificial application, in mg/kg of dry weight
(according to Bonner at al., 1939)

Plants	Treated Plants	Control Plants
Brassica alba Schmalh.	15.8	6.0
Brassica nigra Koch	6.4	3.9
Agrostia tenuia L.	8.6	5.8
Poa trivalis L.	7.2	4.4
Cosmos Gay	6.0	5.0

Radioactive vitamins are being used in recent years for the detection of their uptake by plants.

Shavlovskii (1954) employed vitamin B_1 containing radioactive sulfur S^{35}. The plants were grown under sterile conditions on an agar medium of Knopp, supplemented with the mentioned radioactive vitamin at a concentration of 0. 1 y/ ml. The activity of this medium was 4,400 counts / minute (cpm) per 1 ml. The amount of vitamin in the tissues of the plant was determined after various periods of time. The number of cpm showed the amount of absorbed vitamin. The results are given in Table 55.

Table 55

Absorption of radioactive vitamin B_1 by plants, from sterile agar substrate
(according to Shavlovskii, 1954)

Plant organs	Buckwheat: cpm per plant after 11 days	Buckwheat: cpm per plant after 38 days	Peas: cpm per planlt after 11 days	Peas: cpm per plant after 38 days	Corn: cpm per plant after 11 days
Leaves	3,458	4,920	1,110	1,326	6,993
Stems	1,655	13,288	1,486	2,912	--
Roots	505	1,601	455	3,360	783

These results show that the radioactive (S^{35}) vitamin B_1 enters the plant is the roots, is initially concentrated in the leaves, and then gathers in the stem and in the roots. Apparently, the increase of the vitamin concentration in the roots and stems in the late growth phase of the plant may be explained by the fact that the vitamin is absorbed in the early phase of the growth when the vitamin synthesis by the shoot is inadequate. Later on the plant begins to synthesize the required amount of the vitamin. and the latter, under favorable conditions, enters the stem and roots. It has been shown that plants can obtain vitamins from microorganisms. This was demonstrated by the use of radioactive isotopes. If bacteria or yeasts were saturated with an amino acid or vitamin labeled with phosphorus P^{32} or sulfur S^{35} and inoculated into a medium where plants were being grown, the radioactive substances were soon detected in the tissues of plants. Shavlovskii (1954) grew buckwheat in sand with a culture of *Ps. aurantiaca* or yeasts *(Torlopsis utilis, T. latvica* and *Rhodotorula rubra)*. All the cultures were previously saturated with radioactive vitamin B_1 (S^{35}). After 11 days the plants were analyzed for the S^{35} content of their tissues. Simultaneously the excretion of the radioactive vitamin by the bacteria was measured. The analyses showed that buckwheat takes up the vitamin excreted by the microbes in detectable amounts. The greatest amount went to, the plant (8.5 % of the total activity) from *Ps. aurantiaca* (Table 56).

Table 56

Transfer of vitamin B1 from microbes to buckwheat
(according to Shavlovskii, 1954)

Part of plants	S^{35} (of vitamin B_1) cpm/plant *Ps. aurantiaca* 8,100 cpm	S^{35} (of vitamin B_1) cpm/plant *T. latvica* 45,000 cpm	S^{35} (of vitamin B_1) cpm/plant *Rh. rubra* 9,100 cpm	S^{35} (of vitamin B_1) cpm/plant *T. utilis* 10,230 cpm
Cotyledons	294	252	144	150
Stems	217	161	96	114

Roots	180	109	77	77
Total per 1 plant	691	522	317	341
Given up by microbes to plant, in percent	8.5	1.2	4.0	3.3

Assimilation of amino acids

The capacity of roots to absorb amino acids has been proven experimentally. Petrov (1912) gives data on the absorption of asparagine by plants (corn). Shulov (1913), Pryanishnikov (1952) and Byalosuknya (1917) confirmed these data. According to them, asparagine is a good source of nutrition for peas, corn, cabbage, mustard, and other plants.

Hutchinson and Miller (1911) showed that pea plants can assimilate leucine, glycine, aspartic acid and tyrosine. Klein and Kisser (1925) have shown that during growth in sterile conditions, oats can assimilate arginine not less than they can assimilate nitrates. According to Virtanen and Lathe (1937, 1946), peas and clover assimilate aspartic acid well, while wheat and barley react negatively to this substance, Steinberg (1947) showed that isoleucine has a harmful effect on the growth of tobacco. Tanaka (1931) has shown that the plant *Sysirinchium bermudianum* utilizes asparagine, glycine, and cystine. Miller (1947) followed the absorption of dimethionine by tobacco and tomatoes growing under sterile conditions. The sap of roots and aerial parts contained 1.0- 2.5 mg methionine per 5 ml. Miller is of the opinion that amino acids are assimilated by plants.

Riker and Gütsche (1948) have shown that some amino acids suppress the growth of isolated tissues of sunflower. The authors think that the deleterious action of the amino acids is caused by their excess, which interferes with the metabolism. Sanders and Burkholder (1948) noted that a mixture of amino acids has a more favorable effect than each of them separately.

Virtanen and Lincol ascribe a stimulating role to the amino acids. According to them, small concentrations of alanine and phenylethylamine (decarboxylation product of phenylalanine) strongly affect the growth of peas, in the same way as the heteroauxin. The plant parts change markedly, becoming stronger and greener.

In order to demonstrate the uptake of amino acids by plants, Shavlovskii (1955) studied the absorption of, radioactive methionine S^{35} by buckwheat, corn, and peas.

This culture excretes more vitamin than the others. Yeasts excreted various amounts of the vitamin depending on the species. The data given below should be considered as approximate and possibly lower than in reality. This amino acid was added to the medium of Knopp in which these plants were grown. After 11 days of growth the plant tissues were examined for the presence of radioactive sulfur. The results are given in Table 57.

Table 57
The absorption of radioactive methionine S^{35} by plants (cpm per plant)
(according to Shavlovskii, 1955)

Plants	Specific activity: Leaf	Specific activity: Stem	Specific activity: Root	Activity, (dry weight) of: Leaf	Activity, (dry weight) of: Stem	Activity, (dry weight) of: Root
Buckwheat	56	81	625	389	759	1,389
Corn	65	--	124	5,814	--	3,452
Peas	46	31	628	828	992	7,530

As can be seen from the table, the absorption of this amino acid by plants is quite vigorous. The highest concentration of the amino acid in in the roots. The intensity of absorption varies according to the composition of the medium and external conditions.

The absorption of methionine is more rapid and more pronounced in the presence of vitamins B_1 and B_6.

Ratner and Dobrokhotova (1956) have shown the activating effect of thiamine pyridoxine and pantothenic acid on the synthesis of glutamic acid and alanine in the roots of sunflower.

The distribution of the radioactive sulfur of methionine is different from that of radioactive inorganic sulfur. In the former came sulfur is concentrated in the roots and in the latter, in the aerial parts (Thomas and Hendricks, 1950, and others). This gives grounds for the assumption that methionine as well as other amino acids are assimilated by plants without any change in their molecules and are being used by them for protein synthesis. Shavlovskii extracted protein from the roots of peas which had grown in the presence of radioactive methionine and showed that they contained the major part of the methionine absorbed by the roots. One gram of fresh roots gave 23,171 cpm, and the proteins isolated from them--13,220 cpm.

Plants absorb the amino acids synthesized and excreted by microbes. This was shown by Shavlovskii who used methionine containing radioactive sulfur S^{35}. As in experiments with vitamins, Shavlovskii grew bacteria $Ps.$ $aurantiaca$ and yeasts $Sacchar.$ $cerevisiae,$ in a medium containing radioactive sulfur S^{35} in the form of $(Na_2 S^{35}O_4)$. The bacteria were then autolyzed and the autolysates containing radioactive (S^{35}) amino acids were added to the medium where plants were grown, Radioactive sulfur was then found in the plant tissues; the roots showed more radioactivity than the aerial parts. In the presence of autolysate of $Ps.$ $aurantiaca$ the activity in the cotyledon tissues was 137 cpm, in the stems--356 cpm, and in the roots--720 cpm; in the presence of yeast autolysate the corresponding figures were: 43 cpm, 199 and 569 cpm.

In his subsequent experiments Shavlovskii grew buckwheat in a medium with living cultures of $Ps.$ $aurantiaca$ previously grown in a medium containing radioactive sulfur. Seeds of the buckwheat were infected with these cells before sowing (700 million cells per seed, with a total activity of 125,000 cpm).

The following activity was found on the second day of growth in the plant: cotyledons--228, stems--181, roots--132 cpms. Consequently, on the second day the plants had taken up about 0.4 % of the bacterial radioactivity. A direct relationship between the amount of bacteria and the absorption of radioactivity was noted. Upon introduction of 3.45 billion cells per seed, about 1% of the total radioactivity of the cells was detected in the plants (after 7 days growth).

The capacity of microorganisms to transfer their metabolic products to the plants was shown in a paper by Akhromeiko and Shestakova (1954). These authors, in their studies, employed radioactive phosphorus P^{32}. They grew $Az.$ $chroococum,$ $Ps.$ $fluorescens$ and yeasts (isolated from soil) in media containing P^{32} as a source of phosphorus nutrition. The cultures of microorganisms thus grown were carefully washed with water and inoculated into the sand in which saplings of oak and ash trees were grown.

Experiments had shown that radioactive phosphorus is taken up by the plants in considerable amounts; it is emitted at a different rate and in different amounts by the various microbial species. The greatest effect is obtained in experiments with yeasts. About 43 % of the radioactive phosphorus from yeast was taken up by plants in the first days of their growth. Oak saplings absorbed more radioactivity than ash-tree saplings.

These experiments show that biotic substances formed by microbes, in addition to the amino acids and other metabolites, are excreted into the substrate, and from the substrate are absorbed by plant roots.

Assimilation of antibiotics

Higher plants absorb not only vitamins, auxins and amino acids but also many other organic compounds present in the soil and formed by microorganisms.

Of all the metabolites which can serve as indicators of absorption by plants, antibiotics, in our opinion, are the most outstanding.

Antibiotics are very specific. They are not present in plant tissues and are not formed by them. They are easily detected and differentiated from other organic substances including phytocides. In our experiments we have employed antibiotics in their native state as well as in the form of chemically pure preparations. Antibiotics produced by different representatives of soil microorganisms were used: penicillin (mold product), streptomycin, globisporin, aureomycin, terramycin, and others (products of actinomycetal metabolism), subtilin, gramicidin, and others (of bacterial origin).

The crude as well as the chemically pure preparations were added, in various concentrations, to substrates where plants were grown. Experiments had shown that antibiotics were taken up by roots rapidly and in considerable amounts and were more or less uniformly distributed in all plant tissues and organs.

In a manner similar to that of biotic substances and amino acids the antibiotics concentrate in the root system more than in other parts. From there they enter the aerial parts.

Antibiotics are absorbed by plants directly from the soil where they are formed by microorganisms. The latter, as it will be shown below, developing in soil under certain conditions, produce and accumulate considerable amounts of these active metabolites, These naturally formed substances enter the plant tissues via the root system in the same way as the chemically pure antibiotics.

The antibiotics are known to be rather complex organic substances of high molecular weight. For example, streptomycin consists of three basic groups: N-methyl, S-methyl and also carbonyl group. Its formula is $C_{21}H_{39}O_{12}N_7$. Its molecular weight is more than 500. No less complex are aureomycin, terramycin, penicillin and other antibiotics.

If such complex organic compounds as antibiotics, amino acids and vitamins are absorbed it may be assumed that many other carbon and nitrogen compounds present in the soil are also absorbed by plants.

A voluminous work was performed in the laboratory of the famous French botanist Bonne to determine the absorption of organic compounds by plants. Laurent J. and Laurent J. (1903). studied assimilation of many organic compounds by plants. According to their data, peas, lentils. corn, rice, and wheat utilize glucose, saccharose, glycerol, dextrin, starch and potassium humate. Lefevre (1905, 1906) noted the capacity of plants to assimilate amino acids and other nitrogenous compounds. Ravin (1913), experimenting with horse radish came to the conclusion that plants can assimilate the organic acids--succinic, citric, malic, tartaric and oxalic.

Maze and Perrier (1904) grew corn in a methanol-sugar solution. In 30 days of growth these plants absorbed 10-14 g of saccharose, their dry weight was 14-21.9 g. Approximately the same amount of glucose was absorbed by plants in another series of experiments.

According to Pryanishnikov, 1952; Shulov, 1913; Byalosuknya, 1917; and others, peas, corn, cabbage, buckwheat and other plants assimilate well glucose, saccharose and lactose. Especially good growth of plants was obtained in the presence of levulose.

Klein and Kisser (1925) had shown that oat cultures assimilate organic compounds well without preliminary clevage of their molecules. Arginine is assimilated less than nitrate nitrogen. Similar data were obtained by a Japanese investigator Tanaka (1931). This author studied the assimilation of organic nitrogenous co pounds by higher plants under conditions of sterile growth and obtained positive results.

Seliber (1944) brings results of his own studies and those of others on the growth of potatoes at the expense of starch and other substances present in the tubers. The sprouts of potatoes grow worse if the mother tubers are removed. The author describes cases of formation of sprouts and tubers inside the mother tuber. The growth of sprouts in those cases proceeds completely at the expense of food reserves of the mother nodule. Organic compounds of the latter pass into the embryo and into the daughter tuber.

Numerous experimental data as well as agrobiological observations exist which confirm the possibility of heterotrophic nutrition of plants.

The assimilation of mineral and organic nutrients by plants proceeds at various intensities and depends not only on the composition and properties of the compounds in question but also on the quantitative and qualitative composition of the microflora around the root system.

Studies show that under conditions of sterile growth of plants these substances are not absorbed to the same extent as in the presence of microorganisms. We have grown wheat, peas and corn in a sterile nutrient solution and followed the absorption of penicillin, streptomycin, aureomycin and other antibiotics in the presence and absence of various bacterial species.

Bacterial cultures were chosen which did not inactivate or decompose the above-listed antibiotics and were at the same time resistant to them for the inoculation of the nutrient solution. We have tested and chosen more than 10 cultures belonging to different species: 3 cultures of root-nodule bacteria *(Rh. trifolii, Rh. meliloti and Rh. phaseoli)*, 2 strains of azotobacter *(Az. chroococcum)*, 4 strains of *Ps. fluorescens*, isolated from the rhizosphere of different plants, and 2 strains of the genus *Bacterium (Bact. denitrificans* and *Bact.* sp.) also isolated from the rhizosphere of wheat a nd peas. The effect of the bacteria was determined by the rate of disappearance of the antibiotics from the solution and their uptake by the plants. The presence of antibiotics in the plants was determined by the conventional microbiological tests. The indicator microbes were cultures of the sporiferous bacillus *Bact. subtilis* and staphylococcus--*Staph. aureus*.

Tables 58 and 59 bring data on the uptake of streptomycin and penicillin by wheat and peas in the presence of 5 bacterial cultures and their mixture. Experiments with corn and other plants gave similar results.

Table 58
The effect of bacteria on the uptake of streptomycin by wheat
(units per g of plant tissue; the initial solution contained 2,500 units-100 units/ml)

Bacteria	Roots after 1 day	Stems after 1 day	Leaves after 1 day	Roots after 3 days	Stems after 3 days	Leaves after 3 days
Az. chroococcum	100	20	10	160	20	75
Rh. trifolii	150	30	40	200	20	100
Ps flourescens No 4	30	0	0	50	20	30
Bacterium sp. 25	20	0	0	50	10	10
Oligonitrophil	100	30	50	150	20	50
Bacterial mixture	200	50	100	200	30	120
Control (without bacteria)	20	10	0	100	10	60

Table 59
The effect of bacteria on the uptake of penicillin by peas
(units per g of plant tissue; initial solution in the vessel contained 5,000 units--200 units/ml)

Bacteria	Roots after 10 hours	Stems after 10 hours	Leaves after 10 hours	Roots after 2 days	Stems after 2 days	Leaves after 2 days
Az. chroococcum	120	30	40	200	120	180
Rh. trifolii	100	20	40	250	100	150
Ps flourescens No 4	10	0	0	50	10	20
Bacterium sp. 25	5	0	0	30	5	10
Bacterial	250	50	100	250	120	210

mixture						
Control (without bacteria)	50	5	0	150	50	100

As can be seen from the table, the bacteria have a considerable effect on the uptake of antibiotics by tissues of plants. The various bacterial species have different effects. Some of them enhance the uptake *(Az. chroococcum* and *Rh. trifolii)* others *(Ps. fluorescens* No 4 and *Bacterium* sp. No 25) inhibit the uptake of antibiotics by the plants. The largest amounts of streptomycin and penicillin were absorbed in the presence of a bacterial mixture and the least absorption was in the vessels containing the culture of *Bacterium* sp. No 25.

The analyses have shown that the increased concentration of antibiotics in the plants is accompanied by decreased concentration of antibiotics in the solution.

Analogous data were obtained in experiments with plants growing in sand. The uptake of antibiotics was highest in the presence of *Azotobacter* and root-nodule. bacteria and the smallest in the presence of the nonsporiferous bacillus *Bacterium* sp. No. 25 (Figure 66).

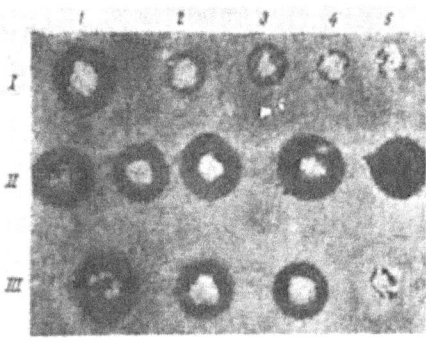

Figure 66. The effect of bacteria on the uptake of streptomycin (from sand) by plants (peas). The analysis was performed 10 hours after the introduction of the antibiotic into the vessel: I--bacteria of the group of oligonitrophils (strain 25 and others); II--*Azotobacter* (the same results with *Rhizobium);* III--control without bacteria. 1--roots, 2--lower part of stem, 3--middle part of stem, 4--upper part of stem, 5--leaves.

Gerretsen (1948) described the enhancement of nutrient uptake by plant roots under the influence of the microflora. He grew oats, sunflower, buckwheat and other plants in vessels with sterile and nonsterile soil, in the presence and absence of bacteria. Phosphates, soluble. insoluble and sparsely soluble in water, were introduced into the soil (bi- and tricalcium, phosphate, phosphoric powder, etc).

The experiments have shown that under sterile conditions, without bacteria, the uptake of phosphates proceeds at a lower rate than in the presence of bacteria. The greatest effect was obtained in the experiments with buckwheat. in the presence of specially chosen bacteria.

Kotelev and Garkovenko (1954) studied the uptake of inorganic radioactive phosphorus by bacteria. Under sterile conditions the radioactive phosphorus was taken up at a slower rate and in smaller amounts. The plants showed only 4% of the total radioactivity when grown under sterile conditions, in the presence of bacteria the radioactivity of the plants increased to 13%.

Experiments with labeled (P^{32}) lecithin gave analogous results. Lecithin was added to the solution used for wetting sand in which barley was grown. The experiments were carried out in the presence and in the absence of bacteria. In the absence of bacteria 1.9% of the radioactive phosphorus was taken up, while in the presence of bacteria 16.6-18.6% of the phosphorus was taken up, i. e., 8-9 times more.

213

Bacteria markedly increase the rate of phosphorus uptake by plants from granules prepared with radioactive superphosphate. Barley sprouts grown under sterile conditions gave 400--500 cpm, but 1,000-1,500 cpm, when grown under nonsterile conditions (cpm. per 10 mg of plant dry weight) (Kotelev, 1955).

We employed the tracer-atom technique to carry out a number of experiments on the uptake of phosphorous compounds from a substrate in the presence and in the absence of different bacterial species. Plants (barley) were grown in a nutrient substrate (solution, sand and soil), previously sterilized and then inoculated with pure cultures of bacteria. Radioactive phosphorus (P^{32}) was then added to the substrate. After 15-20 days of growth the plants were analyzed. We determined the presence of organic and inorganic phosphorus, as well as sparsely soluble (in water) phospholipides, and phosphoproteins. In addition, amino acids were determined (by the use of paper chromatography) in the fraction containing organic water-soluble compounds (Krasil'nikov and Kotelev, 1956). Bacteria isolated from the rhizosphere of corn, grown in Moldavia, were used in these experiments. All of them belonged to the nonsporiferous bacteria of the genera *Bacterium* and *Pseudomonas.* One culture (10-A) was isolated from podsol soil of the Moscow Oblast'. Each vessel filled with sand was supplemented with 465 mg P205 (P^{32}). The total activity per vessel was 6,400,000 cpm. Counting was performed after 17 days growth. The results of the first series of experiments are given in Table 60.

Table 60
The effect of soil bacteria on the uptake of radioactive phosphorus by plants

Experiment	cpm/100g of dry plant substance	Uptake of P^{32}, %	Fraction of organic P, cpm/100g plant mass in aqueous solution	Fraction of organic P, cpm/100g plant mass in lipides	Fraction of organic P, cpm/100g plant mass in proteins
Control (without bacteria)	3,000	0.7	375	450	1,700
Infected with *Bacterium* sp.	3,830	1.4	475	575	2,300
Infected with *Bacterium* sp	4,300	1.3	578	675	2,300
Infected with *Bacterium* sp	--	1.2	650	700	2,900
Infected with *Bacterium* sp	--	0.6	300	400	2,000

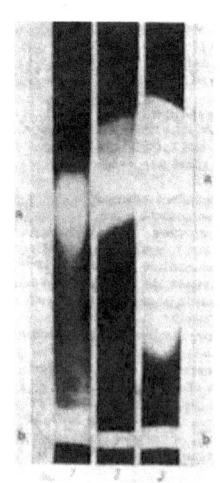

Figure 67. Radiochromatograms of organic and inorganic phosphorus compounds in the soil and in plant tissues. The plants were grown in sterile and nonsterile soil (in the presence of natural microflora): 1--phosphorus compounds in the soil (sterile soil); 2--phosphorus compounds in plants (barley), grown in sterile soil; 3-- phosphorus compounds in plants grown in nonsterile soil; a--organic phosphorus; b--inorganic phosphorus.

The radiochromatogram, (Figure 67) represents the process of accumulation of organic phosphorus in the tissues of barley grown under sterile conditions, and in the presence of bacteria. It can be seen from the figure and Table 58 that bacteria have a considerable effect on the uptake and accumulation of phosphorous compounds in the tissues of plants. Under the influence of the bacteria a formation of quite different phosphorous substances takes place.

The results of the determination of amino acids are not loss demonstrative. We have determined the amino acids present in extracts of 70% alcohol by paper chromatography. Five hundred-mg samples of the dry mass of plants were extracted for 4 hours at 45° C with 25 ml of alcohol. The extracts were then filtered, the filtrate was evaporated to 1-2 ml and subjected to chromatography. The results are given In Table 61.

Table 61
The effect of bacteria on the composition of amino acids in plant tissues

Experiment	Dry weight of plants, mg	Uptake of P^{32} per 100 g	Lystine	Aspara-gine	Aspartic acid	Serine	Glycine	Glu-tamic acid	Alanine	Valine
Control (without bacteria)	545	465	++	+	+/-	-	-	-	-	-
bacteria 1p	573	700	++++	+++	+++	+++	+/-	+/-	++	-
bacteria 2p	710	786	+++	++	++	++++	+	+/-	+++	-
bacteria 125	--	--	+/-	-	-	+++	++++	++++	+/-	+
bacteria 151-A	--	--	+	++	++	++	-	+++	+/-	+/-

Legend: 4 pluses--very intense color; 3 pluses--intense color; 2 pluses--moderate color; 1 plus--weak color; plus--minus-color hardly detected, on the border of accuracy; minus--absence of the amino acid.

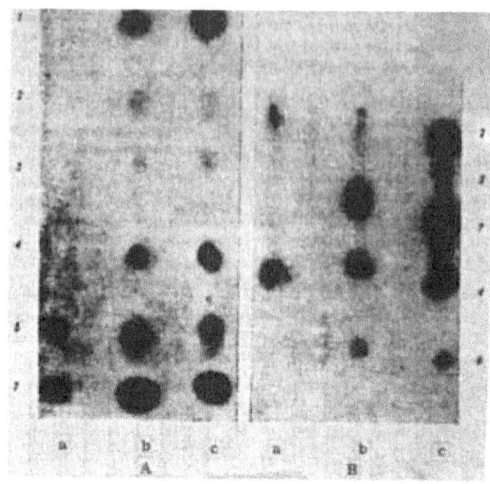

Figure 68. Radiochromatograms of the amino acid composition of plants grown in sterile conditions after artificial contamination with bacteria: A) amino acids of plants grown in sand; B) amino acids of plants grown in the soil (chernozem). a-- amino acids of plants grown in the absence of bacteria; b--amino acids of plants grown in the presence of *Bacterium* sp 1; c--amino acids of plants grown in the presence of a strain of *Bacterium* sp. 2; 1--lysine; 2-3--asparagine and aspartic acid; 4--serine; 5--glycine; 6--glutamic acid; 7-- alanine; ?--unknown compound.

The radiochromatogram (Figure 68) shows the distribution of amino acids in plants grown in the presence of 2 bacterial cultures (lp and 2p). It can be seen from the figure, that the presence of bacteria in the substrate causes formation and accumulation of amino acids of a different nature to those formed in the absence of bacteria. Some amino acids (serine, glycine, alanine, valine, cysteine) cannot be detected, in the control sprouts of the barley grown under sterile conditions, only their traces can be found. These amino acids are present in considerable amounts in plants grown in the presence of bacteria. Different bacterial cultures have a varying effect on the formation and concentration of amino acids in plant tissues. In our experiments cultures 125 and 151-A favored the accumulation of glutamic acid, and bacteria lp and 2p favored the accumulation of alanine, lysine and asparagine.

The effectivity of the same bacteria to plants growing in soil (heavy loam chernozem), although considerable. was somewhat different to that noticed when the plants were grown in pure sand. The quantitative and qualitative relationship of amino acids was different. Neither lysine nor alanine could be found in the barley sprouts grown in the presence of lp and 2p bacteria, asparagine and glycine were found only in traces (Figure 68 B).

According to Akhromeiko and Shestakova (1954), microorganisms inhibit the uptake of phosphorus (P^{32}) by woody plants. Saplings of oak and ash tree were grown in sterile and nonsterile soil and the uptake of radioactive phosphorus was determined. The authors have noticed that bacteria of the rhizosphere at first take up the tracer phosphorus and subsequently release it.

Ratner and Kozlov (1954) grew corn under sterile conditions, according to the method of Shulov, in the presence and in the absence of bacteria in the solution. They found that when bacteria were present in the solution, organic compounds of phosphorus and nitrogen were found in larger amounts than when the plants were grown without bacteria. In the sterile vessels the plants contained the following: amides and amines--1.300 mg; organic nitrogen--1.191 mg (32.33%); in nonsterile vessels the respective figures were 1.562 mg and 1.960 mg (44.22%).

The difference in composition of the amino acids is well defined on the chromatogram. The addition of microbial metabolic products, or a dead culture, to plants in their early growth stage on sterile medium, causes an increase in the phosphorus and nitrogen content of their exudate. The microbial metabolic products increase not only the uptake of these substances by the roots but also the synthetic capacity of the roots. The increased incorporation of the radioactive phosphorus P^{32} into the lipides and nucleoproteins, as well as the increased content of amide and amine nitrogen in the exudate confirms these assumptions.

Microorganisms and their metabolic products affect the process of nitrogen transformation in the root. In the presence of microorganisms the rate of metabolism of amino acids in the roots increases as well as the process of transformation of the inorganic to organic nitrogen. In the presence of microorganisms the uptake of inorganic and organic compounds--microbial metabolites--is increased.

Besides, microorganisms promote the transport of nutrients in the soil. They are carriers of nutrients, supplying the root system with various nutrients.

Numerous papers have stressed the point that mere contact of roots with the soil is not sufficient to secure the nutrient requirements of the plant. Mediators between nutrient sources of the soil and the root system exist in soils. Such mediators are the microbes.

Khudyakov (1953 b) has shown that molds can transport nutrients in their hyphae. It is known that protoplasm moves along the hyphae at high speed. In mucor fungi it is as much as 50-80 μ and more per minute. Various inorganic and organic compounds, including nutrients move together with the protoplasm from the site of their concentration to the site of demand in the growing mycelium of the fungus. It is well known that many fungi grow abundantly on roots and around them, forming the mycorhiza.

Not only fungi but also bacteria promote the transportation of nutrients in the soil. Kotelev (1955) employed tracer techniques in the following manner. He introduced grains of radioactive P^{32} in the form of superphosphate into the soil and followed the diffusion of the phosphorus in the presence and absence of bacteria. The diffusion of phosphorus in sterile soil was very slow and uptake by roots was either lacking or negligible. The diffusion of phosphorus in the presence of bacteria was much more rapid.

The above indicates that it is not permissible to consider the problem of plant nutrition only from the point of view of autotrophy. Alongside the assimilation of inorganic compounds, plants also assimilate various organic-carbonaceous and nitrogenous substances. Some of these are utilized to meet the plant's energy requirements, others serve as biocatalysts. The latter are taken up by the roots and aerial parts of the plant and increase the intensity of the biochemical and biological processes in the cells and tissues. They enhance the growth of plants and increase the absorption capacity of the roots the assimilation of absorbed substances and other vitally important functions.

The biologically active substances of the soil not only enhance the growth and increase the yield of plants but also confer on the plant better nutritional qualities.

Plants which obtain vitamins and other organic compounds from the soil in adequate amounts, yield crops of higher quality and their seeds are of a higher vitality.

The fact that plants can grow in pure mineral nutrient media in the absence of microorganisms cannot serve as proof of the uselessness of the latter in the nutrition of plants.

Plants can indeed be grown in mineral media and yield seeds without the participation of microbes. Is it possible, however. to secure the vitality of such plants in subsequent generations?

We have presented our observations on the growth of green algae and duckweeds. Grown under sterile conditions, in mineral media without the addition of composts or metabolic products of bacteria, these plants lose their viability and eventually die out. Plants grown in the same media but supplemented with composts or metabolic products of bacteria, as well as plants grown in the presence of living bacterial cells (nonsterile conditions) have been kept in our laboratory for more than 20 years without any visible decrease of vitality.

The assumptions of some authors concerning the fact that soil contains only small amounts of organic compounds of phosphorus and nitrogen (in the form of vitamins, amino acids and phytin) which cannot, therefore, be considered as nutrients of any great importance, are also groundless. Our knowledge of forms of organic compounds and of the dynamics of their transformations in the complex microbial coenoses, is inadequate. The knowledge we do possess, however, allows us to assume that the processes of synthesis of various organic compounds proceed incessantly in the soil. Owing to such uninterrupted synthesis (be as it may in small amounts) the total production of these compounds may be sufficient to meet the needs of plants.

We cannot understand the assumption of certain authors that microorganisms by their assimilation of inorganic substances preclude the utilization of the latter by plants. This, according to some authors, leads to plant starvation. The saprophytes of the soil are looked upon as a harmful factor. They do such harm to plants that, according to some specialists, they should be separated from plants because of their competition for soluble salts (Peterburgskii, 1954).

Bacterial life in the soil is known to be very short; it is counted in hours. Even during their life, the dying cells are subjected to autolysis. At the end of the enzymatic lytic processes, processes of solubilization of their residues by enzymes of other microbes ensue. The process of bacterial cell destruction is rapid, The metabolism of living cells is an endless sequence. The elements absorbed from the substrate are soon excreted. Experiments with radioactive elements have shown that phosphorus (P^{32}), for example, appears in the substrate after a few minutes.

The microflora of the root zone is of great importance in plant nutrition. Growing near or on the roots, microorganisms, together with the plants, create a special zone--the rhizosphere. Soil in this zone differs in its physical, chemical and biological properties from the soil outside the root zone. It possesses different conditions, for the absorption and excretion of substances by the roots.

The interaction between microorganisms and plants on one hand, and between individual microbial species and their metabolites, on the other hand, is the basis for the different transformations of inorganic and organic compounds. As a result, compounds which serve as nutrients for plants are formed. These are absorbed by the roots.

Substances present in the soil are subjected to a greater or lesser extent of processing before their absorption by the roots. The plants do not absorb those compounds which are characteristic for this or other soil, but metabolic products of the rhizosphere. The rhizosphere microflora prepares various organic and inorganic nutrients for the plants.

The role of the rhizosphere microflora reminds one of the digesting organs of animals. Microorganisms in the final account serve the same function in the plant nutrition as the digestive system of animal organisms (Krasil'nikov, 1940).

The same point of view is held by the American specialist Prof. Clark (1949). He considers that microorganisms living in the rhizosphere perform the same work as do the intestines of animals. Academician Lysenko (1955) is even more definite in this respect. He thinks that the microflora of the root zone acts as the digestive organ of plants. We can agree with such a comparison if we recall the function of the microflora of the intestines. In recent years, the microflora of the intestines has more and more come to be considered as a factor of supplementary nutrition for animals and human beings. Many intestinal bacteria are known to produce substances which enter the organism of the animal and play an essential role in biological processes as biocatalysts.

In many animals, microorganisms of the digestive tract participate directly the digestion of food. For example, the cellulose bacteria in the intestinal tract of ruminants, decompose cellulose into digestible products.

We have already noted the importance of microorganisms in the life of plants as being among the biological factors of soil fertility. Between the two there exist special interrelationships which express themselves to some degree by the total productivity of the soils.

Investigations show that the vegetative cover is a powerful factor in the life of microorganisms. The peculiarities of the plant species leave their imprint on the quantitative and qualitative composition of the microflora biocoenoses of soils. Plants create and form microbial societies, which effect the microbial population of the entire root system and harvest remains. During their life, plants excrete through their roots various organic and mineral substances which attract microorganisms. During the growth cycle of the plant, roots constantly form and shed root hairs, lose necrotic epidermal cells, etc. All these elements are then taken up by the microbes and become the source of their nourishment.

The species composition of the microflora, just as the soil, has a definite and considerable influence on the growth and development of plants, and consequently on the crop yield. The study of the nature of the interaction between microorganisms and plants is one of the main and most interesting problems, not only of microbiology, but also of the study of soil and plant cultivation.

EFFECT OF PLANTS ON THE SOIL MICROFLORA

The effect of the vegetative cover on the microflora of the soil was studied long ago by various investigators. Plants not only have an influence on living microorganisms (through their root system) but also influence them after their death and after the harvesting of the crops (in cultivated fields). In the former case, this effect is caused by root excretion and by the dying particles of the roots themselves. In the latter case, the active factors are the remnants of the roots and the aerial parts. In both cases, great changes take place in the composition of the soil microflora. In addition, the roots exert a beneficial effect on the physical and chemical properties of the soil, improving the conditions for the existence of microorganisms. One can judge the importance of plant roots in the biology of soils by the great bulk of the root mass and its extension.

Root mass of plants

The studies made by Kachinskii (1925), Chizhov (1931), and others showed that the root system is a great mass by weight with a vast surface. This was observed by Tol'skii (1904-1911) when he studied the root system of forest plantations in the Buzuluk forest. The author gives numerical data on the development of the root system and its dispersion in woods on different soils and with differing densities of planting. At a density of 77 trees per 0.2 hectares, the length of the roots of one plant is 0.21 km, and at a density of 260 trees per 0.2 hectares, it is 0.42 km.

L. P. and L. L. Golodkovskii (1937), on the basis of observations made on the development of the root system of lucerne under differing soil-climatic conditions of its growth, obtained the following data (Table 62).

Table 62
Development of the root system of a 3-year-old lucerne
under various soil and climatic conditions
(Weight of the air-dried mass of roots in 1 cubic meter of soil, in grams)

Horizon, cm	serozem, watered	meadow, watered	weak podsolized	typical podsol	southern chernozem
0-10	343.1	253.7	315.0	348.0	692.0

10-20	149.2	157.0	386.0	386.0	181.3
20-30	95.1	57.0	100.0	95.0	47.3

According to data obtained by Dittmer (1937, 1938), one plant of winter rye has a total root length (without hairs) of 623.4 km. The thinner the roots, the greater their length. The main roots have a length of 0.07 km; the secondary roots 5.4 km; the roots of the third order, 175 km, and those of the fourth order, 442.8 km. The largest mass consists of the active part of the root system.

The area of all the roots of one plant is 237.4 m^2; the area of the roots of the first order, 0.1 m^2 ; the root area of the second order 4.2 m^2; the root area of the third order, 70.5 m^2; and the area of the roots of the fourth order, 162.8 m^2.

The root hairs on the roots of one plant number to 14.3 billions. They are mainly located on the roots of the third and fourth orders (14.1 billions). The length of all the hairs of one plant is more than 10,000 km. Their total area (one hair is 700-1,000 μ in length) is double the surface area of the root, i. e., more than 400 m^2. Consequently, the roots of one rye plant in the fourth month of growth have a total length of 11,250 km and an area of 6,388 m^2.

In the textbooks written by Kursanov, Keller, and Golenkin, smaller figures are given for the length of roots. For instance, according to their data, the total length of melon roots without hairs is 25 km; in wheat, about 20 km; and in spring rye, 623 km. The surface of the root system of one rye plant is 237 m^2. The surface of the roots of rye is 130 times greater than the surface of the aerial organs (Kursanov, 1940).

According to Pavlichenko (1938), the length of roots of the wild oat *(Avena fatua)* is 87.8 km, and of wheat, 71.2 km. In one cubic decimeter of soil, the total length of roots in as follows: in oats, 6 5 cm; in rye, 90 m; and in meadow grass, 553 m. The length of the root hairs is as follows: in oats, 11.6 km; in rye, 24.3.km; in meadow grass, 73 km.

Detailed studies of the root system of fescue plants, conducted by Savvinov and Pankova (1942) in the Volga steppes, showed the following: in a two-meter 2 layer of soil there are 1.75 kg dry mass of roots per m^2, of which one third to one quarter are living. The weight of the aerial part of these plants comprises 0.48 kg, i. e., about one quarter of the total weight of the roots.

The nature of the distribution of the roots in the soil is illustrated in Table 63.

Table 63
Distribution of roots in soil layers, per 1 m^2
(according to Savvinov and Pankova, 1942)

Type of plants and soils	Depth, cm	Root length, cm	Root surface, m^2	Weight of dry root mass, gm, living roots	Weight of dry root mass, gm, total
Fescue vegetation, chestnut soil	0-25	18.5	12.3	278	1,112
	0-50	28.0	18.09	344	1,419
	0-100	32.0	20.77	418.4	1,735
	0-300	36.0	22.40	436.3	1,757
Cereal-grass vegetation, chestnut soil					
	0-25	61.6	10.3	186.8	1,547
	0-50	80.4	13.96	251.1	1,792.5

	0-100	94.5	16.97	310.9	2,002.3
	0-200	100.0	18.10		

Shalyt and Kalmykova (1935) studied the formation of the roots of steppe vegetation in chernozem soils ("Askaniya-Nova" National Reserve). According to their data, the air-dry mass of roots of the *Festuca-Stipa* vegetative association is 3,002.5 g (30 tons/hectare). and in the Basaltic solonets of the same reserve 1,175.8 g per 1m³ (11.75 tons/hectare). The total surface of the *Festuca* roots is 225 m², and for the *Stipa*, 126.2 m². Savvinov and Pankova consider these numbers to be somewhat exaggerated. One can assume that the cited differences have a purely ecological cause and are not the result of a methodical error. It is obvious that under the conditions of the southern Ukraine, which has chernozem soils, the relationship a between roots and their aerial parts will differ from those of the chestnut soils near the Volga. Moreover, the differences discussed by the authors are not very striking and fundamental. As in the data submitted by Savvinov and Pankova, as wen as that of Shalyt and Kalmykova, the numerical indicators are of about the same order.

Muntz and Girard (1888) measured the length and diameter of the roots of plants grown on the experimental fields of the Paris Agricultural Institute and obtained the following data. In 1 m³ of soil, there was a 5.18 m² total area of clover roots; in the case of meadow grasses, 7.58; oats, 10.70; winter wheat, 11.30; and poppies, 2.17m².

According to Belyakova (1947), the root mass of the dry roots of lucerne in the soils of the Vakhsh valley is as follows: in the first year of growth, 11.12 tons, in the second year of growth, 22-24 tons, and in the third year, 30 tons and more per hectare.

Nad"yarnyi (1939) found that mixtures, several years old, of leguminous and cereal grasses accumulate more roots than pure cultures. In the upper layer of the soil (0-20 cm), over a two-to three-year period, up to 40-75 centners* roots per hectare were found. [*One Russian centner equals 100 kg.] A greater accumulation of root mass was observed by Belyakova and Parishkura (1953) in soils having mixed crops of grasses. In other studies, many tons of dry root mass per hectare are given.

The main mass of roots is concentrated in the surface layer (0-25 cm) or somewhat deeper, depending on soil-climatic conditions and on the type of flora. Sometimes, in the deeper layers of soil, a second maximum (less pronounced) of root concentration is observed. The nature of the distribution of the root mass along the horizons of the soil is given in Figure 69.

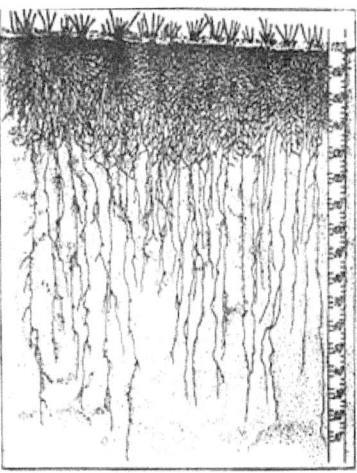

Figure 69. Distribution of wheat roots in the soil (according to Kachinskii, 1950)

221

According to their distribution, the importance of the roots will be greatest in the upper horizons. The biological significance of the root system is determined by its activity, i e., by its ability to absorb elements of nutrition from its surroundings and excrete products of metabolism into the environment, Studies have shown that the active part of the root system is its largest part. According to some date, it comprises 50 to 75 percent of the total root mass.

The substances which are excreted by the root system are utilized by microorganisms as nutrient sources. A great number of microbes concentrate around the roots, growing, multiplying, and excreting their metabolites, many of which are assimilated by the plant roots. The whole root system during the life of plants, as well as after their death, exert an immense influence on the growth and development of microorganisms.

The effect of the root system on the composition of the microflora may be direct or indirect, positive or negative.

The utilization of the nutrient elements by the plants is connected to some degree with the metabolism of microorganisms. A greater influence of plants on microorganisms is that the former enrich the soil with organic substances.

Root excretions

Roots are no longer looked upon as mere suctorial organs through which plants absorb various nutrient elements from the soil. As early as the 18th century, the ability of roots to excrete certain substances, which affect the properties of the soil and determine its fertility, was noted.

The presence of CO_2 in the excretions of roots was noted by many authors including Sossur (1804), Trevisan and Meis (1839), Pollaci (1858), and others. Sachs in 1860-1865 experimentally demonstrated that the root systems of various plants excreted CO_2 (Pryanishnikov, 1952; Konstantinov, 1950).

Lundergardh (1924) determined the amount of CO_2 liberated by the roots of wheat grown in sterile sand and in sand containing bacteria. He obtained the following results: one gram of a dry mass of roots excretes 3.05 mg of CO_2 per hour in sterile sand and 5.57 mg in the presence of bacteria.

The considerable excretion of CO_2 by the roots of plants was observed by Zaikovich. According to his observations, the roots of well-developed corn excreted 0.24 g per day, and, according to Knopp, 0.25 g. In the experiments carried out by Kossovich, mustard roots excreted, on the average, 27.3 mg of CO_2 per day. Barakov observed the excretion of CO_2 by the roots of different plants and concluded that the maximum amount of CO_2 excretion occurs during the period of the most active metabolism of the plant, during its flowering (according to Konstantinov, 1950).

Chesnokov and Bazyrina (1934) grew flax in vessels with podsol soil or sand and determined that the respiration of the soil with the plants growing in it greatly exceeded the sum of the respiration of the same roots and soil taken separately.

The more bacteria present in the rhizosphere, the more intense the formation of CO_2 by the roots of plants (Table 64).

Table 64
The influence of plants on the formation of CO_2 in soil at a temperature of 20i C
(according to Waksman, 1952)

Plants	Number of bacteria, millions / g	pH of soil	CO_2, mg per kg, per day
Triticum vulgare L.	49	6.75	69.4
Secale cereale L.	42	6.44	68.2
Avena sativa L.	45	6.42	79.0
Beta vulgaris L.	78	6.89	74.3

Medicago sativa L.	120	6.89	86.8
Trifolium pratense L.	--	6.66	82.4

The intensity of CO_2 formation depends on the species of the plants, their age, the season of the year, and other factors.

The approximate volume of the root respiration of grain cultures under conditions of growth in the field comprised 25-30 percent of the respiratory volume of the soil as a whole (Konstantinov, 1950).

During their life, plants excrete different mineral and organic compounds via their roots. Compounds of phosphorus, potassium, calcium, sodium and other elements have been found in root excretions.

Sabinin (1940, 1955) and his associates (Minina, 1927) have shown that the excretion by roots of elements of mineral nutrition is accomplished by exosmosis and is regulated by the concentrations of these substances in the external substrate. Tueva (1926) established that the exosmosis of calcium and potassium from roots takes place until an equilibrium state of these elements is established in the surrounding medium. Such a regularity was found by Osipova and Yuferova (1926) in relation to the absorption and excretion of sulfur and phosphorus by the roots of corn and wheat.

Avdonin (1932) observed the loss of ash elements in cultivated oats under field conditions, These losses differ in quantity depending on the conditions of the growth of the plant.

Akhromeiko (1936) decided that some plants excrete mineral substances via their roots, while others do not. He observed phosphoric acid in the root excretions of lupine, peas, buckwheat, mustard, and rape. The amount of phosphorus excreted attained 14-34 per cent of all the phosphoric acid taken up by the plant.

There are some writers who deny that it is possible for roots to excrete mineral compounds. The authors of these works assume that the substances found are the decomposition products of root residues.

Of the root excretions, the organic substances are of the greatest importance, The presence of these substances was observed for the first time at the end of the last century. Dyer (1894) established the presence of acidic compounds in the root excretion of plants of barley, wheat, oats, foxtail, and others. Acids were detected in root excretions by Lemmerman (1907), Künze, (1906), Schreiner and Reed (1907), Doyarenko (1909), and others.

Stoklasa and Ernest (1909) found that plants excrete acetic, formic, and oxalic acids through their roots. Maze (1911) and Shulov (1913) found organic acids and sugar in root excretions. Organic substances were found by Kostychev (1926), Truffaut and Bessonoff (1925, 1927), and others.

Mashkovtsev (1934) found that the roots of germinating seeds of rice excrete sugars, aldehydes, ethyl alcohol, and other compounds precipitating with lead acetate.

Minina (1927) detected organic substances in root excretions of lupine, beans, corn, barley, oats, and buckwheat, upon cultivating them in Knop's nutrient solution, The excretion in most of these cultures reached its maximum during the fourth week of growth, and in buckwheat, at a somewhat earlier period. Upon the ripening and aging of the plants, the amount of root excretions decreases and toward the end of the growth period stops altogether.

Lyon and Wilson (1928) found nitrogenous and nonnitrogenous organic compounds in the root excretions of corn. The amount of nitrogenous substances in root excretions, according to these authors, decreases with the age of the plant.

Winter and Rümker (1952) observed phosphatides, amino acids, thiamine, biotin, meso-inositol, paraaminobenzoic acid, carbohydrates, tannins and alkaloids in the root excretions of plants. Harley (1952) found sugars, amino acids, vitamins, and other organic compounds in root excretions.

Virtanen and his associates (Virtanen and Laine, 1937) observed in the root excretions of young sprouts of leguminous plants; peas, clover, etc, aspartic and glutamic acids, tryptophan and ß-alanine.

Cereals, oats and barley, grown in the same vessel with leguminous plants in the complete absence of nitrogen sources in the medium, grew normally and developed at the expense of the nitrogen excreted by the leguminous plants. Similar experiments were conducted by Lipman as early as 1912.

The possibility of transferring metabolic products from certain plant species to others was confirmed by the experiments of Preston and his associates (Preston, Mitchell and Reeve, 1954). Plants sprayed with methoxyphenylacetic acid were grown in the same vessel with plants which were not sprayed, After some time, the given substance was detected in all the tissues of the unsprayed plants, in larger or smaller quantities: the nonsprayed plants absorbed the methoxyphenylacetic acid excreted by the roots of the sprayed plants.

Many investigators have detected a considerable amount of nitrogenous organic compounds in the roots of cereal plants when they were grown together with legumes (Scholz, 1939; Wyss and Wilson, 1938; Madhok, 1940; Nicol, 1934; Isakova and Andreeva, 1938).

Sabinin (1940) found that the roots of pumpkins excrete from nine to eleven different amino acids. These acids were determined and differentiated by paper chromatography (Kursanov, 1953).

Other investigators deny the presence of nitrogenous substances in the root excretions of leguminous plants (Bond, 1937, and others).

Wilson, Wyss, and others (1937, 1938) assumed the possibilty of the excretion of nitrogenous compounds by roots. However, these compounds, according to these authors, are metabolic products of the nodular tissues and not of the roots of the leguminous plants themselves.

Engel and Roberg (1938) in order to verify Virtanen's data, cultivated alder which was inoculated with cultures of proactinomycetes, forming nodules, and observed in the substrate (sand) a considerable quantity of organic nitrogenous substances excreted by the roots.

Virtanen, in reply to Wilson, Bond, and others, noted that the process of the excretion of organic substances is closely linked with external conditions--sunshine, aeration, nutrition, and the pH of the medium. Confirming earlier data by new experiments, the author stated that the detected nitrogenous compounds are products of the fixation of free nitrogen, which were not utilized in the formation of protein and plant tissues and not products of protein decomposition (Virtanen and Torniainen, 1940).

The organic compounds excreted by the roots of various plants are not identical. In leguminous plants, one detects more nitrogenous compounds--amino acids, amide compounds, and others (Virtanen and others, 1937, 1938, 1940). In cereals. the root excretions are richer in carbon substances--sugars, organic acids, and others. According to our observations, peas, broad beans, beans, lupine, and other leguminous plants excrete substances having a neutral or weakly alkaline reaction, and cereals--corn and wheat--secrete substances having an acid reaction, According to the data of some investigators, the roots of peas excrete nucleotides and flavins (Lundegardh and Stenlid, 1944).

West and Wilson (1939) observed biotin and thiamine in the root excretions of flax and sugar in the excretions of certain cereals, Brown and others (1949) proved the presence of pentoses or closely related compounds (alpha-ketoxylose) in the root excretions of grasses.

Brown and Edwards (1944) found special substances which stimulate the growth of other plants in root excretions.

Groh (1926) studied the root excretions of lupine, broad beans, wheat, oats, barley, and rye. In some plants, substances were detected which have an acid reaction, and others which have an alkaline one. On the basis of these findings, the author divides plants into two groups: the acid group, including peas, broad beans, lupine and wheat; and the alkaline group, including oats, rye, barley, and mustard. According to Pryanishnikov (1905), lupine excretes substances of an acid nature. Due to these excretions, this plant dissolves the highly soluble phosphates, transforming them into an easily assimilated form. Other plants, like mustard and buckwheat, are not able to excrete substances of an acid nature and cannot dissolve the mineral compounds essential for nourishment.

The studies made by Fred (1918, 1919), conducted under strictly sterile conditions, clearly showed the presence of substances of an acid nature in root excretions, which dissolve marble plates. The author pointed out in this connection that, in the presence of bacteria, the process of the dissolution of marble is considerably faster.

The roots of Italian rice excrete a substance which fluoresces with a blue light upon ultraviolet irradiation. This substance is so characteristic that it may, according to the author, serve as an indicator of the given plant (from Audus,1953)

The difference in chemical composition between the root excretions of different varieties of the same species of plant was also noted by other investigators. Timonin (1941) established the presence of substances in a variety of flax resistant to fusariosis (Bizon var.), which activate the growth of the fungus *Trichoderma viridis* an antagonist of the organism causing fusariosis. In the strain which was sensitive to fusariosis (Novel var.) substances were found in the root excretions which stimulate the development of the fungus *Fusarium,* the cause of fusariosis of flax.

Chemical analysis has shown that in the root excretions of the resistant variety of flax, there is a great amount (25-30 mg per plant) of hydrocyanic acid, which possesses antimicrobial properties. In the excretions of the roots of the sensitive variety of flax, this acid is either absent or present only in traces.

Eaton and Rigler (1946) observed an analogous situation in the root excretions of the cotton plant. In the variety resistant to root decay more carbon compounds were found than in the sensitive variety. According to the authors, the given substances attract microbe antagonists, which inhibit the development of the organism causing root rot.

It should be noted that the problem of plant-root excretions only comparatively recently became a subject of thorough study. Therefore, we possess only scant information on the qualitative composition of root excretions. In addition, our knowledge of the quantities of the substances excreted by the roots in even more scant.

The few studies available show that roots excrete considerable amounts of organic substances. Dyer (1894), in determining the amount of acids excreted by the roots of plants, established that 100 ml of nutrient solution from barley contained 0.38 mg of acids; from wheat, 0.58; oats 0.65, foxtail, 0,86; timothy grams, 0.80; orchard grass. 0.81; white clover, 1.28; red clover, 1.55 and from broad beans, 1.11 mg of acids.

According to Maze (1911), one corn plant in a sterile nutrient solution excretes 57 mg of sugar and 84 mg of acids in twenty days of growth. Shulov (1913) found in the root excretions of this plant, after a two-month period of growth in a nutrient solution, 94 mg of nonreducing and 34 mg of reducing sugars and 80 mg of malic acid. The roots of peas excreted, during the same period, 140 mg of sugar. According to the observations of the author, when plants were cultivated on ammonium nitrate, there were more root excretions than when the plant was given calcium nitrate.

Pfeifer, (1912, 1917) investigated the root excretions of wheat and buckwheat. According to this author, 0.27 g of wheat roots excreted 0.134 mg of organic acids and 0.110 g of buckwheat roots excreted 0.155 mg of organic acids, which comprises 1.3 per cent of the total weight of the plants. According to Shulov (1913), the root excretions of corn comprise 0.6 per cent of the plant's weight.

Demidenko (1928) grew corn and tobacco in solutions which were either changed or unchanged. The corn roots of one plant, grown in a solution which was not changed excreted 486 mg of organic substances during the whole vegetation period and, when the solution was changed seven times. the roots excreted 1,136 mg of organic substances, Roots of tobacco, for the same period, excreted 158 mg in an unchanged solution, while in a changed one, it excreted 439 mg of organic substances. In summarizing these observations, the author concluded that the total root excretion comprised 27 per cent of the plant mass.

Mashkovtsev (1934) found that seeds of rice upon germination lose 20-30 per cent of dry weight, with about one fourth of the loss consisting of root excretions of organic compounds.

Virtanen and his associates (1933) found that the roots of peas grown in vessels along with cereals excrete 126.4 mg mineral nitrogenous compounds in 58 days, of which 77.4 per cent is comprised of the nitrogen of amino acids, 3.3 per cent amide compounds, 2.05 per cent of melanin*, and 2.73 per cent of other nitrogenous compounds. *["Melanin" appears in Russian text but it may erroneously refer to "humin."]

When barley was grown together with peas, it grew normally and developed, although no nitrogen was introduced into the vessel with sand; in the tissues of experimental plants of barley, 32.3 mg of nitrogen were found, and in the tissues of the control plants. which were grown without peas, the nitrogen found amounted to only 0.7 mg and these plants developed very poorly.

The barley in these experiments did not utilize all the nitrogen excreted by the peas. A considerable part of it, up to 89.0 mg, remained in the substrate in the form of these or other organic nitrogen compounds. (Virtanen, Synnöve Karström, 1933; Virtanen, 1937).

Virtanen and Laine (1936, 1937) found that in the root excretions of clover and other leguminous plants in the period preceding flowering, mainly (75 per cent of the total bound nitrogen) aspartic acid, gluconic acid, tryptophan and ß-alanine were detected. During flowering, the major part of the nitrogenous root excretions consisted of tryptophan.

Lyon and Wilson (1921) calculated that for the whole vegetative period, the roots of plants excrete up to 5 per cent of the total weight of the plants' organic substances.

Engel and Roberg (1938) determined that, during a two-month period of growth, the roots of one alder plant, inoculated with proactinomycetes, excreted 27.7 mg of nitrogenous compounds and uninoculated plants excreted 23.6 mg of nitrogenous compounds (Table 65).

Table 65
Excretions of nitrogenous substances by alder roots, mg
(from data supplied by Engel and Roberg, 1938)

Experiment	Nitrogen in the initial substrate (sand)	Nitrogen in substrate after 2 months growth of alder	Increase in nitrogen
Inoculated (with nodules)	3.3	31.0	27.7
Uninoculated	2.5	26.1	23.6

Meshkov (1953) investigating the root excretions of peas and corn grown in a sterile nutrient solution, obtained the following results: during twenty days of growth, the roots of peas excreted into the solution 2.87 mg of reducing sugars in experiments performed during 1946, and 4.28 mg in the experiments carried out during 1947. The weight of the dry mass of the vegetable crop comprised 1.92 g in 1946 and 1.85 g in 1947. The roots of corn for the same period excreted into the solution 8.4 mg in 1946, and 8.17 mg in 1947. The weight of the dry mass was 3.69 and 2.35 g respectively. According to the observations of the author, the amount of root excretions depends to a considerable degree on the weight of the roots, rather than on the weight of the green parts, leaves and stems. The total weight of the latter amounts to 2 per cent in the case of peas and 1.3 per cent in the case of corn, of the total weight of the mass of the plants.

In our investigations (1934 b) we determined the growth of microorganisms in the media to which these substances were added. For this purpose, of a great number of species tested the following cultures were chosen: two cultures of yeast; *Torula rosea* and *Sporobolomyces philippovi*, and two bacterial cultures, *Pseudomonas fluorescens* and *Ps. denitrificans*.

These microorganisms were grown in a solution in which wheat and corn were grown, and also in a pure nutrient solution with various concentrations of glucose. After certain time intervals, the cells were counted in a Thoma counting cell and plated on liquid and solid nutrient media. The results are given in Tables 66 and 67.

Table 66
Growth of microorganisms in a rhizosphere solution of wheat
(in thousands per 10 ml of medium)

Time of action of solution--->	3 days	8 days	15 days	25 days	40 days
Torula rosea	75	1,500	1,000	1,500	1,100
Sporobol. philippovi	150	2,200	2,000	2,000	1,500
Ps. flourescens	1,200	100,000	150,000	7,000	1,000
Ps. denitrificans	1,800	150,000	160,000	7,500	1,500

Table 67
Growth of bacteria and yeasts in a pure initial solution with various concentrations of glucose (in thousands per 10 ml on the eighth day of growth)

Glucose concentration, mg per 100 ml	Torula rosea	Sporobolom. philippovi	Pseudomonas flourescens	Pseudomonas denitrificans
5	400	650	20,000	1,500
10	680	900	35,000	2,500
20	1,000	1,500	60,000	50,000
50	2,000	3,000	150,000	100,000
100	3,500	5,000	250,000	200,000

In comparing the maximum numbers of microbial calls which had grown on the rhizosphere solution with the corresponding numerical indicators of growth in glucose-containing medium, we obtained the following results: the maximum number of *Torula rosea* cells, which attains 1,500,000 in the rhizosphere solution, is equal to the same number in the case of a glucose concentration which slightly exceeds 20 mg. Approximately the same amount is also necessary for *Sporobolomyces philippovi*. In order to accumulate 150 million bacterial cells in the rhizosphere solution, we evidently require approximately 50 g of glucose or some other equivalent substance.

Consequently, according to the data of this analysis, the roots of wheat (the vessel contained three plants) excreted in 15 days of growth about 50 mg of organic substances, utilizable by bacteria, and about 20 mg of substances utilizable by yeast.

In experiments with corn, similar results were obtained. Substances utilizable by bacteria were excreted in a larger amount than substances suitable for the nourishment of yeast.

It became known recently that the roots of vegetating plants excrete various enzymes into the medium. The presence of enzymes in root excretions had already been suspected when the problem of the saprophytism of higher plants and the problem of their growth and nutrition on organic media was investigated (Kamenskii, 1883; Lyubimenko, 1923, 1935; Keller, 1948).

Eckerson (1932) has shown that plant roots are able to reduce nitrate to nitrite with the aid of excreted nitrate reductases. Thus, the roots formed up to 2 mg of nitrite nitrogen during 17 hours at 37° C. Klein and Kisser (1925), growing plants in a sterile nutrient solution, detected after some time in this solution an enzyme which reduces nitrate to, nitrite.

Kuprevich (1949) investigated the root excretions of 23 species of plants, belonging to 16 families: oats, wheat, barley, vetch, clover, flax, heather, camomile, willow herb, dandelion, nettle, knotweed, sorrel, tea, oak, birch, poplar, willow, pine, spruce, the common brake, and others. Various enzymes were detected: catalase, tyrosinase, phenolase, asparaginase, urease, invertase, amylase, cellulase, protease, and lipase.

The amount of excreted enzymes and their activity varies among the different species of plants. For example, the activity of amylase was expressed by indices from one to four, i. e., from barely detectable activity to the full decomposition of the substrate. Lipase was detected in traces in only four species of plants (dandelion, touch-me-not, nettle, and pine).

We studied amylase (1952 a) in the root excretions of wheat, corn, and peas, grown under sterile conditions. It was observed that when small samples of roots were placed in a vessel with starch, the latter was comparatively quickly decomposed (Table 68). For example, 0.2 g of wheat roots decomposed 20 mg of starch in 60 minutes.

Table 68
Decomposition of starch by enzymes excreted by plant roots

Plants	Root sample, grams	Reaction for 0.5 hours	Reaction for 1 hour	Reaction for 2 hours	Reaction for 4 hours	Reaction for 8 hours	Reaction for 24 hours
Wheat							
	0.5	-	-	-	-	-	-
	0.2	+	-	-	-	-	-
	0.1	+	+	-	-	-	-
	0.05	+	+	+	-	-	-
Corn							
	0.5	+	+	+	+	-	-
	0.1	+	+	+	+	+	+/-
Peas							
	0.5	+	+	+	+	+	-
	0.1	+	+	+	+	+	+

A plus stands for the presence, and a minus designates the absence of starch, plus-minus designates an undetermined reaction.

Roots of plants grown in the field decomposed starch even more intensively. In one hour, 25 mg were decomposed by a suspension of 0.1 g of wheat roots and by a suspension of 4.5 g of corn roots.

Starch was most quickly decomposed by roots which were not detached from the plant. Young wheat plants, extracted from the soil and washed with water, when submerged in a starch solution, decomposed 25 mg of starch in 30 minutes, while corn plants decomposed 5 mg.

The enzyme amylase was also detected in the water in which wheat plants which were taken from the soil were immersed for some time.

One wheat plant, two years old and grown in the field, excreted into the solution an amount of amylase which decomposed, on the average, 20-25 mg of starch in one hour at room temperature.

It can be seen from the above that starch is most actively decomposed by the root excretions of wheat and to a lesser degree by the excretions of peas and corn.

Ratner and Samoilova (1955) detected in the root excretions of corn and sunflower, enzymes which break down glycerophosphate and saccharose. The amount of these enzymes, according to these writers, changes with the growth phase of the plant. The maximum excretion of enzymes by corn is observed at the period preceeding flowering and during the period of the formation of the spadix (Table 69).

Table 69
Excretion of enzymes by corn roots during various phases of growth, per gram of roots (according to Ratner and Samoilova, 1955)

Phases of growth	Phosphorus liberated from glycerophosphate	Reducing sugars, mg
Initial vegetation period	0.06	4.06
Middle vegetation period	0.106	4.28
Formation of pseudo-ears	0.052	3.66
Beginning of formation of spadices	0.102	6.06
Formation of spadices	0.124	4.32
Ripening of seeds	0.068	0.51

The roots of these plants form enzymes which, in addition to glycerophosphate and saccharose, also split glucose phosphate and ribonucleic acid. Thus, one gram of sunflower roots splits 0.338 mg of ribonucleic acid in three hours, while one gram of corn roots splits 0.048 mg of ribonucleic acid.

A similar picture is observed in relation to the splitting of glycerophosphate and glucose phosphate. Sunflower roots split 0.375 mg of glycerophosphate and 0.208 mg of glucose phosphate in three hours, while corn roots split 0.129 mg and 0.095 mg, respectively.

The authors concluded that, due to the enzymatic activity of the roots, the latter can supply the plants demand for phosphorus at the expense of the organic phosphorous compounds if they are present in the medium.

In addition to enzymes, the plants excrete into the soil a number of other biologically active compounds-- various biotic substances (vitamins, auxins, toxins, etc. The amount of these substances in soils may be quite considerable.

All these substances are sources of direct or supplementary nutrition for soil microorganisms and enhance their growth and accumulation in the soil.

Root residue

In addition to root excretion, microorganisms utilize, as sources of nutrition, dead root cells, hairs, epidermis, etc.

The chemical composition of roots varies in different plants. The roots of some plants contain more water-soluble substances (proteins, sugars. etc), the roots of other plants contain more hemicellulose, cellulose, and lignin. Belyakova and Parishkura (1953) found the, following chemical composition of roots (Table 70).

Table 70
Chemical composition of various roots of plants, in per cent of dry weight
(Belyakova and Parishkura, 1953)

Substance	Lucerne	Eragrostis	Dactylis glomerata (orchard grass)	Labium multiflorum (rye grass)
Ash	5.19	11.67	12.89	13.69
Carbon	43.75	42.42	41.19	43.05
Nitrogen	2.38	0.75	1.22	1.89
C:N ratio	18.4	56.56	38.8	22.7
Protein	14.87	--	--	11.81
Water-soluble part in percent of	2.36	3.29	4.62	1.39

carbon				
Fats	3.01	9.0	1.3	2.0
Hemicellulose	18.21	11.26	15.82	12.58
Cellulose	20.67	20.19	18.21	19.04
Lignin	20.0	30.0	20.0	31.0

It is, obvious that the roots of different plants may attract different types of microflora and may be decomposed by the various species of this microflora.

Upon the decomposition of roots of different plants by microorganisms, different products of intermediate decomposition and different products of the metabolism of the microbes themselves are formed. The latter are also sources of nutrition for other species of microorganisms and, as such, together with the decomposition of roots, attract another population of microbial biocoenoses. The shift of the microbial population continues until the organic substances of the plant, animal, and microbial residues are decomposed into their final products. These final products may be quite versatile, depending on the composition of the organic residues, on the microflora forming them, on soil and climatic conditions, and on the processes of synthesis taking place in the soil outside the cells.

Rhizosphere

The role of roots in the life of microorganisms is not only limited to the supply of nutrient substances. Around the roots more favorable physicochemical and biological conditions for the existence of microbes, as well as for the plants themselves are created.

In regions where there is an abundant accumulation and development of roots the physical properties of the soil improve. The soil particles have more structure in the rhizosphere of plants (see chapter on the structure of soils). With the structure of the soil particles, the respiration process of roots and microorganisms improves, moisture is better conserved, temperature is kept at a more constant level, etc.

The soil around roots is distinguished by a higher moisture content. According to our observations, during the vegetative period, the soil in the rhizosphere of wheat under conditions of the Volga steppes had a higher (by one-two per cent) moisture content, and its moisture capacity was three to five per cent higher, than that of soils outside the root region (Krasil'nikov, 1940), This is evidently connected with the change in the structure and composition of the rhizosphere soil and with the capacity of the root systems of plants to actively change the moisture content of the surrounding soil.

Breazeale and McGeorge (1953) have shown that when soil dries out beyond a certain threshold, the plants moisten it in the vicinity of their roots with water transported from their aerial parts. The latter utilize the atmospheric moisture.

The authors grew tomatoes in soil which was gradually dried. When the plants began to wither, the vessels were transferred to a room with a high humidity (80-90 per cent of full humidity). The plants soon recovered, turgor was reestablished in the leaves, and the soil around the roots became more moist.

The soil in the vicinity of roots also varies considerably with respect to acidity. Around the roots of clover, lupine, and certain other plants, the strongly acidic podsol soils became less acidic. If the control soil had a pH of 4.5. then the pH in the region of the lupine roots increased to 5-5.4. In less acidic soils, the neutralization in the zone of the roots is much more noticeable. (Table 71).

Table 71
Change in the acidity of the soil in the root area of different plants

Soil	Clover; control	Clover; rhizo-	Lupine; control	Lupine; rhizo-	Wheat; control	Wheat; rhizo-

		sphere		sphere		sphere
Strongly podsolized deforested, the first year after plowing	4.5	4.9	4.7	5.4	4.5	4.7
Cultivated, 15 years	5.1	5.8	4.9	5.8	4.9	5.1
Intensively cultivated garden soil	5.6	6.4	5.6	6.5	5.3	5.9

The change which takes place in the environment of the root zone of plants was observed by Kaserer (1940) and Eklunde (1923, 1930). According to Thom and Humfield (1932), the neutralization of acidic as well as alkaline soils takes place in the root zone. For instance, acidic clay soils have a pH of 4.5 and, in the vicinity of roots--6.1. Alkaline soils of Colorado with a pH of 7.9 have a pH of 7.5 in the vicinity of the roots of cereals.

Heller (1953) has shown that plants reduce the redox potential of the soil around the roots. This lowering of the potential, according to his data, is caused by the presence of root excretions and the microorganisms attracted by them. Intense photosynthesis of the green parts of beets lower the rH_2 value (redox potential) of the soil at a distance of one cm from the root surface. Cessation of photosynthesis is immediately accompanied by an increase of the rH_2 in the soil of the root zone. The introduction and the growth of bacteria in the zone of the roots lowers the rH_2 of the best tissues.

The soil around the roots is richer in organic substances. As noted above, it possesses greater quantities of various products of microbial metabolism, products of the decomposition of root hairs, epidermal cells, and root excretions. In this zone, one also notices higher concentrations of enzymes, vitamins, auxins, certain amino acids and other biotic compounds,

Nitrate nitrogen is absent from the root zone or is only present in small quantities. We analyzed the soil around roots of different plants during the whole vegetation period in the fields of the Volga area. Nitrates have only been detected in the rhizosphere during the early stages of the growth of plants and at the end of vegetation (Table 72).

Table 72
Nitrate content in the rhizosphere of plants
(in mg per kg of soil)

Date of analysis	Wheat; rhizo- sphere	Wheat; control	Soy; rhizo- sphere	Soy; control	Sunflower; rhizo- sphere	Sunflower; control
31 May	11.4	29.86	0	12.3	4.65	21.4
5 June	15.9	35.22	0	12.68	4.85	21.1
12 June	22.5	33.9	0	14.2	4.12	22.1
15 June	7.7	37.7	0	11.03	3.1	18.5
19 June	0	24.34	0	9.67	0	15.9
26 June	0	23.96	2.9	14.19	0	6.1
2 July	0	22.54	0	2.3	0	3.5
5 July	0	16.34	0	4.58	0	13.33
10 July	0	9.11	0	12.65	0	12.45
16 July	0	11.27	3.0	13.54	0	11.4
22 July	0	10.8	0	8.65	0	19.4
27 July	0	10.3	0	9.54	0	12.54
31 July	-	-	0	7.24	0	7.86
6 August	-	-	3.4	6.9	0	5.7
10 August	-	-	5.4	6.4	0	4.2
15 August	-	-	3.6	4.2	3.4	5.2
19 August	-	0	5.4	4.7	4.6	7.2

| 25 August | - | 0 | 4.1 | 4.2 | 3.6 | 5.2 |

Katznelson and Richardson (1943) have found that the soil in the root area is less subject to the sterilizing effects of chemical substances. On processing soil with formalin and chloropicrin, the authors detected a much greater decrease in the number of microorganisms outside the zone of the root system. In the root region of certain plants (tomatoes and others), the microbes did not react at all to these chemicals and their number did not decrease. Living organisms, the root region and the root system of plants seem to be less accessible to chemical action.

In our experiments, plants of corn and beans were grown under sterile and unsterile conditions, in growth containers (9 kg) filled with garden soil from the Moscow area. When the plants reached the stage of flowering or bud formation, antiseptic substances were introduced into the soil: 0.5 liter of a 30 per cent solution of formalin and two g of chloropicrin per vessel. In the sterile soil the plants perished and, in the unsterile soil, they continued to grow normally,

It can be assumed that, in the rhizosphere of plants, a protective barrier is formed in the form of metabolic products of microbes, which are much more numerous here than outside the rhizosphere. Evidently, the chemical substances are directly decomposed in the rhizosphere by microbial organisms.

The metabolism of microorganisms is more intense in the root region, as are many chemical and biochemical processes, as well as the transformations of various organic and mineral substances. In the rhizosphere, various minerals, rocks, limestone, marble, etc are decomposed at a faster rate. This process is not only caused by root excretions (CO_2 and other acids) but also by the microflora of the rhizosphere. The more intense the growth of microbes, the faster the decomposition process of substances. Certain compounds, for instance, tricalcium phosphate, do not dissolve in the sterile rhizosphere of plants, but when soil microbes are added to the vessel the substance becomes available to the plants (Gerretsen, 1948). One of the tasks of agricultural microbiology is the enrichment of the root region with microbes, which transform nonsoluble phosphorus compounds into the soluble compounds available to the plant.

Under the influence of the microflora in the rhizosphere, one notices an increase in the solubility of iron and manganese compounds. According to Starkey (1955), this increase is caused by the change in the redox potential, which in quite different here than outside the rhizosphere. In the rhizosphere, iron, manganese. and other metals occur in combination with organic compounds formed by microbes. According to the author, amino acids, organic acids, and other metabolites of microorganisms form stable complex compounds, which are preserved in the soil for a long time. They are utilized by the plants and used as a source of iron, manganese, and other elements. The quantities of these organometallic compounds are greater in the rhizosphere than outside this region.

Weinstein and others (1954) experimentally confirmed this data. They grew plants (sunflowers) in solutions both with and without the addition of microbial metabolites and they followed the absorption of the mineral salts of iron. In the presence of metabolites or ethylenediaminetetraacetic acid, the uptake of iron was faster, while in the absence of these substances and of microbes, the applied elements were not taken up by the plants. These observations showed that plants evidently take up iron, not in the form of mineral compounds, but in the form of organomineral substances formed under the influence of microorganisms.

All the above data show that in the vicinity of the roots of vegetating plants, a special zone is formed in which more favorable conditions prevail for the existence, not only of microorganisms, but also of the plants themselves.

The microflora of the rhizosphere

Increased accumulation of microbes in the root soil was first observed by Hiltner in 1904. He proposed the term "rhizosphere." In investigating the root system of various plants, Hiltner came to the conclusion that the accumulation of microbes in this area was not accidental and that it was caused by the biological activity of the roots.

It should be noted that some investigators before Hiltner observed the accumulation of certain species of bacteria in the root region, but they looked upon this phenomenon from a narrowly specialized point of view.

Epstein (1902) observed the development of special bacteria on the roots of beets, which differ from those of the soil. Welich (1903), Grüber (1909), and Maassen (1905) found large numbers of *Clostridium gelatinosum* cells in the rhizosphere of the sugar beet.

After Hiltner, the phenomenon of the accumulation of microorganisms around roots was observed by Layon (1918). He pointed out that the root systems of plants strongly change the natural microbial associations of the soil. The ratios between species in the rhizosphere microflora, according to this author's observations, differ from the ratios prevailing in the microbial biocoenoses of nonrhizosphere soil. Hoffman (1914). Maaasen and Behn (1923), Stoklasa (1926) and Richter and Werner (1931) observed the increased development and accumulation of fungi in the rhizosphere.

A more detailed and systematic study of root-region microflora began in the 19301's.

Detailed studies made by Starkey (1929, 1934) showed that the roots of plants have a considerable influence on the accumulation of microorganisms in the soil. According to his observation, the number of microorganisms in the rhizosphere is several times higher than that in ordinary soil outside the root area. For example, on the rhizosphere of beets, 427 million bacteria per gram were found, while in the control soil only 8.2 million per gram were found; in the rhizosphere of clover, he found 11,320 million per g and in the control soil only 6.6 million per g; in the rhizosphere of wheat, 653.4 million per g were found; and in the control soil, only 22.8 million bacteria per g were found. Pochenrider (1930) studied the microflora of the soil in the root region of cruciferous plants and also observed the intensified development of certain bacterial species *(Azotobacter* and others).

The most numerous and detailed investigations were performed by the Soviet specialists, Isakova (1934-1940), Krasil'nikov and others, (1934, 1945), Sidorenko (1940), Obraztsova (1936), Berezova (1941, 1945), Korenyako (1942), and many others. Berezova published data on her study of the microflora of the rhizosphere of flax; Isakova on that of rubber plants, tangerines, tung trees, and the tea plant, and Obraztsova, on the tea plant.

Large accumulations of microbes in the region of the roots of forest plants were found by Samtsevich and his associates (1949) and by Kozlova (1953).

Romeiko (1954) observed the development and accumulation of microflora in the rhizosphere of hops. Noting the concentration of microbes in the root region the author emphasized the connection between this phenomenon and the developmental phase of the plant.

Popova (1954) obtained data on the intensified accumulation of microbes on the root systems of vines which grow in the serozem soils of Uzbekistan.

Our investigations (1934-1945) of the soil of the root area have shown that plants in any soil under different climatic conditions, regardless of the zone, possess the capacity of concentrating soil microorganisms in the vicinity of their roots, We studied the microflora of cereals (wheat, rye, oats, barley) grasses (orchard grass, rye grass, etc), leguminous plants (beans, peas, vetch, clover, lucerne, etc), commerical crops (cotton, tobacco, flax, hemp, etc), and several varieties of fruit trees and decorative plants (acacia, poplar, lime tree, oak, apple, plum, pear, etc). Studies were performed in various parts of the Soviet Union: in the South--in Crimea, in the Caucasus, Armenia, and Georgia, on the coast of the Black Sea, in Central Asia, and on the fields of the Tadzhik, Uzbek, and Kirgiz Republics, in Kazakhstan, in the lowlands of the Volga area and in the Astrakhan steppes; in the Ukraine, Moldavia, Siberia, on the Sakhalin; in the northern regions of Yakutiya, Igarka, and the Kola Peninsula. The microflora of plants of the moderate region was also studied: Leningrad, Moscow, Yaroslavl', Kuibyshev, and other regions, and the plants of Belorussia, Latvia, Estonia, and other places,

According to our observations, the total number of microorganisms in the zone of the plant roots exceeds the number of microbes in ordinary soil by tens, hundreds, and thousands of times. The differences in the quantitative relationships depend on the species of plant and the soil-climatic conditions (Table 73).

Table 73
Total number of bacteria in soils
(in thousands per gram)

Soil and plants	Rhizo-sphere	Control	Soil and plants	Rhizo-sphere	Control
Kola Peninsula			**Central Asia, Serozem**		
Potatoes	12,500	320	Cotton	3,500,000	74,000
Clover	21,000	400	Lucerne	7,100,000	60,000
Oats	7,800	270	Orchard grass *(Dyctylis)*	3,800,000	35,000
Mixture of grasses	19,500	350	Oat grass *(Arrhentherum)*	4,200,000	47,000
Severnaya Zemlya			**Armenia, Serozem**		
Saxifrage	7,600	120	Wheat	1,800,000	12,000
Poppies	5,200	210	Lucerne	3,600,000	14,000
Cereals	12,000	400			
			Chestnut Soil		
Moscow Oblast' Podsol			**Wheat**	2,400,000	10,000
Wheat	750,000	1,500			
Flax	500,000	1,150	Georgia, Krasnozem		
Clover	1,000,000	1,200	Tobacco	500,000	500
			Tea plant	800,000	600
Trans-Volga region, Chestnut soil					
Wheat	2,100,000	14,000	**Crimea, Southern shore**		
Lucerne	3,700,000	13,800	Tobacco	1,200,000	8,000
			Vineyards	1,700,000	3,000

Note: The numerical data given in the table was obtained from the study of soils by the dilution method, by inoculation into synthetic media (Chapek, Gil'taya).

In recent years, the microflora of the rhizosphere has gained more of the attention of many foreign specialists. In addition to the above-mentioned studies, made by Starkey, others have been made by Lockhead and his associates, and by Jensen, Katznelson, Timonin, and others, on the bacteria of the rhizosphere.

All of these investigators obtained data on the considerable accumulation of microorganisms in the root region. Katznelson (1946) found 100-1,000 times more microbes in the rhizosphere of beets than were found outside the root zone. Thom and Humfield (1932) obtained the following data: in the rhizosphere of corn, there are 136 million bacteria per one gram of soil, while outside the rhizosphere (control) only five million.

According to Starkey (1955), in the rhizosphere of beans, there are 200 million bacteria per gram of soil; in beets, 427 million per gram; and in grain crops, 653 million per gram. The control samples of soil contained one to five million bacteria per gram of soil.

The poorer the soil in organic substances, the less fertile it is, and the weaker is the influence of the root system on the quantitative composition of the microflora of the rhizosphere. The quantitative relationship of the microflora of the rhizosphere (M.R.) to that of the control (M.C.) is most strongly expressed in poor soils. It is more marked in the primary, slightly fertile podsol soils of the Kola Peninsula than in the chernozem soils of Moldavia or Kuban,.

In the primitive noncultivated or slightly cultivated soils of the Kola Peninsula the total number of bacteria reaches from 50,000 to 300,000 cells per gram. In the rhizosphere of clover in its first year of life, there are 150-400 million per gram. The M.R. to M.C. ratio is 1000:1. In the well-cultivated chernozem soils of the Khar'kov Oblast, the number of bacteria in the rhizosphere of Lucerne is from 2,500 to 5,000 million per one gram, and outside the rhizosphere, from 150 to 300 million per gram. The M.R. to M.C. ratio is 17:1.

Central Asia's slightly cultivated serozem soils contain comparatively small numbers of bacteria, within the range of five to ten million per gram. In the root region of lucerne in its first year of life, we counted up to 1,000 million bacteria per gram. This same soil, after five to eight years of cultivation and the introduction of the corresponding fertilizers, contained 7,000 million bacteria in the rhizosphere of one-year-old lucerne, and, outside the roots, 60 to 100 million organisms per gram. The M.R. : M.C. ratio was 100 to 200) in the first case, and 70 to 100 in the second.

Such a shift in the M.R. : M.C. ratio was observed by us in chernozems, podsols, serozems, chestnuts, and other soils, regardless of the geographical region, Only the quantitative expression differed.

The quantitative ratios between the number of microbes in the rhizosphere and outside it increase, with the depth of the penetration of the root system. In the lower layers, the numerical indices are expressed more strongly (Table 74).

Table 74
Number of microorganisms in the rhizosphere and outside it at different soil horizons
(in thousands per 1 g of soil)

Soil	Plant	Depth, cm	Rhizosphere	Control	MR:MC ratio
Moscow Oblast' podsol					
	Rye	0-25	350,000	1,200	300
		40-60	250,000	300	800
		80-100	5,000	3	1,700
	Clover	0-25	950,000	1,500	630
		50-70	300,000	300	1,000
		90-110	10,000	5	2,000
Moldavia chernozem					
	Lucerne	0-25	5,000,000	100,000	50
		40-60	700,000	3,500	200
		80-100	80,000	300	270
		120-150	10,000	20	500
	Wheat	0-25	1,500,000	75,000	20
		40-60	300,000	2,000	150
		80-100	30,000	100	300

At a depth of 90-110 cm in the rhizosphere of clover, growing in podsol soil, the number of bacteria is 2,000 times higher than outside the rhizosphere, and in the rhizosphere of lucerne in chernozem soil at the same depth, it os 270 times higher. Approximately the same ratios are observed in the case of wheat.

In the deep layers of soil there are usually very few bacteria, while in the rhizosphere of plants, even at the depth of two to three meters, they grow abundantly.

Under conditions of deep growth, the root systems of plants create favorable conditions for the growth of microorganisms, not only by their excretion of nutrient substances, but also evidently due to such factors as the improvement of their conditions of respiration and metabolism.

The number of microbes growing in the root area varies with the ago of the plant. As should be expected, the maximal number of bacteria is observed during the period of the most active growth of the plant. The more intense are the life processes, the more organic substances are excreted by the root, and the more intense the multiplication of microbes in the rhizosphere. Observations show that an abundant growth of microorganisms takes place in the early stages of the plant's growth; however, the most vigorous growth of microbes ensues during the period of flowering and in the period directly preceding it. (Figure 70). Sometimes one also observes three elevations in the growth curve of microbes: the first small elevation is seen in the early stage; a second great elevation ensues before and during flowering; the third elevation occurs before ripening. The last is usually barely noticeable.

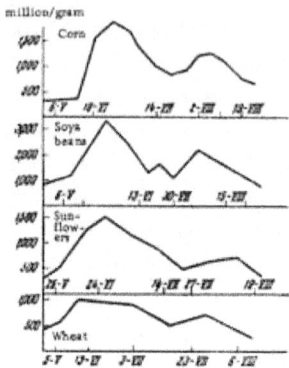

Figure 70. Quantitative composition of the microflora of the root area during different phases of the plants' growth

Under conditions of irrigation, small rises in the growth curve of microbes are observed after each irrigation (Krasil'nikov, Kriss, and Litvinov, 1936b). Under these conditions, they are caused by the increase in soil moisture. The question of whether the moisture was the direct cause of the enhanced growth of the microbes or whether this growth was strengthened as a result of an increased amount of nutrient substances excreted by the roots due to the higher intensity of metabolism in the plant may probably be answered as follows: both mechanisms are operative.

When soils in the rhizosphere are analyzed after harvest, one can observe that the activity of microorganisms does not cease. In the presence of moisture and higher soil temperatures, a sharp increase in the number of microbes in the root zone is observed. In this case, the dead roots are subjected to intense decomposition. The number of microbes during this period may quite often exceed that of the rhizosphere of living plants during rapid growth.

In cases when there is little moisture in the soil (in arid regions), the root residues after harvest decompose slowly and there is no noticeable increase in the number of microbes. The roots remain in the soil undergoing no major changes for long periods.

Differences in the quantitative composition of microbes in the rhizosphere of different plants are slight. In our investigations we could not detect regular and significant differences in the numbers of bacteria in the rhizospheres of different plants of the cereal and leguminous families. In the same field sector, under the same agrotechnical and soil-climatic conditions, the total number of bacteria in the rhizospheres of wheat, oats, and barley are approximately the same. When comparing the microflora of the rhizospheres of cereal and of leguminous plants, one notices a slightly higher number of microbes in the leguminous type. However, this difference is not always noted.

As to the nature of microbe distribution in the rhizosphere of vegetating plants, we noticed regular diffuse growth of microbes, as well as the existence of more or less isolated colonies.

In the literature, one finds indications that microbial cells are present not only on the surface of roots but also inside them, having penetrated the tissue of the epidermis, the intercellular substance, and also the protoplasm of the cells themselves (Berezova, 1953; Rempe, 1951; Hennig and Villforth, 1940). According to the data obtained by Schanderl (1939, 1940), the cells and tissues of plants are not sterile and always contain bacteria. In his last paper, the author claims that plant cells grow in symbiosis with bacteria. The latter are present in the protoplasm in greater or smaller amounts.

Our investigations did not confirm Schanderl's data. Neither in the tissues nor in the cells of healthy plants could we detect bacteria, fungi, or actinomycetes. Upon microbiological analysis, the plant tissues always remained sterile. Even the roots of leguminous plants did not have bacteria in their tissues outside the nodules. Other scientists also, (Burcik, 1940; Schaede, 1940) did not detect bacteria in tissues of healthy plants. Detailed studies were recently made by Stolp (1952). This author attempted to verify Schanderl's data. He investigated various plants, leguminous, cereals, and others. in not one case did he find microbial cells in plant tissues. Bacteria can penetrate into the dying tissues and cells when the re sistance of the latter is weakened. For example, the root hairs, upon dying, are completely filled with bacterial cells. Obviously, Schanderl and others studied such dying cells and tissues, or they accepted as bacteria the intracellular structures.

In order to elucidate the nature of the distribution of microorganisms on and in the vicinity of roots, we employed the method of growing vegetation on glass plates mounted on special vegetative frames, set up in such a way that the roots of the growing plant spread on the glass, leaving their imprints on them. For these frames we used flat boxes with glass walls, 5-6 cm wide, 20-25 cm high and 50 cm long.

The frames were filled with soil or sand and the plants were grown in them. In order to make sure that the roots would stick to the glass, the frames were arranged at a slope of 40-50° (Figure 71). The roots, due to geotropism, grew in a downward direction and touched the glass, sticking tightly to its surface.

Figure 71. Frames for growing plants: A--frame in a sloping position, for obtaining imprins of roots and microflora on glass; B--nature of the growth of roots on the lateral side of the frame (glass).

For the sake of the convenience of microscopic examinations, we placed microscopic slides on the inner side of the glass surface of the vegetation frame. After certain time intervals, the glass with the roots growing over it was taken out and examined microscopically after having been previously stained or left unstained,

When the growth of microorganisms was abundant, one could see with the naked eye a zone of film, measuring three to five mm in radius, around the root, This zone was especially noticeable on the glasses of overgrowth which had been immersed in sand (Figure 72). The sand was easily removed from the surface of the glass while the microbial cells remained. One found by simple microscopy that the microorganisms grow diffusely in the zone of the roots in the form of isolated colonies or small foci. The colonies are located between the hairs on the surface of the roots, or in their vicinity (Figure 73). In cases where there was a great deal of moisture covering the roots and the root hairs, the bacterial cells spread. often occupying considerable segments along the roots (Figure 74).

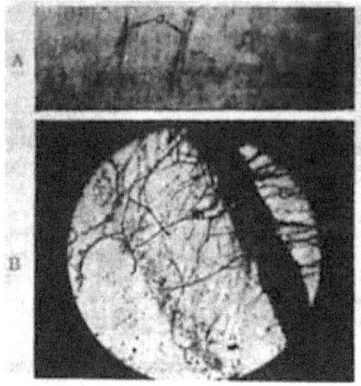

Figure 72. Imprints of root branches on glass with surrounding microflora: A--natural size of the slide; imprints of roots in the form of hairs (a) are seen; B--the same preparation magnified 1:100; branch of the root overgrown by hyphae of fungi and actinomycetes.

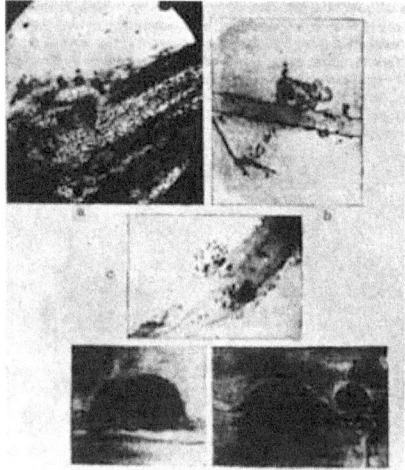

Figure 73. Colonies of bacteria and actinomycetes on the surface of roots: a--colonies of actinomycetes (marked by +), b--bacterial colonies, c--colonies of mycobacteria (our own observations); d and e--bacterial colonies (after Linfold, 1942).

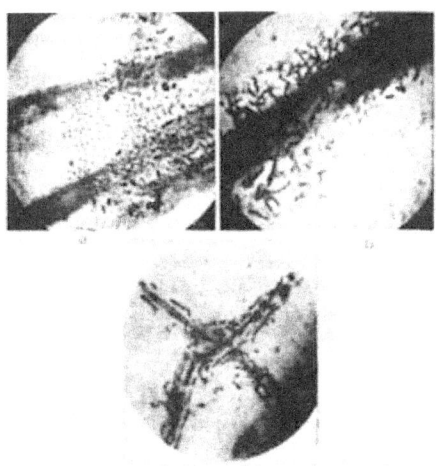

Figure 74. Diffuse distribution of bacteria on root surfaces: a--nonsporiferous bacteria; b--mycobacteria; c--nonsporiferous bacteria and mycobacteria.

Isolated colonies, as wen as profuse areas of microflora, mainly consist of nonsporeforming bacteria and mycobacteria. One also finds colonies of actinomycetes and mycelia of fungi. Very seldom does one see small colonies or single cells of sporeforming bacteria.

Nonsporeforming bacteria and mycobacteria form well-defined, compact colonies, located most often on the surface of the roots, between the root hairs, and also at a certain distance from them (Figure 75). Colonies of actinomycetes often consist of proactinomycete elements, (Figure 75), or of entangled mycelia. Fungi also grow around roots in the form of single hyphae and mycelia.

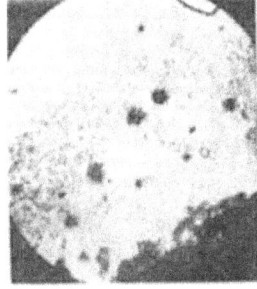

Figure 75. Colonies of actinomycetes around roots. The zone of massive growth of bacteria, at a certain distance from the root, is clearly seen

If one prepares imprints of the root area of microflora growing in sand, and not in soil, one does not encounter this picture. Upon the microscopic examination of these imprints, one can observe only mycelial threads of fungi and actinomycetes, occasionally with sporangiophores on branches. One seldom observes single colonies of actinomycetes, Around the root branches and hairs bacterial cells are also encountered, but in limited numbers and, as a rule, they appear as single cells or in pairs and very seldom in colonies. When such a preparation is being grown, by covering it with a thin layer of an agar medium, a considerably larger number of bacteria and mycobacteria appear than can be observed under direct microscopy. The great majority of cells are not detected by microscopy. This in due to the fact that the cells of bacteria and mycobacteria, and possibly certain actinomycetes, exist in a fragmented state in the form of small granules hardly discernible from soil particles. These granular elements, stained with erythrosin, are encountered in the soil of the root zone in great numbers. They probably form the major part of the mass of the rhizosphere microflora. These elements, when transferred to a nutrient medium, grow out, giving rise to cells of normal size, which are detectable under laboratory conditions.

The methodof overgrowing the root zone ofplants on glass was employed by Linfold (1942). He grew seeds of corn, lettuce, pineapple, and the cowpea *(Vigna sinensis)* in soil in a special chamber. The imprints obtained on glass plates were microscopically examined. The author observed an abundant growth of bacteria in the form of large colonies around the roots and root hairs; colonies were often located at the hair tips (Figure 76). At a certain distance from the roots, according to the author, amoebae, Infusoria, and nematodes grew.

Figure 76. Bacterial colonies on the tips of root hairs (according to Linfold, 1942)

Starkey (1938) employed the method developed by Rossi-Kholodnii for the study of root-area microflora. He found that the bacteria grew on the roots in clusters, and fungi and actinomycetes, in the form of threads. The author buried the slides in the soil. under the root system. Of course, he could not see the greater part of the bacteria on the overgrown glass. As in our experiments, the bacteria were probably in a fragmented state and could not be detected among the numerous soil particles without being cultured.

Stille (1938), using this method, found an abundant growth of bacteria around the root hairs.

These data show that, in the root zone, the microflora probably grows in the same way as it generally does in the soil, in colonies or in aggregates. Only when there is a high moisture content in the substrate do bacterial cells grow in extensive areas.

Group composition of the microflora of the root area. Studies have shown that around and on the roots of vegetating plants one finds various representatives of microorganisms--bacteria, actinomycetes, fungi, algae, yeasts, protozoa, phages, and other living organisms.

However, the prevailing group in the rhizosphere, regardless of the conditions of growth and the age of the plants, are the nonsporiferous bacteria,

The second place (largest group) among rhizosphere microflora is occupied by mycobacteria.

The remaining forms of microbes, actinomycetes, fungi, sporiferous bacteria, etc are encountered in much smaller numbers. The quantitative ratios of these microorganisms can be found in any plating of rhizosphere soil on agar media containing protein or on synthetic media. In Table 75 data are given on the of different groups of microbes growing in the rhizospheres of various plants immediately before flowering. They were obtained by plating one drop (0, 05 ml) of a 1:1,000 dilution.

Table 75

Quantitative relationships between microorganisms in the rhizosphere of plants
(number of colonies per plate)

Soil	Plant	Total number of colonies	Nonspore-forming bacteria	Spore-forming bacteria	myco-bacteria	actono-mycetes	fungi
Trans-Volga region	Wheat	1,200	950	10	200	38	2
Chestnut soil	Oats	1,800	1,450	5	300	35	10
Moscow Oblast'	First year clover	2,200	1,675	8	500	15	2
Podsol	Rye	2,200	1,675	6	50	20	4
Central Asia	Lucerne	3,200	2,700	15	450	34	1
Serozem	Cotton	2,500	2,150	12	305	30	3

Approximately the same group-composition ratio of rhizosphere microflora is also found in other plants. Nonsporeforming bacteria are the prevailing group among the rhizosphere microflora of wheat, oats, clover, lucerne, cotton, and other plants either growing in the southern regions on chernozem, chestnut, and serozem soils, or in the northern regions in podsol or other soils. In some cases, deviations in the number of bacteria of the different groups may exist, the percentage of this or that microbial group may change, but the general nature of the ratio between the different groups of microbial biocoenoses in preserved.

The group relation of the microflora in the root area varies considerably with the age of plants.

It was noted above, that upon the ripening of plants, the total number of microorganisms in the rhizosphere decreases. During this period the quantitative ratio between the different representatives and groups changes; the number of sporeforming bacteria, fungi, and actinomycetes increases, and new organisms appear. At the same time, the total amount of nonsporeforming bacteria decrease, some species disappearing altogether, etc.

On the diagram in Figure 77, data are given on the analysis of the rhizosphere microflora of oats during different periods of its growth, obtained in studies of the plants of the Trans- Volga region (Krasil'nikov, Rybalkina, Gabrielyan, Kondratleva, 1934).

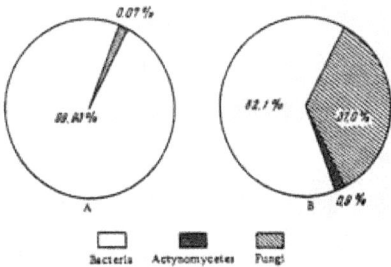

Figure 77. Quantitative relationship between microorganisms at different stages of the plant's growth (oats): A--flowering period; B--period of ripening.

In later works we noted the same changes in the ratios of the representatives of the rhizosphere microflora of many other plants, growing in different areas on differing soils (Krasil'nikov, Kriss, and Litvinov, 1936).

The change in the quantitative composition of rhizosphere microflora with the age of the plant was also noted by Starkey (1938), Isakova (1939), and some other investigators.

Nonsporeforming bacteria comprise the main, and the most numerous and versitile group of soil microflora, in general. This group embraces representatives of various families, genera, and species; these include such organisms as *Azotobacter*, rhizobia, thiobacteria, photobacteria, *Azotomonas, Sulfomonas,* nitrifying bacteria, denitrifying bacteria, and others.

All these representatives are encountered in the rhizosphere of plants. Most frequently, organisms of the genera *Bacterium* and *Pseudomonas* reach the largest numbers, These organisms, comprise the main microflora of the root area,

The species composition of these organisms is scarcely known. We have no adequate methods for the diagnosis and differentiation of these bacteria and we are not in a position to distinguish between the nonsporeforming bacteria in the rhizosphere of one genus of plants and those of another. Only a few genera can be accounted for: *Azotobacter, Rhizobium,* and some others.

Kostychev and others (1926) were of the opinion, that *Azotobacter* adapted itself to the rhizosphere of certain plants: tobacco, rice. and others and is their essential companion.

Later, many investigators found *Azotobacter* in the rhizosphere of plants. Sidorenko (1940b, c) studied its growth in the root region of wheat, barley, beets, lucerne, sudan grass, soybean and other plants grown on the chernozem soils of the Khar'kov Oblast'. Popova (1954) found *Azotobacter* in the rhizosphere of grape vines: Obraztsova (1936) and Isakova (1935) found it in the root zone of subtropical plants: lemon, tangerine, and tung tree; Daraseliya (1950) found it in the root zone of the tea plant. Raznitsyna (1947) and Tanatin (1949, 1953) observed the distribution of *Azotobacter* in the soils of Central Asia under cotton, lucerne, wheat, and various vegetable crops. Pavlovich (1953) studied the distribution of *Azotobacter* in the rhizosphere of various plants in the Latvian Republic; Banak (1956), in Moldavia; Gerbart (1952), in the western Ukraine districts, Petrosyan and his associates (1949) and Afrikyan (1953) in Armenia.

According to our observations, the growth of *Azotobacter* takes place in the root area of various plants under differing climatic conditions and in different geographical regions, from the northern regions to the southernmost points, on mountain tops and in valleys. The growth of *Azotobacter* is observed in the rhizosphere of forest trees (Samtsevich and others, 1952); Krasil'nikov, 1945), fruit trees, bushes, and in other plantations (Kanivets, 1951; Kulikovskaya, 1955, and others).

The number of *Azotobacter* organisms in the rhizosphere of plants may reach considerable dimensions. We counted from 0 to 200 million *Azotobacter* cells in 1 g of soil in the root area. Tanatin (1953) counted tens of thousands and hundreds of thousands in a gram of rhizosphere soil. Approximately the same numbers were found by other investigators (Pavlovich, 1953; Babak, 1956; Raznitsyna, 1947).

No less frequent are the rhizobia. Their identification in the rhizosphere is not only of diagnostic. value but also of practical significance, Depending on the abundance of their growth in the root area of leguminous plants, their effectiveness will differ under the same conditions of activity.

Root-nodule bacteria, as is well known, grow abundantly on the roots of those plants on which they can form nodules. However, they can also grow on roots of other plants. For instance, the nodule bacteria of lucerne grow luxuriously on the roots of lucerne and also in the root zone of cotton; the nodule bacteria of clover also grow well in the rhizosphere of lucerne, peas, and certain other plants. The nodule bacteria of peas grow well in the root zone of peas, clover, wheat, etc. The total number of nodule bacteria in the rhizosphere of plants may be considerable. Korenyako (1942) counted them in hundreds of thousands and in millions per gram of soil. Similar data is given by Raznitsyna (1947), Petrosyan and his associates (1949), Petrosyan (1956), and others.

Ammonia-producing bacteria, denitrifying bacteria, and certain other groups of nonsporeforming bacteria grow abundantly in the rhizosphere. Of the first two groups many millions and billions are present per gram of soil in the rhizosphere (Krasil'nikov, Rybalkina, and others, 1934; Krasil'nikov, Kriss and Litvinov, 1936, Raznitsyna, 1947; Starkey, 1928; and others).

Nitrifying bacteria, cellulose bacteria, nitrogen-fixing anaerobic bacteria of the genus *Clostridium,* and certain other groups do not grow well, comparatively speaking, in the rhizosphere (Rempe, 1252; Krasil'nikov, 1934, 1936).

The second place, with respect to their quantitative growth in the rhizosphere, is occupied by mycobacteria. Their number in the root area reaches hundreds of thousands and millions. Their species composition was not studied. Most often one encounters the nonpigmented forms of *Mycob. album, M. mucosum,* and others.

One finds sporeforming bacteria in the rhizosphere of plants much less frequently and in smaller numbers. In general, they comprise only a fragment of one percent of the microflora. They are especially scarce during the period of vigorous vegetative growth of plants. Usually these bacteria begin growing abundantly at the end of vegetation, especially on dead, decomposed roots.

Actinomycetes also occupy a small place among root-area microflora during the early stages of the development of plants. This group of organisms is very widespread and versatile in its species make-up. In the rhizosphere, there are various representatives of actinomycetes. Toward the end of vegetative development their number increases considerably. They multiply with special intensity on semidecayed dead roots. One often sees rootlets covered completely with the mycelia of these organisms in the form of a fluffy or a mealy-white coating.

Fungi are detected in the rhizosphere by the conventional analyses of small quantities.

It is known that certain plants have a well-developed fungal coating on their roots, coalescing with the root tissue. This: fungal coating is called mycorhiza. Depending on the nature of its relation to the roots one can distinguish between endotrophic and ectatrophic mycorhiza. Mycorhiza fungi are widespread in the root systems of many plant species, both woody types and grasses. Some investigators are of the opinion that all plants have mycorhiza. The fungi participating in the mycorhiza belong, according to their systematic positions, to different classes and orders, families and genera. It is supposed that these organisms are of great importance to the plants. However, this problem has been only slightly studied (Lobanov, 1953; Reiner and Nelson-Jones, 1949; Kelly, 1952).

Many plants do not grow well without mycorhiza fungi, and some do not grow at all, However, one seldom encounters roots with an abundant growth of these organisms in grassy field crops.

It should be noted that upon an ordinary microbiological analysis of the rootlets, one observes, as a rule, single mycelial hyphae. Fungi are detected by special studies with the use of special analytical methods.

The question of algal growth in the root area of vegetating plants has been only slightly studied. The research done by Katznelson (1946) and Shtina (1953, 1954 b) showed that various algae live in the rhizosphere of plants in considerable quantities., Their total number reaches tens and hundreds of thousands in one gram of soil.

Shtina studied the growth of algae in the rhizosphere of rye, timothy grams, clover, lupine, potatoes, barley, and oats. Among some of these plants, the number of algae in the root area was two to three times higher than outside it (rye, timothy grass, clover, lupine). For example, in the root zone of clover, 149,000 cells were found in one gram of soil, and in the control area (outside the rhizosphere), only 99,000 cells per gram were detected.

Qualitatively, the algae composition within the rhizosphere is approximately the same as outside it. It consist mainly of diatoms, green and blue-green algae (Shtina 1954 a and b). They probably also have a certain importance in the life of root-area biocoenoses.

In the rhizosphere of plants, one observes invertebrate animals: protozoa, nematodes, insect larvae, etc. The number of these organisms in the root area of healthy plants is small and does not exceed a few thousands per 1 m^2 under various plants. Shilova (1950) studied the soils under grasses, rye, and lupine, in long fallow, and fallow soils. According to her calculations, in 1 m of the surface horizon (0-10 cm) of soil, there are from 7,000 to 187,000 invertebrates, and the number of certain members of Apterygota in this area, under grasses, reached 760,000.

The composition of this fauna is extremely diverse. One finds in the rhizosphere members of Acarina (mainly under forest cultures), and representatives of Apterygota. Enchytraeidae, and others (Shilova, 1950; Gilyarov, 1949, 1953). Katznelson (1946). Linfold (1942), and Brodskii (1935) have described the distribution of the protozoa, amoebae, ciliates, flagellates, and others in the soil of the root area. These organisms are frequently encountered when studying the soil of the root area in ordinary laboratory studies.

Nikolyuk (1949) has established that there are two to three times more protozoa in the root zone of lucerne than outside it.

The greatest number of these organisms are accumulated in the rhizosphere of cotton during its first year of cultivation (up to 100,000 in 1 g soil). The author ascribes this accumulation of protozoa to the abundant growth of bacteria in the rhizosphere, which serves as nutrient material for the protozoans.

Ressel (1955), Brodskii (1945). Nikolyuk (1949), and others ascribe great importance to this group of organisms, as a factor affecting the composition of microbial biocoenoses in the soil.

Quite often one encounters worms and nematodes in the root zone of plants. Certain nematodes, as is well known, grow well on roots and, in penetrating into plant tissues, cause diseases. Arkhipov (1954) noticed the death of nematodes under certain plants.

Pathogenic microbial forms--bacteria, actinomycetes and fungi--often grow and accumulate in the root area. Under certain conditions, these organisms penetrate the tissues of plants. in agricultural practice, cases of mass infections of plants, as a result of the growth of pathogenic fungi and bacteria in the root zone are not infrequently encountered. The development of microbial antagonists inhib iting phytopathogenic organisms in the rhizosphere is also possible.

In general, many different forms of organisms may grow in the rhizosphere of plants, both useful and harmful; those which facilitate the nourishment and develop ment of plants, and, an the contrary, those which inhibit and poison them. The prevalence of these other organisms depends on soil-climatic conditions, on the manner in which the farm is handled, and on the whole agrobiological complex.

The specificity of the microflora of the root zone

As indicated above, the chemical composition of root excretions, as well as that of the dying root hairs and cells of the root epidermis, varies in different plants. Consequently, root production will attract different soil microflora. Plants which excrete carbon compouonds attract to their roots a microflora which differs from the microflora attracted by plants which mainly excreted nitrogenous substances. With the presence of sugar in root excretions one kind of bacterial and fungal species will develop and, in the presence of organic acids, others will develop. These microorganisms which utilize the available nutrient substances more quickly and fully will predominate in the plant rhizospheres.

The qualitative composition of root excretions and dying root residues determines the characteristics of the quantitative and species makeup of the microflora in the soil of the root region.

The first to have observed the capacity of plants to affect the microflora of soils was the writer, S.T. Aksakov.

In 1896 he wrote: "The mushroom is the child of the forest. . . . As is well known, if one sows, plants-- in a word, if one plants a forest in a bare field--the mushroom types characteristic of the varieties of the planted forest will undoubtedly start to grow there. However, the shadow (as many have thought) cast by the branches of trees is not the only secret force by which trees cause mushrooms to grow around them. It is true that the shadow is one of the primary reasons for this phenomenon. It protects the soil from the burning rays of the sun, and it create a moisture in the soil and even the dampness which is essential for the forest, as well as for the mushrooms. However, the main reason for the formation of the mushrooms, in my opinion, are the tree roots which, in their turn, having moistened the surrounding soil, give it the sap of the tree. This, in my opinion, is the secret of mushroom growth." (Notes and Observations on how to Collect Mushrooms, vol. V, 1896).

Later, data on this subject were published In the scientific literature (Galakhov 1929; Danilov, 1943, 1949; Vasil'kov, 1953, and others). Detailed information on the interrelationship between edible mushrooms and higher plants in the forest phytocoenoses of Latvia is given in the dissertation presented by Mazelaitis (1952), and information on the distribution of mushrooms in the forests of the Moscow area in given in the book by Shiryamov "The Search and Collection of Mushrooms in the Forests of Moscow Area," (1948).

An analysis of the obtained observation and studies shows that the distribution of edible mushrooms is closely connected with the composition of phytocoenoses. When the forest and plant species of an area are known, one can determine beforehand which species of mushroom will grow in the area and the distribution of the mushroom types of interest to us. For instance, the white mushroom is encountered in heather, mountain-cranberry, spruce-sorrel, and oak-bilberry forest. The ordinary brown mushroom is found in pine forests with birch and heather; *Boletus lutens* is found in young pine forests having a certain grassy vegetation, etc (Vasil'kov, 1953).

We demonstrated experimentally the selective action of plants. Different plants were grown under sterile conditions, in sand or cotton, and soaked with Knop's nutrient solution. Bacteria were introduced into the medium, either separately, or in the form of mixture. The results are given in Table 76.

Table 76
The effect of plants on the qualitative composition of bacteria
(number of cells in thousands per ml on the 20th day of growth)

Plant	Root-nodule bacteria-- clover	Root-nodule bacteria-- lucerne	Az. chroococcum	Pseudomonas flourescens strain No 1	Pseudomonas flourescens strain No 2
Wheat	10	100	0	200,000	1,000
Corn	0.1	200	0.01	15,000	4,000
Cotton	10,000	100,000	3.0	100,000	30,000
Sugar beets	--	1,000	1.5	100	100,000
Flax	0.1	10	0.01	300	100
Clover	100,000	10,000	6,000	1,000	1,000,000
Lucerne	1,000	100,000	5,000	1,000	100,000
Peas	10,000	1,000	3,000	1,000,000	1,000

As can be seen from the given data, some bacteria grow well, others grow only slightly, while others show intermediate growth on the same plant. *Azotobacter,* for instance, did not grow at all or grew very poorly in vessels in which wheat, corn, and flax were planted, showed intermediate growth on cotton, and abundant growth on clover, lucerne, and peas.

The reaction of two strains, Nos. 1 and 2, of the same species of nonsporeforming bacteria. *Ps. fluorescens,* to the root excretions of plants differed. Strain No. 1 grew most abundantly in the rhizosphere of wheat, corn, and especially cotton, while strain No. 2 grew much better in the presence of the root excretions of clover and lucerne. The root-nodule bacteria reacted positively toward the roots of cotton, clover, lucerne, and peas, and less so to the action of the root system of sugar beets.

We also grew a series of other sporeforming and nonsporeforming bacteria, mycobacteria and yeasts. The sporeforming bacteria *Bac. mesentericus, Bac. subtilis, Bac. megatherium,* as a rule, did not grow in a medium with the root excretions of the experimental plants, but did not die in it. The mycobacterium, *Mycob. album,* grew fairly well in the rhizosphere of corn and wheat and somewhat better in that of clover and peas. Brewer's yeast did not grow at all with corn, wheat, and beans, but wild yeast of the genus *Torula,* on the other hand, grew beautifully and multiplied.

Similar data were obtained by Metz (1955) in experiments with sterile cultures, nonsporeforming bacteria and, on other plants, their growth stopped. According to his data, the growth of *Azotobacter* is strongly suppressed in the rhizosphere of celandine, buttercups *(Ranunculus acer* L., *Ran. repens* L), peonies *(P.officinalis* L.), and fumitory, *(Fumeria officinalis* L.), and is less suppressed in the rhizosphere of *Viola tricolor* Wittr., *Allium schoenoprasum* L., *Rumex patientia* L., and *Epillobium montanum* L.

Plants such as *Crepis virens* K., *Hieracium pilosella* L., *Armoracia rusticana* Gaertn, and others, which strongly suppress the growth of sporeformong bacteria, do not have any deleterious effect on the growth of *Azotobacter*. On the contrary, many of them stimulate the growth of this microbe.

A similar effect on the growth of *Azotobacter* by these or other plant species prevails under conditions of natural growth (Table 77).

Table 77
Accumulation of azotobacter in soil with different plants
(number of cells per gram of soil)

Region	Plant	Cultivated soil	Fallow soil
Moscow Oblast', podsol			
	Wheat	0-10	0-10
	Rye	0-10	0-10
	Barley	50	60
	Oats	70	50
	Potatoes	100	50
	Clover	1,000	80
	Flax	0	10
Kola Peninsula, podsol			
	Clover	1,000	0
	Cow parsnip	50	0
	Barley	10	0
Kuibyshev Oblast' Serozem			
	Wheat	200	200
	Barley	500	200
	Lucerne	5,000	250
Moldavian SSR, Serozem			
	Wheat	500	800
	Lucerne	600	800
	Sudan grass	1,200	500
Central Asia Uzbek SSR, Serozem			
	Wheat	20	100
	Corn	50	100
	Lucerne	3,500	250
Vakhsh Valley, Serozem			
	Cotton	450	400
	Lucerne	10,000	800
	Rye grass	6,000	600
	Orchard grass	300	400
	Rice	12,000	1,200
Kirgiz SSR, Serozem			
	Wheat	380	250
	Beets	500	300
	Potatoes	180	150
	Cotton	10	50
Trans-Volga region.			

Chestnut soil			
	Wheat	0-10	0
	Millet	0	0
	Sunflower	40	50
	Corn	0-20	20
	Lucerne	2,000	200
	Sweet clover	1,500	200
Crimea, Southern coast			
	Tobacco	250	200
	Vineyards	150	180

The effect of grass plants on the growth and accumulation of *Azotobacter* in the soil is especially well demonstrated in monocultures. The longer a plant is cultivated, the more bacteria and fungi accumulate in the soil. Such a long-term accumulation of microbes was observed by us in the soils of Central Asia, in the lucerne-cotton crop rotation, Usually, after lucerne is grown for three years whether in the pure form or in a grass mixture, cotton is cultivated for several years (six to nine). After this type of crop rotation, the microflora changes considerably in relation to *Azotobacter,* and to other species as well. (Figure 78). Under lucerne, the number of *Azotobacter* cells increases and, under cotton, it decreases. However, *Azotobacter* does not completely vanish under cotton.

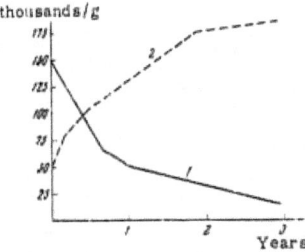

Figure 78. Growth of *Azotobacter* in soils with a crop rotation of lucerne-cotton in the fields of the Vakhsh valley, Tadzhik SSR: 1-- cotton, 2--lucerne.

'This regularity of change in microbial forms is observed in monocultures of plants, for example, under lucerne and wheat in the chestnut soils of the Trans-Volga region, or under clover-flax and oats in the podsol soils of the Moscow Oblast' (Figure 79).

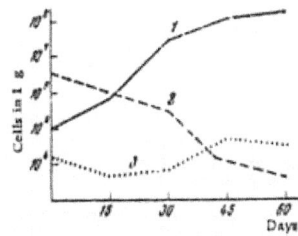

Figure 79. Growth of root-nodule bacteria of clover on various plants: 1--clover; 2--wheat; 3--oats (monoculture) on the experimental fields of the Agricultural Academy im, Timiryazev.

247

In the crop rotations of an ordinary farm with industrial crops, one observes thesame results, but with less markedly expressed numerical indices (Krasil'nikov, 1940a).

Sheloutnova (1938) and Wenzl (1934) observed the growth of *Azotobacter* under a culture of tobacco and vineyards; Mashkovtsev (1934) and Uppal and coworkers (1939) observed its growth under rice.

We have studied the growth of *Azotobacter* in forest plantations in the fields of Kirgizia, Moldavia, Ukraine, and the Moscow Oblast'. The arboretum of a seven- to ten-year-old plantation, in sectors previously under lucerne or other vegetation possessing a considerable number of *Azotobacter* cells was investigated. The growth of *Azotobacter* was quantitatively measured on strips sown with different tree varieties: maple, ash, poplar, acacia, spruce, pine, oak, etc (Table 78).

Table 78
Growth of *Azotobacter* in soil under forest trees
(number of cells in 1 g of soil)

Soil	Plant	Azotobacter in forest strip	Azotobacter in plowed-up strip
Moldavian SSR, Chernozem			
	Oak forest	80	2,700
	Acacia	150	2,800
	Lime	0	3,000
Moscow Oblast', Podsol			
	Spruce	0	100
	Birch	0	250
	Oak	0	100
Kirgiz SSR, Serozem			
	Maple	60	6,000
	Ash	500	5,000
	Poplar	0	4,200
	Acacia	1,500	4,000
	Elm	0	6,000
	Birch	0	3,500
	Grass mixture	--	100,000

As can be seen from the data given, afforestation, as a rule, removes *Azotobacter* from the soil. Only under certain species in it preserved in small numbers,. Under certain plants, the acacia and the ash tree, *Azotobacter*grows moderately well, although it does not reach the numbers found in plowed-up soils.

Artificially planted birch, aspen, oak, and especially spruce and pine trees in podsol soils quickly remove *Azotobacter*. *Azotobacter* also perishes in Southern chernozem soils, if the latter are afforested with oak, pine, hornbeam, etc.

Fruit trees, such am apple, pear. plum, cherry, and others, do not suppress the growth of *Azotobacter* and many of them are even favorable to its accumulation in the soil. The soils of the orchards investigated by us in Moldavia (chernozem). in the central belt of RSFSR (podsol), in Crimea, and in other regions of the USSR contain larger quantities of Azotobacter than the soils of fields which are intensively cultivated (Table 79).

Table 79
Growth of Azotobacter in soils under fruit trees
(number of cells in 1 g soil)

Region and soil	Azotobacter in orchard	Azotobacter in field
Moldavia, Chernozem		
Kishinev region	2,500	450
Kalarash region	7,500	1,200
Slobodzei region	7,000	1,500
Rezinski region	5,000	650
Moscow Oblast', Serozem		
Chashnikovo, experimental field	1,200	120
Snegiri	2,400	60
Kragnyi Mayak	1,400	0
Crimea		
Alupke, schist	1,500	0
Yalta, humus	3,700	450
Sudak valley		
Redish-brown, sepia brown	12,000	1,500
Steppe zone	4,200	540
Koktebel'. Heavy aluvium loam	2,100	250
Old Crimea. Gravel chernozem with low humus content	1,800	80
Perekop region. chestnut solonchak	3,600	240

As can be mean from the above-mentioned, certain species of plants enhance the growth of *Azotobacter* in soil, others suppress it, and others neither enhance nor suppress its growth.

This type of plant classification is only relative and only holds true for sterile cultures, where the microbes are subjected to a one-sided action by root excretions, with the exclusion of other external factors.

Under conditions of natural growth in the field, the effect of root excretions is to a large extent annulled by many factors and, first of all, by microorganisms. Therefore, the summary effect in such cases will express itself in a weaker form and sometimes will not be observed at all. A comparison of the effectiveness of plants should be performed under the same soil-climatic conditions.

The same plant in different soils and in differing climatic and geographical zones may act differently on *Azotobacter*. As was noted above, wheat acts negatively on the growth of *Azotobacter* under conditions of sterility, in open ground in the Trans-Volga region on chestnut soils, and in the central belt on podsol, but does not suppress the growth of *Azotobacter* under the conditions prevailing in Kirgizia on chernozem soil, on the chernozem soil of the Kuibyshev Oblast', and in other places. Even in the same region and in the same soil, the effect of wheat may differ, depending on the extent of the cultivation of the soil. Soils of the podsol zone, where cultivated, often contain a sufficient number of *Azotobacter* under wheat and other plants. Of great importance for the growth of *Azotobacter* is the agrotechnical cultivation of soil.

The root excretions of young plants differ from those during the period of ripening and, therefore, the microflora during these periods of growth will also differ.

Only by consideration of a multitude of factors affecting the growth of *Azotobacter* can the contradictory indices, obtained by different investigators, on the distribution of this microorganism be explained.

These considerations not only pertain to *Azotobacter* but to all other microbial species in the rhizosphere of plants.

The selective effect of plants is well demonstrated in the case of root-nodule bacteria. Korenyako (1942) tested three species of bacteria, *Rh. trifoli. Rh. meliloti* and *Rh. leguminosarum,* by introducing them into containers in which clover, lucerne, peas, wheat, corn, and cotton were grown on sand. Root-nodule bacteria of lucerne grew equally well under lucerne and under cotton; *Rh. trifolii* grew abundantly under clover, lucerne, peas, and wheat, less well under flax, and hardly at all under corn. Root-nodule bacteria of peas were found in large numbers under peas, clover, and wheat.

These data were confirmed by experiments performed in open soil (podsol). Root-nodule bacteria of clover were introduced under clover, lucerne, peas, wheat, and corn. The growth of the bacteria in the rhizosphere was followed through the entire vegetative period, The results are given in Table 80.

Table 80
Growth of *Rh. trifolii* in the rhizosphere of various plants
(in thousands in 1 g soil)

Date of analysis	Clover	Lucerne	Peas	Wheat	Corn
23 June	100	10	10	1	1
2 July	100	100	10	0.1	0.1
23 July	100	1,000	10	10	1.0
30 July	100	100	100	100	0.1
21 August	10,000	100	1,000	1,000	10
20 September	1,000	100	190	100	10
20 October	1,000	10	100	0.01	0.1

The intense growth of root-nodule bacteria under leguminous plants was also noted by other investigators (Wilson and Wagner, 1936, and Lewis, 1938).

Rudin (1956) noted the activating action of corn sap, and especially that of pea sap, on root-nodule bacteria. According to his observations, the sap of inoculated peas having nodules on their roots is more active than the sap of noninoculated peas, The sap of corn roots enhances the virulence of the root-nodule bacteria of peas.

According to our observations, the root-nodule bacteria of lucerne grow well under timothy grass, cotton, and rye grass. Their number in the serozem soil of Central Asia, reached hundreds of thousands in one gram of soil, i. e., almost the same as under lucerne. In the chestnut soils of the Trans-Volga region, these bacteria in the rhizosphere of timothy grass amount to tens of thousands in one gram of soil, and under orchard grass, oats, and millet, their number was considerably smaller.

The number of root-nodule bacteria in the soil, after leguminous plants, decreases, if the subsequent culture in the crop rotation is not favorable to their growth this decrease in the number of bacteria in the soil takes place at different rates, depending on the soil and on the plant species. Under flax and wheat, the number of root-nodule bacteria of clover in the podsol of the Moscow Oblast, decreases comparatively rapidly.

The experimental data obtained by Chailakhyan and Megrabyan (1955) on the specific action of the roots of leguminous plants on root-nodule bacteria are of interest. The authors found that the ground roots or the root sap of leguminous plants do not negatively affect the growth of root-nodule bacteria of their own species, but suppress the growth of bacteria of other foreign species (Table 81), The maximum expression of this selective action by the roots of leguminous plants is observed during budding and flowering stages, becoming less marked later.

Table 81
Suppressing effect of the roots of leguminous plants on the growth of root-nodule bacteria.
Zones of inhibition of growth around the roots, in mm
(according to Chailakhyan and Megrabyan, 1955)

Roots of plants	Root-nodule bacteria, Vetch	Root-nodule bacteria, Onobrychis	Root-nodule bacteria, Lucerne	Root-nodule bacteria, Clover	Root-nodule bacteria, Peas	Root-nodule bacteria, Beans	Root-nodule bacteria, Soy	Root-nodule bacteria, Broad beans	Root-nodule bacteria, *Trigonella*	Root-nodule bacteria, Lupine
Vetch	0	4	3	4	6	3	3	7	4	4
Onobrychis	4	0	3	5	7	4	4	8	5	3
Lucerne	2	5	0	4	6	6	4	6	3	3
Clover	5	6	4	0	6	5	5	5	4	4
Peas	2	2	3	3	0	3	2	3	3	3
Beans	6	5	5	6	6	0	3	6	7	5
Soy	5	4	6	5	5	4	0	7	6	5
Broad beans	3	3	2	2	3	3	2	0	3	3
Trigonella	4	4	3	7	4	3	4	6	0	6
Lupine	4	5	4	4	7	4	3	7	4	0

In our investigations we were unable to find such specificity in the selective antibacterial action of either the roots or the aerial parts of leguminous plants.

Torne and Brown (1937) tested the bacterial effect of the sap of a great number of plants, leguminous and others. According to their data, the extracts of leaves of many plants inhibit the growth of root-nodule bacteria. These plants include clover, cabbage, carrots, turnips, and others. The authors did not observe any specificity in the inhibitory action of the sap. Lucerne sap inhibited the bacteria of its own species to the same extent as it inhibited that of clover, beans, and other plants. Root sap was less bactericidal than the sap of the aerial parts and, in many plants, the root extract had no toxicity for bacteria whatsoever.

Among the other microbes which grow in the root zone of vegetating plants we also studied mycolytic bacteria. These bacteria are characterized by their ability to dissolve the mycelia, of fungi (Khudyakov, 1935; Novogrudskii, 1936). They differ in classification and comprise a mixed group of bacteria. They also possess a more or less defined specificity. Each species in this group of bacteria dissolves certain forms of fungi--saprophytes and phytopathogenic forms.

According to Korenyako (1939). Raznitsina (1942), and Kuzina (1951, 1955), mycolytic bacteria grow abundantly and accumulate under certain plants. Under the conditions prevailing in Central Asia, in serozem, soils under lucerne, their number varies from 100,000 to one million in one gram. Under cotton, on the other hand, their number decreases. In soil which has been plowed for six to nine years, under this culture [cotton], the number of mycolytic bacteria is minimal, and under three-year-old lucerne it is maximal (Figure 80).

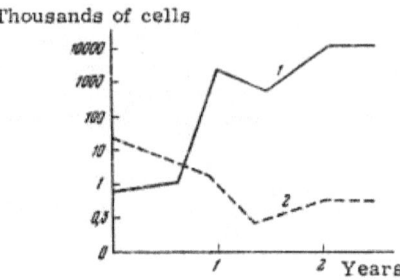

Figure 80. Accumulation of mycolytic bacteria in soil under lucerne and cotton (Central Asia) 1--under lucerne; 2--under cotton.

Under the conditions prevailing in the central belt of the Soviet Union, many mycolytic bacteria grow well and accumulate in podsol soils under clover and under certain other leguminous plants. Some of these bacteria dissolve fungi of the genus *Fusariam,* and others dissolve fungi of the genus *Helminthosporium.* Mycolytic bacteria do not grow uinder wheat and especially under flax, and they are removed from the soil relatively quickly.

Daraseliya (1949) found a large accumulation of mycolytic bacteria in the rhizosphere of the tea plant. According to the author, the limited distribution of phytopathogenic fungi of the genus *Fusarium* in the soils of tea plantations is caused by the abundant growth of these bacterial antagonists.

Nitrifying bacteria also grow in the rhizosphere of plants, where, under certain species, for instance under leguminous and certain nonleguminous plants, their number is considerably higher than under cereals and certain vegetable cultures. In studying the species characteristics of denitrifying bacteria, which were isolated from the rhizosphere of lucerne, wheat, and millet at the Ershov Station (Saratov Oblast'), we successfully determined some of their characteristics. The strains which were isolated from the root zones of wheat were in the most cases more active than the strains growing under lucerne and millet. Nitrate reduction was completed by the former in three to five days, while the latter completed this reaction under the same conditions in five to ten days. In the former case, the process was accompanied by the violent evolution of gas (nitrogen), while in the latter, the formation of gas was not observed or was negligible.

This example of the growth of denitrifying bacteria is probably not general and was observed only under the soil-climatic conditions of the given region. In the serozem soils of Central Asia and in the podsol soils of the Moscow Oblast' we have not observed such specificity. Certain differences in the properties of denitrifying bacteria, growing in the rhizosphere of lucerne and wheat, were observed in the chernozem soils of Moldavia.

Among other species of nonsporulating bacteria growing in the rhizosphere of plants, specific forms also probably exist which have better adapted themselves to the conditions of life under certain species of plants. Unfortunately, with the methods available we are unable to detect specialized forms under these or other plants. Certain investigators have observed different concentrations of physiological groups of bacteria under different plant cultures.

According to observations made by Starkey (1929, 1931), the process of the formation of nitrate from ammonium sulfate in the rhizosphere of certain plants, was more intensive than in others.

According to our observations, the root system of peas activates the process of the decomposition of cellulose. In our experiments we used special growth containers, filled with turfy podsol soil. One wall of the container (made of glass) was covered with filter paper. The plant grew from the container, which had been placed in a sloping position, and the roots were established so that they grew on the surface of the paper. The consecutive stages of the process of the destruction of the paper could be followed through the glass. Wheat, corn, peas, vetch, and beans were planted. The experiments showed that the roots of cereals do not exert a noticeable effect on the process of cellulose decomposition. Among the leguminous plants, beans proved to be ineffective, vetch only slightly stimulatory, while the roots of peas strikingly enhanced the process of the decomposition of paper (Figure 81).

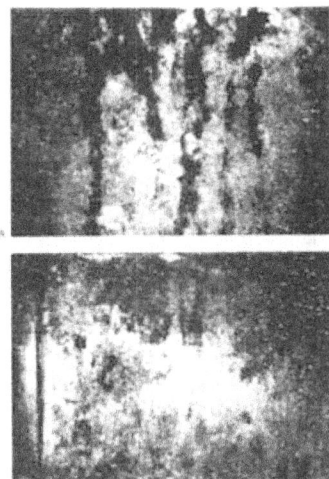

Figure 81. Decomposition of the cellulose in soil under the influence of the root system of plants a) destruction of paper around the roots of peas; b) destruction of paper in control soil, outside the root area.

Lockhead and his associates (1950, 1955) demonstrated that the composition of the microflora of the rhizosphere of various plants differs with respect to vitamin requirements. In the root area of some species of plants, bacteria prevail which require vitamin B_1 or B_2 and, in the rhizosphere of other plants, bacteria requiring biotin, vitamin B_{12}, cysteine, methionine, etc are prevalent.

There are indications in the literature, that algae also differ in their growth in the root area of plants. According to data obtained by Shtina (1954a, 1955), rye, timothy grass, and potatoes mainly enhance the accumulation of diatom algae; in the rhizosphere clover and lupine enhance the growth of green algae and perennial grasses, and potatoes partially enhance the growth of blue-green algae (Table 82).

Table 82
Growth of algae in the root zone of plants
(in thousands per 1 g of soil)

Plant	Diatoms in rhizosphere	Green algae in rhizosphere	Blue-green algae in rhizosphere	Diatoms in control	Green algae in control	Blue-green algae in control
Ry e	42.0	79.8	8.2	18.0	65.0	8.0
Timothy grass, first year	43.6	63.6	8.6	28.6	78.1	6.6
Timothy grass, second year	73.2	147.6	8.6	28.6	78.1	6.6
Clover, first year	15.4	90.0	11.0	28.6	40.0	6.0
Clover, second year	37.4	105.0	6.8	39.5	58.4	1.1
Lupine	19.2	93.6	2.4	37.2	69.6	1.0
Potatoes	27.6	46.8	7.2	15.6	45.6	3.6

There are many indications that if unfavorable conditions prevail during the growth of the plant, more or less great numbers of phytopathogenic organisms grow and accumulate in their rhizosphere.

Timonin (1941) showed that under a culture of flax considerable numbers of the following fungi often develop: *Fusarium lini, Alternaria, Cephalosporium,* and certain others causing plant diseases. Sanford and Broadfoot (1951) observed the growth of the phytopathogenic fungi *Helminthosporium sativum* and *Fusarium culmorum* in soil under wheat and oat monocultures. The fungus *Ophiobolus graminis,* under the conditions which prevail in certain localities in Canada, is suppressed by oats and clover and, according to Winter (1940), it grows better in the rhizosphere of wheat, than in the soil outside the root zone. Martin (1950) found an accumulation of the fungi *Fusarium solani, Pyrenochaeta sp.,* and other species in the soil under citrus plantations. The author is of the opinion that the weak growth of the seedlings and saplings of citrus plants in the soil of citrus groves is caused by the deleterious effect of the microflora. Cotton is favorable to the accumulation in the soil of the phytopathogenic fungi *Verticillium dahliae* and *Fusarium vasinfectum,* causing it to wither, At the same time, lucerne suppresses the growth of these fungi. The accumulation of phytopathogenic fungi in soil under the influence of vegetative cover was also noted by other investigators (Lockhead and others, 1940-1950; Weindling, 1946; Eaton and Rigler, 1946; Timonin, 1946; Grammer, 1955, and others).

Under certain conditions, plants can accumulate microbial antagonists in the soil, which inhibit the growth of such useful species of microorganisms as *Azotobacter,* root-nodule bacteria, and mycorhizal fungi, which are the producers of various biotic substances: vitamins, auxins, amino acids, and other products of microbes.

Plants may also favor the growth and accumulation in the soil of the microbial antagonists of phytopathogenic bacteria, fungi, actinomycetes, and even viruses.

Hildenbrand and West (1941) observed a lower rate in the root-rot disease in strawberries when they were sown after soybean, and an increased incidence of the disease when they were sown after clover. According to data obtained by Cooper and Chilton (1950), in the root zone and on the roots of sugar cane, actinomycetes, antagonists of the fungus *Pythium arrhenomonas,* which is the causative agent of the root disease of this plant, grow abundantly. Their number in the rhizosphere reaches 77,000 and more, while outside the rhizosphere, away from the root, it does not exceed 2,000 in one gram of soil.

Similar data wam given by Sanford (1946, 1948), Weindling (1948), Fallings (1954), and others.

Plants have a beneficient effect on soil. not only in relation to phytopathogenic microbes, but also in relation to pathogenic microbes affecting men and animals.

Bogopol'skii (1948, 1950) studied the effect of the root system of various plants on the viability of bacteria of the colon group in the soils of urban plantations, in the parks and squares of Kiev. According to his observations, certain grasses used for lawns considerably hastened the death of these bacteria. For instance, in a plantation of sweet clover, after 60 days, 25 organisms out of one and a half million originally introduced were found, 45 were found under clover, and 110 were found under oats, of the total number originally introduced in the soil. Under orchard grass the colon bacillus almost totally disappeared (10 organisms per gram), while In the control zone without plants, 180 bacteria per gram were observed.

According to our observations, the colon bacillus and a pyogenic staphlococci die in the soil under some plants quicker than under others (Table 83).

Table 83
Death of *Staph. aureus* and *Bact. coli* under the influence of plants
(number of cells per one gram of soil)

Bacteria	Time of stay in soil, days	Fallow soil	Two-year-old clover	Grass mixture
Staph aureus				
	0	2,500,000	2,500,000	2,500,000
	5	100,000	10,000	50,000
	10	10,000	100	10

	20	1,000	0	0
	30	100	0	0
	50	0	0	0
Bact. coli				
	0	1,500,000	1,500,000	1,500,000
	5	500,000	20,000	30,000
	10	25,000	1,000	5,000
	20	1,500	100	600
	30	200	0	10
	50	10	0	0

On the tenth day after the introduction of the staphylococci, it could not be found under a grass mixture of lucerne and rye grass, nor under clover with timothy grass. Under clover it disappeared after 20 days and in fallow soil, after 50 days. The colon bacillus disappeared more quickly under clover than under a grass mixture or even under fallow soil. The same results were obtained by Mishustin (1954) in vegetation experiments with orchard grass, rye grass, brome grams, clover, and fescue grass. Two cultures were introduced into the soil, *Bact. coli* and *Bact. coli aerogenes.* They died sooner under clover and fescue grass. Under other plants these bacteria died much later.

Arkhipov (1951, 1954) studied the extent of the growth in soils of the bacterium which causes anthrax under different plants, under conditions of vegetation experiments, and under field conditions. Wheat, rye, clover, vetch, lucerne, barley, Euagropyrum, potatoes, buckwheat, millet, lupine, flax, garlic, onions, etc were grown. Experiments showed that certain plants (garlic, winter wheat, rye, onions, rhubarb, and vetch) completely remove the anthrax bacillus from the soil. Lucerne, spring wheat, hemp, and the castor-oil plant exert a weak inhibitory effect. *Ornithopus sativus,* carrots, radishes, rape, water cress, and others have no effect whatsoever. Finally, such plants as potatoes, Eusgropyrum, horseradish, radishes, and turnips stimulated the growth and accumulation in the soil of the above microbe. On the basis of the results of his own studies and data obtained from the literature, V. V. Arkhipov recommended the sowing of winter wheat, rye, vetch, clover, rhubarb, garlic, and onions for the more rapid removal of anthrax bacilli from the soil.

The selective action of plants on the microflora is not only caused by the specificity of the nutrient substances excreted by the root system, but also by special antimicrobial compounds. In the chapter on toxicity of soils we shall give data showing that many plants form and excrete different toxic substances into the environment. Among them there are many compounds which strongly inhibit the growth of certain species of microbes.

Earlier we noted (1934b, 1939) that the root excretions of wheat, corn, flax, and some other plants visibly supress the development of certain kinds of bacteria--*Azotobacter,* sporiferous bacteria, and other groups both under sterile experimental conditions and directly in the soil in their natural surroundings. It was shown that the root excretions of corn act differently to those of wheat. Under the influence of these excretions, the cells of the bacteria undergo a considerable deformation, degeneration, involution, and subsequently die. Sometimes this degeneration of the culture is accompanied by the birth of new forms and species.

The presence of antimicrobial substances in the root excretions of plants was observed by Sidoranko (1940b), Meshkov (1953) and others. The soil studies carried out for many years in the Wareham forests (in England) by Rayner and Nelson-Jones (1949) showed that the obstacle standing in the way of afforestation of certain soil zones is not the lack or insufficiency of nutrient elements but the presence of special organic substances. These substances, according to these authors, retard the growth not only of young saplings of various wood varieties but also that of many microorganisms. Rayner has shown that by introducing organic fertilizer manure, or compost into these poisoned soils, the toxicity diminishes or even vanishes. According to her observations, this decrease in toxicity takes place due to the activation of the microflora in the presence of fresh organic substance.

Extensive studies of the antimicrobial properties of toxins excreted by plants were made by Metz (1955). The root excretions and root sap of 100 species of plants were studied. It was shown that 10 species out of 100 (*Armoracia rusticana* Gaertn, *Chelidonium majus* L., *Crepis viren* K., *Hieracium. pilosella* L., *Hypericum perforatum* L.., *Lampsana communis* L., *Ranunculus acer.* L., *Viola silvatica* Schm., *V.*

tricolor Wittr.,*Pulmonaria officinalis* L.) greatly inhibited the growth of sporiferous bacilli or micrococci; 15 species--*Brassica campestris* L., *B. napus* L., *Campanula rapunculoides* L., *Capsella bursa pastoris* Mad., *Fumaria officinalis* L., *Ranunculus repens* L., *Paeonia officinalis* L., *Sinapis alba* L., *Galium verum* L., *Allium schoenoprasum* L., and others inhibited their growth moderately; 31 species: *Achillea millefolium* L., *Aethusa cynapium* L., *Avena sativa* L., *Chrysanthemum lencanthemum* L., *Cichorium intybus* L., *Datura stramortium* L., *Hordeum sativum* Gessen and others inhibited the growth of these bacteria only slightly. The remaining species of plants did not have any inhibitory effect.

Stiven, (1952) found that in the soil under the plants *Tragopogon plumosus* L., and *Pentanisia variabilis* Harw., the processes of nitrification and the growth of *Bac. subtilis*, *Bac. coli*, and others are inhibited considerably. Pure substances obtained from the root excretions of these plants have the same effect.

The action of root toxins is not specific, according to Metz. The roots inhibit the growth of bacteria, both when they were isolated from the rhizosphere of the given plant and when they were isolated from that of another species. For example, the roots of *Chelidonium majus* L. inhibit, to the same degree. the growth of bacteria isolated from their own root zone and those from the root zones of goutweed, violet, hawkweed, and others, and also bacteria from fallow soil. However, the author notes, the root-zone microflora is less sensitive to the action of the roots than organisms from outside this zone. The growth of bacteria and mycobacteria, isolated from fallow soil or from soil outside the root zone, in the majority of cases, is inhibited by the roots of many plants, while most of the bacteria of the rhizosphere are not inhibited at all.

According to our observations, the roots of lucerne and peas, under conditions of growth in a sterile nutrient solution. excrete substances which inhibit the growth of *Azotobacter chroococcum* and *Pseudomonas fluorescens,* certain species of rootnodule bacteria, and others.

The growth rate of the bacteria from two-month-old plants grown in a solution was determined (Table 84).

Table 84
Effect of root excretions of peas and lucerne on growth of bacteria

Bacteria	Peas	Lucerne
Az. chroococcum		
strain No 54	++	-
strain No A	++	-
halophylic strain	-	-
garden strain	-	+
strain No 6	-	-
Az. Vinelandii	++++	+++
Ps. flourescens No 8a	+	-
Ps. aurantiaca	+++	+++
Root nodule bact. of:		
kidney beans	+++++	++
peas	+	-
vetch	+	-
lucerne	+	-
Lathyrus	+	_
broad beans	+++++	+++
soy	-	+
lupine	-	-
sweet clover	+++++	+++
clover	++++	-

The longer plants have been growing in a solution, the more strongly the solution affects the bacteria. After peas grew for three months in the solution, *Azotobacter* strain No 54 and strain A did not grow at all, nor did the root-nodule bacteria of clover and sweet clover.

Antibacterial substances are also excreted by isolated roots if the latter are grown in an artificial nutrient medium. We cultivated the roots of peas. lucerne, lupine, and certain other plants in Bonner's nutrient solution for one to three months and, after various periods of time, we determined the presence of antibacterial substances in it. The bacteria were introduced into the solution and, by means of plating, their viability and degree of multiplication were established, Cultures of *Azotobacter* strain No 54, a halophilic strain and a garden strain, and, in addition, root-nodule bacteria of peas, vetch, lucerne, and others, were sown into the solutions.

The results of the experiments showed that the excretions of roots grown in an isolated form act on the same species of bacteria as do the excretions of non-isolated roots. Only the degree of their inhibition was weaker. The antibacterial spectrum of root excretions was the same in both cases. This indicated the fact that isolated and nonisolated roots excrete the same antimicrobial substances.

Thus, by this experiment, it was observed that certain antimicrobial substances excreted into the medium, are directly synthesized in the root system of the vegetating plant.

Microflora of decomposing roots

Investigations have shown that plants not only determine the microflora of the soil during their growth but also determine them by means of their dead residues, especially those of roots. It was established that these residues, depending on the species of plant or, more accurately, on their chemical composition, are decomposed by various forms of microbes. The qualitative composition of the microflora of the decaying roots of wheat and clover, cotton and lucerne, differ considerably.

We performed microbiological studies of the root flora and of the flora of the area around roots which were rotting in the soil after the harvesting of crops. Studies have shown that around the roots of wheat, corn, sunflower, soybean, and other plants, the number of microorganisms is considerably higher than in the soil outside the root zone. If by ordinary calculation (plating on an agar medium) in the soil outside the root zone one finds four to eight million bacteria in one gram, then around decaying roots there are 20 to 140 million. The number of bacteria varies according to the degree of the decomposition of the root tissue.

The group composition of the microflora of decaying roots is given in Table 85, which shows that the increase in the total number of microorganisms paralleled the multiplication of the cellulose bacteria.

Table 85
Qualitative composition of the microflora of corn roots at various stages of decay
(number of cells in millions per gram of soil)

Time of sampling after harvesting of crops, in days	Nonsporiferous bacteria	Sporiferous bacteria	Coccoid forms	Myco-bacteria	Actino-mycetes	Fungi	Cellulose bacteria
5	52	6.5	3.5	17.0	9.0	0.3	6.8
12	46	8.5	3.0	12.0	12.0	0.8	5.0
18	120	12.5	2.5	17.5	4.0	0.3	13.0
22	180	15.0	2.8	15.0	5.0	0.6	12.0
27	120	14.0	12.0	27.5	15.0	0.3	8.0
32	100	10.0	20.0	30.0	25.0	1.5	4.0
38	60	5.0	20.0	23.0	22.0	1.6	3.0

Actinomycetes, mycobacteria, and coccoid forms appeared somewhat later, after the development of the nonsporiferous bacteria. Coccoid forms are basically mycobacteria, actinomycetes. and, to a certain degree, mycococci, i.e., organisms belonging to the group of actinomycetes.

Our subsequent studies were made with root residues which were introduced into the soil. We studied the microflora of the decaying roots of lucerne, clover, wheat, corn, and Euagropyrum, introduced in podsol soil in glass vessels.

At the same time, the microflora of a root-mass compost was studied under the conditions of a laboratory experiment. The results are given in Table 86.

Table 86
Quantitative and qualitative composition of the microflora of decaying roots
(number of cells in thousands per one gram of soil)

Roots	Root-nodule bacteria	Azoto-bacter	Nonspori-ferous bacteria	Mycolytic bacteria	Actino-mycetes	Fungi	Sporiferous bacteria
IN COMPOSTS:							
Lucerne	1,000,000	0	1,000,000	1,000,000	0.05	1	0.01
Clover	1,500,000	0	1,000,000	10,000	0.1	5	0.001
Wheat	0	0	100,000	0.05	15	35	35
Corn	0.03	0.3	400,000	0.1	250	15	150
Euagropyrum	0.05	0.01	5,000,000	1,000,000	150	20	0.001
IN SOIL:							
Lucerne	500,000	2,500	1,200,000	10,000	150	40	0.5
Clover	150,000	1,700	750,000	1,000	300	30	0.8
Wheat	0	0	15,000	0	1,500	360	280
Corn	0	0	13,000	0	4,500	120	340
Euagropyrum	10	0	70,000	100,000	2,800	400	450

In the table, data are given on a number of microbes during the initial stage of root decay. During the subsequent stages of decomposition, the quantitative ratios of microorganisms changed; the number of actinomycetes increased considerably, while the number of sporiferous and nonsporiferous bacteria and fungi decreased. In a half-decayed mass of roots, actinomycetes often cover the root particles with a white coating of aerial mycelium.

Nonsporiferous bacteria prevail during all stages of root decay, but their species composition varies. The number of cellulose bacteria during the root decay of various plants differs. Filter paper spread on a soil plate containing the roots of clover or lucerne decomposes more quickly and more effectively than on soil containing the roots of wheat. In the former case, 500-700 eroded spots were counted on the paper and in the latter case, only 130-270 spots were found. A compost of the roots of Euagropyrum contains fungi which are absent in the decaying roots of clover, and vice verse. In some cases, there is a mass multiplication of mycolytic bacteria and, in other cases, these bacteria are scarce or not detected at all (Krasil'nikov and Nikitina, 1945).

Bodily (1944) introduced residues of clover and wheat straw into the soil and he observed that in the presence of clover the number of microorganisms in soil was greater than in the presence of wheat straw.

Other plant residues, in addition to roots, possess, to a certain degree, a determining effect on the composition of the soil microflora. Birch leaves decompose differently and possess a different microflora than leaves of aspen and oak. The composition of the microflora of Euagropyrum straw and wheat straw also differ. Consequently, all the plant residues of the roots, stems and leaves upon decomposition, enhance the growth of different species of microorganisms in the soil.

Al'bitskaya (1954) obtained data on fungi from her study of the decomposition of plant residues of forest and steppe vegetation. According to these data, the roots of steppe vegetation are mainly decomposed by fungi of the genera *Penicillium, Cephalosporium , Fusarium,* and also, to a limited extent, by members of the genus *Mucorales.* Roots of oak are decomposed by fungi of the genus *Trichoderma--T. lignorium, T. koningii,* and rarely, by members of the genus *Penicillium.*

Plant residues of both steppe and forest vegetation are more intensively decomposed after being inoculated by a mixture of the natural microflora where, in these cases, the greatest loss in water-soluble organic substances is observed, which indicates a most complete decomposition of the residues. Upon the decomposition of steppe vegetation, there is a greater CO_2 evolution and a decrease in water-soluble substances and in lignin. Upon the decomposition of the roots and leaves of oak, the water-soluble substances are depleted in carbon, and upon the decompo sition of steppe vegetation, their car bon content increases.

Thus, plants, while alive, differentiate, select, and accumulate certain compounds. In other words, the vegetative cover, as a whole, is a powerful determining factor in the microbial biocoenoses of soils. In the zone of the root system, only those organisms which assimilate the root excretions of the plant in question more quickly, can develop, supplanting other, less well adapted, species.

In different plants, the dominating microbial forms differ in their systematic position as well as biologically. It may be said that each species of plant, or group of closely related species, concentrates a more or less specific microflora.

Unfortunately we are not yet in the position to determine this specificity in a precise way. Our methods of recognizing and classifying microbes, especially bacteria, are far from being completely developed. We cannot exactly say in what manner the nonsporiferous bacteria, which dominate in the rhizosphere of wheat differ from the bacteria in the rhizosphere of clover, oats or potatoes. By their appearance, cell size, motility, colony structure, and nature of growth on media, they do not differ in the majority of cases, nor do they differ in their generally accepted physiological properties.

Only a more thorough study of the biochemical activity of microorganisms will enable us to differentiate between them. However, such a method of study has not yet found wide use in laboratory investigations of rhizosphere microflora.

The selective action of the vegetative cover may be directed toward the selection not only of useful, but also of harmful microflora. Under unfavorable conditions, when agrotechnical rules are not observed, with an incorrect choice of crop rotation, the fields are contaminated by phytopathogenic bacteria and fungi and other harmful microbes--weeds. Especially, after the prolonged repeated cultivation of plants on the same field in monocultures, one observes this effect. The accumulation of an undesirable microflora under monocultures is most often caused by the insufficient growth of microbial antagonists in the rhizosphere, which are characteristic of the given plant under conditions of normal growth.

Many outbursts of epiphytotic diseases, as for example fusarial infections of cereals, cotton, saplings of woody plants, as well as cotton wilt are, in our opinion, caused by the above phenomenon. This was proved by microbiological studies made of cotton fields which were afflicted by the fungi *Verticillium dahliae* and *Fusarium vasinfectum.* These fungi are removed as soon as mycolytic bacteria begin to grow in the soil. These bacteria, as was shown above, grow under lucerne and certain grass mixtures.

In agricultural practice from time immemorial, crop rotation has been used as one of the methods of increasing crop production. It was empirically found that with certain crop rotations, not only crop production increased, but disease decreased. It is known that lucerne ameliorates the soil in cotton farms and inhibits the growth of the organisms causing cotton diseases. On this basis, one can select plants for rational crop rotation.

Noting the great influence of the vegetative cover on the formation of microbial biocoenoses in soil, one should not forget the importance of the soil itself as a substrate and the effect of the activity of man on external conditions. The physicochemical state of soil determines to a great extent the direction of the microbiological processes, and to a similar extent the development of different microbial species. The distribution of *Azotobacter*in soil not only depends on the plants. on their root excretions and decomposition products, but also on soil acidity and the presence of phosphorus, calcium, molybdenum, and other nutrient elements.

The cultivation of soil, the liming of soils, the use of fertilizers and other factors favor the growth and accumulation of *Azotobacter*. In arid regions the growth and accumulation of *Azotobacter* in soils depends to a considerable degree on irrigation.

EFFECT OF SOIL MICROORGANISMS ON PLANTS

In the foregoing chapter, the considerable influence of the microbial population of the soil, and the accumulation of individual species and groups of microorganisms in the root zone, was shown. The importance of root microflora for the life of plants has been studied only a little and the information available on the action of different microbes on the growth of plants is meager. We will not dwell here on the activity of root-nodule bacteria,*Azotobacter,* and mycorhizal fungi, since data on this subject are abundant in scientific literature. In this chapter, data are given only on those organisms which exert a beneficial or harmful effect on plants, due to the products of their metabolism; these are microbial antagonists, activators, inhibitors, etc.

Microbial activators

It was noted above that certain soil microorganisms are capable of producing various biotic substances-- vitamins, auxins, amino acids, and other biocatalysts. Such microorganisms activate the biological processes and, therefore, we named them microbial activators.

There is much data in the literature on the positive effect of pure cultures of bacteria, fungi, and actinomycetes on the growth and development of plants. Microbial activators increase the percentage of germinating seeds, enhance the growth of the young plants, and often change the nature of the biochemical processes.

Already toward the end of the last century, Geier (1882), and later Zimmermann (1902), described the bacteria living in the tissues of plants which had a certain activating effect on their growth. Such bacteria could form special nodules in the leaves of subtropical and tropical plants.

According to their systematic position, these bacteria differ from each other. In members of the genus *Ardisia Thunb.*, the nodules in the leaf tissues are formed by nonsporiferous bacteria of the genera *Bacterium* and*Pseudomonus.* In *Pavetta* L., *Chomelia* L,, *Psychotria* L., and certain other genera, mycobacteria were isolated from the nodules, in *Dioscorea* L., and others, bacteria of the *Rhizobium* type were isolated (Krasil'nikov, 1940 a, b).

Miehe (1911, 1918) studied in detail the bacteria which are members of the genera *Ardisia* Thunb. and *Pavetta* L. According to his data, they formed special substances which cause the stimulation of the tissues (Reizwirkung). They do not fix nitrogen.

Jongh (1938) found that *Ardisia* does not grow or grows poorly without bacterial symbionts, and does not bloom nor bear fruit. When bacteria were introduced into the tissue of the plant, its growth and development improved drastically, the growth of branches was enhanced, leaves acquired normal form, and flowers and fruits appeared.

The role of the symbionts of the root-nodule bacteria group is widely known. Forming nodules on the roots of leguminous plants and on certain nonleguminous ones, under certain conditions they considerably improve the growth of plants and increase crop yields.

The biological role of root-nodule bacteria for plants is widely known. However, the mechanism of the action of these organisms is still obscure. It is assumed that root-nodule bacteria fix molecular nitrogen and supply it to the host plant. There is no clear-cut experimental data to support this assumption.

There are reasons to believe that root-nodule bacteria, as well as the bacteria from the nodules on the leaves of the above-mentioned plants, act favorably through their metabolites. According to our data, leguminous plants, in symbiosis with the nodule hacteria, fix molecular nitrogen for themselves from the air. The bacteria, due to their metabolic products, act as biocatalysts, activating the nitrogen-fixing ability (Krasil'nikov and Korenyako, 1946a).

The positive action of mycorhizal fungi has already been mentioned. These fungi are widespread in nature. One view has been expressed that all plants have mycorhizae, but differ as to the nature of the co-habitation. In some plants the mycorhiza is endotrophic and in others ectotraphic. In the former, the fungal hyphae grow almost exclusively in the root tissues and only a few extend into the soil, outside the root, The endotrophic mycorhiza, in its turn, includes two types of mycorhizae the phycomycetal and vesicular type, wore often encountered in grassy and woody plants, and the orchid mycorhiza found in the plants of the orchid family These two types of mycorhiza differ in the nature of the structure and development of their mycelial hyphae. In phycomycetal mycorhizas, the hyphae are not septate and often form characteristic swellings in the root tissues--vesiculae. The mycelium in the orchid mycorhiza is septate; the hyphae form characteristic entanglements only within the root cello. As a rule, they do not have vesicles.

In the endotrophic mycorhiza of both types, the hyphae develop only in the cortex of roots in the intercellular space, or penetrate into the cells. The mycelia of the fungus do not penetrate into the central part of the root.

Ectotrophic mycorhiza is characterized by the growth of mycelial hyphae on the surface of the root tips, which surrounds them with a thick and quite dense cover. From this cover the hyphae extend into the soil. There are no root hairs on this part of the root. A small part of the hyphae penetrates inside the root, but not very deeply, their growth being usually limited to the intercellular space of the epidermis, where the hyphae interweave, forming he dense Hartig net. The hyphae seldom penetrate into the upper two to three layers of the cortical cells. The hyphae do not penetrate the cortical cells, and, in the few cases where this does happen, they soon die inside the cells. Entotrophic mycorhizae are most often encounter ed among woody coniferous and leaf-bearing plants.

Between the endotrophic and ectotrophic mycorhizae, there are intermediate formations--the ecto-endotrophic mycorhizas.

Some authors (Lobanov, 1953) are of the opinion that an absolutely ectotrophic mycorhiza does not exist at all. They maintain that in woody plants, especially during the early stages of their development, there is always an endotrophic mycorhiza. Only later the external cover on the root tip develops.

The systematic distribution of the mycorhiza fungi varies. Fungi-forming vesicular mycorhiza probably belong to Phycomycetes of the Endogenaceae family, The endotrophic mycorhizas in orchids are formed by the Fungi Imperfecti of the genus *Rhizoctonia,* while in certain orchids, the mycorhiza is ferried by higher fungi, Basidiomycetes with well-developed fruiting bodies--*Armillaria mellea, Xerotus javanicus,* and *Marasmius coniatus.*

Ectotrophic and endotrophic mycorhizae of woody plants are mainly formed by mushrooms, most often by the imperfect fungi of the genus *Phoma.* There are mycorhiza-forming fungi, among the Ascomycetes and other groups of fungi. For instance, in beech, 12 different species of fungi have been described as participants in the mycorhizae in pine 17 species were described; and, in spruce, up to nine species, etc (Kursanov, 1940; Yachevskii, 1933; Kelly, 1952; Magru, 1949; Lilly and Barnett, 1953, Goimann, 1954, Lobanov, 1953 and others).

These data clearly show that there is no strict specificity among mycorhizal fungi, as in root-nodule bacteria. The first to observe the positive effect of mycorhizal fungi on the growth of plants was Kamenskil (1880) and after him, Voronin (1886), Vysotskii (1902), and others. They all looked upon mycorhizal fungi as symbionts, which exert a great influence on the growth of plants. Baraney (1940) presents extensive data confirming this point of view. One of his tables is given below (Table 87).

Table 87
Effect of mycorhizas on the growth of pine seedlings

Indexes of growth	With mycorhizas	Without mycrohizas
Length of shoots of seedlings, cm	35.5	17.5
Increase in length after 2 years, cm	18.0	3.0
Length of sprouts of 2nd order, cm	10.0	0.3
Weight of shoots part, gm	17.0	3.1
Weight of roots, gm	11.0	4.5

Number of leaves	42	12
Total area of leaves, cm^2	591.0	96.0

The essence of the action of fungi consists in supplying the plants with nitrogenous and carbonaceous elements of nutrition in some cases and, in others, in the supply of auxiliary nutrients or biotic substances, and more correctly with both. There is a great deal of data in the literature on the significance of mycorhizal fungi in the nutrition of plants, which was already mentioned In the previous chapters of this work. Recently, by the use of direct experiments with labeled atoms, it was shown that mycorhizal fungi take up find transmit various nutrient elements.

Kramer and Wilbur (1949) and Mellin and Nilson (1950, 1952) have shown that fungi transmit P^{32} and N^{15} from the external solution into the tissues of the roots and stems of pine *(Pinus taeda* L., *P. resinosa, P. silvestris* L.). Morrison (1954) found that in the presence of the mycorhizas, there in enhanced transfer of P^{32}, not only to the roots and atoms of *Pinus Rediata,* but also to the leaves. Herley and MacCready (1950, 1952) have shown that mycorhizal fungi take up labeled phosphorus, accumulating up to 90% in their mycelia.

The data on studies made with labeled atoms does not disclose the nature of the compounds by way of which the labeled phosphorus and nitrogen are transmitted to the mycorhizal fungi, one must assume that these elements when entering the cell of the fungus, take part in the general process of building its substance in the form of one of the organic compounds of metabolic products, The labeled elements are released from the cell as metabolites which enter into the medium and, from there, into the roots and green parts of the host plant. Such it process was demonstrated in Shavlovskii's experiment with rhizosphere bacteria (see above).

The free-living soil bacteria also have a considerable effect on plants. Clark and Roller (1931) studied the action of pure cultures of the following bacteria on the growth of duckweed: *Bact. coli, Clostridium sporogenes, Clastrid welchii, Ps. fluorescens liquefaciens, Bact. aerogenes, Staph. aureus, Bac. subtilis, Bact. prodigiosum* etc. Some of these bacteria stimulated the growth of buckwheat, while others had no visible effect on it.

Kozlowski (1935) observed the action of pure bacteria cultures on the growth of barley and apples, The tested cultures of the sporferous bacteria, *Bac. cereus, Bac. mycoides,* and *Bac. subtills* and of the nonsporiferous bacteria, *Bact. denitrificans, Bact. putidium,* and *Ps. pyocyanea* and also cultures of fungi. The sporiferous bacteria had no effect on the growth of plants. Of the nonsporiferous bacteria, *Ps. pyocyanea* and *Bact. putidum* inhibited their growth, while *Bact. denitrificans,* stimulated the growth of barley, but not that of apples.

We tested 130 different bacteria, mycobacteria, and actinomycetes isolated from various soils. Among them were the following; 32 strains of *Azotobacter,* 33 strains of root-nodule bacteria, 40 of *Pseudomonas,* 10 of*Bacterium,* four of mycobacteria, six of actinomycetes, three of *Bac. mycoides,* and two of *Bac. subtilis.* Cultures of these organisms and products of their metabolism were added to the medium in which wheat seedlings were grown in one series of experiments, and to that of isolated roots of peas and wheat in another series of experiments.

Isolated roots are a convenient object for the study of the requirements of biotic substances, since they are incapable of the independent synthesis of the whole gamut of these substances.

We grew isolated roots of peas, wheat, rye and other plants on Bonner's synthetic medium of the following composition:

Distilled water 1,000 ml
Ca (N0$_3$)$_2$ · 4H$_2$ O, 0.23 g
MgSO$_4$ · 7H$_2$O, 0.36 g
KNO$_3$,0.81 g
KCl, 0.65g
KH$_2$PO$_4$,0.12 g
Iron, Traces
Saccharose, 20 g

Extracts and filtrates of bacterial cultures were added in various amounts, as auxiliary substances. The bacterial cultures were grown in liquid media and filtered with bacterial filters. The results are given in Table 88.

Table 88
Influence of metabolic products of soil microorganisms on the growth of isolated roots
(increase in cm on the 30--40th day of growth)

Microorganisms	Increase in length of roots of peas	Increase in length of roots of wheat
Control (no metabolites of microbes)	0.3	0.1
Az. chroococcum		
strain 54	15.0	17.0
strain 37	5.0	0.8
strain A	27.0	15.0
Rh. terifolii		
leguminosarum	31.0	30.0
phaseoli	50.1	45.0
lupini	7.0	15.0
meliloti	0.0	5.0
sojae	3.0	10.0
Ps. aurantiaca	65.0	55.0
Ps. flourescens		
strain 4	40.0	35.0
strain 15	10.0	0.5
strain 30	25.0	12.0
strain 69	38.0	12.0
Ps. denitrificans	6.0	38.0
Ps. mycolytica	10.0	31.0
Ps. nonflourescens		
strain 3	4.0	30.0
strain 10	10.0	31.0
strain 12	23.0	11.0
strain 15	0.1	0.2
Bact. denitrificans	25.0	12.0
Bact. album	37.0	37.0
Bact. mycolyticum	6.0	25.0
Bact. vulgaris	0.0	0.0
Bac. mycoides	0.0	5.0
Bac. mesentericus		
strain 3	0.0	3.0
strain 11	0.0	0
strain 27	5.0	0
strain 29	3.0	0
Bac. subtilis		
strain 7	0.0	0
strain 17	2.0	0

strain 21	3.0	0
A. violaceus	0.0	0
A. aurantiacus	3.0	0
A. globisporus		
strain 160	65.0	25.0
strain 187	37.0	30.0
strain 375	10.0	45.0
A. grisus		
strain 17	0.0	0.0
strain 57	0.0	0.0
strain 1067	0.0	25.0
A. albus	37.0	56.0
A. alboflavus	71.0	45.0

As can be seen from the table, a large increase in the length of the roots was observed in the presence of filtrates of *Azotobacter,* root-nodule bacteria, and bacteria of the genera *Pseudomonas* and *Bacterium* (Figure 82) Metabolic products of certain actinomycetes were quite active. Sporiferous bacteria often showed a negative action, inhibiting the growth of roots.

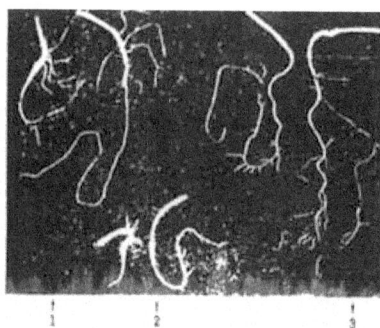

Figure 82. Effect of products of bacterial metabolism on the growth of isolated roots of peas: 1--metabolites of Ps. fluorescens ; 2--metabolites of Ps. aurantiaca; 3-control (no metabolites added).

In the same microbial group and even in the same species, different strains showed different effects on the roots. For example, among cultures of *Ps. fluorescens,* strain No 4 strongly activates the growth of wheat roots, while strain No 15 only slightly activates or does not activate these roots at all; strainNo 15 of *Ps. nonfluorescens* is inactive, while strains Nos 5 and 10 are active; the root-nodule bacteria of lucerne did not promote the growth of these roots and even suppressed them, while the root-nodule bacteria of peas and especially those of beans greatly enhanced root growth. The same was observed in all other groups of organisms.

The roots of different plants reacted differently to the action of filtrates of the same culture. The roots of wheat reacted more intensively than the roots of peas to the metabolites of *Ps. nonfluorescens,* strain 10, while with the filtrate of strain 12, the picture was reversed.

Filtrates of microbial cultures show a positive effect only when used in small amounts. When the amounts are large, their effect on the growth of isolated roots is negative. The roots do not grow or grow very poorly, deviating from the normal, they thicken, swell, do not branch, become brown too soon, and die.

The metabolites of many organisms in small concentrations strongly suppress the growth of roots. Such organisms which produce toxic substances are found among various groups of microorganisms but they are especially numerous among the sporiferous bacteria. This group of bacteria is in general the most toxic in relation to plants and many microbes (see below).

Similar data were obtained in experiments with plant seedlings. The filtrates of certain microorganisms noticeably activated the germination of seeds, while the filtrates of others had an inhibitory or no effect.

As in the experiments with isolated roots, filtrates of *Azotobacter,* nonsporiferous bacteria of the genera *Pseudomonas, Rhizobium* and *Bacterium,* and many species of actinomycetes most actively enhanced the growth of seedlings, In the genus *Azotobacter,* the strains of *Az. vinelandii* and *Az. agile var. jakutiae* were active. The strains of *Az. chroococcum* differed considerably in activity. Some possessed strongly and accentuated activating properties with respect to the growth of wheat, others had only weak activating properties and still others had no effect whatsoever on the growth of plants. Under the influence of certain strains, the suppression of wheat growth was observed.

We studied the activating effect of bacteria on different plants under field conditions for a period of three years (1945a). The seeds were treated with a culture of bacterial activators and were sown on kolkhoz fields in various regions. Altogether more than 100 experiments were performed, not including those with A*zotobacter.* The summary of the results is given in Table 89,

Table 89
Effect of bacterial activators on plant crop

Crop	Bacteria	number of experiments	positive experiments	increase in crops, %
Wheat				
	Ps. flourescens			
	strain No. 14	16	12	12-27
	strain No. 30	9	7	15-25
	strain No. 25	12	6	10-14
	Bact. sp strain No. 106	10	8	15-29
	Oligonitrophiles	10	7	12-18
	Az. chroococcum strain No. 103	20	12	10-23
Oat				
	Ps. flourescens			
	strain No. 14	10	7	14-28
	strain No. 30	--	--	--
	strain No. 25	5	5	10-16
	Bact. sp strain No. 106	12	8	12-30
	Oligonitrophiles	2	2	14-22
	Az. chroococcum strain No. 103	25	15	13-19
Clover				
	Ps. flourescens			
	strain No. 14	6	5	18-23
	strain No. 30	4	2	12-25
	strain No. 25	4	3	10-30
	Bact. sp strain No. 106	6	4	18-26

	Oligonitrophiles	--	--	--
	Az. chroococcum strain No. 103	12	8	14-27

The data in Table 89 show that bacterial activators exert a similar effect on crops as do azotogen*, nitrogen, and other bacterial compounds. In our experiments, *Azotobacter* was used in the form of peat azotogen. *[*Azotogen--Russian commercial name for *azotobacter* field-inoculating preparation.] In Table 89 are given only those cases where a positive effect was obtained when *Azotobacter* was absent from the soil; it had perished during the first few days after its introduction into the soil. Therefore, the effect was caused, not by *Azotobacter,* but by other microbes.

Akhromeiko and Shestakova (1954) successfully tested bacteria, which had been isolated from the rhizosphere, on the growth of oak and ash tree seedlings. With oak, there was a 24-34 per cent increase in the increment of dry matter, and with ash a 40 per cent increase. Samtsevich and others (1952) used an *Azotobacter* culture for the inoculation of oak seedlings in a steppe zone. According to their observations, this microbe increases the percentage of acorn germination and enhances the growth of oak seedlings. Similar results were obtained by Runov and Enikeeva (1955), Mishustin (1950b) Smali (1951) and others.

Shtern (1940 a, b) tested radiation strains of *Azotobacter* on oat seedlings, grown in vegetation containers. Certain radiation strains were more active than the initial culture.

Afrikyan (1954a) studied a large collection (more than 200 strains) of sporiferous bacteria isolated from Armenian soils. The wheat seeds which were inoculated with these cultures were allowed to germinate either in Koch dishes on cotton or in sand in containers. The experiments showed that there are very few activators among the sporiferous bacteria of the *Bac. subtilis* and *Bac. mesentericus* group. More often one finds bacterial inhibitors in this group which suppress seed germination and the growth of plants. These data are in agreement with our observations (see below).

Popova (1954) employed cultures of bacteria isolated from the rhizospere of grape vines for the enhancement of the germination of grape seeds and grape stalks. Certain species of *Ps. sinuosa* increased the percentage of germinating seeds to 80% while in the control plants, only 10-12% of the seeds germinated by the 45th day. These bacteria also enhanced the growth of seedlings and roots. In the control plants, the buds swelled on the 16th day and, in those treated with bacteria, on the fourth day. The highest activity was shown by *Azotobacter chroococcum* and the nonsporiferous bacterium *Bact. album,* strains 2 and 3.

Pantosh (1955) found that nonsporiferous bacteria of the Pseudomonas and Bacterium groups have an activating effect on the growth of the plant from the rhizosphere of which they were isolated. In experiments performed in vegetation containers, on quartz sand, the inoculation of the seeds with bacteria before sowing increased a crop of wheat by 20-65%above that of the controls, as follows:

Inoculated with	Weight of dry plants, g	Total nitrogen content, mg
In the control (uninoculated)	9.9	376
Bacterium sp.	11.1	430.7
Ps. radiobact.	16.5	466.5
Flavobacterium solare	11.2	408.9
Bact. parvulum	14.3	430.3

Petrosyan (1956) studied the effect of bacterial activators on leguminous plants, on the formation of their nodules and on their accumulation of nitrogen. The experiments were performed in containers and in plots under field conditions. In the former case, the following results were obtained:

	Weight of plants, g	Number of nodules	Total nitrogen, %
In experiments with vetch			

Control (no bacteria added)	34 (100%)	110	4.86
Inoculated with root-nodule bacteria	45 (132%)	121	5.11
Inoculated with rood-nodule bacteria and activator	65 (190%)	161	5.29
Inoculated with a pure culture of activator	56 (163%)	129	5.88
In experiments with lucerne			
Control (no bacteria added)	34 (100%)	87	4.39
Inoculated with root-nodule bacteria	38 (109%)	129	4.96
Inoculated with rood-nodule bacteria and activator	53 (153%)	177	5.24
Inoculated with a pure culture of activator	48 (139%)	155	5.07

Similar results were obtained in field experiments.

The increase in crops due to bacterial activators was as follows: after vetch, 172.6%--control crop of 11.6 kg; after lucerne, 141.3%--control crop of 10.2 kg; after Onobrychis, 150%--control crop of 8.4 kg from one plot.

In our experiments, we obtained similar results. In vegetation containers certain bacterial activators (Ps. aurantiaca No 1, Pseudomonas No 145, Bacterium sp. No 160 produced an increase in leguminous plants as follows: clover, beans, lucerne and lupine, by 30-80% above that of the control plants. In field experiments the crop increase of these plants after inoculating them with the activators, was 24-30% above the control levels. The number of nodules increased considerably when activators were employed.

On the roots of one plant, the following number of nodules were found: in control plants without inoculation of bacteria, 8; on plants inoculated with root-nodule bacteria of lucerne, 9-12; and after inoculation of bacterial activators (Ps. aurantiaca), 28. On the roots of beans, the number of nodules was 6. 8 and 16; on the roots of lupine, 0.2, 0.5 and 1.2 (Krasil'nikov and Korenyako, 1945c).

Inoculation of seeds	Dry mass of clover obtained, g
Uninoculated	26
Active initial culture	28
Avirulent strain I	35
Avirulent strain II	21
Avirulent strain A	40
Avirulent strain D	42
Avirulent strain B	26
Avirulent strain A_1	36
Avirulent strain A_3	26
Avirulent strain E	18

Bacterial activators stimulate the activity of root-nodule bacteria, enhancing the formation of nodules and, by means of the latter, the growth of the plants. Activators also act positively on leguminous plants without nodule bacteria directly stimulating their growth by their metabolic products.

In one series of experiments, we employed avirulent strains of the root-nodule bacteria of clover and vetch, obtained experimentally. The seeds were inoculated with cultures of these bacteria and sown in sterile sand in glass containers. Crops were obtained after two months.

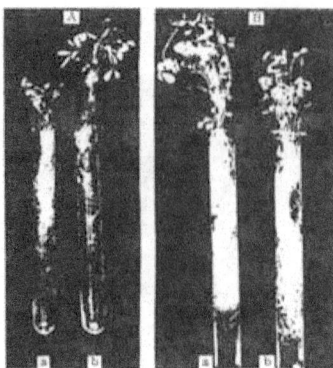

Figure 83. Organotropic action of bacteria on lucerne:

Exp. A: a--stimulation of root growth with the simultaneous suppression of the growth of the aerial parts; b--control plants; exp. B: a--stimulation of the growth of aerial parts; b--control plants.

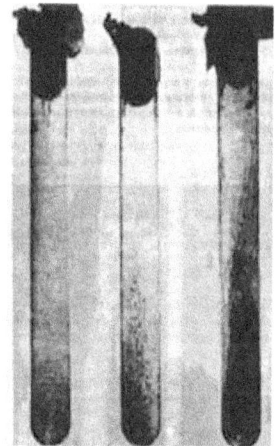

Figure 84. Effect of metabolic products of bacteria on the growth of *Phycomyces blakeslseanus:*

1--control growth of the fungus in the absence of metabolites; 2--mass formation of zygotes (dark spots) in the presence of metabolites of bacterial stimulators *(Bacter.* sp. No 2); 3--abundant growth of aerial mycelium with conidia in the presence of metabolites of bacterial activators *(Azotobacter chroococcum).*

Similar results were obtained in experiments with vetch. Using avirulent experimental strains of root-nodule bacteria, Nos 1, A, D, and A_1, the crop was considerably greater (123-161%) than when the inoculation was performed with the initial virulent culture (107-115%). In experiments with avirulent bacteria, no nodules were found on the roots of the plants while when the initial culture was used for Inoculation, 9-58 nodules were found on each plant.

Certain chemically pure substances obtained from cultures of actinomycetes and other fungi, as for instance gibberellins and gibberellin-like substances, activate the growth of plants (Figure 85).

Figure 85. Effect of antibiotics on the growth of corn: 1--seeds treated with antibiotic solution before sowing, 2--control, seeds treated with water.

Dorosinskii and Lazarev (1949), Dorosinskii (1953), Lazarev and Dorosinskii (1953) grew oats in sterile, well-washed, loamy soil in the presence of bacteria and in their absence.

The plant crop in the absence of bacteria but with full mineral fertilization was, on the average, 2.6 g; in vessels with bacteria, 7.1 g; in vessels with bacteria, but without a mineral fertilizer, 6.5 g.

Fomin (1951) grow certain melon cultures, fruit, and wood varieties of plants, treating them with preparations of *Azotobacter, Psaudomonas,* and "silicate" bacteria. After such treatment, the crops of all these plants increased.

Certain microorganisms exert an organotropic action on plants: they activate the growth of individual organs or tissues. For instance, *Ps. tumefaciens* stimulates the multiplication of the cells of the root tissue or stem tissue in tomatoes, carrots, and other plants, as a result of which swellings are formed, There are microbes which, by their metabolic products only, activate to a considerable extent, the growth of roots or serial parts.

In our investigations, we have observed bacterial cultures which, under the conditions of vegetation experiments (in sand) only stimulated the growth of roots, not affecting the aerial parts at all. In other variations of the experiment some bacteria enhanced the growth of the aerial parts without influencing the root system (Figures 83 and 85). Observations were also made of bacterial cultures which activate the process of fertilization in fungi (Figure 84) and yeasts, In Table 90, data are given on the activity of the substances which stimulate this process.

Table 90
Activity of substances stimulating the sexual process of fungi and yeasts
(number of units in 1 g of dry substance)

Substrate	*Phycomyces blakesleanus*	*Zygosacchar.* sp.
Compost inoculated with bacteria	150	80
Soil humus	60	60
Az. chroococcum strain A	180	120
Ps. flourescens strain 30	150	30
Extract from an aspen raceme	260	80

Molliard (1903) observed the stimulation of the formation of apothecia in the fungus *Ascobolus* under the influence of the bacteria within. Sartory (1916) obtained perithecia in aspergilli only in those comes where the fungi grow together with the sporiferous bacillus *Bac. mesentericus.*

Stimulation of the copulation of the fungus *Phycomyces blakealesanus* when grown together with bacteria or their metabolic products (factor Z), extracted from agar-agar, was observed by Robbins and Schmidt (1939, 1945). This "factor", according to their data, is present in the tissues of many plants and is also formed by microbes.

Nickerson and Thimann (1941, 1943) found that a certain substance among the metabolic products of the fungus *Aspergillus niger* enhances the copulation of the yeast *Zygosaccharomyces*. The active principle of this stimulant is soluble in water and 90% ethyl alcohol and consists of two substances: an acid closely related to glutamic acid, and riboflavin, Burnett (1956) caused an increased copulation and the formation of zygotes in mucor fungi by the addition of ß-carotene to the medium.

In our collection, there are actinomycete-antagonists, the metabolic products of which have a stimulating effect on the fruiting processes in higher plants. One such culture enhanced flowering and fruition in the cotton plant. Under conditions of the experiment in growth containers, plants treated with the native liquid of the given actinomycete had 10-12 bolls, while the control plants had 5-6 bolls on each bush. Correspondingly, the cotton crop approximately doubled in quantity. We treated plants with nonpurified substances. It should be assumed that chemically purified preparations will have an even stronger activating effect on plants.

In recent years great interest has arisen over gibberellins and, especially, gibberellic acid, as a stimulant to the growth and development of plants. These substances are obtained from the fungus *Gibberella fujikoroi* (a conidial stage of *Fusarium moniliforme)*. This fungus was first isolated in Japan by Kurozava, in 1926, from the tissues of diseased rice. The rice disease caused by this fungus is quite widespread in Japan, and expresses itself in an extreme elongation of the stems, in the yellowing of leaves, and in the death of the plant. Yabuta isolated the active substance from a culture of the fungus and he found that it stimulated the growth of many plants. Later, this substance was isolated in the crystalline form and studied in greater detail. In England and in America, three substances, gibberellin A_1, gibberellin A_2, and gibberellic acid (gibberellin A_3) were obtained from the culture fluid. The last substance is the most active and is of the greatest interest. It was, therefore, more thoroughly studied and more fully elucidated, Chemically defined, it is a dihydroxylacetone acid, a tetracyclic compound, with the general formula $C_{19}H_{22}O_6$.

The activating effect of gibberellic acid expresses itself in extremely small concentrations. Only 0.01 microgram/ml is enough to stimulate the growth of peas; at higher concentrations the stimulation of growth increases. The greatest effect is observed at a concentration of 10 micrograms/ml. According to the data obtained by Brian (1957), the increment in the growth of peas exceeds by 10 times that of the control plants. Its height was 42 cm after treatment with gibberellic acid, while in the control, nontreated plants, it was 7 cm. This stimulating effect was also observed by Brian in experiments with wheat, but to a lesser degree than in the experiments with peas. Gibberellic acid does not show an activating effect on the growth of roots, neither in peas nor in wheat.

Of special interest is the effect of gibberellic acid on the growth of biennial plants: cabbage, rape, carrots, sugar beets, etc. It is known that these plants give off flower shoots and bear fruit during their second year of life; during their fire, year of growth only a rosette of leaves and roots is formed. The flowering and formation of seeds may also be achieved during the first year of growth, but only after yarovization [a Russian term for vernalization; translator].

Experiments have shown that these plants form flower shoots and flowers and produce seed during their first year of growth after being treated with gibberellic acid, without previous yarovization, This acid has the same effect as the one obtained by yarovization. The enhancement of flowering and fruition is observed in longday plants. Lang (1956) obtained flowers and fruit on henbane *(Hyoscyamus niger)* during the first year of its growth. After the introduction of 300 micrograms/ml in the tissue, the plant soon formed flowering, or main shoots, on which flowers developed.

During the last two or three years, a vast amount of material has accumulated showing the strong activating effect of gibberellic acid on the growth and flowering of various plants. The strongest effect is seen with the long-day plants.

It should be noted that the stimulating effect of gibberellic acid was so far only obtained under experimental conditions in a hothouse. Under field conditions in soil, there have either been no results, or a weak stimulation of only certain plants has been observed. A small increase of crops, within the range of 11-25 per cent has been observed in meadow grasses.

In plants which react to gibberellic acids, one often observes that there is a decrease in the amount of chlorophyll, that the leaves have a yellowish tint, that their total nitrogen content decreases, and that in the tobacco leaves the percentage of nicotine decreases and , in rice, a decrease in the amount of sugars. It is assumed that this is caused by a lack of nutrition. With appropriate fertilization this has not been observed.

The mechanism of the action of gibberellic acid is not clear; however, all the investigators note that it differs from that of auxins. The latter affects the processes of growth in a different way. These substances also differ in their chemical composition.

The data given in this chapter show that there are species of microorganisms which form very active substances, which stimulate the growth and certain processes and functions in plants. Gibberellic acid is the first metabolic product of microorganisms obtained in a chemically pure form. It should be assumed that in the near future many other substances will be obtained which possess the capacity to activate the growth and development of plants. Similarly, as takes place among microbial antagonists. The stimulating substances of the microbial activators belong to various classes of compounds. Here, a great research study lies ahead in the isolation of these substances and in the study of their nature.

One must note that among the antibiotic substances and activating compounds, there is often much in common in the way they act on organisms. Many antibiotics show a stimulating effect on the growth of plants and animals; they enhance the increment of live weight of the latter, and sometimes enhance the process of fruition in both higher and lower plants. On the other hand, activating substances quite often possess clearly expressed antimicrobial properties.

Distribution of microbial activators in the soil

Very little is known on the distribution and growth of microbial activators in soil. At the same time, in daily laboratory practice, one very often encounters these bacteria, We have in mind the auxoautotrophs, which have been mentioned before. These organisms synthesize all the substances necessary for growth and development, and, therefore, grow well on simple synthetic media. For the purpose of their study, we used the following medium:

Twice distilled water	1,000 mm
KNO_3	1.0 g
KH_2PO_4	1.0 g
$MgSO_4 . 7 H_2O$	0.2 g
$CaCl_2$	0.1 g
NaCl	0.1 g
$FeCl_3$	traces
Glucose	20.0 g

There were no vitamins or auxins in the medium.

The total number of auxoautotrophs growing on this medium was quite large. We counted from several tens of thousands to many hundreds of millions of bacteria in one gram of soil.

The number of auxoautotrophic bacteria changes, depending on the properties of the soil and climatic conditions. As a rule, their total number is greater in fertile soils than in nonfertile ones. When the auxoautotrophs from the chernozem soils of Moldavia, Crimea, and Kuban are plated on an agar medium, 30 to 150 million appear in one gram. In the turfy podsol soils of the nonchernozem belt of the Moscow and Leningrad Oblast's and of Latvia, the number of these organisms varies within the range of 100,000 to 4.5 million per gram. In the soils of the Kola Peninsula (primary soils, loose calcareous soils, etc) there are 5,000 to 30,000 cells in one gram,

In cultivated soils their number is greater than in noncultivated soils, and in gardens it is greater than in fields (Table 91).

Table 91
Quantitative ratio of auxoautotrophic and auxoheterotrophic types in soils
(in thousands per one gram of soil)

Soil and region	On a vitamin-free medium	On a meat-peptone agar MPA
Podsol. Arctic Circle. Virgin soil	5	3.5
Podsol. Arctic Circle. Cultivated soil	300	360
Podsol. Moscow Oblast'. Virgin soil	200	180
Podsol. Moscow Oblast'. Cultivated soil	2,500	2,000
Podsol. Moscow Oblast'. Garden soil	45,000	60,000
Krasnozem. Caucasus. Cultivated soil	3,500	1,500
Podsol. Latvia. Virgin soil	150	100
Podsol. Latvia. Cultivated soil	1,800	1,500
Chernozem. Moldavia. Virgin soil	1,200	800
Chernozem. Moldavia. Cultivated soil	30,000	50,000
Chernozem. Crimea. Virgin soil	20,000	35,000
Chernozem. Crimea. Cultivated soil	160,000	200,000
Chernozem. Kuban'. Cultivated soil	220,000	280,000
Chernozem. Kuban'. Cultivated soil	450,000	600,000

Comparing the data on the number of bacteria growing on a synthetic vitamin-free medium and on an ordinary protein medium, one can observe that the number of auxoautotrophs and auxoheterotrophs is almost the same. In humus soil the ratio between these two categories is approximately 1:1. In certain cases, there are even more auxoautotrophs (see above).

According to Schmidt and Starkey (1951), about 30 per cent of soil bacteria synthesize biotic substances and excrete them into the soil.

Lochhead and Chase (1943) found from 10-14 per cent of auxoautotrophs among the microflora of soil. These authors divide soil microbes into seven groups, according to their ability to grow on vitamin-free media with the addition of a few auxiliary substances. To the first group belong the microbes growing on media completely devoid of vitamins. In the second group are included organisms growing on a vitamin-free medium with the addition of a few amino acids (cysteine, alanine, proline, asparagine, arginine, leucine, glycine, lysine, etc). To the third group the authors relate those microbes which require for growth certain vitamins: pantothenic and nicotinic acids, thiamine, riboflavin or others. The fourth group includes organisms growing on a medium to which both amino acids and vitamins were added. Those microorganisms growing on a basal vitamin-free medium with an admixture of yeast extract comprise the fifth group.

A basal vitamin-free medium with an admixture of soil extract reveals the bacteria of the sixth group. This medium, with a small admixture of yeast and soil extracts, is suitable for the bacteria of the seventh group, the most numerous one. Microbes of the first group comprise approximately 10-14 per cent of all soil bacteria; the second group, 10 per cent; the third, 12-14 per cent; the fourth 16-17 per cent; the fifth, 18-20 per cent; the sixth, 3-7 per cent; and finally, the seventh, 40-50 per cent of all soil bacteria. This subdivision of bacteria is quite arbitrary.

A great number of vitamin-producing bacteria are found in the rhizosphere of plants. According to our calculations, they comprise 40-80 per cent of all soil bacteria and their number varies depending on the species of the plant, the stage of its growth, and upon external conditions. In analyzing the rhizosphere of wheat and clover, growing in the fields of the Moscow district Dolgoprudnoe), we obtained the data given in Table 92. In the root zone of these two plants and in all other plants studied by us, the number of auxoautotrophs was no smaller and often larger than the number of auxoheterotrophs.

Table 92
Number of auxoautotrophic bacteria in the rhizosphere of plants
(in thousands per gram of soil)

Plant	Soil	Autxoauto-trophs in early phase of growth	Auxohetero-trophs in early phase of growth	Auxoauto-trophs in the fruit-bearing phase	Auxohetero-trophs iin the fruit-bearing phase
Wheat	Rhizosphere	800,000	100,000	300,000	500,000
Wheat	Control	5,000	3,500	3,000	3,000
Clover	Rhizosphere	1,500,000	1,800,000	1,000,000	1,600,000
Clover	Control	40,000	30,000	20,000	20,000

West and Lochhead (1940a, b), and Wallace and Lochhead (1949), in their detailed studies also found that auxoautotrophs prevailed in the rhizosphere of flax and other plants. The same was noted by Katznelson and Richardson (1943), Stolp (1952) and others (Table 93).

Table 93
Quantitative ratios of auxoautotrophs in the rhizosphere and outside it, by percentage
(according to Wallace and Lochhead, 1949)

Plant and soil	Early growth stage	Flowering stage
Wheat / Rhizosphere	42.3	48.5
Wheat / Outside the rhizosphere	14.4	9.7
Oats / Rhizosphere	44.1	47.4
Oats / Outside the rhizosphere	14.8	9.7
Clover / Rhizosphere	55.5	42.8
Clover / Outside the rhizosphere	9.3	2.4
Lucerne / Rhizosphere	40.5	37.2
Lucerne / Outside the rhizosphere	9.3	2.4
Flax / Rhizosphere	35.8	32.9
Flax / Ouitside the rhizosphere	9.7	8.6
Timothy grass / Rhizosphere	25.3	15.4
Timothy grass / Outside the rhizosphere	9.3	2.4

As is well known, the ability of microorganisms to produce biotic substances is, in most cases, not connected with their taxonomic division into groups. Auxoautotrophs and heteroauxotrophs are found among various microbial groups: among sporiferous and nonsporiferous bacteria, micrococci, mycobacteria, actinomycetes, fungi, etc. The largest percentage is encountered among the bacteria of the genus *Pseudomonas.* According to our calculations, the number of these bacteria reaches 0-60 per cent of all the bacteria growing on an agar medium devoid of vitamins. A smaller, but quite a high percentage (30-40 per cent) is encountered among the genus*Bacterium* and mycobacteria. Considerably rarer are the auxoautotrophs among sporeforming bacteria of the genus *Bacillus* (Table 94).

Table 94
Number of auxoautotrophs of different groups of microorganisms,
living in the rhizosphere of plants and outside the root zone
(by percentage)

Bacteria	Outside the rhizosphere	In the rhizosphere, wheat	In the rhizosphere, clover
Pseudomonas	30.0	40.0	46.0
Bacterium	40.0	34.0	23.4
Mycobacterium	18.0	25.0	30.0
Bacillus	6.0	0.3	0.1

Oligonitrophiles are active producers of biotic substances. These organisms are widely encountered in soils and their characteristic feature is that they grow abundantly in a vitamin-free and nitrogen-free medium. Evidently, they all belong to the auxoautotrophs.

Among the actinomycetes one rarely finds cultures which will not grow on Chapek' s synthetic medium. They grow well on a vitamin-free medium.

Yeasts of the genera *Torula, Mycotorula,* and some others are widespread in the soil. These organisms are powerful producers of vitamins and various other biotic substances. Therefore, they grow well on mineral, vitamin-free media, forming large slimy or semislimy colonies. Some of them, as, for instance, *Torulopsis pulcherrima* are widespread in the soil and grow abundantly on the nitrogen-free medium of Ashby, forming large colonies similar to *Azotobacter* colonies. They are of special interest as activators of plant growth and of the life processes of microorganisms.

Microbial inhibitors and their action on plants

These data show how large and versatile is the microflora which produces biotic substances. With the aid of these substances, the microorganisms growing in the soil activate the growth, nourishment, and many other vital processes of both higher and lower organisms.

In soils, as well as in other natural substrate, there are microorganism-inhibitors. which, in course of their metabolic activity, suppress the growth and development of higher and lower plants. They form special substances which are toxic to plant tissues and organs. Toxins, or phytotoxins formed by phytopathogenic fungi and bacteria, were studied long ago by numerous investigators (see Kuprevich, 1947; Sukhorukov, 1952; Bilai, 1953, Gorlenko, 1953; Goiman, 1954). However, the question of whether these organisms produce toxic substances directly in soil remained unsolved in the literature. It is known that many species of fungi and bacteria form toxins which act on animal organisms. Growing on food products, fodder, and on various plant residues, they excrete toxic substances. Upon feeding these products to animals, one often observes a strong case of poisoning (Pidoplichko, 1953).

Investigations show that microbial inhibitors may poison plants with their toxins under conditions of their natural growth in soil, if favorable conditions for such growth are formed. They suppress germination of seeds, the growth of sprouts and plant growth in general and decrease the total crop. Consequently, when there is a massive growth of these organisms, they may become an important factor in determining the fertility of soil and the crop yield of plants.

The suppressing action of microbial inhibitors is also manifested in the growth of lower plants--fungi, bacteria, algae and others. In such cases, the microbes are called antagonists.

Microbial inhibitors are found among various groups of lower forms: bacteria, fungi, and actinomycetes. Greig-Smith (1911) established the fact that toxic substances formed by certain bacteria, suppress the growth of plants. Ressel (1933) ascribed great importance to Protozoa, which consume microbial cells. Hutchinson and Thaysen (1918), Lewis (1920), and Laudenberger (1952) noticed in certain nonsporiferous bacteria of the genus*Pseudomonas* the ability to synthesize potent toxic substances. Johnson and Murwin (1931), and later Braun (1950), discovered this ability in *Pseudomonas tabaci,* the causative agent of tobacco disease. The ability to form toxic substances was also found in other bacterial groups.

Among the group of nonsporiferous bacterial inhibitors, representatives of the genera *Bacterium* and *Pseudomonas,* are comparatively often encountered and lose often those of the genus *Rhizobium.* They are often found in the rhizosphere of vegetative plants. We studied more than 300 cultures of these organisms isolated at different times from different soils, from the chestnut soils of the Trans-Volga region, the serozem soils of Central Asia, and the podsol soils of the Moscow and other regions.

Of this number of cultures which were studied, about 100 suppress to a greater or smaller degree the growth of plants and the germination of seeds, Strongly expressed herbicidal properties were possessed by certain strains of *Ps. flourescens, Ps. pyocyanea* and *Bacterium* sp. They completely or almost completely inhibited the germination of needs of clover, vetch, and wheat (Figure 86). The seeds merely sprouted and died, or did not show any signs of germination.

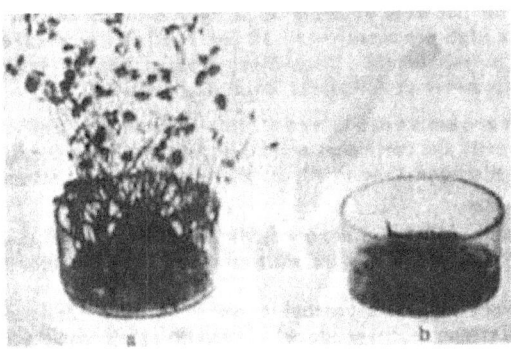

Figure 86. Suppression of the germination process of clover seeds by nonsporiferous bacterial inhibitors, *Posudomonas* sp. a--control; b--in the presence of bacterial inhibitors.

The toxic properties of many sporiferous bacteria are sharply expressed, We studied more than 350 cultures, isolated from various soils of the Soviet Union. The bacteria were grown in liquid nutrient media. The seeds were treated by soaking them for several hours in the culture fluid. Seeds of plants treated with culture fluid were germinated on cotton or on paper which had been wetted with water.

The toxic or herbicidal effect of the bacterial fluid revealed itself in the suppression of growth and the lowering of the percentage of germinating seeds, Analyses have shown that approximately 20-30 per cent of the cultures investigated possessed inhibitory proportion. Among the bacteria isolated from turfy podsol soils, the number of inhibitors was large (about 34-45 per cent).

The nature and strength of their effect vary in different cultures. Some organisms completely or almost completely inhibit the germination of seeds (Figure 87), others are less inhibitory, and still others do not show any inhibitory effect whatsoever.

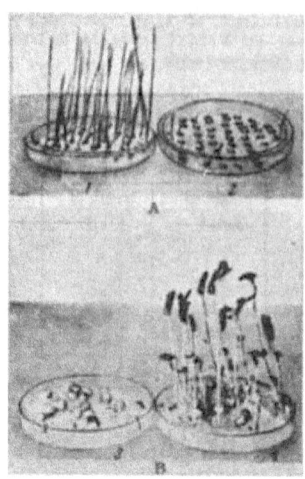

Figure 87. Effect of sporiferous bacterial inhibitors on the germination of plant seeds. The seeds were soaked in the bacteria-culture fluid and germinated on cotton or on paper wetted with water:

A--effect of *Bac. mesentericus* (strain 50) on germination of wheat seeds: 1--control, seeds soaked in water; 2--seeds soaked in culture fluid; B--effect of *Bac. mesentericus* (strain 67) on the germination of seeds of peas: 3--seeds soaked in culture fluid; 4--control, seeds soaked in water.

The species of the inhibitors studied by us mainly belonged to *Bac. mesentericus* and *Bac. subtilis*.

The capacity to suppress the germination of seeds and the growth of seedlings is revealed in various degrees among strains belonging to the same species. Among cultures of the bacterium *Bac. messentericus,* we found more than 180 strains isolated from various soils, including 100 strains from the podsol soil of the Chashnikovo Experimental Station. Among these were some very strong inhibitors, while others did not inhibit plants growth at all. Some of them inhibited the germination of wheat seeds, others those of peas, vetch or clover, while some of thorn inhibited the germination of wheat, peas, vetch and clover seeds. In Table 95 data are presented from an experiment with vetch and clover.

Table 95
Effect of metabolic products of bacteria on the growth of plants
(calculated for the 30th day of growth, in cm)

Bacterial cultures	Clover: height of shoots	Clover: length of roots	Vetch: height of shoots	Vetch: length of roots
Control	5.0	4.0	24.5	12.0
Bac. subtilis strain 7	4.5	0	26.0	6.0
Bac. subtilis strain 15	5.5	0.5	22.0	3.5
Bac. brevis. strain 3	4.8	0-0.2	25.0	2.5
Bac. mesentericus	5.2	0	25.5	1.0

Among sporiferous and nonsporiferous bacteria, there are sometimes strains possessing an organotropic or selective herbicidal action. They either suppress the growth of only this root system or of only the aerial parts. We found cultures which completely inhibited the growth of the roots of vetch and wheat. The seeds sprouted without the formation of roots, while the latter were very much reduced (Figure 88). The aerial part developed more or less normally as long as the seeds contained nutrient substances.

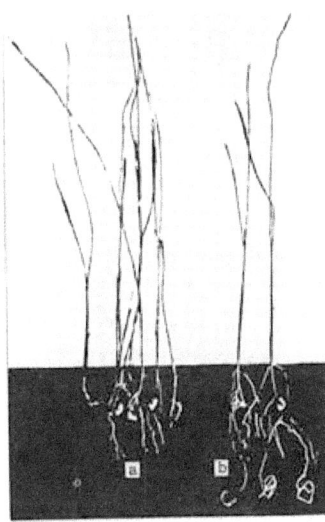

Figure 88. Suppression of growth of wheat roots by bacterial inhibitors: a--experiment; b--control.

Some bacterial strains in our collection (three nonsporiferous and four sporiferous strains) inhibited the growth of the aerial parts, but did not affect the root system. The needs germinated a root, while the aerial part was strongly reduced (Figure 89).

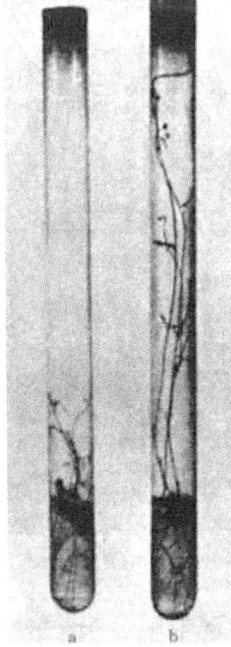

Figure 89. Inhibition of growth parts of vetch by cultures of bacterial inhibitors: a--experimental, b--control plant.

Certain strains of bacteria suppress the sporulation process of lower organisms, the formation of zygotes in phycomycetes and the formation of spores in yeasts (Krasil'nikov, 1947 a). It is possible that there are microbes which inhibit the fruiting process of higher plants as well.

In our collection of actinomycetes there are strains which cause chlorosis of higher plants by the action of their metabolic products. Chlorosis appeared in corn and wheat after treating the seeds before sowing with a culture fluid of certain species of actinomycetes and, even more markedly, with purified preparations of antibiotics. If the seeds of these plants are kept in a solution of an antibiotic for two, to four hours before sowing, the seedlings are completely colorless, without the slightest sign of the formation of chlorophyll. The growth of much plants is suppressed and, soon ceases altogether. In some cases the plants, recover, become green, and continue to grow more or less normally,

If the seeds are treated with weaker solutions of the antibiotic, one obtains seedlings which are slightly green and somewhat etiolated. Strongly etiolated plants are obtained upon the treatment of seeds with streptomycin. Soaking seeds for two hours, in a solution of one microgram per ml causes the complete etiolation of the seedlings. The latter do not become green for a period of 15-30 days and finally die. A suppression of chlorophyll synthesis is caused by aureomycin, terramycin and other antibiotics.

We obtained the etiolation of duckweed by growing it in a nutrient solution to which an antibiotic had been added. Depending on the concentration of the antibiotic, the growth of the plants was inhibited to a greater or smaller degree. The extent of the appearance of the green color also varies, from slight chlorosis to full colorlessness.

Certain toxins of microbial origin cause the phenomenon of chlorosis in grapevines, According to our observations, this phenomenon may be due to fungi of the genus *Fusarium*. We found certain strains, the toxins of which caused the etiolation of shoots, of cuttings, and grape stock, when treated before planting in the soil. The plants that grew from them had light green leaves with a yellowish hue, their development was slow, and other deviations were observed which are characteristic of chlorosis of grape vines (Krasil'nikov and Kublitskaya, 1956).

This picture of the etiolation of cuttings was observed by us after the treatment of the vine with antibiotics of actinomycete origin. Certain strains of gray and pigmented actinomycetes synthesized substances which inhibit the formation of chlorophyll in the leaves of grapevines. Cuttings, when immersed with their basal ends in the crude fluid culture and subsequently planted in the soil developed and showed obvious signs of etiolation.

The inhibition by antibiotics of the formation of chlorophyll in plants has been mentioned by certain other investigators. Provasoli, Huntner, and Schatz (1948) obtained colorless cultures of *Euglena* sp. under the influence of streptomycin. The antibiotic was added to the nutrient solution in small quantities; under its influence the chloroplast of the cells was destroyed, as a result of which completely etiolated forms of organisms were obtained.

The phenomenon of chlorosis as an effect of streptomycin was observed in cereals (wheat, corn, etc) by Von Euler (1947) and Hagborn (1956). They wetted the seeds of plants in the antibiotic solution and planted them in the soil. The seedlings were devoid of green color.

Berezova and Sudakova found that the death of the growing tip of flax is not connected with boron starvation, but is the result of poisoning by toxins formed by bacteria.

Kugushova has shown that on the roots of lucerne, bacteria may grow which, by their excretions, cause the failing off of the buttons (according to Berezova, 1953 a).

Inhibitors, suppressing the growth of plants and the germination of seeds, are encountered in great numbers among actinomycetes. In this group of microorganisms, cultures with strong herbicidal properties are most often found among the orange *A. aurantiacus,* among the gray *A. griseus,* and among other species and groups (Table 96).

Table 96
Effect of the culture fluid of actinomycetes on the germination of plant seeds

Number of germinated seeds by % of control----->	beans	corn	clover	lucerne	wheat
A. aurantiacus, strain 1149	86	60	66	77	12
A. aurantiacus, strain 1306	44	60	50	88	12
A. griseus, strain 2283	142	60	100	100	100
A. griseus, strain 293	86	120	83	111	100
A. globisporus, strain 070	114	80	50	100	87
Control	100	100	100	100	100

Experiments with seedlings have shown a more or less similar picture. Some cultures of actinomycetes strongly suppress growth, while others only slightly or not at all (Table 97).

Table 97
Effect of filtrates of actinomycetes on plant seedlings
(in length of plant parts in cm)

Actinomycetes	Wheat, rootlets	Wheat, sprouts	Corn, rootlets	Corn, sprouts	Beans, rootlets	Beans, sprouts
A. aurantiacus, strain 1149	1.5	2.0	12.0	7.0	15.0	13.7
A. aurantiacus, strain 1306	1.5	7.5	0.7	3.5	1.7	3.8
A. griseus, strain 2241	12.4	10.5	13.7	15.0	11.0	8.8
Control	14.0	15.0	15.0	12.0	18.5	14.0

Toxic substances of actinomycetes and other microorganisms exert a suppressing effect on single isolated organs or parts of plants; on leaves, cuttings, etc. If one immerses the cuttings, or cuts off the leaves, then, after a certain period of time, they wither and die. By the speed of the withering and death of these parts one may judge the strength of the action of the toxic substances.

In our experiments we used cuttings of various plants, of beans, peas and corn, and branches of lemon, apple, pear, and apricot trees, etc.

If one puts on the surface of an uncut leaf a piece of cotton which has been wetted with toxin, after a few hours spots appear of a necrotic nature. The stronger the poison, the more sharply the necrotic spots on the leaf are expressed. This method was used by us in testing the toxic substances formed by microorganisms.

Among the inhibiting factors of great importance are the phages: bacteriophages and actinophages. The studies of Rautenshtein (1955), Khavina (1954), and certain others show that these agents are widely distributed in soils, where they are detected in considerable numbers. There is reason to believe that they suppress and lyse cells of bacteria or actinomycetes as readily as under conditions of pure cultures.

For example. root-nodule bacteria become inactive when phages multiply abundantly in the soil. Under conditions of the experiment, they multiply to a considerable extent. We have counted tens and even hundreds of thousands in one gram of soil. According to Demolon and Dunez (1934), phages of root-nodule bacteria of clover and lucerne, under certain conditions, saturate the soil to such an extent that the soil becomes much less fertile for these plants, root-nodule bacteria do not develop in it, and there is only a slight or no formation of nodules on their roots, and when they do develop they have an abnormal appearance. The authors are of the opinion that the observed clover-lucerne soil exhaustion is caused by the accumulation of phages. According to certain data, phages penetrate the plant, and, by interfering with plant metabolism, lower crop production (Vandecaveye and others, 1940).

There is data in the literature on the formation of toxic substances by fungi. Leng (1949) has shown the poisoning effect of the Penicillium fungi on the seedlings and of cereals. The most active inhibitors in these experiments were *P. notatum*, and *P. oxalicum*. Monnaci and Torini (1932) and Diachum (1934) note the formation of toxins by fungi, which act on cereals under conditions of their growth in soil.

Producers of toxic substances are known among various groups of soil microflora. An important place is occupied by representatives of the genus *Fusarium*. The substances formed by them were obtained in a chemically pure form having a known structure; for example, lateritin, $C_6H_{46}O_7N_2$; avenacein, $C_{25}H_{44}O_7N_2$; fructigenin, $C_{26}H_{44}O_7N_2$; sambucynin, $C_{24}H_{42}O_7N_2$, and enniatins, lycomarasmin, yavanicin, etc.

These substances act differently on plants and animals. Some of them are specific (Goiman, 1954).

Fusaria are very widespread in nature. The probably play an important role in the toxicoses of soils. Their inhibitory effect on the growth of plants was observed by many authors (Rehm, 1953; Laundoldt, 1952; Sukhorukov, 1952). The significance of these fungi for the fertility of soils is not only determined by their ability to synthesize toxins and excrete them into the soil but also by their phytopathogenic properties.

Bilai (1955) described in his monograph many strains of the genus *Fusarium* which have a deleterious effect on the germination of seeds and on the growth of seedlings of rye, oats, and barley. The products of their metabolism, obtained in the form of filtrates, were tested under various conditions. The results of the author's experiments are given in Table 98.

Table 98
Effect of filtrates of Fusarium cultures on the germination of plant seeds
(in length of plant parts in cm)

Fungal culture	Rye, rootlets	Rye, sprouts	Barley, rootlets	Barley, sprouts
Control	21.5	4.25	29.8	3.6
Fus. poal., strain 2	3.8	1.9	--	--
Fus. poal., strain 5	8.3	2.6	16.0	3.6
Fus. poal., strain 9	11.7	2.5	11.8	2.3
Fus. poal., strain 41	2.4	1.6	--	--
Fus. poal., strain 45	15.0	5.4	8.4	1.2
Fus. sporitrichioides, strain 28	6.1	1.4	18.4	2.1
Fus. sporitrichioides, strain 30	11.3	3.2	6.0	1.5
Fus. sporitrichioides, strain 51	15.3	6.3	11.2	1.5

As can be seen from the table, the filtrates of some strains affect the seedlings of rye, while others act predominantly on the growth of barley. Certain strains suppress the growth of rye and wheat to the same extent as that of barley or oats.

Klechetov (192 6) in studying the phenomenon of the flax exhaustion of soils found the growth of the fungi *Fusarium*, *Thielaviopsis basicola*, *Cladosporium herbarum*, *Alternaria*, and *Macrosporium* in these soils; these fungi, according to the author, form toxic substances and are the reason for the death of the sown flax.

A considerable role in the exhaustion of soils and in the lowering of plant yields is attributed in the literature to the fungi of the genus *Fusarium*. Kvashina (1938), Kurtesova (1940), and Ioffe (1950).

Kublitskaya (1955) studied the degree of the distribution of fungi of the genus *Fusarium* in the soils of Central Asia (Uzbek SSR) under grapes. She isolated 52 cultures and many of them proved to be toxic for grapevines, causing poisoning and death to the cuttings and stock under the conditions of growth in soil. Certain strains caused chlorosis under experimental conditions.

Strongly expressed herbicidal properties are exhibited by fungi of the genus *Pythium*. According to Likais (1952), *Pythium debaryanum* forms toxins in the soil which inhibit the root systems of plants.

Mirchink (1950) studied a large collection of fungi isolated from turfy podsol soils of the Moscow district and found among them many toxigenic forms. The most toxic and the most widespread fungi in these soils are representatives of the genus *Penicillium* and, secondly, *Fusarium* and *Trichoderma*. Fungi of the genus *Trichoderma (T. lignorum)* and certain representatives of the genus *Fusarium* strongly suppress the germination of wheat seeds, as a result of which the number of germinating seeds decreases by 68 per cent and more. The length of sprouts in the presence of the metabolic products of *Trichoderma* is 3.5 cm; in the presence of the fungus *Fusarium*, 4.0 cm; and in the control, 4.6 cm. The Penicillia inhibitors are often found in the turfy podsol soil in a great number of species. Some of them are very toxic for wheat, which can be seen in Table 99 and in the photograph (Figure 90).

Table 99
Toxic effect of fungi of the genus *Penicillium* on wheat seeds

Fungi	Per cent of germinated seeds	Mean length of sprouts, cm
Control (nutrient medium)	100	4.6
Control (water)	100	4.6
P. cyclopium	0	--
P. paxilli	54	2.6
P. ochro-chloron	74	1.5
P. martensii	74	3.0
P. nigricans, strain II/14	100	1.0
P. nigricans, strain II/35	87	0.6
P. nigricans, strain VIII/8	90	1.0

Figure 90. Effect of the culture fluid of the fungus Penicillium nigricans on the germination of wheat seeds: a--control; b--treated seeds.

Active toxin producers in soil are fungi of the genera *Trichoderma, Trichothecium, Botrytis,* and others. From cultures of *Helminthosporium (H. victoriae),* the toxin victorin was isolated, which inhibits the growth of roots and seedlings of oats at a dilution of 1:1,000,000. This substance is formed by the fungus directly in the soil (Weeler Luke, 1954; Tyler, 1948). Toxic substances harmful to plants were found among the representatives of *Verticillium.* The most well studied among them is *V. alboatrum.* Its toxic substance was found by Bewley (1922), It causes the withering of tomatoes, cotton, tobacco, and other plants. Green (1954) discovered two substances in this fungus-- a protein and a polysaccharide. The former is excreted into the medium and the latter enters the tissues of the plants, The poisoning effect of this fungus was also noted by Sukhorukov (1952) and others,

Among members of the genus *Trichothecium* were found the toxic substances trichothecin and others, which inhibit plants and certain microbes, Similar substances were found in *Deuterophoma tracheiphilus,* causing "malsecco" in citrus plants, They were also encountered in many other fungi (Hossayon, 1953; Freeman and Morrison, 1949, Gelman 1954).

It is obvious that the importance of microbial inhibitors in soil toxicosis will be mainly determined by the degree of their growth and activity.

The distribution of microbial inhibitors and their accumulation in the soil has been but occasionally studied; as were microbial activators. Monnaci and Torni (1932) found about 60 per cent of the soil fungi isolated and investigated by them, to be inhibitors.

According to our data, there are a great number of inhibitors among the fungi, bacteria, and actinomycetes in soils. Out of 1,500 cultures of actinomycetes, more than 200 inhibited, to a larger or smaller degree, the germination of beet or wheat seeds and 16 strains completely suppressed their germination; 21 cultures strongly suppressed and 58 weakly suppressed the growth of clover and lucerne, The total number of inhibitors among actinomycetes is comparatively small, on the average 5-15 per cent,

One finds inhibitors among sporiferous bacteria considerably more often, Out of 560 strains studied, belonging mainly to three or four species, *Bac. mesentericus, Bac. subtilis, Bac. cereus,* and *Bac. brevis.* 178 strongly suppressed the germination of clover seeds, more than 200 cultures suppressed to some degree the germination of peas. According to our data, there are about 40 per cent inhibitors among the sporiferous bacteria of *Bac. mesentericus* and *Bac. subtilis* isolated from the turfy podsol soils.

Inhibitors among nonsporiferous bacteria are encountered much less frequently than among sporiferous bacteria, According to our calculations, their number can be expressed in a tenth part of one per cent. Some species of the genus *Bacterium* and *Pseudomonas* possess, however, strongly expressed toxic properties in relation to plants and microorganisms.

It should be noted that certain microorganisms among bacteria and fungi react to toxic substances in the same way as do higher plants, which enables us to use them as test organisms in the screening for and the study of phytotoxins. Microbial tests have a number of advantages, With them one can more quickly determine and solve a number of problems related to the toxicosis of the soil and the poisoning of plants, In mass studies we often use both tests; the microbiological and the plant test.

We carried out the quantitative evaluation of microbial inhibitors in different soils, but went into greater detail in the turfy podsol soils of the Moscow Oblast', the Kola Peninsula, and in other regions of the USSR. Virgin and cultivated soils, forest and swampy soils, meadows, ate were investigated,

We counted from 5,000 to 450,000 inhibitors in one gram of soil depending on the properties of the latter (Table 100). In slightly cultivated soils, the absolute number of inhibitors is smaller, but its percentage may be higher than in wellcultivated soils.

Soil	Bacteria inhibiting azotobacter	Actino-mycetes inhibiting	Fungi inhibiting azotobacter	Bacteria inhibiting beet	Actino-mycetes inhibiting	Fungi inhibiting beet

Table 100
Number of inhibitors in podsol soils of Moscow area
(Number of cells in 1 g of soil)

		azotobacter		seedlings	beet seedlings	seedlings
Dolgoprodnoe						
Virgin soil	15,000	23,000	1,300	8,000	3,000	500
Plowed fields	45,000	17,000	2,000	15,000	7,000	1,000
Agricultural Academy Timiryazev						
Forest	40,000	80,000	17,000	25,000	10,000	4,000
Virgin soil	10,000	32,000	1,500	10,000	3,000	500
Plowed fields	120,000	150,000	2,300	50,000	35,000	3,000
Chashnikovo						
Forest	120,000	82,000	12,000	20,000	12,000	7,000
Virgin soil	40,000	16,000	1,600	10,000	5,000	1,000
Plowed fields	450,000	160,000	1,400	150,000	60,000	500

Inhibitors which suppress the growth of *Azotobacter* in podsol soils are much more numerous than microbes which suppress plant growth. Mirchink (1958) studied the fungal flora of soils of the experimental station Chashnikovo (Moscow Oblast') and found that 11-38% were inhibitors which suppress plant growth. They were distributed in the following manner: in forest soil--13%, in glades--11%, in cultivated soils, 15-38% of the total microflora, detected by by existing methods (Table 101).

Table 101
Number of fungi in podsol soils
(thousands in 1 g of soil)

Soils	Total	Inhibitors, %
Control soils without fertilizers	60	32
Fertilized with mineral nitrogen	138	38
Calcium-containing fertilizers + manure	36	24
Calcium-containing fertilizers + manure + P.K.	18	15

As can be seen from the data given, the greatest number of inhibitors was found in soils cultivated to a limited extent. Mineral fertilizers do not diminish but, on the contrary, they noticeably increase the content of inhibitors.

It was experimentally established that microbial inhibitors form toxic substances directly in the soil in which they grow.

If these organisms are introduced into nontoxic or inactivated soil and the soil is incubated under certain conditions of humidity and temperature, then after a certain time it will become toxic for these or other plants or for certain species of microorganisms, depending on the peculiarities of the inhibitor.

Rybalkina (1938 a) observed the appearance of toxicosis in flax-exhausted soil upon growth of the fungus (Fusarium lini).

Mirchink (1956) incubated soil (podsol) with fungi-inhibitors and she observed the appearance of toxicosis. In soils in which the fungus *Penicillium cyclopium* grew abundantly if artificially introduced, seeds of wheat did not germinate at all or germinated in small numbers (Figure 91). Other species of fungi isolated from podsol soils also poisoned the soil but to a lesser degree. On such soils germinating wheat seedlings constituted 15-60% of the number of seedlings in normal control soil.

Figure 91. Poisoning of soil by cultures of fungi upon artificial infection. Germination of wheat seeds: a--in control (noninfected) soil; b--in soil in which *Penicillium nigricans* grew; c--in soil infected with *Penicillium cyclopium*.

Clover-exhausted soil inactivated by heating regains its toxicity by growing the appropriate microbial inhibitors in it. In such soil with regenerated toxicity, clover and root nodule bacteria grew much more poorly than in normal soil. There was either no nodule formation on the root of clover or it was considerably suppressed (Table 102). Root-nodule bacteria in such soil became avirulent and lost the capacity to penetrate the roots and form nodules in them. Cultures of some fungi of the genera *Penicillium, Fusarium, Trichoderma* and some sporeforming bacteria, when growing abundantly in inactivated forest or field soil restored the soil's original toxicity in relation to wheat and *Azotobacter* (Table 103).

Table 102
Growth of clover on soil with restored toxicity

Expermiental conditions	Number of sproutings	Number of sprouts on the 30th day	Number of nodules per plant (average)
Inactivated soil not infected with inhibitors (control)	46	46	23
Inactivated soil, infected with inhibitors:			
Ps. pyocyanea	32	26	0.05
Ps. tumefaciens	39	22	0.0
Fusarium sp.	41	31	0.5
Mixture of all bacteria	30	19	0.0

Note: Each vessel contained 50 seeds. Inoculation was performed with active cultures of *Rhizobium trifolii.*

Table 103
Accumulation of toxic substances in podsol soil an a result of growth of inhibitors

Name of inhibitor	Length of wheat rootlets, cm	Length of wheat sprouts, cm	Survival of *Azotobacter* cells, hours

Control: inactivated soil (not infected)	12	8	240
Infected with:			
Bacillus strain 12	3	5	6
Bacillus strain 23	6	2	12
Bacillus strain 8	4	4	16

Toxins produced by inhibitors may, under certain conditions, accumulate in considerable quantities and endow the soil with toxic properties. The extent of accumulation of toxic substances depends upon the intensity of their formation by microorganisms, the rate of destruction and leaching, and also upon the degree of adsorption.

The phenomenon of toxicosis of soils has been known for a long time in agricultural practice and has always attracted the attention of many investigators.

Toxicosis expresses itself in the suppression of the growth and development of higher plants, and in the lowering of crop yields. The phenomenon of toxicosis is frequent under monocultures. In such cases, one speaks of the soil exhaustion as the reason for the suppressed growth of plants.

Tiring of soils was observed by agriculturists and scientific specialists. Plank (1795), De Candol (1813), Daubeni (1845), Uzral (1852), and others, indicated the lowering of soil fertility under monocultures and explained it by an accumulation of toxic substances. Later more attention was devoted to this phenomenon by many investigators: Kossovich (1905), Pryanishnikov (1928), Timiryazev (1941), Vorob'eva and Shchepetil'nikova (1936), Krasil'nikov and Garkina (1946) and others (of Krasil'nikov and Mirchink, 1955, Grummer, 1955).

Soils toxicosis expresses itself in relation to both higher plants and lower plants--bacteria, fungi, actinomycetes, algae, etc. Much date has been accumulated on fatigue and toxicosis of soils and the significance of this factor for the fertility of the latter. However, the essence of this phenomenon remained obscure until now. There are different points of view concerning its causes, but they can all be reduced to two basic ones.

According to one opinion, soil toxicosis is caused by the accumulation of special toxic substances as a result of growing plants against the rules of agrotechnique (Ishcherekov, 1910; Whitney and Cameron, 1914; Greig-Smith, 1913, 1918 and others). According to the other point of view, the existence of toxins in the soil is denied, and the fatigue of soils is explained by lack of nutrient substances, as a result of their unbalanced withdrawal from the soils in case of monocultures (Ressel, 1933; Pryanishnikov, 1928; Kossovich, 1905; Hutchinson and Thaysen, 1918).

Whitney and Cameron (1914) during their studies of soil fatigue found that plants in such soils do not suffer from lack of food but from accumulation of large amounts of toxins. Ishcherekov noted the possibility of removing toxic substances from the soil by washing with water. After washing, the plants grow better and produce normal crops. Thorough studies by Greig-Smith show that in Australian soils toxic substances accumulate in considerable amounts. Their concentration depends upon the type of soil, season of the year and other external factors. According to his observations, the toxins are thermolabile and are destroyed upon boiling and by drying.

Hutchinson and Thaysen (1918) studied European soils and found that the toxic substances which accumulate in them are thermostable, unlike those formed in Australian soils.

Many investigations have shown that soil toxicosis in relation to microorganisms is observed much more often and expressed more strongly than toxicosis in relation to higher plants.

It has long been known that microorganisms, pathogens and saprophytes which enter the soil do not grow and sooner or later perish. The bactericidity of soils was noted by Garre (1887), Freudenreich (1889) and others, They showed that pathogenic bacteria of the colon group, pyogenic cocci and diphtheria bacilli perish in the soil. Microbes such as the tubercle bacillus, the bacillus of anthrax and many others also perish in the soil (cf. Mishustin and Perteovskaya, 1954).

The soil possesses the ability to rid itself of pathogenic bacteria entering it. The rate of autoliberation from these microbes differs in various soils (Table 104).

Table 104 Survival of pathogenic bacteria in various soils (in days)		
Bacteria	minimum	maximum

Bact. typhi	15-20	360
Bact. dysenteriae	6-10	270
Vibrio cholerae	6-12	120
Mocob. tuberculosis	60	210
Bact. necrosis	10	75
Mact. melitensis	3-10	90
Bact. pestis	3	30
Bact. tularense	--	75

Many pythopathogenic fungi and bacteria cannot remain in soils for prolonged periods of time. The survival of certain bacteria of this group upon being introduced into the soil is as follows: *Bact. armeniaca* --6-8 days,*Bact. citri* --6-40 days, *Bact. aroideae* --3-15 days and *Bact. tabacum* --7-14 days.

Bact. malvacearum, Bact. citriputeale, Bact. amylovorum and others die relatively quickly in the soil (cf. Gorlenko, 1950). The death of the fungus *Pythium ultimum* in forest humus was observed by Peitsa (1952). According to this author, humus from under various woody plants possessed different toxicity; the strongest toxicity was found in extract of humus from under pine, next from under beech and the the weakest of all, from under birch. The antimicrobial properties of soil are not less sharply expressed in relation to saprophytic bacteria, fungi, actinomycetes and other microorganisms. Members of the soil microflora, which come from other soils often also perish in the soil.

Root-nodule bacteria introduced into clover-tired soil do not grow, and die comparatively quickly. Thus, from 65,000 bacteria introduced per one gram of a soil:

after two days 15,000 bacteria remained
after three days 1,000 bacteria remained
after five days, 10 bacteria remained.

After ten days the bacteria disappeared almost completely and only a few cells were found.

Kazarev (1907) observed, that the fungus *Pyronema confluens* grew well in sterilized soil and did not grow in nonsterile soil. Extract of nonsterile soil added to the sterile soil made the latter unsuitable for the fungus. A similar phenomenon was also observed by Novogrudskii (1936 a). The toxic substance causing the death of the fungus is thermolabile; it can be destroyed by heating to 120° C.

Numerous data are available concerning the adaptability of *Azotobacter,* rootnodule bacteria and certain sporeforming bacteria to soil.

Most strongly expressed and most widespread toxicosis is observed in soils of the podsol zone. According to our observations, there is either no *Azotobacter* growth in these soils, or it dies fairly quickly.

During our many years of study of soil microflora, we have investigated thousands of samples of podsol soils taken from various places of the Soviet Union.

Selected indexes of soil-toxicosis distribution in the various districts are given in Table 105.

Table 105 Toxicosis of turf-podsol soils		
Soils and region	Total No of samples studied	Samples toxic to Azotobacter
Kola Peninsula		
Forest	43	0

Humus-ferruginous	105	5
Swamps	22	0
Cultivated garden	215	87
Leningrad Oblast'		
Fields, acid	28	0
Fields, neutral	25	5
Garden	17	15
Arkhangel'sk Oblast'		
Virgin	35	0
Cultivated	40	8
Forest	27	0
Garden	23	16
Kaluga Oblast'		
Forest	17	0
Virgin	25	0
Cultivated	32	10
Garden	35	28
Ryazan' Oblast'		
Forest	13	0
Virgin	18	0
Cultivated	21	8
Garden	18	15
Yaroslavl' Oblast		
Virgin	30	0
Cultivated	30	6
Garden	30	27
Karelo-Fin AASR		
Forest	21	0
Field, virgin	30	0
Field, cultivated	50	10
Garden	20	18
Kalinin Oblast'		
Fields, virgin	47	0
Fields, cultivated	53	12
Garden	23	16

Gor'kii Oblast'		
Virgin	20	0
Cultivated	20	6
Garden	20	12
Moscow Oblast', Volokolamak Region		
Forest	13	0
Virgin	33	0
Cultivated	40	5
Garden	50	42
Dmitrovsk Region		
Forest	33	0
Virgin	53	0
Cultivated	67	12
Garden	67	53
State Farm "Krasnyi Mayak"		
Virgin	150	0
Cultivated	200	12
Garden	70	52
Chashnikovo		
Virgin	250	0
Cultivated	350	18
Garden	50	48
Forest	70	0

As can be seen from the data given, podsol forest and virgin field soils are not suitable for *Azotobacter*. All the 717 fields soils and 237 forest soils were toxic. Very seldom does one encounter samples of weakly-cultivated field soils (116 out 2,100 of studied samples), where *Azotobacter* grows. Well-cultivated and fertilized garden soils are less toxic or not toxic at all. Of 1,863 samples 1,244 contained *Azotobacter* at a greater or lesser density and 23 samples proved to be toxic for it.

Of all the podsol soils studied by us, those which were studied in greatest detail were the soils of the Moscow district on the fields of the experimental station in Chashnikovo and the Academy of Agricultural Sciences im. Timiryazev. These soils are loams with considerable leaching, Soils from under forests, with different woody species (spruce grove, birch wood, aspen grove, oak grove, etc), and soils of glades covered by grassy vegetation, soils weakly -cultivated which were plowed one or two years ago, soils cultivated for a long time (15-20 years and more) and soils of a renewed forest were studied.

In all cases the investigations were conducted all the year round; the samples for analysis were taken at 6-10 day intervals during the summer months and once a month during the winter. The soils were analyzed while fresh.

We determined the toxicity of soils by the viability of *Azotobacter* and by germination of seeds of plants (wheat, beet, etc). At the same time a total count of the microflora was made, including microbial inhibitors, which form toxic substances.

Studies have shown that many soils contain toxic substances. In forest soils, as a rule, there are more of these substances than in forest-free soils, there are less in plowed soils and still less in well-cultivated ones.

The toxicity of forest soils is determined by the varieties of trees growing in it. The greatest amount of toxic substances is found under spruce grove, and in a smaller amount, under pine and aspen grove. Soils under birch wood and oak grove are weakly toxic or nontoxic at all.

In Table 106 data are given on the toxic action of soils on germination of seeds of beet and wheat and on *Azotobacter*.

Table 106
Toxic action of forest podsol soils of the Moscow Oblast'

Soils	Germination of beet seeds, %	Germination of wheat seeds, %	Time of death of *Azotobacter* cells (in hours)
From under a spruce-grove	1	5	2-4
From under a pine-grove	5	25	20
From under a birch-grove	50	80	72
From under an oak-grove	80	90	82
Fallow cultivated soil	72	90	30 days

After clearing a forest and plowing, the soil becomes less toxic; *Azotobacter* does not perish for several days or even weeks. With renewal of the forest, the soil's toxicity is also restored (Krasil'nikov, Mirchnik and others, 1955).

The formation of toxic substances in chernozem soils under an artificially planted forest in the region of the southern steppes was observed by Runov (1953).

Plots covered by forests in the chestnut-soil zone of the Trans-Volga region, lose their *Azotobacter*. We observed a similar picture upon afforestation of soils in Central Asia, Kirghizia, Vakhsh valley of the Tadzhik SSR, Moldavia and other places. Soils rich in *Azotobacter*, lose them as soon as certain varieties of trees start growing in them.

Formation of toxic substances in soils is observed under certain grassy plants, particularly often in monocultures.

While studying the clover-exhausted soils of the experimental fields of the Agricultural Academy im. Timiryazev, we observed that they were obviously toxic for *Azotobacter* and for root-nodule bacteria, as well as for plants (Krasil'nikov and Garkina, 1946).

Toxicity changes considerably with the seasons of the year, as do many other properties of soil. It is most apparent during the summer-autumn months (July-September); in the late autumn and in winter it decreases and approaching spring it reaches its minimum. *Azotobacter* perishes more quickly in summer soils than in winter ones, while in spring soils, it even grows (Table 107).

Table 107
Toxicity of soils in relation to the season of the year
(Survival of *Azotobacter* in different months, hours)

Soils	July	Aug.	Sept.	Oct.	Dec.	Jan.	Feb.	April	June
Forest	2	6	2	2	24	24	72	216	96

Weakly cultivated	24	24	24	72	72	120	--	216	96
Well cultivated: under Timothy grass	24	24	2	6	24	40	24	24	--
Well cultivated: under clover	72	96	96	96	120	216	--	216	--

Similarly, seeds of beet germinate with greater energy and in greater numbers in spring soils (April-May) than in summer-autumn ones (August-October).

According to Rybalkina, toxicity of soils in relation to *Bac. mycoides* is lost after cool rainy weather.

According to the observations of Reiner and Nelson-Jones (1949), the greatest toxicity in the forest fields of Wareham (England) is detected in the autumn-winter months, with a decrease beginning in March.

One may assume that the increase and decrease of soil toxicity is caused by quantitative fluctuations in the toxin content. Toxic substances are either washed out by rain waters in the autumn and thaw waters in the spring, as it was assumed by Reiner and Nelson-Jones, or they are inactivated by the low temperature in the winter. For the verification of these assumptions we conducted special experiments.

In one series of experiments the soil (forest) was thoroughly washed with water and studied for toxicity. In washed soil *Azotobacter* perished at the same rate as in nonwashed soil. As can be seen, the toxic substances present in the soil which we studied are not washed out with water or only a part of them is washed out, as may be assumed, the part that is not adsorbed by the soil particles. The majority of the toxic substances are probably in the adsorbed state. Therefore, the decrease of the soil toxicity in the spring is not caused by washing out by rains and thawing snow but, one may assume, by the action of winter frosts.

In another series of experiments we subjected forest soil to freezing at minus 15-20° C for two months, and after thawing, an *Azotobacter* culture was introduced into it and the soil was incubated at 25° C. The results showed that in control soil, maintained at room temperature, *Azotobacter* died after one and a half hours while in soil which had been subjected to freezing, it did not die for 96 hours.

The soils investigated by us were not inactivated by heating at 100° C for 30 minutes, Inactivation was not attained even after autoclaving at 120° C for 30 minutes. In "exhausted" soils as we have shown earlier, the toxic substances are destroyed and disappear, at a temperature of 100° C maintained for 30 minutes, Obviously, the nature of the inhibitory substances in these soils is different.

Cultivation of podsol soils exerts a large influence on their toxic properties. *Azotobacter* grows better in chalked soils than in nonchalked soils (cf. Sushkina, 1940).

Many investigators (Christenson, 1915; Gainey, 1918, 1940 and others) ascribe the absence of *Azotobacter* in podsol soils to their acidity. They think that *Azotobacter* cannot grow in soils which have a pH of 5.5 or lower, If one even occasionally finds this microbe, it is considered to be of a special acidophilic species *(Az. indicum)*. According to these authors, as soon as one neutralizes acid soils, the ordinary *Azotobacter (Az. chroccoocum)* will start to grow and accumulate in them.

There exists another opinion, according to which proliferation of *Azotobacter* is conditioned, not by acidity, but by the ratio of the anions $CO_3 : PO_4$ (Niclas, Poshenrider and Mock, 1926), by the condition of the oxide and suboxide salts of metals (aluminum, iron, etc).

There are indications that, *Azotobacter* does not grow in many acidic soils after chalking, when the pH comes close to neutral (6.5--7.0).

Levinskaya and Malysheva (1936) observed that introduction of $CaCO_3$ into acidic soil (podsol of Murmansk Oblast') does not improve the growth of *Azotobacter*.

Brenner (1924) chalked acid soils of Finland and introduced *Azotobacter*. This measure did not decrease the soil toxicity. The introduced microbe perished as fast as in the nonchalked soils. According to the author. toxicosis of podsol soils is caused by special toxic substances. formed upon decomposition of plant residues (moss, etc) and also by toxic iron compounds.

Katznelson (1940) studied the viability of *Azotobacter* in acidic soils of America and tested various organic and mineral fertilizers with and without neutralization of the soil. The author reached the conclusion that there was no strict correlation between soil acidity and viability of *Azotobacter*. At pH 5.9 its number may be higher than at pH 6.6. At the same pH value of soil, *Azotobacter* proliferates in some cases and does not grow in others.

In our experiments on neutralization of soils by CaCO₃, MgO, NaOH the conditions of growth of *Azotobacter* in podsol forest soil improved considerably, but its toxicity was not completely eliminated. Under field conditions chalking also failed to reduce toxicosis of soil. As in the experiments of Brenner, we did not find *Azotobacter* in podsol, sparsely cultivated and well-chalked soils. We did not find it either in the year when chalking was applied, or 1-3 years later. *Azotobacter* was not detected in these soils even after a single introduction of manure. *Azotobacter* introduced in such soil died out at a slower rate than in the control, surviving 6-10 days and more, however its multiplication was not observed. Upon introduction of potassium nitrate or potassium phosphate the death of the introduced *Azotobacter* was speeded.

Therefore, the suppressing factor consists, not only of soil acidity, but also of many agents: physical, chemical and mechanical ones. For instance suboxide salts of aluminum, iron and other compounds, which, by the way, are in direct relationship to the pH of the environment, may inhibit growth of *Azotobacter*.

However, the main factors which cause soil toxicosis are, in our opinion, in many (if not in all) cases, excretion products of plants and microbial metabolites.

Formation of toxic substances by plants

Schreiner and Reed (1907) found that roots of certain plants excrete toxic substances. When they grew wheat repeatedly in the same vessel containing sand or soil, they observed a decrease in crops after each new sowing; this differed in various soils (Table 108).

Table 108
Wheat crops upon repeated sowing in the same soil (after Schreiner and Reed)
(in % of the first sowing)

Soil and region	First sowing	Second sowing	Third sowing	Fourth sowing
Clay--Cecille	100	68	57	44
Loamy--Leonardstown	100	30	37	23
Clay--Tacoma	100	53	53	46
Sandy--Portsmouth	100	64	30	--

Application of fertilizers causes a certain increase in crops but only after the first sowing. The authors obtained the strongest effect after the application of lime and manure. The crop of the first sowing was as follows (per cent of control):

Control (without fertilizer), 100
Mineral fertilizer, 130
The same plus lime, 205
Manure, 200
Manure plus lime, 238

Upon repeated sowing with the same amount of fertilizers the crop decreased. The solution from under wheat was not suitable for this plant: seeds did not germinate well in it, and seedlings were clearly retarded. The inhibition of root growth of flax seedlings was especially strong in this solution. When nutrient substances were added to this solution, only the growth of the aerial parts improved, but the roots remained undeveloped.

Toxic substances obtained from the substrate (from the solution or from the soil or sand) in which wheat grew, are thermolabile, inactivated by boiling, and absorbed by charcoal and chalk. If a toxic solution In filtered through charcoal or chalk, or subjected to heating, plants grow normally in it. The addition of pyrogallol to the soil extract removes its toxic properties. The same effect is obtained with the use of naphthylamine. These substances differ in their chemical composition and belong to the picolinic acid, salicylaldehyde, vanillin, and dihydroxystearic acid.

The experiments of Schreiner and Reed were repeated by Periturin (1911, 1912) with the same results. According to his data, the root excretions of wheat suppress not only the growth of the wheat seedling but also that of the oat. Schmuck (1911) grew cereals in sand after wheat; in some cases he cut the wheat at the root, while in others he let it grow to the end. After the harvest of crops, wheat or oats were sown again. In some containers roots of wheat were introduced into the sand. These experiments showed that the presence of wheat roots lowered the crop of oats to 76.8% and that of wheat to 45.2%.

Molliard (1915) tested the effect of root excretions of peas and corn, grown under sterile conditions, on seedlings of the same plants. The data obtained by him agree with those of Schreiner and Reed.

Hedrick (1905) observed an inhibitory effect of root excretions of oats on the growth of young apricot trees, The effect of root excretions of potatoes and tomatoes was less strongly expressed. A still smaller effect was that of roots of mustard and rape. Root excretions of beans and clover did not suppress growth of the above-mentioned trees. Similar phenomena were observed in the horticulture department of the Experimental Station of Woburn (USA), where it was shown that root excretions of grass suppressed the growth of young peripheral root tips of apples and pears which constitute the most active part of the root system.

Jones and Morse described the inhibitory action of the nut tree (gray nut--*Jualans cinerea* L.) on the growth of the creeping cinquefoil shrub. The latter does not grow in the vicinity of this tree to a distance of approximately twice the diameter of the foliage. Jensen has experimentally established that the root excretions of maple, cornel, cherry, tulip, and pine suppress the growth of wheat; the strongest suppression was observed in the summer, during the period of plant growth, when root excretions were more abundant than in autumn (after Schreiner and Reed, 1907).

Pickering (1903-1913) observed the deleterious effect of grasses on growth of fruit trees. He grew grasses in baskets with soil, hung under the foliage of an apple tree. The water coming through during downpours irrigated the soil around the apple tree. The roots of the latter were poisoned and the plants perished.

Fletcher (1912) observed the high sensitivity of *Sesamum indicum* L. to root excretions of *Andropogon sorghum* Brot. According to his observations, this plant cannot ripen in the vicinity of sorghum. Sewell (1923) found that roots of sorghum are toxic to wheat. Shull (1932) did not confirm the results of Fletcher and Sewell. According to his data, sorghum has no toxic effect on plants.

Mashkovtsev (1934) observed the thinning out of rice plantations after 2- 3 years of monoculture. The author thinks that the reason for this was the presence of toxic substances in the soil.

Ahlgren and Aamodt (1930) studied the interaction between different plants by growing them separately and together in containers. When Timothy grass was grown together with spear grass, or meadow grass with spear grass or with Timothy grass, much lower yields were obtained than when they were grown separately. The dry weight in grams of the plants in isolated cultures was as follows:

Spear grass, 0.396
Meadow grass, 0.343
Flat meadow grass, 0.450
Timothy grass, 0.577
Mixture:

Spear grass, 0.244
Timothy grass, o.360
Mixture:
Meadow grass, 0.195
Spear grass, 0.311
Mixture:
Flat meadow grass, 0.260
Meadow grass, 0.264

Waks (1039) found toxic substances in the root excretions of *Robinia pseudoacacia* L.

Many investigators connect the formation of toxic substances in soil with the growth of plants (Jakes, 1937; Rippel, 1936, Winter and Bublitz, 1953, Dimond and Waggoner, 1953, Nutman, 1952; Grümmer, 1955 and others).

It has been found that in many plants there are special substances--cholines and blastocholines (blastanein--germination and cholycin--to prevent), inhibiting germination of their own seeds and also of seeds of other species of plants. The nature of those substances differs in different plants. They may be excreted by the roots of seedlings and suppress growth of neighboring plants. Root excretions of seedlings of birch suppress the growth of rye grass and lychins: root excretions of seedlings of wheat and rye grass suppress germination of moods of certain weeds of *Anthemis arvensis* L. and *Metricaria inodora* L; seedlings of beans suppress germination of seeds of flax and wheat and seedlings of violets inhibit germination of wheat seedlings (after Audus, 1953).

Benedict (1941) observed the death of *Bromus inermis* Leyss, after its repeated sowing for many years in the field. The soil of such fields was toxic for the plant itself and for certain other plants. an well. Introduction of fertilizers did not abolish the toxicosis but only somewhat diminished it.

Bode (1940) described the poisonous effect of root excretions of *Artemisia absinthium* L. on the growth of fennel, caraway, sage and other plants sown in its vicinity. The height of the anise stem at a 70 cm distance from absinth was 5.7 cm at a distance of 100 cm--17 cm and at a distance of 130 cm--39 cm. A toxic substance was isolated, which was found to be a glucoside. This glucoside, called absinthin, is formed in the leaves of absinth and in easily extracted with water; it is washed out by the rains and enters the soil under the plant crown. It is preserved for prolonged periods in the soil.

There are indications that accumulation of toxins in the soil also takes place under fruit trees. Proebsting and Gilmore (1940) have shown that soil from under old peach trees is toxic for young peach saplings. Martin (1950-1951) found the same to be true for soils that have been under lemon trees for a long time. Plants planted on plots that have never before been under citrus grove a grew 9% faster than plants planted on old citrus-grove soils. Tomatoes and other vegetables grew quite satisfactorily on soils of old lemon groves and showed no signs of repressed growth.

According to Martin, the toxicity of much "exhausted" soils was not eliminated by washing with water for six weeks. Only by treatment with 2% sulfuric acid or 2% KOH and subsequent saturation with calcium did he succeed in removing the toxicity and restoring the fertility of the soil.

The inhibitory effect of root excretions of grassy plants, mustard, tobacco, tomato, and others in noted by many authors. Their effect in expressed in grassy plants and woody varieties as well. The degree of their inhibitory effect varies from 6 %to 97 % depending on plant species and on external conditions (Livingston, 1023; Breazeal, 1924; Conrad, 1927 and others, cf. Grammer, 1955).

Guyot (1951) ascribes a decisive role in toxicosis of soil to root excretions. He investigated a great number of plant species, and detected in many of them the ability to form toxins and to excrete them into the soil. These plants can be divided into groups according to the intensity of their toxic action. If one were to use *Brachypodium pinnatum* P. B. as a control plant which does not form toxins, with an index of 100, the remaining plants will have the following indexes:

Helianthemum vulgare Gärten., 85, weakly toxic
Barkhausia foetids Mönch., 81, weakly toxic

Thymus serpyllum L., 75, moderately toxic
Hieracium pilosella L., 70, moderately toxic
Origanum vulgare L., 70, moderately toxic
Asperula cyanchica L., 69, moderately toxic
Teucrium chamaedrys L., 68, moderately toxic
Picris hieraciodies L., 65, moderately toxic
Papaver rhoeas L., 61, moderately toxic
Achillea millefolium L., 59, moderately toxic
Hieracium umbellatum L., 39 strongly toxic
Solidago virga aurea L., 30, strongly toxic
Hieracium vulgatum Fr., 29, strongly toxic

Upon repeated sowings of the same plants, their seeds progressively germinate less well, and the mature plants yield smaller and smaller crops. According to Guyot and Massenot (1950), *Hypericum perforatum* L. under the same sowing conditions had in the first year a density of 4,200 plants per plot, the year after--1,100 and on the third year--500 plants on the same plot. Similar results are given by Curtis and Cottham (1950) with various species of sunflower. Hurtis (1953) tested the effect of root excretions of corn, peas, wheat. oat, rye, and lucerne, on the germination of mustard seeds. Excretions of barley roots caused a strong inhibition of seed germination, root excretions of other plants showed no such effect. According to Schilling (1951), cauliflower grows better if there in celery in its vicinity. Piettre (1950) noticed a strong poisoning of soils under many-year plantations of coffee plants. He isolated a substance belonging to the fatty acids from these soils. The most ubiquitous among them is lignoceric acid ($C_{24}H_{48}O_2$) which strongly suppresses the growth of plant seedlings.

The Swedish scientist Oswald (1947) studied the toxicity of soils under certain grassy plants. According to his data. seeds of rape--*Brassica napus,* and *B. rapa* L. germinated very weakly or not at all in soil from under*Agropyrum. repens* P. B. or *Festuca rubra* L. The author succeeded in isolating a substance inhibiting germination of rape seeds from roots of the couch grass. The toxicity of soils on which couch grass or Festuca rubra L. have been grown was removed by heating at 80-90° C (Oswald, 1949, 1950).

According to Schuphan (1948), lettuce suppresses growth of radish seed and radish is toxic to the development of lettuce seeds. Lettuce excretes, according to this author, saponin, and radish excretes mustard oil. The toxic substance was extracted from the leaves with ether and obtained in crystaline form (0. 5 mg from 1 g of leaves). It proved to be 3-acetyl- 6-methoxybenzaldehyde (Bonner 1950)

Lyubich (1955) tested the interaction between various woody plants. Planting various species in pairs on the same hole it was found that certain species inhibit the growth of others (Table 109).

Table 109
Interaction between woody plants

Plants	English oak	Green ash *(Fraximus viridus)*	Box elder	Indian bean	Japanese pagoda tree	Elm *(Ulmus pinnato-ramosa*
English oak	+	+	+			-
Green ash *(Fraximus viridus)*	+	+	-	+	-	-
Box elder	+	-	+	-		
Indian bean		+	-	+		
Japanese pagoda tree		+	-			
Elm *(Ulmus pinnato-ramosa)*	-	-				

Note; a plus sign means good growth (no suppression), a minus sign means suppression of growth.

An is evident from the table, each of the tested plants suppressed the development of shoots of any other type.

According to Guyot, root excretions of *Hieracium pilosella* L., even at low concentrations, are toxic to many plants. Vigorov has found that by its root excretion *Agropyrum repens* P. B. inhibits the growth of seedlings of pine and Caragana (after Grümmer, 1955).

Rodygin (1955) observed a 40-80% cancer morbidity of the lime tree when it was grown in the vicinity of asp. In the neighborhood of other plants (pine, spruce, fir) the percentage of cancerous lime tree did not exceed 2-3%.

According to Bordukova (1947) the kind of plants that grow in the vicinity of potatoes is important. In the vicinity of sunflower, tomatoes, apple, cherry, raspberry, pumpkin or cucumber, the resistance of potatoes to*Phytophthora* was lowered. Potatoes grown in the vicinity of a birch wood rot more easily than potatoes grown in the vicinity of pine.

One cannot combine narcissus and lily-of-valley flowers together in one bunch since they would soon wither; similarly, mignonette increases the withering of flowers in a vase.

The prolonged studies of Bonner and his co-workers (1938-1948) have shown that guayule roots excrete a toxic substance--trans -cinnamic acid. Ten milligrams of this acid in 1.5 kg of soil completely inhibits germination of guayule seeds. The higher the concentration of this substance, the longer it remains in the soil. One milligram of cinnamic acid endures 14 days in nonsterile soil and in sterile soil for an even longer period of time.

The shrub *Encelia parinosa* Adana, (of the compositae) grows in deserts. It is characterized by the fact that no grass grows for a certain distance around it. Investigations have revealed that the soil under the crown of this shrub is poisoned by the toxic substances, formed by its leaves. An extract of the leaves, or even better, the leaves themselves when introduced into the soil arrest growth of other plants--tomatoes, pepper and rye. These toxins do not act on the growth of *Encelia*, barley, oats and sunflowers.

It is well known that a great number of plants produce various compounds possessing toxic properties in relation to bacteria and to plants. Substances such as glucosides, saponin and coumarin are very widespread among plants. Inhibitors of the saponin group were found in hay stacks which suppress the germination of seeds and growth of algae (Moewus and Bonnerjee, 1951, Lindahl, Cokk et al, 1954), These substances reach the soil with plant residues. Upon decay of the latter they re liberated and may exist in the soil, in active form for a certain length of time, affecting the microflora and the higher plants. According to Benedict (1941). dead roots of brome grass release toxic substances upon their decomposition which suppress germination of brome-grass seeds. The above-mentioned absinth and *Encelia* release their toxins upon the decomposition of their aerial parts.

Golomedova (1954) studied the effect of aqueous extracts of many grasses and shrubs on plant growth. Tartar honeysuckle, maple, ash tree, buckthorn and amorpha inhibit growth of fescue and Euagropyrum, Extracts of couch grass, root and green parts of Austrian absinth inhibit oak saplings.

Feldmeier and Guttenberg (1953) obtained alcoholic and ether extracts from seeds and seedlings of beans, which inhibited growth of oat coleoptiles.

Bublitz (1954) tested the effect of extracts of pine branches, decomposing in soil, on the germination of *Lepidium sativum* L, seeds and on the growth of certain bacteria in compost. The extracts noticeably inhibited growth of bacteria and germination of seeds, more so in an acid medium (pH-5.6) than in a neutral one.

Lindahl, Cook et al, (1954) obtained a toxic substance of the saponin type from lucerne hay, According to Mishustin (1956) these substances are excreted by the roots of lucerne during vegetation,

Callison and Conn (1927) and McCalla (1948, 1949) found toxic substances of the soil in the form of decomposition products of plant residues. They established the presence of the following compounds among these substances: vanillin, coumarin, dehydrostearic acid, salicylic acid and other compounds. Small doses of these substances, noticeably inhibited growth of plants.

Toxic substances are present in plants of many species of the umbellate family: poison hemlock, poisonous cicuta, water dropwort, marshwort, *Anthriscus* and others. In their tissues one finds phthalides, various tars, esters, acids and other compounds. In Cruciferae plants one finds mustard oils.

Various volatile substances are present in many odorous plants. In some plants they are formed in the seeds and fruits, while in others in the leaves and stems or in the roots. Essential oils of a series of plants: citrus plants clove, mint, *Satureia,* thymes, germander, eucalyptus, etc and the resin of coniferous trees, poplar and others inhibit the germination of seeds of various plants to various degrees (Weintraub and Pricae, 1948, Grümmer, 1955, Molisch 1937; Madaus, 1936, Clausen, 1932).

Lebodev (1948) has found that absinth inhibits growth of flax, peas, beans, sage and clove. Roots of ash excrete volatile substances which inhibit growth of the oak.

Golubinski (1946) observed the stimulation of pollen germination by volatile substances formed by plants.

Solov'ev (1954) found that volatile substances from certain plants *(Agropyrum pectiniforme* R. et Sch.) stimulate the germination of lucerne pollen, those of other plants *(Bromus intermis* Leyss, *Phelum Pretense* L.) inhibit and still others have no effect at all.

The volatile substances excreted by onion, garlic and horse radish are well known (cf. Tokin, 1951).

In agricultural practice plants forming volatile substances have, since ancient times, been used as preservatives against spoilage of foodstuffs, For instance peasants put pieces of garlic into the cornbins for protection against the weevil, and against *Agrostis segetum* they use branches of the bird cherry tree (Grimm, 1950).

Volatile substances excreted by plants have a definite effect on phytopathogenic microflora and may play a protective role, Fahlpahl (1949) observed a protective effect of hemp, The volatile substances which it excreted inhibited growth of certain pathogenic microorganisms, due to which plants growing in the vicinity of hemp were less subjected to diseases. Schilling (1951) gives data on the protective effect of celery. When cabbage grows in the vicinity of this plant it is less affected by microorganisms.

Pirozhkov (1950) noted the lethal effect of the volatile substances of tomatoes on certain insects attacking the gooseberry shrub, as Tenthrodinodea and Pyralididae. According to this author's observations gooseberry shrub in the vicinity of tomatoes do not suffer from these insects.

Certain organic acids of plant origin are also toxic for seedlings of a number of plants, They are often found in the fruits. Malic and citric acids, in apples; 3.4-dihydroxycinnamic acid and 3-methoxy-4-hydroxycinnamic acid in tomatoes, transcinnamic acid, in guayule, etc (Akkerman and Veldstra, 1947),

The alkaloids are very widespread among plants. Some of them inhibit growth of plants. The most well known are cocaine, physiostigmine, aconite, caffein and quinine,

It should be noted, that the nature of the toxic substances excreted by plants is unknown in the majority of cases. Grümmer (1955) relates these substances to a special group of specific substances--the cholines,

Recently, artificially produced substances have penetrated more and more into agricultural practice; these substances exert a certain inhibitory effect on plants, Substances have been obtained that put put potato tubers "to sleep." To these belong a series of compounds--methyl esters of alpha-naphthylaceatic acid, ß-naphtyldimethyl ester, isopropylphenylcarbonate and certain other esters of phenylcarbonic acid (Krylov, 1954). Small doses of these substances (0.1% of aqueous solution) keeps tubers from sprouting in storehouses (Moewus and Schader, 1951). Preparation M-1 (methyl ester of ß-naphthylacetic acid) is offered for use in the preservation of potatoes, A dosage of 1.5- 3 kg of this substance is enough to protect 1 ton of potatoes from sprouting, increasing their yield by 10-14%, and decreasing weight losses 2.5-5 times, it also protects the starch and vitamin C, decreases the accumulation of the glucoside and solanin in the tubers (Krylov, 1954).

Pteroylglutamic acid (4-amino-9-methylpteroylglutamic acid) and indolyl-3-acetic acid suppress growth of roots and, to a lesser extent, inhibit the growth of the aerial parts of the plants.

Maleic hydrazide causes a long-lasting inhibition of plant growth. At a concentration of 0.01% this compound suppresses the growth of the raspberry for 24-38 days and ripening of berries for 16-33 days. This substance is recommended for use on lawns, i. e., it strongly checks the growth rate of grass.

Suppression of plant growth by chemical preparations in widely used in agriculture, In a number of cases it is necessary to arrest the development of buds and especially the beginning of flowering in fruit trees (apples, pears, aprocits, peaches, etc), These substances can also be used in decorative plant breeding, when it is necessary to arrest the growth of plants, etc.

It should be noted, that many substances of the auxin group may act as inhibitior or herbicides and as stimulants as well. For example, the well-studied compound- 2.4-dichlorophanoxyacetic acid (2.4-D) sometimes stimulates and sometimes suppresses seed germination, depending on the concentration used. Para-aminobenzoic acid has a stimulating effect at a concentration of 0.001% and strongly inhibits seed germination and plant growth at a concentration of 0.05%. Nicotinic acid at a concentration of 0.01% is also a strong poison for plants.

The toxic substances differ in their stability. Some of them are quite stable, are not destroyed upon prolonged stay in the soil, and may be concentrated to a greater or smaller extent. Other substances are easily destroyed and vanish from the soil. In such cases, plowing is enough to remove the toxicity of the soil. The vegetative cover, fertilizers and other measures also change actively the toxic substances in the soil,

Certain substances are easily leached by rains and are removed from the soil, while others are in the adsorbed state and are not eluted by water.

Studies show that there is a direct relationship between the concentration and length of time of preservation of toxins in the soil, and the qualitative and quantitative composition of the soil microflora,

In sterile soils the active substances are preserved for considerably longer periods of time than in nonsterile soils. When sterile soil is inoculated with a particle of nonsterile soil, the toxins in it soon become inactivated as in nonsterile soil (Brown and Mitchell, 1948, Audus, 1953, Jorgensen and Hammer, 1946),

Stapp and Spicher (1955) and Jensen and Peterson (1932) have described bacteria which strongly destroy herbicides. Their activity aids in the liberation of soil from toxins.

As can be seen from the above-mentioned data, the accumulation of toxic substances under natural conditions may vary, depending on the type of soil, the nature of the substance and on external conditions. Some substances accumulate in considerable quantities, while others remain in small concentrations or are not accumulated at all. The concentration of these substances determines the degree of toxicosis and fatigue of soils.

Under natural conditions. in soils and other substrates into which toxic substances constantly enter there is some degree of accumulation of the latter. At certain concentrations of toxins in the soil, poisoning of plants may take place.

The question of whether toxins can themselves enter plants is answered in the affirmative.

It was experimentally shown, that a number of toxic substances penetrate the plants via the roots and spread to the tissues. The mere fact of the poisonous effect of inhibitors in the above experiments show that these substances penetrate the plants. There are also direct experiments with chemically pure substances obtained from bacteria, fungi and actinomycetes. Gramicidin--a preparation obtained from the sporeforming bacterium--*Bac. brevis*, possesses strongly toxic properties; it poisons the tissues of animals as well an those of plants. Small doses of it in the nutrient medium cause rapid browning of roots, withering of aerial parts, and death of the whole plant.

Pyocyanin, a substance formed by the blue pus bacillus--*Ps. pyocyanea,* is endowed with strongly toxic properties. Plants-- wheat and clover--perished within a few hours under its influence. Substances obtained from many actinomycetes are also toxic for plants: mycetin, lavendulin, actinomycetin, longisporin, etc. Among the fungal metabolites, notatin, glyotoxin, and some others possess herbicidal properties.

Toxins and antibiotics may enter the plants through their leaves. If a drop of the solution of a toxic substance is placed on the surface of the plant, after some time one observes symptoms of poisoning, not only in the tissues that have been in direct contact with the solution, but in distant parts as well. Sometimes the whole branch withers away. This phenomenon of poisoning of branches and leaves was observed by us in birch, due to the action of the toxin produced by the fungus *Botrytis cinerea* (Krasil'nikov, 1953b).

MICROBIAL ANTAGONISTS

Among the soil microorganisms there are forms that inhibit the growth of other microbes. They are usually called antagonists.

There is no fundamental difference between inhibitors and antagonists. Both groups act with the aid of special metabolites in their metabolic products: the inhibitors affect cells of higher organisms and the antagonists act on lower organisms. Even this distinction is not always sharp, since there are inhibitors, which also suppress microbes and, on the other hand, there are antagonists that are also toxic to plants.

As was already mentioned, substances formed by inhibitors are called toxins or phytotoxins, and substances produced by antagonists are called antibiotics. This differentiation is of a purely formal or conventional nature. As is well known, there are many substances among the antibiotics which are highly toxic to plants and animals.

Regardless of this relativity of concepts and designations the antibiotics and their producers are a special branch of science and are considered as special substances with specific manifestations.

Microbial antagonism has caught the attention of scientists for many years. Pasteur, Mechnikov and their contemporaries noticed the ability of some species of microbes to suppress the growth of others (cf. Nakhimovskaya, 1937; Waksman, 1947; Krasil'nikov, 1950 and others). Pasteur observed this ability in the anthrax bacillus in relation to the microbe causing chicken cholera. Mechnikov found it in the lactobacilli in relation to the putrefactive bacteria and certain colon-type bacilli. On this basis he devised a method of changing the intestinal flora and sanitation of the human and animal intestines. The phenomenon of antagonism was observed in various groups of microorganisms, among bacteria, fungi, actinomycetes, algae, protozoa, etc. Microbial antagonists acting upon various pathogenic bacteria against cocci (staphylococci, streptococci, pneumococci and diplococci), against organisms causing intestinal infections (dysentery, parathyphoid, typhoid and cholera) against the tubercle bacillus, diphtheria, peat and anthrax brucellosis, tularemia and gas gangrene have been described. A great number of antagonists were described acting against pathogenic fungi, yeasts, protozoa, etc.

Recently, the efforts of investigation have been concentrated on the detection of antagonists acting against viruses and malignant tumors, Antagonists of viruses and tumors can be found among actinomycetes and bacteria (Waksman, 1953; Kashkin, 1952; Kurylowicz and Slopek, 1955).

Much data are available on the suppressing action of antagonists of phytopathogenic bacteria, actinomycetes and fungi. Antagonists were found acting against various organisms. The greatest attention is paid to antagonists--actinomycetes.

Among 20 cultures of actinomycetes isolated from rhizosphere soil, Lochhead and Lauderkin (1949) found 11 cultures that suppressed the growth of phytophathogenic strains of *Actinomyces scabies*.

Meredith and Semeniuk (1945-47) found that among the actinomycetes which they studied 21% inhibited growth of the fungus *Pythium graminicola,* which causes root necrosis in a number of plants. The actinomycete-antagonists of *Chalaria quercina* which cause wilt in the oak, and actinomycete-antagonists of *Cerasto mella ulni,* which affect woody plants, such as the elm, etc have been described (Stallings, 1954).

From the soils of the southern shore of Crimea, Petrusheva (1953) isolated 31 cultures of actinomycetes. Twenty-two of these inhibited the growth of the fungi *Thielaviopsis basicola* which causes root rot of tobacco plants, and *Fusarium* sp. which causes the "black foot" disease of citrus saplings, Leben and Keitt (1948) had a collection of actinomycetes that inhibited 33 species of phytopathogenic fungi.

Cooper and Chilton (1950) have done much work on the detection of actinomycete-antagonists of the phytopathogenic fungus *Pythium arrhenomonas* which is widely distributed in the soils of Louisiana. Out of 8,302 strains which they isolated, 18.5 to 31.5% were antagonists. Kublanovskaya (1950) noted actinomycete - antagonists to the agent of wilt of the cotton plant-- *Verticillium dahliae* and *Fusarium vasinfectum*. Lechevalier et al., (19 5 3) tested 197 strains of actinomycetes for antagonism to *Cerastomella ulni* and found only one active strain. Actinomycete-antagonists were found in soils which were active against phytopathogenic fungi-- *Helminthosporium sativum, H. victoriae, Coletotricum circinans, Verticillium albo-atrum* and others (Stevenson, 1954; Stessel et al., 1953).

In our studies we found actinomycetes in soil, which suppress phytopathogenic fungi., *Fusarium lin, F. solani, F. vasinfectum, Helinthosporium sativum, Alternaria humicola, Rhizoctonia solani, Botrytis alii, Deuterophoma tracheiphilus, Trichoderma lignorum, Monila fructigena,* and also fungi of the genera *Penicillium, Aspergillus, Cladosporium, Verticillium* and others.

Among fungi there are many antagonists. Antagonists have been described against agents of various diseases: *Fusarium, Peziza, Rhizoctonia, Ophiobolus, Botrytis, Monilia, Sporotrichum, Pythium, Phymatotricum, Phytophthora* and *Sclerotium*.

Porter (1924) described antagonistic relations between fungi and bacterial antagonists, and the phytopathogenic fungi *Helminthosporium* and *Fusarium*. Sanford and Broadfoot (1931) isolated from the soil 6 species of fungi which inhibit the growth and activity of the fungus *Ophiobolus graminis*. Weindling (1932, 1948) described a case of parasitism of the fungus *Trichoderma lignorum* on the fungi of the genus *Rhizoctonia* and others. The same was observed by Novogrudsk (1936). The latter given a long list of fungi and bacterial antagonists, indicating the species of fungi and bacteria inhibited by them.

In this list appear more than 30 fungal species which inhibit over 50 fungi belonging to different genera and families. Many other investigators also found antagonists among the fungi (Allen and Haenseler, 1935; Joseph, 1952; Vasudeva, 1950, 1953 and others).

Stemmel, Leben and Keitt (1953) isolated 170 fungi antagonistic to other fungi from soils.

Anwar (1949) found that among soil fungi and bacteria approximately half (out of 86 studied) suppressed the growth of the fungus *Helminthosporium sativum*, and about 12% inhibited *Fusarium lini*. From various soils, Gregory at al., (1952) isolated 14 cultures of fungi, 29 strains of actinomycetes and 31 strains of bacteria, which actively suppressed the growth of the fungus *Pythium debaryanum*. Three actinomycete strains and 1 bacterial strain suppressed growth of the root-nodule bacteria, *Rh. meliloti* and *Rh. trifolii*.

Among the sporeforming and nonsporeforming bacteria, it is not uncommon to encounter antagonists to phytopathogenic microbes. Among the sporeformers antagonists to various phytopathogenic fungi have been described. Porter (1924) mentions a number of bacteria which inhibit the growth of *Helminthosporium*. The latter did not grow in the presence of *Bac. capoulatus, Bac. mesentericus* and *Bac. mycoides*. Bamberg (1931) indicated inhibition of growth of the fungi *Tilletia tritici, Ustilago zeae, U. levis, U. avenae* by the sporeforming bacillus *Bacillus* D.

The afore-mentioned bacteria showed similar action against phytopathogenic fungi such as *Penicillium* sp., *Helminthosporium, Ophiobolus, Acrostalagmus, Fusarium. Sclerotinia gleosporium, Alternaria* and others (Novogrudskii, 1936; Weindling, 1946; Stessel et al., 1953; Stallings, 1954; Krasil'nikov, 1953 a and others).

Many antagonists have been described among the nonsporeforming bacteria. Studies show that they are most frequently encountered among the *Pseudomonas* and *Bacterium* species and also among myxobacteria. Bisby (1919) described the species *Pseudomonas phaseoli* which inhibited the fungus *Fusarium oxysporium*. Fawcett (1931) indicated *Ps. juglandis* as an antagonist of the fungus *Dothlorella gregaris*. Johnson and Marvin (1931) detected the antagonistic action of certain nonsporeforming bacteria *(Bacterium* C-1. *Bacterium* X, etc) on growth of the fungi *Ustilago zeae, U. avenae, Alternaria solani, A. brassicae* and *A. Tenuis*.

Khudiyakov (1935) isolated and studied in detail bacteria which dissolve the mycelium of the fungi *Fusarium graminearum, F. culmorum, F. scirpi, F. lini. F. herbarum, F. equiseti, Sclerotinia libertiana* and others. These bacteria were called mycolytic bacteria. Subsequently, many other investigators detected these bacteria in the soil (Raznitsyna, 1942; Berezova, 1932; Korenyako, 1939; Kublanovskaya, 1953, etc).

Ark and Hunt (1941) had cultures of the bacteria *Bac. vulgatus* and *Bac.* sp. that inhibited the growth of many phytopathogenic bacteria--*Bact. amylovorum Bact. aroideae, Bact. carotovorum, Bact. phytophthorum, Ps. campestris, Ps. lachrymans* and others. These bacteria also inhibited certain fungi--*Fusarium graminearum, F. lycopersici, Phytophthora* sp. and others. Similar data are given by other authors (Johnson, 1935, Weindling, 1946; Christensen and, Davis, 1940; Vasudea, 1952 and Skinner, 1956).

Antagonists to phytopathogenic fungi have been described among the myxobacteria (Johnson and Marvin, 1931; Kononenko, 1937).

Fungi which attack nematodes are described in the literature. They were first described by M. S. Voronin in his work "Mycological Studies" published in 1869 and by Sorokin in 1871. In a series of papers Soprunov (1954) showed that these fungi differ in their species composition and are very widespread in soils. The majority of them belong to the *Hyphomycete,* to the genera *Trichothecium, Arthrobotrys, Dactylaria, Dactylella,*etc.

These fungi "catch" the nematode with their hyphae and poison it with their metabolites. Attempts were made to use these fungi in the struggle against phytopathogenic nematodes. The introduction of this fungus into the soil lowers the incidence of plant disease. In the struggle with nematodes which affect cucumbers, the fungi-antagonists, or an they are called the predatory fungi noticeably decrease the incidence of disease; in the control group there were 23 galls per each plant and in the treated plants, an average of 0.6 galls (Soprunov, 1954).

According to Gorlenko (1955), artificial enrichment of soil by predatory fungi lowers the morbidity of cucumbers 1.5-7 times. Tendetnik (1957) used the predatory fungi for exterminating pathogenic larvae, the ancylostomes in mines and also to destroy strongyles in the manure of infected animals.

On the basis of existing evidence, one may say that there are no species of bacteria. actinomycetes, proactimomycetes, micromonospores, protozoa, algae, etc against which no antagonist can be found. In laboratory practice, one usually gives the name antagonist only to a microbe which suppresses the widely used test organisms, which usually comprise a narrow range of known cultures of bacteria or fungi. One does not take in account the fact that the so-called inactive forms (nonantagonists) in these tests might be active against other organisms. On the basis of our own experience and data from literature, we can say that the property of antagonism is characteristic of all species of microorganisms, but is expressed differently and to various degrees depending on the natural properties of the antagonist on the one hand, and the sensitivity of the test organism on the other hand, and also upon the type of substrate and other external conditions.

The antimicrobial action of the antagonists manifesto itself not only under laboratory conditions on artificial media but also under natural conditions of habitation, in the soil. In sterile soil, where there are no antagonists the growth of microbes and the biochemical processes which they cause take place at an intense rate. However, it is sufficient, to introduce an antagonist in such soil in order to stop or to slow down the growth and biochemical processes of microbe.

In his experiments, Afrikyan (1951) directed the action of antagonists of the group of sporeforming bacteria: *Bac. subtilis* and *Bac. mesenterics* against the organisms: *Bac. mycoides* and *Az. chroococcum.* The result are given In Table 110.

Table 110
Intensity of *Azotobacter* growth as depending on
growth of the antagonistic bacterium--*Bac. mesentericus*
(experiment with sterile soil; table shows number of cells, in thousands,
in 1 g of soil after passage of days)

Experimental conditions	Initial number	1	2	3	5	7	10
No antagonist (control)	80	150	380	520	760	800	1,000

With antagonist	80	100	30	0.05	0	0	0.05
No antagnist (control)	200	500	600	800	800	800	500
With antagonist	200	100	50	0.5	0	0	0.05
Antagonist	80	400	1,200	1,500	1,600	1,000	800

Bac. mesentericus was introduced into sterilized soil together with cultures of *Azotobacter* to which it is an antagonist, The growth of *Azotobacter* was suppressed by *Bac. mesentericus*. Under conditions of growth in sterilized soil when grown alone, *Azotobacter* reached 1 million cells per 1 gram of soil and in the presence of *Bac mesentericus, Azotobacter* grew very slowly and only during the first day; then the number of its cells decreased rapidly, dropping practically to zero. Only at the end of the experiment, after 10-15 days, when growth of the antagonist and formation of the antibiotic stopped, did the *Azotobacter* resume growth although at a very slow rate.

The same was observed in experiments with *Bac. mycoides*. This microbe in very sensitive to the antibiotic action of the potato bacillus. In an isolated condition, in sterilized soil, its growth is excellent but it ceases to grow completely or almost completely in a mixture with the antagonist (Figure 92).

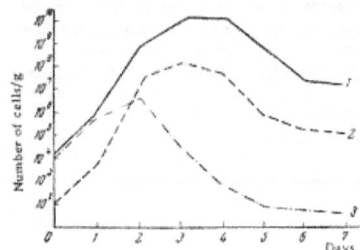

Figure 92 Growth of *Bac. mycoides* in soil (sterile) in presence of the antagonist--*Bac. mesentericus:* 1--growth of *Bac. mycoides* without antagonist; 2--growth of *Bac. mesentericus* in soil; 3--growth of *Bac. mycoides* in a mixture with *Bac. mesentericus.*

Mikhaleva (1951) studied the growth of the root-nodule bacteria of clover, peas, kidney, beans, lupine, soy, lucerne, etc in the presence of antagonists and actinomycetes, The most resistant, according to their data, were the soy bacteria In podsol soils these bacteria are more strongly suppressed by actinomycetes than in chernozem soils. Root-nodule bacteria of clover perish after 4 days in the soil in the presence of the antagonist.

The activity of the root-nodule bacteria decreases markedly under the influence of the antagonist; the number of nodules on the roots is usually smaller than in the controls. According to our observations, the root-nodule bacteria of clover vetch and peas react noticeably to the action of antagonistic *Bac. subtilis, Bac. mesentericus* and others. When the soil is artificially enriched with these bacteria the nodules do not develop on the roots of the above-mentioned plants or they appear in small numbers only, with a degenerated appearance, an unusual form and a small size.

Stolp (1952) studied the growth and activity of root-nodule bacteria of peas in the presence of *Pennicillium expansum*. Robinson (1946) observed cultures of antagonists among actinomycetes, bacteria and fungi, which actively suppressed the growth and virulence of root-nodule bacteria in the laboratory and under field conditions as well.

The general importance of bacterial antagonists is determined not only by the nature and strength of their activity but also by their number in the soil. The more intensely they grow in the soil, the higher their concentration and the stronger the effect they exert.

It is difficult or almost impossible to assess the total number of antagonists in the soil. As was already mentioned, all microbes possess antagonistic proprties against some microbe. However, it is practically impossible to detect antagonism against all existing microorganisms.

In our studies we give quantitative indexes of distribution. of antagonists in relation to certain species of bacteria or fungi. Among the bacteria, cultures of *Azotobacter*, root-nodule bacteria, mycobacteria, micrococci and nonsporeforming bacteria (colon bacillus, etc) were used.

According to our calculations, there are very large quantities of actinomycete-antagonists with clearly expressed antimicrobial properties in different soils.

In the previous chapter (microbial inhibitors) data were presented on the number of microbes which suppress the growth of *Azotobacter*. The number of bacterial antagonists ranged between 10,000 and 450,000; fungi, between 1,300 and 17,000 and actinomycetes, from 10,000 to 160,000 in 1 g of soil, depending on the properties of the latter.

In relation to certain other species of bacteria *(Bac. mycoides, Bac. subtilis)* and fungi *(Fusarium lini, Fusarium* sp., etc) the total number of antagonists is much higher. In our analyses of various soils of the Soviet Union we found tens and hundreds of thousands and often millions of them in 1 g of soil.

In the podsol soils of the Moscow area one may find 40,000- 1,000,000 actinomycetes-antagonists or even more, per 1 g of soil; sporeforming bacteria in numbers of 20,000-500,000 may be found in 1 g of soil. In the chernozems of the southern districts of the USSR the total number of microbes is higher and, as a rule, the number of antagonists is also higher. However the percentage of antagonists may be lower. For example, the soils of the Crimean steppes contain a smaller percentage of antagonists than the northern soils of the Kola Peninsula or the soils of the central districts. There are relatively few antagonists in the red soils of the Caucasian coastal area (Table 111).

Table 111
Number of actinomycetes-antagonists in soils (in thousand/1 g soil)

Soil and region	Actinomycetes	Antagonists
Serozem, Tashkent	560	200
Serozem, Vakhah valley	1,300	360
Chernozem, Kuban'	2,000	400
Chernozem, Kar'kov	1,000	400
Chernozem, Kuibyshev	1,500	800
Chernozem, Crimean steppe	1,200	120
Krasnozem, Batumi	200	10
Chestnut soil, Armenia	600	200
Brown soil, Armenia	1,200	600
Chernozem, Armenia	800	200
Chernozem, Georgia	1,800	350
Podsol, Moscow, Oak forest	1,200	350
Podsol, Moscow, Spruce forest	360	0
Podsol, Moscow, Birch forest	1,200	420
Podsol, Moscow, tilled, a	1,500	800
Podsol, Moscow, tilled, b	600	200
Podsol, Leningrad, tilled	1,500	800
Podsol, Kola Peninsula, ferruginous	0.4	0.3
Humus, Kola Peninsula, ferruginous	2.0	1.5

Tundra, Kola Peninsula, forest	0	0
Swampy soil, Kola Peninsula	0.2	0.2

In certain soils one may find 2,000-400,000 bacterial antagonists per 1 g of soil; they belong to two groups-- *Bac. subtilis* and *Bac. mesentericus*. Among the nonsporeforming bacteria and especially among species of the genera *Pseudomonas* and *Bacterium,* there are many antagonists. A special place in relation to their quantity among the antagonists of this group is occupied by the mycolytic bacteria. These bacteria grow well in many soils and in the rhizosphere of various plants. According to our figures, their total number approaches tens of millions and more in 1 g of soil (Krasil'nikov, 1940 a,b, c; Kusina, 1951; Kublanovskaya, 1953).

Rasnitsyna (1947) studied the soils of the Vakhah valley and counted 100,000 to 100 million mycolytic bacteria (which lyse the fungus *Fusarium vasinfectum)* in 1 g of soil, depending upon the vegetative cover. The greatest number was found in the rhisosphere of lucerne, spear grass and the Euagropyrum; they were less numerous under rye grass, brome grass and peas. Kuzina (1955) gives more or less similar indexes for light serozems of the Uzbek SSR.

Novogrudskii (1949 a) isolated mycolytic bacteria from Kazakhstan soils which lyse the fungi: *Fusarium culmorum, F. graminiarum, Verticillum dahliae, Colietotrichum lini, Alternaria tenuis, Amblyosporium botrytis, Pyronema confluens,* and *Mucor racemosus.*

In the same soils the author counted from 100 to 100 thousand mycolytic bacteria in 1 g of soil. These are active solely against *Fusarium graminiarum* and were more numerous under lucerne than under oat or millet.

Above were given data on growth and accumulation of these bacteria in soils under different plants.

In nature microorganisms do not live alone, but in associations which contain many foreign species-- competitors and noncompetitors. In these associations definite and quite complicated relations are established among the species, of both a symbiotic and an antagonistic nature.

In microbial populations or coenoses, each species in its struggle for survival throughout a long history of evolution, elaborated certain means of struggle with its competition. These means are versatile. Microbes may displace their competitors by abundant multiplication or they may form during their metabolism various specific and nonspecific substances which suppress microbial growth. These nonspecific substances are, for example, organic acids, alcohols, peroxides and other compounds. These metabolic products are characteristic of many microbial species.

Examples of the nonspecific metabolites are certain inorganic substances, such as ammonia, hydrogen sulfide, ions of chlorine, SO_3, aluminum, iron and other elements, Plentiful emission of H_2S, formed as a result of the vigorous metabolic activity of the appropriate microbes, may cause the poisoning of the milieu, making it inadequate for the growth of many foreign species of microbes and even for higher plants, as was observed by us in soils with high ground-water level in the Trans-Volga region (Krasil'nikov, Rybalkina and others, 1934).

Microbes producing acids, alcohols and other organic compounds suppress the growth of organisms which are sensitive to these substances, regardless of the species to which they belong.

The most characteristic and outstanding reactions are those which are caused by particular specific substances, the so-called antibiotics, which act against microbes. These substances have specific effects. The microbial antagonists which form these substances suppress growth of certain specific species only. Some antagonists inhibit only gram-positive bacteria and others, inhibit both grarn-positive and gram-negative bacteria. Some of them act on cocci and others on bacilli.

Some antagonists have an inhibitory effect only on fungi or only on phages and viruses, etc.

Antibiotic substances produced by antagonists are a potent weapon in the struggle with competing microbes.

A characteristic feature of such antagonists is the fact that as a rule, they act only upon foreign species. *A. streptomycini,* the producer of streptomycin, does not inhibit cultures of its own species. The producer of aureomycin, *A. aureofaciens* does not inhibit cultures of its own species, regardless of from where they were isolated or under what conditions they lived previously. Similarly, other antagonists which produce antibiotic substances, terramycin, chloromycetin, actinomycin, sulfactin, etc do not suppress the growth and development of strains which belong to the same species.

These characteristics of specific or selective action of antagonists determine to a large extent the species makeup of microbial populations in natural substrates.

Experience shows, that there are microorganisms that form antibiotics under conditions of solitary growth, on artificial nutrient media. Their ability to form these antibiotic substances is hereditarily fixed and expresses itself in the absence of competitors.

These organisms are looked upon as potent antagonists. They are often found among various microbial species. They comprise the main group of producers of antibiotic substances obtained to date in various laboratories and in the antibiotic industry. In other species this property is not fixed by heredity and appears only in the presence of competitors, i.e., under conditions of mixed populations. In pure isolated cultures these organisms do not produce antibiotic substances. For example, the colon bacillus, *Bact. coli* suppresses the growth of *Bac. anthracis* only in those cases where both organisms are together. In a pure culture the colon bacillus does not produce antibiotic substances which are active against the mentioned microbe. Apparently these microbes only produce antibiotic substances under pressure, out of necessity.

Shiller in 1914 was the first to pay attention to the forced nature of the formation of antibiotics by microorganisms. He was working in Mechnikov's laboratory on the interrelations between the acidophilic bacilli, sporeforming bacteria, streptococci, pneumococci and other species (Shiller, 1952).

The phenomenon of forced antagonism was also noticed by other investigators (Peretts and Slavskaya, 1934; Izabelinskii and Soboleva, 1934 and others). Streshinskii (1949, 1950) has demonstrated the formation of antibiotic substances in mixed cultures of a fungus and a sporeforming bacillus. According to his observations, *Bac. subtilis* forms an antibiotic substance against the *Penicillium* fungus only when the two grow together. The fungus too becomes a more active antagonist in the presence of the bacillus. We observed forced antagonism in a number of actinomycetes. Cultures grown separately on nutrient media, do not form antibiotic substances, but in the presence of certain microbes (fungi or bacteria) these substances which suppress the growth of competitors are formed. Two inactive species of actinomycetes when grown together, form antimicrobial substances.

All such organisms possess a latent antagonistic capacity which is not fixed hereditarily.

The majority of microbial antagonists readily lose their ability to produce antibiotics, active strains turning inactive in the process of variation. This property is encountered in many organisms used in industry and causes many difficulties in the antibiotic industry and in laboratory practice.

Antibiotic substances should be classified an potent weapons in the struggle of microbes with neighboring competitors; as a biologically important attribute formed in complex populations during phylogenesis and as a property determining the degree of development and distribution of the microorganism in nature. The biological role of antibiotic substances and, therefore, the phenomenon of antagonism as a whole, is quite important in the life of both higher and lower soil organisms.

Through use of their metabolic products antagonistic microbes suppress their competitors, removing them from the substrate and thus exerting a definite selective action. To a certain degree, microbial antagonists regulate the formation of microbial coenoses (colonies) in the soil in general. They play an important role in the improvement of soils, in the so-called process of self-purification of soils. The removal of harmful pathogenic and phytopathogenic flora and fauna is accomplished by microbial antagonists.

Microbial antagonists suppress not only the growth and propagation of competing organisms, but also many of the functions of the latter. There are among microorganisms those which inhibit certain processes which occur in microbes.

Inhibitors were observed to suppress nitrification, denitrification, nitrogen fixation, decomposition of cellulose, etc. Appropriate cultures inhibit the decomposition of organic substances, fermentation of sugars, etc. Microbial cultures are known which inhibit the formation of certain metabolites, various acids, enzyme biotic substances, auxins, vitamins, amino acids and other biocatalysts. There are organisms in the soil which, with their metabolic products, suppress the formation of toxins and antibiotics and often inactivate them in the medium, if they are formed there.

There are data in the literature dealing with the suppression by inhibitors and their metabolites of processes of cell multiplication, spore formation, budding and the sexual process. Inhibitors often inhibit the process of respiration.

In the presence of antagonists or their metabolic products many active substances: biocatalysts, antibiotics, toxins, and enzymes lose their peculiar functions, and become inactive. Toxins become harmless, antibiotics no longer suppress the corresponding microbes, enzymes no longer cause the decomposition of organic matter, etc. In general, one may say that for any metabolite, living nature creates an antimetabolite (Woolley, 1954). The metabolite penicillin becomes nonbactericidal for staphylococci and certain other bacteria, which produce an antimetabolite penicillinase (Abraham and Chain, 1940; Woodruff and Foster, 1945). Sulfonamide compounds are inactivated by bacteria which produce paraaminobenzoic acid (Woods, 1940; Sevag and Green, 1944, 1950) or the factors H.P. etc (Moller and Schuerz. 1941; Green 1940). Mucin suppresses the antibacterial action of tyrothricin; tannic acid neutralizes actinomycin (Waksman, 1947). In our bacterial collection there were strains, which completely inactivated the antibiotic substances, formed by the sporeforming bacteria: *Bac. subtilis, Bac. mesentericus* and *Bac. cereus*. We also possessed bacterial cultures, which annuled the toxic effect of *Ps. pyocyanea* metabolites. The toxin of the fungus *Botrytis cinerea* is neutralized by metabolites of certain actinomycetes.

A solution of mycetin at a 1:10,000 dilution completely suppresses the growth of *Staph. aureus* and in the presence of filtrate or cells of a *Bac. proteus* culture the antibacterial effect is either nonexistant, or becomes very weak. A culture of the colon bacillus and certain *Pseudomonas* strains had such an inactivating effect in our experiments. An inactivating effect of microbes was also found in relation to other antibiotics (Krasil'nikov and Nikitina, 1951).

As seen from the above-mentioned data, antagonists exert a definite suppressing effect on various microorganisms. Due to their action, they have a considerable influence on the formation of microbial coenoses in soil in general and determine, to a certain degree, the distribution and accumulation of the various species of soil microflora.

Antagonists, therefore, can be considered as one of the powerful factors governing soil fertility and plant-crop abundance.

It is only natural that this group of microorganisms attracts the attention of specialists in many fields-- microbiologists, phytopathologists, plant breeders and others. The possibility of practical utilization of antagonists is one of the most important reasons for the study of the phenomenon of antagonism.

The protective role of microbial antagonists

As was noted before, in soils in which antagonists grow abundantly (bacteria, fungi or actinomycetes). microbes, sensitive to them, saprophytes as well as phytopathogens, grow much more slowly, or not at all. This served as a basis for the use of microbial antagonists in the struggle against harmful microflora, and against organisms causing plant disease.

The first attempts in this direction were made by Porter (1924). He treated wheat seeds with bacterial antagonists and then infected them with *Helminthosporium* fungi. The seeds either did not become infected and germinated normally, or they were slightly affected. Bamberg (1931) infected wheat seeds in order to protect them against smut. Weindling (1940, 1948) used the fungus *Trichoderma lignorum* for the protection of citrus saplings from *Rhizoctonia*. This fungus according to other authors, also protects cucumbers and peas from *Rhizoctonia* and wheat from fusariosis (Allen and Haenseler, 1935; Bisbu, James and Timonin, 1933).

Milliard and Taylor (1927) observed a protective effect of the actinomycetes-antagonist---*A. praecox* in relation to the agent of scab in potatoes *A. scabies.* No scab was observed upon prolific growth of the antagonist.

A similar phenomenon was described by Kissling (1933). He experimented with bacterial antagonists in subduing potato scab. Under conditions of a field experiment, upon sufficient growth of the antagonist the disease either did not appear at all or had a limited occurrence only.

Gregory, Allen, Riker and Peterson (1952) demonstrated the protective role of actinomycete-antagonists in their experiments with the phytopathogenic fungi *Pythium ultimum* and *Pythium debaryanum.* These fungi were quite virulent for lucerne. The introduction of certain species of actinomycotes-antagonists into the soil *(Actinomyces* sp.) protected the plants from the disease. A positive result was also obtained through use of antagonists--fungi *Trichoderma lignorum* and *Penicillium patulum,* and the sporeforming bacteria, *Bacillus* sp. No 6.

Rehm (1953) treated wheat and rye seeds with actinomycetes-antagonists in order to fight the organisms which cause fusarlosis *(Fusarium nivale* and *Fus. culmorum).* As a result, seed germination Increased by 30 %.

Gorrard and Lockhead (1938) compiled a long list of microbial antagonists which suppress the development of phytopathogenic organisms in soil. Among them there are sporeforming and nonsporeforming bacteria, fungi, actinomycetes and protozoa. Cordon and Haenseler (1939) have shown experimentally the inhibitory effect of the sporeforming *Bacillus simplex* on the growth and development of the fungus *Rhizoctonia solani* which affects cucumbers and peas. A culture of the antagonist was introduced Into the soil in which the plants were grown. As a result, the morbidity of the latter dropped. In the control without bacterial antagonists 65 % of the total number of cucumber plants and 48% of peas were affected; after introduction of the antagonist the mortality role was 35 and 45% respectively.

Ark and Hunt (1941) mention two cultures of sporeforming bacteria--*Bac. vulgatus* and *Bacillus* sp. which suppressed the growth of many phytopathogenic bacteria in soil and protected the plants from disease. Other investigators also speak of the antagonistic action of microbes in soil against phytopathogenic bacteria and fungi (Johnson, 1935; Eaton and Rigler, 1946; Allen and Haenseler, 1935; Feeney and Garibaldi, 1948; Kenknight, 1941 and others).

In the Soviet Union, as already mentioned above, Khudyakov (1935) and Novogrudskii (1936) established the lytic effect of mycolytic bacteria on phytophatogenic fungi. These bacteria were subsequently exclusively tested fighting plant infections under laboratory conditions, and in open fields as well. The results were positive in many cases.

The percentage of mortality noticeably decreased. For example, the experiments of Berezova (1939), performed on kolkhoz fields with flax, have shown, that mycolytic bacteria lower the incidence of fusariosis by an average of 40% and in isolated cases their action was even more effective.

Raznitsyna (1942) used mycolytic bacteria which are isolated from the soil against fusariosis of pine saplings (Figure 93). Under open-field conditions on a sector strongly affected by fusariosis, inoculation of bacteria gave an outright positive effect, and the lowering of morbidity reached 80 % and more (Table 112). The plants looked considerably healthier on sectors treated with mycolytic bacteria, than in the control. The needles were longer, stronger and greener, the stem of the saplings thicker and taller than in saplings not treated with bacteria (Figure 94).

Figure 93. Protective action of bacteria in fusariosis of pine saplings: 1--plants infected with the fungus *Fusarium;* 2--plants also infected with *Fusarium,* but treated with mycolytic bacteria.

Table 112
Protective action of bacterial antagonists in fusariosis of pine saplings sown in May 1940

Experimental conditions	Seeds germinated per 100 planted	Number that survived until September	Height of plants in September, cm
Control (not inoculated with bacteria	40	5	3.0
Inoculated with bacterial culture No. 30	75	61	4.9
Inoculated with bacterial culture No. 77	65	46	3.9
Inoculated with Euagroypyrum compost	70	46	4.1

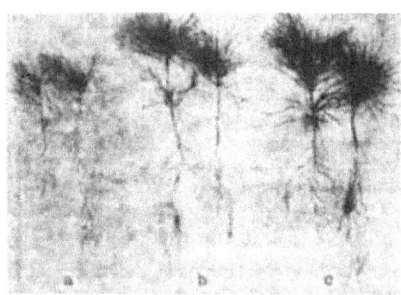

Figure 94. Pine saplings growing on sectors affected by fusariosis: a--not treated with bacteria; b and c--treated with mycolytic bacteria.

Korenyako (1940) studied mycolytic bacteria in the soils of the Uzbek SSR for a number of years. She isolated cultures of bacteria of the genus *Pseudomonas,* which showed antagonistic properties against the agents causing wilt of the cotton plant *Verticillium dahliae.* When these bacteria were tested on the experimental fields of STAZRA SOYUZNIKHI (Tashkent) which were strongly affected by the mentioned fungus, positive results were achieved. The mycolytic bacteria suppressed the growth of the fungus and lowered the morbidity of the cotton plant by 60- 80%. Kuzina (1951) and Kublanovskaya (1953) corroborated these results. They showed that mycolytic bacteria in mixtures with other bacteria suppress phytopathogenic fungi more vigorously.

Davydov (1951 used mycolytic bacteria against the mildew organism *(Sphaerotheca mors uvae)* on the gooseberry shrub. The parasitic fungus disappeared upon introduction of the bacterial antagonists and the plant recovered and developed normally.

Petrusheva (1953) used cultures of actinomycetes as antagonists against tobacco-seedling rot caused by the fungus *Thielaviopsis basicola,* and against blackleg of citrus cultures, caused by *Fusarium* sp. In soil treated with antagonists the plants grew normally; when the actinomycetes were not introduced into the soil, the mortality of the plants reached 70%.

Gurinovich (1953) used cultures of actinomycetes and bacterial antagonists in the struggle against the vascular disease of cabbage caused by nonsporeforming bacteria *Pseudomonas campestris.* The antagonists were introduced into the soil affected by the soil microbe, and prevented the plants from becoming infected.

Seiketov (1951) tested four cultures of the fungus *Trichoderma--T. glaucum T. lignorum, T. koningii* and *T. album* as antagonists of *Rhizoctonia,* which affects potatoes. The percentage of mortality of the "early rose" variety was considerably lower (7-8%) in the experimental plants than in the controls (54 %).

Kuzina (1955) used mycolytic bacteria against wilt of the cotton plant caused by verticillium. She treated the seeds with bacteria before sowing. The morbidity in the control sectors was 54% and after treatment with the bacterial antagonists it was 8--9%. Correspondingly, the crop yield was: in the control- 1, 030 bolls and the total weight of seed cotton was 342 g. In the experimental plants there were 1,630 bolls and the weight of the seed cotton was 514 g, i.e., 57% higher than those of the control.

Kublanovskaya (1953) used actinomycetes as antagonists to fight cotton-plant wilt. She prepared a special composted preparation of cotton cake with actinomycetes. This was introduced into the soil in which the cotton seeds were sown. The actinomycetes introduced in this manner proliferated abundantly in the soil and suppressed the growth and activity of the fungi *Verticillium dahliae* and *Fusarium vasinfectum,* protecting the plants from the disease. At the end of the growth period the number of plants in the cotton plants of the 8517 variety affected by the wilt was: 52.6% of all control plants and 18.1% in sections treated with actinomycetes; in the cotton plants of the 108-F variety there were 23.3 % diseased control plants and 3.3 % treated plants. In the treated sections all the plants looked stronger, the bushes were larger, the leaves wider and flowering was more abundant, etc (Figure 95).

Figure 95. Protective effect of actinomycetes-antagonists against cotton-plant wilt caused by *Verticillium:* a) plants affected by wilt, not treated with the actinomycete: b) plants whose seeds were treated with a culture of actinomycetes (according to Kublanovskaya (1953).

The crop in these experiments also increased noticeably in sectors treated with the antagonists. The 8517 variety yielded: 9.4 centners/ha in the control sector and 22.6 centners/ha in the experiment. The variety 108-F yielded: 17.8 centners/ha in the control sector and in the experiment 22.8 centners/ha.

Chinese scientists Yin, Chen, Yang et al., (1955) corroborated Kublanovskaya's data. They prepared compost from soil with cotton cake and grew actinomycetes in it which were antagonists of the wilt fungus*Verticillium dahliae.* The ripened compost was used for the treatment of the seeds. The latter were sown together with the compost. The incidence of the wilt disease decreased by 50-75 %, the cotton crop increased by 13-45%.

Mitchell et al, (1948) observed vigorous growth of antagonists-actinomycetes and bacteria, when those were introduced in the soil together with a composted plant mass. The phytopathogenic fungus causing root rot in the cotton plant perished. The positive role of composted preparations saturated with microbial antagonists was noted by Sanford (1946-1948). He observed in his experiments a drop in the morbidity of potato tubers, pine saplings, etc. Similar data are given by some other investigators (Weindling, 1946; Garret, 1946; Winter and Rümker, 1950). They all noted that the favorable action of composted preparations was not due to an improvement in plant nutrition but to the more intense development of antagonists in the soil.

Schaffnit and Neuman (1953) used peat composts, enriched with bacterial antagonists on the snowy mold caused by *Fusarium nivale*. The preparations were made from various types of post. The best results were achieved with bog peat of mass origin. The percentage of morbidity in wheat treated with composted peat was considerably lower than in the control.

Wood and Tveit (1955), on the basis of data from the literature and their own observations, came to the conclusion that microbial antagonists play an important role in soil improvement. Microbes of local origin are much more effective. In order to enhance their growth and activity the authors suggest the introduction of certain plant residues into the soil as sources of nutrition.

There are many other studies in the literature which confirm the positive effect of microbial antagonists in the struggle against phytopathogenic fungi, bacteria and actinomycetes. All these studies show that microbial antagonists can be used in agricultural practice for the improvement of soils. For this purpose the soil should be enriched with the appropriate antagonists.

This can be achieved by various methods. As may be seen from the above data the antagonists mentioned may be introduced into the soil in the form of pure cultures (which is not very effective), or in the form of composts.

In the enrichment of soils with antagonists, the vegetative cover plays an especially important role. It was noted above that the plants are a powerful selective factor. Some of them, under certain conditions favor the growth and accumulation of phytopathogenic microbes in the soil, and others favor the growth of the antagonists of these microbes. By selecting, by means of special experiments, those plants in whose rhizosphere the needed antagonists grow abundantly, and by using these plants in the crop rotation, one may remove or suppress the growth and the harmful activity of the pathogenic microbe.

If the selected plants are inoculated with microbial antagonists before sowing, their accumulation in the soil can thus be markedly enhanced.

Upon introduction of pure cultures of antagonists, one must take into account their adaptability, their growth in the soil, and their activity. If the antagonists lose their activity in the soil or in a substrate which is not suitable for them which often happens), or if they do not grow or grow but little, their effect will be small or will not express itself at all.

Microbial antagonists, an noted above, suppress their competitors with their metabolic products, among which a special place is occupied by the antibiotics. The latter may accumulate not only in artificial nutrient media but also directly in the soil, being concentrated there in smaller or greater amounts.

It is also known that plants can absorb various organic substances from the medium, including antibiotics, through their roots. If the antibiotic substances enter the plants, it should be assumed that they may produce a certain effect in the tissues, namely, suppress the activity of microbes that have penetrated the cells and increase the toxicity of the cell sap and, consequently, also increase the resistance of the plants to infections. In other words, an assumption is made that the entry of antibiotics is reflected in the immunobiological properties of the plants.

It should be noted that the problem of immunity in plants, regardless of the numerous studies done on the subject, is still very little known.

Literature data show that immunity, nonsusceptibility to infections in plants is determined by various, quite complicated internal and external factors.

One of such internal factors is the toxicity to bacteria of the call sap. It was noted long ago that the sap of plants possesses the ability to suppress growth of bacteria and fungi. Wagner (1915) observed death of bacteria in sap from the tubers and stems of potatoes. Precipitating these juices with ammonium sulfate and washing the precipitate with water, he obtained a substance of a protein nature with strong antibacterial properties. The substance is thermolabile, decomposes quickly in light and under the influence of oxidases and peroxidases. The toxicity of potato juice was observed by Cholodnyi (1939), he found that the toxicity of the juice increases upon sprouting of potato tubers.

Antimicrobial properties were observed in the juice of onions, corn, tomatoes, cotton, orchids and other plants. In the literature there are descriptions of the inactivation of fungal toxins by the sap of plants resistant to the diseases (Yachovskii, 1935, Vavilov, 1919, Naumov, 1940, Carbone and Arnaudi, 1937). There are indications, that the cell sap from varieties which are resistant to the infection is more toxic than that of susceptible varieties (Kramarenko, 1949; Gorlenko, 1950).

The antimicrobial properties of vegetative juice are explained by the presence of various substances in the cells.

An opinion was voiced that plant immunity is caused by the presence of elements of mineral nutrition in the sap. Some of them, potassium copper, cadmium, etc create a nonfavorable milleau in the plant tissues for growth of microbes. According to some authors, they are favorable to the accumulation of organic acid in the cells. Other investigators tried to explain the resistance to infections by the osmotic pressure of the sap, by the presence of alkaloids, enzymes, agglutinins, and lysis, dissolving the microbial cells, etc.

Israil'skii (1952) expresses the opinion, that the nonsusceptibility of plants to certain infections is caused by bacteriophages which saturate the tissues of the plants.

Winter and Ruemker (1952) concluded that the nonsusceptibility of plants is caused by the presence of a special "resistance factor" in the roots.

Many investigators ascribe considerable importance in plant immunity to the redox system (Rubin, 1050, et al.) or to the permeability of the protoplasm of the cells Kuprevich, 194 7, Kokin, 1948; Sukhorukov, 1952), Recently, a connection was noted between plant resistance to infections and the production of pigments of the anthocyanin group.

On the basis of him extensive studies Tokin (1951) attaches great importance to the phytocides in the immunity of plants. In phytocides are included different chemical antimicrobial substances which are formed by plants. Essential oils and organic acids, aldehydes, alcohols, phenols and also various specific compounds may be included here.

One would assume that during the course of the history of their development plants acquired various means of protection against infections, mechanical, chemical and biological.

While ascribing a certain importance to all these, in our opinion, the investigators have not yet touched one of the most important protective factor" the antimicrobial antibiotics, formed in the soil by microbes and absorbed there from the plants.

Formation and accumulation of antibiotics in soil

It was recently established that in the soil there are antibiotic substances which are formed by the microbial antagonists. The formation of antibiotic substances by microbes in the soil was experimentally proved by many investigators. Gottlieb and Siminoff (1952) have shown the presence in soil of chloromycetin, formed by *Actinomyces venezuelae*. The substance was isolated from the soil and chemically purified. Owing to favorable conditions of growth of the actinomycetes, 25.0-27.8 mg/kg of chloromycetin accumulate in the soil. The formation of chloromycetin in the soil by *A. venezuelae* was also noted by Jefferis (1952). The greatest amount is formed upon the addition of peptone or lucerne hay to the soil. In the presence of starch or oat meal the antibiotic is not formed.

Gregory et al., (1952) observed the formation of antibiotic substances in soil by cultures of bacteria-- *Bacillus* sp. B-6, by actinomycetes--*Actinomyces* No 67 and by fungi *Pencillium patulum*. In the presence of appropriate nutrient sources in soil (soy meal) the following amounts of antibiotics were formed:

	In sterile soil, units	In nonsterile soil, units
Bacillus sp. B-6	300	30
Actinomyces sp. No 67	5-10	3
Penicillium patulum	100	10

Hessayon (1951, 1953) showed the formation of the antibiotic trichothecin by the fungus *Trichothecium roseum*. Large quantities of the antibiotic were detected in loamy soil, and smaller quantities in sandy soils.

According to Gottlieb (1952) the antibiotic actidione is formed in the soil in considerable quantities (10 µg/g and more) in the presence of soy meal. Grossbard (1940) has grown *Penicillium patulum* in the soil in the presence of various sources of nutrition. The formation of patulin took place in the presence of beet cake, glucose, fresh wheat straw, timothy grass and certain other plant residues. According to this author, antibiotics are also formed in the soil by *Aspergillus terreus* and *A. antibioticus* in the presence of wheat straw and certain other sources of nutrition.

In a later work Grossbard (1952) has shown the formation of antibiotics in the soil by the fungi: *Aspergillus terreus*--70 units; *Penicillium* sp.--80 units; *Aspergillus clavatus*--110 units and *Penicillium clavatus*--100 units in 1 g of soil.

On the third day of growth of the fungus *Aspergillus clavatus,* in the presence of brown sugar, Gottlieb and Siminoff (1952) noticed the formation of clavacin in the soil in the amount of 16 µg/g. Taking into account the extent of adsorption and inactivation of the substance by soil particles, the authors determined its total production as not less than 50 µ g/g.

Stevenson and Lochhead (1954) studied the formation of antibiotic substances by the fungus *Penicillium* sp. and by the actinomycetes K-1 and V-27 in sterile as well as in nonsterile soil. In sterile soil the fungus forms as much antibiotics as on Chapeks medium (about 20--30 µ g/g and in nonsterile soil 3 times less (7 -l0 µ g/ g). The actinomycetes form up to 16 units/g and more of the active substance. As sources of nutrition both organisms use various plant residues--the green bulk of clover, orchard grass, brome grass and wheat.

Wright (1955) observed the formation of griseofulvin in the soil by the fungus *Penicillium nigricans* in the presence of other soil fungi. According to his data, a pure culture of *Penicillium nigricans* forms the following amounts of the antibiotic (in sterile soils), in podsol at pH 5.3--0.2 μ g on the 7th day and 2.4 μ g on the 14th day, in podsol with Ca(OH)$_2$--2.4 μ g on the 7th day and 20 μ g on the 14th day; in orchard soil at pH 6.2--0.6 μ g on the 7th day and 10 μ g on the 14th day--in l g of soil. Upon infecting the soil with other fungi the production of griseofulvin either decreases or increases, depending on the species of the fungi. On the 14th day of incubation griseofulvin was detected in the soil in the following quantities:

	Amount in μ g
Penicill nigricans at 40 μ g/g (Control)	100
Penicill nigricans + Trichod. viride	3.2
Penicill nigricans + Trichod. viride	3.2
Penicill nigricans + Trichod. viride	0.4
Penicill nigricans + Mucor ramannianus	12
Penicill nigricans + Cladosp. herbrum	100
Penicill nigricans + Penic. expansum	1.5
Penicill nigricans + Penic. frequentans	0
Penicill nigricans + Penic. albidum	200
Penicill nigricans + Penic. stolonifer	200

In sterile soil which is contaminated with nonsterile soil griseofulvin accumulates much less, namely about 6 μ g/g.

The author notes that in sterile and nonsterile soils antibiotics are also formed by many other fungi. On the 14th day of incubation in soil infected by fungi he found the following quantities of antibiotics:

	Antibiotic	In l g of soil, μ g
Trichoderma viride	gliotoxin	40
Trichoderma viride	viridin	0
Penicill. expansum	patulin	200
Penicill. frequentans	frequentin	20
Penicill. stolonifer	mycophenolic acid	0.16
Penicill. nigricans	griseofulvin	40

Our studies show that antibiotic substances are formed in the soil by various antagonistic organisms--bacteria, actinomycetes and fungi, if the conditions are suitable.

In podsol soil in the presence of nutrient sources (soy meal, lucerne or clover straw, sugar or other substances) bacterial antagonists form 40-180 units/g in sterile soil and 10-80 units/g in nonsterile soil. (Table 113). As can be seen from the table, the formation of antibiotic substances is more intense in sterile soils. As an artificial nutrient, here too, the presence of special substances is essential to the formation of antibiotics, and often those are not the same substances as those necessary for the nutrition and growth of the antagonistic bacteria. Of many organic substances tested only some of them were suitable for the synthesis of antibiotics, although growth of the antagonists occurred with any one of the nutrient sources used in the experiments.

Table 113
Formation of antibiotics in soil by bacteria
(Units/ g on the 5th through 7th day of growth)

Bacteria	Sterile soil, antibiotic formed	Nutrient source used	Nonsterile soil, antibiotic formed	Nutrient source used

Bac. mesentericus	180	Starch, soy meal	80	Soy meal, meat-peptone
Bac. subtilis	120	Meat-peptone broth, soy meal	40	Meat-peptone broth
Ps. flourescens	40	Lucerne	10	Lucerne
Ps. nitrrificans	60	Meat-peptone broth	20	Meat peptone broth, lucerne

Similar data were also obtained upon testing the actinomycetes-antagonists. The latter, when growing in soils where appropriate organic sources of nutrition are available, form antibiotic substances in detectable amounts (Table 114). We detected these antibiotics directly by the use of microbiological assay methods in the soil, and after extraction with organic solvents. Extraction as a rule, causes lower quantitative results. Where the microbiological assay method indicates 100units/g by extraction one succeeds in isolating not more than 10-30 units/g, i.e., about 10-30% of the total amount present in soil. Depending on the nature of the soil and the properties of the substance itself, the amounts of the extracted substances may vary (Krasil'nikov, 1954c).

Table 114
Formation of antibiotics by actinomycetes under various soil conditions
(units / g)

Soil	Actino-mycete 290, soy meal	Actino-mycete 290, corn extract	Actino-mycete 287, lucerne straw	Actino-mycete 287, starch	Actino-mycete B, lucerne straw	Actino-mycete B, corn extract
Sterile podsol	80	120	80	10	100	100
Nonsterile podsol	30	40	10	0	20	0
Sterile serozem	0	20	20	0	10	0
Nonsterile serozem	0	0	10	0	0	10
Sterile chernozem	30	60	20	0	50	80
Nonsterile chernozem	0	20	10	0	20	20

Korenyako, Artamonova and Letunova (1955) studied the formation of antibiotics in soil by actinomycetes belonging to different species. In all more than 100 cultures were studied, 13 of them in great detail. The actinomycetes were grown in soil with the addition of various organic nutrients. All of them, except a very few, formed antibiotics in soil in greater or smaller amounts. The most active ones are presented in Table 115.

Table 115
Formation of antibiotic substances by actinomycetes in various soils
(units / g)

Actinomycetes	Podsol	Serozem	Cherno-zem	Krasno-zem	Podsol	Serozem	Cherno-zem	Krasno-zem
	Sterile				**Non- sterile**			
A. Aurantiacus, 1149	200	50	100	80	80	10	30	0
A. globisporus 81-B	120	100	60	60	60	60	40	20
A. globisporus 2302	80	30	100	20	30	0	20	0
A. globisporus 2570	150	20	60	50	50	20	20	0
A. griseus 2535	170	100	100	120	80	40	30	40

The data given in the table show that microbial antagonists form antibiotic substances with different intensities depending on the type of soil. The largest amount of these substances is formed in podsol soil, less in chernozem and chestnut soils and the smallest amounts in krasnozem.

The antimicrobial properties of antibiotics express themselves differently in the soil than in artificial media. Not all the antibiotics are active in the soil. Martin and Gottlieb (1955) have shown, that circulin, neomycin and biomycin do not suppress the growth of the sporiferous bacterium *Bac. subtilis* in the soil even at very high concentrations (500 µ g/g), whereas on laboratory nutrient media they inhibit its growth at 0. 1 µ g/g. At the same time subtilin noticeably inhibits the growth of *Bac. cereus* in these very same soils. In these author's experiments antinomycin was the most active. Five µ g/ g actinomycin in soil was a sufficient amount for the suppression of *Bac. subtilis* growth.

The effectiveness of antibiotics in the soil is determined not only by their properties but also by their concentration. In turn the latter depends on the rate at which these substances are formed and enter, the soil on the one hand, and by the rate of their inactivation on the other hand. As is known, the majority of antibiotics disappear from the soil. Some are inactivated by the soil solution or destroyed by microbes, others are washed out by water and a considerable portion is adsorbed by soil particles.

in our experiments (Krasil'nikov, 1954c) we followed the degree of activity of antibiotic preparations, introduced artificially into various soils under different conditions. The rate of inactivation of antibiotics, the extent of their being washed out by water and their adsorption by the soil were studied. The average indexes chernozem soil are presented in Table 116.

Table 116
Changes in the antibiotic content of soil (chernozem)
two hours after the introduction of the former
(in units/g)

Antibiotics	Inactivated	Washed out with water	Remained in active and absorbed state
Streptomycin	850	30	1,120
Globisporin	800	120	1,080
Terramycin	850	250	800
Pennicillin	300	1,320	380
Preparation No 1609	10,000	0	0

Up to 2,000 units/g of chemically pure preparations were introduced into the soil. Preparation 1609 was introduced at a concentration of 10,000 units/g.

As is seen from these data, streptomycin, globisporin and terramycin are inactivated in chernozem to more or less the same extent (about 25%), penicillin, to a lesser extent and preparation No 1609 is completely inactivated. In other soils the inactivation of these antibiotics presents a different picture (Table 117). The least inactivation of these antibiotics is observed in podsol soils and in krasnozem, and the greatest in chernozem.

Table 117
Minimal doses of antibiotics, at which the soil still exhibits antibacterial properties
(in units/ g)

Antibiotic	Serozem	Chernozem	Krasnozem	Podsol
Streptomycin	350	850	300	80
Globisporin	100	600	400	80
Terramycin	600	850	200	400
Preparation 1609	10,000	10,000	10,000	50,000
Penicillin	120	300	150	60

In order to determine what part of antibiotics are adsorbed by the soils, we washed the latter with water. Upon introduction of antibiotics into podsol soil in the amount of 2,000 units/g the following portions could be washed out with water: penicillin, 1,650 units/g; streptomycin, 40 units/g; globisporin, 120 units/g; terramycin, 350 units/g, and the antibiotic No 1609--none.

Penicillin more than other antibiotics can also be washed out of other soils; e, g., krasnozem, 1, 800 units/ g and from serozem, 1, 500 units/ g.

By comparing the numerical indexes of inactivation of the antibiotics and their proneness to be washed out by water, the amount of antibiotics in the adsorbed state can be determined and therefore also the adsorbing capacity of the soil. According to our data in podsol soil the adsorption picture was as follows: globisporin--1,800 units/g; streptomycin--1,880 u/g, terramycin--1,350 u/g, penicillin--280 u/g per gram. In serozem the figures were 1,670, 1,600, 1,200 and 380 units/g respectively. Streptomycin is adsorbed by chernozem at the rate of 1, 120 units /g and by krasnozem--1,650 units/g; globisporin--1,080 and 1,540 units/g and terramycin--900 and 1,620 units/g respectively.

As seen from this data the amounts of antibiotics detected are considerably smaller than those introduced. In order to create antibacterial activity in 1 g of serozem soil for example, it is necessary to use a minimum of 350 units streptomycin; 100 units globisporin; 600 units terramycin; 130 units penicillin and more than 10,000 units preparation No 1609. Upon introduction into the soil of smaller antibiotic doses than those indicated, the antimicrobial properties of the soil will not be detected either by a qualitative test or by extraction with various solvents (water, alcohol, ether, acetone, etc). All these data show that to the numerical indexes obtained upon quantitative determination of antibiotics formed in the soil by microbial antagonists, one should add the amounts of antibiotics adsorbed and inactivated. If 20 units/g of antibiotic substance were found in the soil, the actual amount produced by the antagonists would be considerably higher.

The rate of destruction of antibiotics in the soil varies. Some of them are inactivated within a few hours and others may be preserved for a few days or even weeks, depending on the nature of the substance and the properties of the substrate.

Antibiotics with basic properties such as streptomycin are very quickly inactivated in the soil, Neutral compounds (chloromycetin) are inactivated slowly, and substances of the acid type occupy an intermediate position.

Adsorption and inactivation of antibiotic substances depend to a considerable degree on soil acidity. Aureomycin and terramycin saturate neutral loamy soils at a concentration of 30,000 μ g/g, while for the saturation of acidic loam 60,000) μ g/g and more are required, i.e., twice as much.

At pH 3.2 soil rich in humus adsorbs 4,000 μ g/g of antibiotics and at pH 5.6-7.6 only 400 μ g/g, i.e., ten times less.

Consequently, the antimicrobial action of antibiotics will differ in these soils. In order to inhibit growth of *Bac. polymyxa* in soil at pH 5.6 a concentration of 5,00 μ g/g of terramycin is required, while at pH 6.2, 200 μ g/g suffice (Gottlieb and Siminoff, 1952; Martin and Gottlieb, 1952).

The stability of antibiotics in the soil also varies with the acidity of the latter.

According to Gregory et al., (1952), the antibiotic actidione is preserved in alkaline soil at pH 7.8 for 8 days and in an acidic soil at pH 5.2--for more than 14 days. Clavacin to completely inactivated in the first day in alkaline soil and in acidic soil it is preserved for 3-4 days. Ninety per cent of fradicin is preserved in alkaline soil for 14 days, in an acid soil it is adsorbed and completely inactivated an the first day.

Chloromycetin is less strongly inactivated in the soil than are streptomycin and terramycin. When chloromycetin is introduced into sterile soil, it remains for more than 14 days without change. In nonsterile soil, with a large number of various microorganisms, the preparation is gradually inactivated; after 3 days only 70% of it is left, after 7 days--about 30% and after 2 weeks only 20% are recovered.

Pramer and Starkey (1052) have found, that streptomycin introduced into the soil at a concentration of 1,000 μ g/g is preserved in sterile soil for over 3 weeks and in nonsterile soil for 2 weeks. About 50% of it is destroyed within the first week. In the presence of glucose the antibiotic is preserved for a longer period of time.

Jefferis (1952) tested many antibiotic substances. He introduced them into various soils and observed the rate of their destruction. The following data have been obtained (Table 118).

Table 118
Preservation time of antibiotics in different soils
(days)

Antibiotic	Introduced μ g/g	Nonsterile podsol	Sterile podsol	Nonsterile orchard soil	Sterile orchard soil
Albidin	30	7	14	2	3
Frequentin	100	10	16	2	7
Gliotoxin	20	40	16	2	7
Griseofulvin	30	20	40	16	17
Patulin	2,000	32	32	2	2
Penicillin	50	3	2	2	2
Streptomycin	400	26	16	6	16
Viridin	100	8	16	1	1

An may be seen, in orchard soil there is a faster inactivation of antibiotics. According to the author, certain substances are destroyed faster in sterile soil than in nonsterile soil, which is puzzling and disagrees with the observations of other investigators.

According to our data (Krasil'nikov, 1954c), the preservation time of antibiotics varies greatly and depends first of all, upon the properties of the substances, and secondly, on the type of soil and external conditions (temperature, humidity, acidity, etc), The same antibiotic is inactivated and parishes at different rates in different soils. For instance, globisporin is preserved for 7 days in podsol, but for only 2 days in krasnozem. Aureomycin is active 10 days in podsol but not more than 3 days in krasnozem (Table 119).

Table 119
Preservation time (days) of antibiotic substances

Soils		Globi-sporin	Prepar-ation No 112	Aureo-mycin	Terra-mycin	Prepar-ation No 1609	Penicillin
Chernozem	Sterile	25	15	30	30	2	10
	Nonsterile	5	2	3	6	0.2	1
Podsol	Sterile	90	80	60	50	5	20
	Nonsterile	7	5	10	8	0.1	3
Serozem	Sterile	35	20	40	30	3	15
	Nonsterile	5	4	8	8	0.1	1
Red soil	Sterile	22	12	20	15	2	5
	Nonsterile	2	2	3	3	0.1	0.5

Antibiotics are most rapidly inactivated in krasnozem and podsol soils.

Preparation No 1609 disappeared immediately in nonsterile soils; other preparations could still be detected after a few days. In sterile soils antibiotics are preserved for considerably longer periods of time than in nonsterile soils. As can be seen from Table 119, the majority of active substances disappear in the first 2-3 days in nonsterile soils, while in sterile soils they are preserved for 2 -3 months.

Similar data are obtained upon introduction of crude antibiotics in the soil. Korenyako, Artamonova and Letunova (1955) introduced active actinomycetal substances in the form of culture liquids into podsol, chernozem, krasnozem and serozem soils. About 700 crude preparations were examined, 12 of them, in greater detail. The results are given in Table 120.

Table 120
Preservation time of crude antibiotics in soils
(days)

Antibiotics	Podsol, sterile	Podsol, non-sterile	Ser-ozem, sterile	Ser-ozem, non-sterile	Cher-nozem, sterile	Cher-nozem, non-sterile	Kras-nozem, sterile	Kras-nozem, non-sterile
A. violaceus strain 1806	20	5	20	5	20	5	10	5
A. aurauntiacus strain 1149	180	8	180	8	180	8	100	8
A. globisporus strain 81-B	180	8	180	8	180	8	5	2
A. globisporus strain 76	180	25	180	25	180	25	180	25
A griseus strain 2535	180	2	180	2	180	2	20	2
A. griseus strain 2392	120	20	120	20	120	20	120	20
Control soils	0	0	0	0	0	0	0	0

The soil solution exerts a slightly inactivating effect. We tested a solution obtained by using a strong press from incubated samples of podsol, serozem and chernozem soils, with 4 antibiotics: penicillin, globisporin, preparation No 15 (grisein) and preparation No 1609. The soil solution was added in various amounts to the antibiotic solution of a known titer and after a certain period (1-5 hours or more) the activity of the preparations were examined. The results were not always clear-cut, however, reliable data were obtained in experiments with grisein and, preparation 1609 and in some cases also with penicillin. The solution from incubated podsol soil inactivated antibiotics to a lesser degree than solutions obtained from serozem and chernozem. One ml of solution extracted from podsol neutralized 20-30 units of preparation No 1609, 5-7 units of grisein and 2-5 of penicillin. Soil solution from incubated serozem inactivated 50-60 units of preparation No 1609, 7-10 units of grisein and 10-12 units of penicillin. Solution of incubated chernozem inactivated 80, 15 and 25 units, respectively. In addition it also inactivated approximately 5-7 units of globisporin.

The inactivating force of soil solution is closely related to the species makeup and metabolism of the microorganisms. If one incubates soil in the presence of small amounts of organic substance and in addition inoculates the soil with certain bacterial species, the solution thus obtained would inactivate considerably more antibiotics. We incubated podsol soil with various bacterial species (sporeformers and nonsporeformers) with an admixture of various sources of organic nutrition (soy meal, clover, hay, corn extract, peptone), The greatest effect was obtained in an experiment with a nonsporeforming bacillus (strain No 6) which was incubated in soil with soy meal. The solution obtained thereof (1 ml) inactivated 100 units of preparation No 1609 and up to 50 units of penicillin, but did not inactivate grisein or globisporin. The latter was inactivated by solution from soil incubated with bacterium No 15 in the presence of clover hay.

The inactivating action of the soil solution was observed by Waksman and Woodruff (1942), They examined the effect of actinomycin in pure solution and with an admixture of a humus extract. A culture of Bac. mycoideswas killed in the former case by 1 μ g and in the latter case, by 10 μ g or more of the substance. A similar effect was observed by Skinner (1956) while studying the antibiotic obtained by him from Actinomyces albido-flavus.

Inactivation of antibiotics in soil is probably mainly by products of microbial metabolism. It has been shown, that with increase of the latter, destruction of antibiotics, is enhanced. In the above-mentioned experiments, in soil incubated together with soy meal, the growth of bacteria was very intense and their composition differed from that in soil with clover or peptone.

Winter and Willeke (1951 a, b) introduced penicillin into composted soil rich in humus and into loamy soil poor in organic substances. In humus soil, where there was abundant growth of microbes, the antibiotic disappeared after 2-3 hours. In soil poor in humus the same antibiotic was preserved for 12 hours. If the soil microflora is removed by sterilization, penicillin in such soil is preserved for more than 3 days, while in nonsterile soil it disappears on the second day.

We have demonstrated the inactivating role of the soil microflora in experiments with pure cultures of bacteria and actinomycetes. It was found, that the antibiotics, penicillin, mycetin and streptomycin are inactivated in different degrees by metabolic products of various species of bacteria and actinomycetes. Some species or strains of bacteria inactivate streptomycin more strongly, while others inactivate penicillin and mycetin more strongly. There are cultures the metabolic products of which enhance the activity of antibiotics (Krasil'nikov and Nikitina, 1951).

The data given here change our concepts on the preservation of antibiotics in soil. Antibiotics are not always fully inactivated in the soil. Depending on the soil-climatic conditions and also an the chemical properties of the antibiotics themselves, they may be preserved and accumulate in the soil.

Entry of antibiotics into plants

The importance of antibiotic substances formed in the soil, for plants, is determined primarily by the extent to which the former enter the plants via the roots and by their activity within the plant.

The question of whether antibiotic substances are capable of entering the plants is now answered in the affirmative. Vegetating plants absorb various organic substances through their roots including antibiotics formed by bacteria, actinomycetes and fungi. The absorption by plants of subtilin, gramicidin, pyocyanine, licheniformin, (antibiotics of bacterial origin), as well as streptomycin, globisporin, aureomycin, terramycin, grisein, griseofulvin, etc (substances formed by actinomycetes) has been experimentally established.

Table 121 shows the degree of entrance of chemically pure antibiotic preparations, from solutions into the plant, via the root.

Table 121
Assimilation of antibiotics by plants
(units per 1 g tissue)

Antibiotics	Wheat, roots	Wheat, stems	Wheat, leaves	Peas, roots	Peas, stems	Peas, leaves
Penicillin	150	60	60	150	80	100
Streptomycin	100	20	20	100	20	30
Grisein (preparation No 15)	120	30	30	150	20	40
Mycetin	100	5	0	100	10	0
Subtilin	300	0	0	80	10	5

Penicillin enters the plant at the fastest rate and in the largest amounts, followed by grisein and streptomycin. Mycetin subtilin and gramicidin enter in small amounts and do not travel high up in the plant.

Aureomycin, terramycin, globisporin and many other active substances enter readily into the plants.

Plants absorb from the substrate not only chemically pure preparations but also crude antibiotics, in the form in which they are excreted by their producers. We added a liquid containing antibiotics to a culture solution and grew plants in it. After a certain time one could detect antibiotics in the tissues of the latter (Table 122).

Table 122
Uptake of crude antibiotic substances from culture fluids by plant roots
(units/ml)

Microorganisms producing antibiotic substances	Introduced into the soil	Found in wheat roots	Found in wheat leaves	Found in pea roots	Found in pea leaves
A. streptomycini	250	60	30	80	40
A. aureofaciens	600	120	50	100	40
A. grisseus strain 80	100	20	25	50	25
Bact. nitrificans	120	45	10	40	10
Bact. fluorescens	200	40	5	40	35
Bact. denitfricans	90	10	10	15	5

Antibiotic substances enter plants from solid substrates, directly from soils.

In our experiments (Krasil'nikov 1951 a; 1952 b, c; 1953 c; 1954 a) we have tested various antibiotics formed by bacteria, fungi and actinomycetes-antagonists. Chemically pure preparations were introduced under adult plants in containers with sand or soil and their concentration was determined after given time intervals on the root tissues and aerial organs. The experiments were performed both under sterile and nonsterile conditions of growth. In Table 123 are given the data from the experiment with the sterile sand substrate. Streptomycin 500 units and grisein (preparation 15), 500 units were introduced per 1 g of substrate.

Table 123
Entrance of antibiotics into plants from sterile substrates
(units/g of tissue)

Antibiotics	Plant origin	Kidney beans in sand	Peas in sand	Wheat in sand	Kidney beans in soil	Peas in soil	Wheat in soil
Grisein	Root	150	250	100	80	70	50
Grisein	Leaves	80	120	60	50	40	20
Streptomycin	Root	220	160	120	100	80	30
Streptomycin	Leaves	100	80	40	40	40	40

As can be seen from the table, antibiotics enter into the plants via the roots in quite large quantities. They are found in roots, stems and leaves for 10 days and more.

In experiments performed with sterile soil (Moscow area podsol) the same results were obtained. The antibiotics, globisporin, 500 units and grisein, 800 units were introduced per 1 g of substrate. Peas, kidney beans and wheat were grown on this soil. Analyses have shown, that globisporin was preserved in soil for 30 days and grisein for 40 days. During this time they penetrated through the roots and into the plants in smaller or larger quantities (Table 123). The rate of entry of antibiotics from sterile soil was only slightly less than the rate of entry from nutrient solutions or from the sand substrate. After a few hours (6-10) the antibiotic substances could be found in the roots and in the lower parts of the stem.

Our experiments have shown, that antibiotics are also absorbed by plants from natural nonsterilized soil. The same antibiotics, grisein and globisporin, were introduced into vessels with podsol soil under adult plants (wheat and kidney beans) in doses of 800 units/g. After 24 hours and sometimes even later we detected these antibiotics in the tissues of the roots and aerial organs (Table 124).

Table 124
Entrance of antibiotics into plants from nonsterile soil
(units/g of tissue)

Antibiotics	Kidney beans, roots	Kidney beans, leaves	Wheat, roots	Wheat, leaves	Peas, roots	Peas, leaves
Grisein	20	10	30	10	30	20
Streptomycin	30	10	20	6	40	20
Aureomycin	--	--	20	4	20	6
Terramycin	--	--	30	10	20	6
Control plants	0	0	0	0	0	0

Plants also absorb crude antibiotics from the soil. We tested three native antibiotics formed by the actinomycetes No 290, 287 and strain "B." In the presence of the appropriate organic matter in the soil, these cultures, as is shown above, form 30-120 u/g of antibiotic substances. When we grew peas and wheat in such soil, we could detect the antibiotics in the tissues in small, but sufficient quantities which were large enough to inhibit growth of microbes sensitive to them (Table 125).

Table 125
Assimilation of crude antibiotics from the soil by plants
(units/g of tissue)

Antibiotics	Peas, roots	Peas, leaves	Wheat, roots	Wheat, leaves
No 290	10	2-3	4	+
No 287	15	3	10	2
"B"	2	+	4	+

The "+" sign designates small quantities detected only qualitatively.

Thus, in pea plants, 2-15 units/g antibiotics were obtained and in wheat somewhat less. In some cases, when there is an abundant growth of antagonists which form large amounts of antibiotics in the soil (100-150 units/g), large quantities of the latter enter the plants-up to 20-30 units/g in the roots and 10-20 units/g in the leaves.

The zones of growth inhibition of the test organism around pieces of tissues of the experimental plants may be seen in Figure 96. Around the shreds of tissue of the control plants no such growth-inhibition zones are formed.

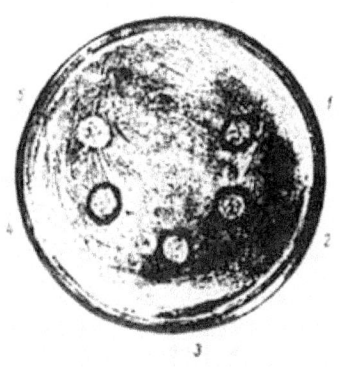

Figure 96. Entrance of antibiotics from the soil into plant tissues. Zones of lack of growth of bacteria around pieces of tissues:

1--roots; 2 and 3--stems (lower and upper parts); 4--leaves; 5--roots of control plant, grown in soil without antibiotics, the zone is missing.

It should be noted that plants may not only absorb the antibiotics that exist in the free state from the soil but also those adsorbed by soil particles. It was stated earlier, that a considerable part of the antibiotics are adsorbed immediately after their formation and become tightly fixed. Adsorbed antibiotics cannot be washed,out with water or with a number of organic solvents, even upon prolonged treatment. However, due to the activity of their root systems, the plants, are in the position to sever this, bond between the antibiotics and the soil particles and to desorb and take up the antimicrobial substances.

We introduced streptomycin, up to 2,000 units/g and more into the soil (podsol, garden) sometimes to full saturation, and then rinsed the soil with water until no more antibiotic was found in the elutions.

After this treatment about 1,500 units of streptomycin per gram remained in the soil. Plant seedlings of peas or wheat were planted in such soil and after some time their tissues were analyzed for the presence of the antibiotic. Usually, after 3-4 days and often even after 30-40 hours streptomycin was found in the roots, stems and leaves in amounts up to 10-15 units/g and more. As a rule, soil analyses have shown the absence of the free antibiotic; the latter was probably in the adsorbed state and was actively absorbed by the roots.

Antibiotic substances entering plant tissues, enhance the bactericidal effect of their sap and thus increase the resistance of the plants to diseases.

The more abundant the growth of antagonists in the soil, the more antibiotic substances they produce, the more of the latter enters the plants and the stronger the bactericidal effect of the sap becomes.

The sap of plants which are grown in a sand substrate without microorganisms and without humus is, according to our observations, less bactericidal than the sap of plants which are grown in nonsterile soil rich in humus.

We enriched the soil artificially with actinomycetes--producers of streptomycin, and we grew in this soil plants--peas and wheat. The sap of such plants was tested for its bactericidal effect on *Bac. mycoides* and *Staph. aureus.* Death of the bacterial cells in the sap of the experimental plants followed after 8-12 hours, and in the sap of the control plants which were grown in soil not enriched with actinomycetes, there was only suppression of growth, but death of the bacteria was not observed.

Plants grown in soil well fertilized with manure or compost had more active sap than the sap of plants taken from unfertilized soil. The sap of plants grown in a greenhouse (corn) was less bactericidal, than the sap of plants, grown in the open ground in the same soil (Krasil'nikov and Korenyako, 1945 a). Eaton and Rigler (1946) observed increased resistance of roots of cotton plant to *Phymatotrichum ornnivorum* upon treatment of the seeds with carbohydrates. In these cases, according to the author, intense growth of bacterial antagonists in the rhizosphere of the plants was observed.

Kublanovskaya and Brailova (1954), studied the bactericidal properties of the sap of the cotton plant in relation to the fungus *Fusarium vasinfectum* and found that its fungitoxicity toward the given fungus was less when the plants grew in soil with antibiotic substances than the fungitoxicity of the control plants growing in soil without antibiotics. The coefficient of multiplication of the fungus in the sap of the control plants was: 13.6 in the germination phase and 11.8 in the phase of cotyledons; in the sap of experimental plants: 7.8 in the phase of germination and 9.8 in the phase of cotyledons. As is seen, the antifungal properties of cotton-plant sap are strengthened at the expense of the antibiotic substances coming from the soil. Accordingly, the plants morbidity due to wilt was less: in the control 96% of the plants were diseased, and among the experimental plants--only 18.4%. Enhancement of antifungal properties of plant sap was also observed by Kublanovskaya in field experiments on plots fertilized with actinomycetal cotton-cake composts.

Stapp and Spicher (1954, 1955) observed the appearance of protective substance a in the sap of the potato in relation to *Bact. phytophthorum* during the development of the plant, when the soils were enriched with microbial antagonists.

The data given show that the plants absorb antibiotic substances from the soil. Antibiotics may be absorbed by the plants not only from solutions of chemically pure substances but also from a complex organic mixture of metabolites, the microbial antagonist.

Actinomycetes, bacteria and fungi which produce antibiotic substances, grow in the soil in the rhizosphere of plants. They saturate this zone or microfoci in the soil with the products of their metabolism, including antibiotics. The latter enter the plants through the roots and exert their action there. It is self-evident that the concentration of antibiotics in soil, when formed under natural conditions, will be lower than the concentrations created upon artificial introduction. However, under natural conditions, these substances are constantly formed and therefore one would assume that their entrance into plants is not stopped during the whole vegetative period.

Having entered the plant tissues the antibiotic substances protect them against the penetration of microbial parasites, suppress the growth of those that have already invaded, produce or elevate the toxicity of the plant sap and thus elevate to a larger or smaller extent the immunological properties of the plant.

In other words, microbial antagonists are factors which increase the resistance and insusceptibility of plants to infections.

Antibiotic substances as a therapeutic means in plant cultivation

Suggestions on the possible use of antibiotic substances for medical purposes were made by Pasteur, Mechnikov and their contemporaries. Scientists attempted to use bacterial cultures together with their metabolic products for curing the sick. Fehlesein (1883) described a case of curing lupus by introduction of *Streptococcus erysipelas* into the patient's skin. Colley (1893) used the same organisms for the treatment of a cancer patient. Pavlovskii (1887) introduced a culture of *Pneumococcus* into the bodies of animals and so prevented them from being infected with anthrax. Bourchard (1889) used the metabolic products of *Pseudomonas pyocyane* against anthrax. Manasein (1871) and Polotebnev (1872) used the green mold *Penicillium* in treatment of patients. Many other specialists attempted to use microbial cultures for the purpose of medical treatment. A special branch of medicine, bacteriotherapy even came into existence (cf. Kashkin, 1952; Ermol'eva, 1946; Waksman, 1947; Waksman and Lechevalier, 1953; Kohler, 1955; Korzybski and Kurylowicz, 1955 and others).

All these investigations had no proper success and recognition and were soon dropped. Only after active substances--penicillin, streptomycin, etc were isolated and chemically purified, did the antibiotic substances produced by microbial antagonists receive general recognition.

Presently, many antibiotic substances, penicillin, streptomycin, aureomycin, terramycin, subtilin, etc are widely used in therapeutical medicine and veterinary science. This successful use of these preparations in medicine naturally served as a stimulus for work on the use of antagonists and their metabolic products against infections of plants.

In the foregoing chapter we have shown the beneficent role of microbial antagonists, their inhibition of phytopathogenic organisms in the soil and then protection of plants against fungal and bacterial infections. We noted there that microbial antagonists remove phytopathogenic forms directly from the soil and by virtue of this alone protect plants from diseases.

However, the antibiotic substances obtained from cultures of antagonists may be used for the removal of phytopathogenic organisms not only from the soil but also within the plant. In other words, these substances should be used as curative remedies.

What then should be the requirements, in this case, from the antibiotics?

As in the case of treatment of human beings and animals, antibiotics used in plant growing should: 1) be active against the agent causing the plant disease and have the ability to inactivate toxins; 2) should penetrate easily into the plant tissues; 3) should not be inactivated too rapidly; 4) should exhibit antibacterial activity within the plant tissue; 5) should not be harmful to the plants at concentrations which are toxic to bacteria.

In addition, the method of use should be technically possible and all the measures should be economically profitable.

The first point was, in principle, proven experimentally. It was established, that phytopathogenic bacteria and fungi are susceptible to the inhibitory action of antibiotics.

As was noted above, for each phytopathogenic microbe it is possible to choose its corresponding antagonist and to obtain its antibiotic substances. Among the immense variety of microorganisms existing in nature, producers of antibiotics against bacteria, fungi, actinomycetes, viruses, etc can always be found.

Concerning the second postulate--the ability of antibiotic substances to penetrate the plants, this question too was answered in the affirmative. In the previous chapter it has been shown that these substances easily enter into the root system and proceed to the aerial parts. Antibiotics can also be introduced through the aerial organs--stems, leaves and seeds.

The possibility of introducing drugs into plants via the stem, was shown by Shevyrev in 1903. He introduced various antiseptics into fruit trees with the aim of killing parasites. The method developed by him is nowadays often used for extrarhizal nourishment of plants. This method is as follows: a hole is bored in the trunk of the tree and one end of a wick (of gauze or cotton) is placed in the hole; the other end is immersed in a bottle which contains the antibiotic solution; the antibiotic enters the trunk of the tree through the wick and spreads to all parts of the plant.

In grassy plants antibiotics may be introduced via the stem by simply wetting it or smearing a paste containing the preparation on it. We used the first method. A wetted piece of cotton or gauze was wrapped around the stem of the plant and covered with wax paper, to prevent rapid desiccation.

We tested the method of introducing antibiotic substances through the trunks of trees on various varieties of fruit-bearing and decorative plants and under different climatic conditions--in Crimea, Caucasus and Moscow (Krasil'nikov and Kuchaeva, 1955). Various antibiotics were introduced into the plants--penicillin, streptomycin, globisporin, aureomycin, terramycin, grisein and other chemically purified preparations. In a few cases we also introduced crude antibiotic substances in the form in which they appear in the culture fluid.

Experiments have shown that all these antibiotics can enter the plant via the trunk, but in different amounts and at different rates. Penicillin enters at the most rapid rate (see above). Most of the experiments on absorbability were performed with it (Table 126).

Table 126
Entrance of penicillin into plants upon introducing it via the stem
(July-August 1954)

Plants	Age, years	Height, meters	Diameter of trunk, cm	Time of introduction of antibiotic, days	Solution adsorbed, ml	Antibiotic units adsorbed	Detected in lower branches and leaves	Detected in upper branches and leaves
Maple (acer platanoides K.)	15	3	8	5	20	200,000	-	-
Ash tree (Fraxinus chinensis L.)	8	4	7	5	15	150,000	-	-
Lime tree (Tilia cordata Mill.)	5	25	7	5	10	100,000	-	-

Plant								
White acacia (Robina pseudo- acacia L.)	8	2.8	7	5	10	100,000	-	-
(Halimodendron argenteum Fisch.)	15	1.8	6.3	5	0	--	-	-
Cherry (Cerasus vulgaris Mill.)	9	4.5	6.5	4	430	4,300,000	+	-
Bird cherry (Padus virginiana Roem)	15	4	5.8	5	210	2,100,000	+	-
Apple (Malus domestica Borkh)	8	3.2	7.3	3	380	3,800,000	+	+
Peach (Persica vulgaris Mill.)	8	1.5	6.0	5	560	5,600,000	+	+
Apricot (Armeniaca vulgaris Lam.)	7	1.6	5.4	5	900	9,000,000	+	+
Sweet cherry (Cerasus avium Moench.)	7	2.1	6.1	5	165	1,650,000	+	+

As seen from the table, various woody plants absorb different amounts of penicillin. Some of them, as for example, cherry, sweet cherry, apple, peach and apricot trees absorb antibiotics in large quantities, others like maple, ash tree and lime tree absorb little of it.

The distribution of the penicillin within the plant also differs. In some plants (cherry, apple, peach, etc) it moves quickly to all parts, into the branches and leaves of the whole crown; in other plants (bird-cherry tree and Halimodendron) it slowly reaches only the lower parts of the branches and leaves and does not reach the upper part of the crown; in still other trees (maple, ash tree and lime tree) it cannot be detected in the leaves at all.

The intensity of the uptake and the distribution of antibiotic substances in the plant changes noticeably with changes in climatic conditions--temperature, air humidity, soil moisture, etc. The lower the temperature and the higher the humidity of soil and air, the slower is the uptake of antibiotics. For example, in May 1954 in the Nikitskii Garden, when the average temperature of the month was 13.4° C the temperature of the soil 14.2°C and the relative air humidity 92%, peaches and apricots absorbed 45-50 ml penicillin solution each on the first day and during 5 days--100-110 ml. In August, the average daily temperature of the air was 24.3° C the soil temperature was 23° C and relative air humidity was 41%, the same plants took up 200-210 ml on the first day and during 5 days--about 1 liter of penicillin solution (Table 127). In May the absorption of penicillin was 3-5 times less than that in August, although in the spring plant suction is usually higher.

Table 127
Intensity of extrarhizal absorption of penicillin by plants under various weather conditions
(the Nikitskii Botanical Garden, 1954) (in ml of absorbed solution of 10,000 unit / ml activity)

Plant	1st day	2nd day	3rd day	4th day	5th day	Total
May: temp of air 13.4° C, soil temp. 14.2° C, relative air humidity 92%						
Apricot	45	30	20	10	5	110
Peach	50	30	10	5	0	95
Sweet cherry	35	15	5	1	0	56
August: sir temp. 24.3° C, soil temp. 23.0 C, relative air humidity 41%						
Apricot	210	200	190	200	100	900
Peach	200	180	100	40	40	560
Sweet cherry	80	50	15	10	10	165

We obtained similar data in experiments with birch (10 years old) under the Moscow climate. A globisporin solution was administered (activity 5,000 u/ml) to the same plant via the stem on dry and on rainy days. Two repeated experiments were performed: the first in June and the second in August. The amounts of antibiotic absorbed during 5 days were: on dry days in June--1,500 ml, in August--900 ml and on rainy days of the same months the corresponding absorption was 350 and 200 ml, i.e., 4.5 times less.

In experiments with lemon trees in the orchard of the Institute of Subtropical Cultures (Anaseuli) in the rainy period in September 1952, the antibiotic grisein was absorbed to such a low degree that the work had to be postponed until drier weather set in.

The rate of distribution of antibiotics within the plant corresponds to the intensity of its absorption. The faster, and the more antibiotic enters, the sooner it is found in the different parts of the plant. In experiments with plants of the Nikitskii Garden we introduced a penicillin solution into trees and the rapidity of its appearance in the leaves 'was measured. Each day after the introduction of the solution 20-30 leaves were removed from the tree and analyzed separately for the presence of the antibiotic in them. Table 128 shows the percentage of leaves in which the antibiotic was detected.

Table 128

Rapidity of the distribution of the absorbed penicillin in the tissues of plants
(the figures designate the percentage of leaves saturated with the antibiotic)

Plants	Quantity adminis- tered in mg	Lower part of crown after one day	Upper part of crown after one day	Lower part of crown after two days	Upper part of crown after two days	Lower part of crown after three days	Upper part of crown after three days
IN MAY:							
Apricot	45	0	0	41.6	33.3	90	13
Peach	50	0	0	50	25	55	18
Sweet cherry	35	0	0	0	0	60	0
IN AUGUST:							
Apricot	200	100	100	100	100	100	94
Peach	210	100	100	100	100	100	100
Sweet cherry	80	100	100	100	100	100	75

In August in dry warm weather, the antibiotic is rapidly distributed throughout the whole tree. It can be found everywhere a few hours after its introduction. In May there was cool rainy weather in Crimea. The uptake of the antibiotic and its distribution in the tissues was very weak. Only 2 days after introduction could one detect the antibiotic in the plant's leaves, and then only in some leaves.

The entrance of the said substances into the plants is connected with the physiological conditions of the latter. The more intense the metabolic processes of the plants, the more vigorous their growth, the faster are the antibiotics absorbed. When the external factors slow down the growth of the plant, the inflow of antibiotics into the roots also slows down. It was noted above, at a lowered air temperature the uptake of active substances is much more sluggish than at a higher temperature in the summer. The same was observed by Stokes (1954) in her work. She determined the rate of uptake of griseofulvin by plants at different temperatures. At 25° C the substance enters the plant 5 times as fast and in larger quantities than at 10° C. She also observed the detrimental effect of excessive humidity on the uptake of the antibiotic. At a 56% relative humidity its concentration in the tissues is 4 times higher than that at a humidity of 91% and a temperature of 25° C.

Antibiotic substances taken up by the root system, are transported via the xylem to the aerial parts, the leaves. If, however, the antibiotics are introduced through the leaf surface, their transportation is accomplished through the phloem, i.e., as in case of substances synthesized in the leaf.

Antibiotic substances entering the plant, penetrate inside the cells and cause a certain effect there. In order to follow the penetration of these substances into the cells we used antibiotics which were luminescent in ultraviolet light. Mycetin and certain other substances belong to these antibiotics.

We allowed the antibiotic solution to pass through the tissues of the plant, we then performed microscopic analyses of microtone slices. Mycetin first enters into the cytoplasm, staining the various granules and rodlike mitochondria and then enters the nucleus, where it reaches higher concentrations than in the cytoplasm

Some of these substances at certain concentrations inhibit nuclear division upon entering the cells.

Pramer (1955) followed the penetration of antibiotics--penicillin, streptomycin and chloromycetin into the cells of the algae *Nitella Clavata*. These algae were immersed in a solution of the antibiotic for some time, were then washed thoroughly, and the cell juice which was squeezed out of the individual cells was collected with a micro-pipette. In this juice the content of the antibiotic which was introduced into the algae was determined.

It was found that streptomycin and chloromycetin penetrate the membrane, reach the inside of the cells, and spread throughout the protoplast giving the latter bactericidal properties.

Penicillin, according to the author, is not detected inside the cells. It either does not penetrate them or, if it does, is immediately inactivated.

Nielsen (1955) found that antibiotics formed by the plankton of water reservoirs suppress the photosynthetic activity of algae of the Chlorella group.

There are theories which state that upon introduction of various substances into the trunk of a tree, they spread in a sectoral fashion, corresponding to the vascular transport system.

Taking this in account, we paid special attention to the distribution of antibiotics in the periphery of the bark of woody plants. The administered antibiotic was determined in the leaves and branches located in various parts of the crownaccording to sectors and circular rings-in the lower, middle and upper parts.

Numerous analyses show quite clearly, that penicillin, streptomycin, grisein and other antibiotics are distributed more or less evenly throughout the crown of the plant. We observed no sectoral distribution of the substances introduced in fruit, ornamental or forest trees.

Certain antibiotics enter the root system from above and travel downward. According to our observations, grisein possesses this property. When it is introduced into the stem or the trunk of a lemon tree it can be found after a certain time in the lower part of the trunk and in the roots. The tissues contained: in the trunk, at the point where the substance was introduced 120 units/g, near the root 60 units/ g and in the roots-30-50 units/g.

The method of administering antibiotics through the intact stem by the use of gauze and cotton bandages, was employed by us in experiments with grassy plants and with shrubs; it was also tested with woody plants. Young branches of garden roses, apple trees and pear trees, stems of peas and wheat were wrapped with cotton (or gauze) wetted with a solution of penicillin, streptomycin, grisein or another preparation; after a lapse of some time, the plant tissues were subjected to analysis. As the investigations have shown, these substances penetrated inside the plants, but never accumulated in high concentrations. This method is thus hardly suitable for wide use. However, it may be used for local therapy.

Introduction of antibiotics through the leaf surface has been performed in experiments with woody and grassy plants. The leaves of the plant were either sprayed with a sprayer or wetted with cotton.

The spraying of the crown of plants with antibiotic solutions, using a sprayer was employed by us in experiments with fruit trees: peaches, apricots and apple trees and with grassy plants; peas, corn, wheat, etc. Preparations of penicillin, streptomycin and grisein in dilutions of 1:1,000-1:5,000 were used. After some time these antibiotics were determined in the tissues of leaves, branches and stems. Before analysis the severed leaves and branches were thoroughly washed in water.

Analysis showed the following amounts of antibiotics (in 1 gram tissue):

	Penicillin in units / g	Streptomycin in units / g
Apple tree	up to 5	2-3
Sweet cherry	15	5-10
Peach	10	10
Apricot	40	20-40 (grisein)
Peas	up to 20-50	10-20
Wheat	5-10	2-10

The greater the distance from the location of the antibiotic administration, the lower its concentration in the organs of the plant.

Upon wetting leaf surfaces with pieces of cotton soaked in a solution, even more convincing results were obtained. Two to five hours after the application of the cotton bandages with the antibiotic, the latter could be detected in the leaf tissue which were quite removed from the spot of its application as well as in the petioles of leaves, and even in the tissues of the branches which bear those leaves.

The American specialists use antibiotics in the form of dust, spraying them on the crown of the plants. The dust particles reaching the surface of the leaves, dissolve and penetrate into the tissues.

It should be pointed out that in all the experiments with the various methods of introduction of antibiotics, the antibiotics move in the direction of the lower parts as well the upper parts of the plant. Upon introduction of a solution of penicillin or grisein through the trunk of an apricot tree, these substances were detected in the branches, leaves and root tissues as well. The same was observed with peas. A preparation of penicillin introduced through the stem surface, was subsequently found in leaves in the upper parts in a concentration of 5-10 units/g and in the roots, in a concentration of 3-5 units/g (Table 129).

Table 129
Distribution of antibiotics in tissues of peas upon their introduction through the stem
(u/g of tissue)

Antibiotics	Roots	Stems	Leaves
Penicillin	3-5	10-20	5-10
Grisein	3-5	10-15	3-5
Streptomycin	1-3	10-30	3-5

Mitchell, Zaumeyer, Andersen et al., (1952, 1954) introduced antibiotics, applying them with lanolin paste. The paste with the antibiotics was spread on the stem surface, and after some time the active substance was determined in the tissues of branches and leaves. Brian, Wright et al., (1951) observed the penetration of the antibiotic griseofulvin into plants. Leben, Arny and Keitt (1953) introduced helixin and antimycin into plants, and Hessayon (1951) introduced trichothecin--an antibiotic obtained from the fungus *Trichothecium.*

Antibiotics can be used for the sterilization of infected seeds. It is known that in plant seeds there are often phytopathogenic bacteria and fungi which are sources of plant diseases. In order to get rid of these agents, various chemical substances are used--antiseptics. However, the antiseptics which inhibit the growth of microbes also act deleteriously on the seed tissues and decrease their ability to germinate.

The antibiotics, unlike the antiseptics, act selectively, inhibiting microbial metabolism without causing any harm to the seed embryo. The sterilizing effect of antibiotics was tested by us on cotton seeds. It is sufficient to immerse the seeds for 4-8 hours in an antibiotic solution in order to kill the microbes in the seed tissues (Krasil'nikov, Mirzabekyan and Askarova, 1951; Askarova, 1951: Mirzabekyan, 1952).

Blanchard and Diller (1951) treated leguminous seeds with aureomycin, allowed them to germinate and then determined the entrance of the aureomycin into the seedlings. The authors noticed a larger accumulation of the antibiotic in the roots than in the upper parts.

The effectiveness of antibiotics depends on their concentration in the tissues. In turn, the concentration depends on the properties of the plants, especially on the properties of the antibiotic and also on external conditions.

It was found that. when the solution of antibiotic is concentrated, more of it penetrates the plant (Table 130).

Table 130
Degree of saturation of plants with antibiotics
(units/g of tissue)

Antibiotics	Introduced into substrate units/ml	Wheat roots	Wheat leaves	Pea roots	Pea leaves	Corn roots	Corn leaves
Penicillin	5,000	3,500	3,000	4,000	3,800	5,000	4,000
	1,000	600	500	500	300	800	300
	500	200	100	300	160	180	80
	100	70	40	50	30	60	25
	50	80	40	50	40	80	70
Grisein No 15	1,000	800	600	500	400	950	600
	500	250	160	300	180	300	100
	100	70	40	80	50	80	50
	50	50	40	60	40	80	85
	10	30	20	50	30	50	30

Very high concentrations of antibiotics (penicillin--5,000 u/ml), grisein--1,000 units/ml) have a toxic effect on plants. They begin to wither and guttation stops. Smaller concentrations are harmless; the plants develop normally.

Winter and Willeke (1951, b) have shown that in tissues of lettuce, penicillin may accumulate up to 500 units/g, and streptomycin, up to 100 units/g and cause no noticeable pathological phenomena. When the penicillin concentration in the tissues reaches 1,000-2,000 units/g, the plant suffers from poisoning.

Brian, Wright, Stubbs and Way (1951) point out that considerable concentrations of griseofulvin in the tissues of oat and lettuce brought about no poisoning effect.

If the concentrations of the antibiotics in the substrate are very low, the plants may accumulate them in their tissues. For example, when the concentration of penicillin in the medium is 50 units/ml, up to 80 units/g accumulates in the roots of or wheat and corn; when the concentration of grisein is 10 units/ml in the medium, up to 20-50 units/g and more are accumulated in the roots of peas.

Pramer (1953-55) tested different antibiotics--streptomycin, aureomycin, chloromycetin, terramycin, neomycin, etc. According to his observations, different substances enter the plants and saturate their tissues to various degrees. In experiments with cucumber seedlings, the most highly absorbed drug was streptomycin, At a concentration in the solution of 500 μ g/ ml, its accumulation in the tissue reaches 100-150 μ g/ g, 18 hours after treatment. Streptomycin penetrates less readily and in smaller quantities the tissues of tomatoes and kidney beans. The author notes that sometimes the concentration of streptomycin in leaves is higher than that in stems and roots.

Chloromycetin penetrates the cucumber seedlings much more weakly than does streptomycin and reaches a concentration of 20-50 μ g/g. Aureomycin, terramycin and neomycin do not penetrate this plant at all.

Pramer (1954), investigated the absorbability of streptomycin, chloromycetin and penicillin into the cells of algae *Nitella clavata* and found that after the algae were immersed for 12 minutes in a streptomycin solution which contained 8 μ g /ml, the algae's cell sap contained the same concentrations of antibiotic as that of the external solution. After immersion in the same solution for 18.5 hours the concentration of streptomycin in the sap was 7 times higher than that of the surrounding solution.

Chloromycetin penetrated the cells of the algae with difficulty. Only after 24 hours was it detected in the sap. Penicillin was not detected at all in the cells of the algae after 25 hours of immersion in a solution with 25 μ g/ml antibiotic concentration. The author believes that penicillin penetrates the cells quickly, but is immediately oxidized.

The concentration of antibiotics in plant tissues depends not only on the amount of the substances entering but also on the rate of their disappearance, or the time of their preservation.

Antibiotics are known to be preserved in the body of animal organisms for a short time only. They are excreted from the body during the first hours after their entrance, which complicates the work with antibiotics in hospitals. Only with the aid of special substances--prolongators--does one succeed in keeping the antibiotic in the animal or human organism.

In plants the antibiotics which are introduced are preserved for much longer periods of time. These very same substances -- penicillin, streptomycin, globisporin, aureomycin, etc are preserved within the plants for several days or even weeks. For example in tissues of sweet cherry, penicillin is preserved for 4 days and in tissues of the apricot tree--16-17 days (Table 131).

Table 131
Time of preservation of penicillin inside woody plants

Plants	Age of plant, years	Antibiotic indroduced, in units	Time of preservation in days	Location where experiments were carried out
Cherry	8	2,400,000	8	Moscow, Central Botan. Garden Ac. Sci, USSR
Apricot	5	2,200,000	16	Moscow, Central Botan. Garden Ac. Sci. USSR
Peach	8	4,800,000	15	Crimea, Nikitskii Botan. Garden
Apricot	7	6,000,000	17	Crimea, Nikitskii Botan. Garden
Sweet cherry	7	1,450,000	4	Crimea, Nikiskii Botan. Garden

In grassy plants antibiotics are preserved for a similar length of time.

Grisein (No 15) when introduced into tissues of the cotton plant and peas, is preserved there for 10-20 days (Table 132).

Table 132
Time of preservation of the antibiotic grisein No 15 in plants
(unit/g of tissue)

Time of analyses, number of days after introduction	Cotton, roots	Cotton, leaves	Peas, roots	Peas, leaves
Initial amount	350	100	180	80
1	200	50	120	60
2	150	30	100	40
3	100	20	70	30
5	80	10	30	10
7	50	8	10	5

10	30	5	10	5
20	10	0	0	0
30	0	0	0	0

Askarova (1951) and Mirzabekyan (1953, 1955) found antibiotic substances inside the cotton plant 20 days after their introduction.

Brian et al., (1951) detected griseofulvin in tissues of lettuce and oats during 3-4 weeks.

Antibiotics first disappear from the aerial parts of the plants and later from the root system.

Analyzing a nutrient solution in which plants saturated with antibiotics have been kept, we could establish that these substances were excreted by the roots. However, the amount of the substances excreted was much smaller than that of the plant. If in the tissues of one pea plant there were about 4,000 units of penicillin then about 600 units were excreted into the solution.

In the experiment with streptomycin, single pea plants were immersed for one day in streptomycin solution. The number of units of the antibiotic which were absorbed from the solution were precisely determined, and the plant was removed from this solution. The roots were washed with water and immersed in Hellriegel' s nutrient solution which did not contain antibiotic. After certain time intervals the amounts of the antibiotic in the solution and in the plant tissues were determined. The results are given in Table 133.

Table 133
Excretion of streptomycin from pea tissues into solution

Times of analyses, number of days after introduction	Units in solution	Units inroots	Units in stems	Units in leaves	Units in in the whole plant
Initially	0	350	45	80	1,707
3	120	270	20	40	1,123
5	240	150	20	20	767
10	320	20	5	5	105.5
15	350	0	0	0	0

In these experiments one pea plant absorbed 1,800 units, after 10 days only 105 units were found, and after 15 days nothing at all was left. During this time only 350 units were excreted into the solution by the roots. Therefore, we assume that the remaining 1,357 units of antibiotic were assimilated by the tissues as sources of nutrition and underwent biochemical changes.

Comparing the degree of absorption of the antibiotic with the nature of the distribution of the latter in the plant, and with the time of its preservation in the tissues, one may conclude that there exists a direct connection between these phenomena. The more the antibiotic was absorbed, the sooner it was found in the leaves and upper branches, and the longer it was preserved there.

However, this is not always so. Quite often, upon intense absorption of an antibiotic the latter is not detected in the tissues or is found there in very small quantities. For example, maple and bird cherry under similar conditions absorb the same amounts of penicillin, but in the bird-cherry tree it penetrates into the upper parts and reaches the leaves, while in the maple it is found neither in the leaves nor in the branches. In the apricot, peach, sweet cherry and apple trees penicillin may be detected in the leaves upon introduction of 350-500 thousand units per tree while in maple, ash tree, lime tree and acacia it cannot be detected even when it is introduced in considerably larger quantities.

We introduced 7-15 million units per tree of globisporin and penicillin into the trunks of a 10-year-old birch and a 7-year-old willow--calculated on the basis of 300-600 units per 1 g woody mass. After 36 hours penicillin and globisporin could be detected in the leaves of the willow; the former was 1.5-2 times more concentrated than the latter (Table 134). In birch the antibiotic was not found either in the leaves or in the branches. However, traces of the antibiotic were detected in the wood of the trunk at a distance of not more than 10-30 cm. upward and downward from the point of introduction.

Table 134
The uptake and distribution of penicillin and globisporin in tissues of birch and willow (solutions introduced: penicillin--15, 000 units/ ml globisporin--10,000 units/ml)

Type of Antibiotic/Type of tree	Amount of anti- biotic solution introduced	Total units per plant	Units per 1 g of wood mass	Anti- biotic found after 10 hours	Anti- biotic found after 20 hours	Anti- biotic found after 36 hours	Anti- biotic found after 48 hours	Anti- biotic found after 72 hours	Anti- biotic found after 120 hours
Birch									
Penicillin	1,000	15,000,000	600	0	0	0	0	0	0
Globisporin	800	8,000,000	250	0	0	0	0	0	0
Willow									
Penicillin	850	13,600	500	0	trace	20	30	20	15
Globisporin	680	6,800,000	300	0	0	10	15	15	10

The absence of antibiotics in the leaves and branches of birch, maple, lime tree and other plants may be explained by their inactivation or by the adsorption by the tissues adjacent to the point of introduction, and, in certain cases, also by the weak uptake of the solution. The last explanation does not apply in the case of the birch. As is seen from the table, the birch absorbed more penicillin and globisporin solution than the willow, but nevertheless, the antibiotic was not found either in the leaves or in the branches.

Investigation of the causes of this phenomenon has shown that the wood of the trunk and branches and the leaf mass of birch possess a clearly expressed inactivating and absorbing capacity in relation to the antibiotics tested. By specially devised methods we have established, that one gram of wood of the trunk of birch absorbs 18,000 units of penicillin and fully inactivates 6,000 units; it absorbs 6,000 units globisporin and inactivates more than 3,000 units. A ground mass of green leaves absorbs 10,000 units penicillin and inactivates 8,000 units; it absorbs 6,000 units globisporin and inactivates 5,000 units. In other words, the wood and especially the leaf mass of the indicated plants almost completely inactivate the absorbed antibiotics--penicillin and globisporin. In order to detect the antibiotics in the given tissues, it is essential that they be administered in a higher concentration (more than 6,000 units globisporin and more than 8,000 units penicillin per g).

Plant tissues probably inactivate all other antibiotics which enter them. We tested streptomycin, penicillin, aureomycin and certain crude preparations of actinomycetal origin on various plant tissues: apples, lemons, peaches, cherries, etc. In all cases a different degree of inactivation was observed. Thus, tissues of lemon saplings (3-5 years old) inactivated aureomycin within the limits of 50--100 units/g, the leaf tissue was a stronger inactivator than the tissue of the trunk. The tissues of the apple tree and even more so, those of ornamental woody plants inactivated aureomycin to the extent of 200-500 units/g.

Inactivation of the crude preparation No 399 (from strain 399) in our experiments, was as follows:

Tissue of the lemon-tree trunk; 70 units/g
Tissue of the lemon-tree leaves; 160 units/g
Tissue of the pear-tree leaves; 220 units/g
Tissue of the apple-tree leaves; 180 units/g

Antibiotic substances may be introduced into the plant through the aerial parts, not only as chemically pure preparations, but also in their crude state, in the form of a culture fluid which is diluted with water to a certain concentration. We introduced the crude antibiotics via the trunk and through the leaf surface of woody plants and grasses. Plant seeds were also soaked in crude substance seeds of wheat, clover, peas, etc.

Antibiotic substances introduced in the form of culture liquid are distributed in the same manner as are chemically pure preparations, but at lower rates.

If seeds treated with antibiotics are immediately germinated, these substances may be detected in the seedlings. This translocation of the antibiotics from the seeds to the seedlings was observed by us in the cotton plant, peas and wheat. Special analyses have shown that the antibiotics completely permeated the cotyledons of the leguminous plants, as well as the endosperm of the cereals. Such a saturation of the food reserves with antibiotics--penicillin, streptomycin, grisein and certain other substances, does not cause any harm to the seedlings. The latter develop normally and utilize the reserves of the endosperm or the cotyledons exactly as the control seedlings but only if the antibiotic is not toxic.

Antibiotics introduced into plants have an antimicrobial action. If plants are artificially infected with a phytopathogenic form of bacteria or fungus, and a corresponding antibiotic is employed, the disease will not appear or will be weaker than it is in the control. We have introduced many nonpathogenic bacteria inside plants--*Bact. coli, Bact. prodigiosum, Bact. album, Ps. fluorescens, Ps. sp., Rhizobium trifolii*, etc. They all died much more rapidly in the tissues of plants to which antibiotics were introduced. For example, the root-nodule bacteria of clover, *Bact. coli and Bact. prodigiosum,* introduced into the stem of peas or kidney beans, die there after 20-30 hours and later, while the plants treated with streptomycin showed no bacteria after only 2-6 hours. The phytopathogenic fungus *Fusarium* sp. spread on seedling of pine, grew well on them, penetrated the inside and caused their death after several days. On plants that were treated with the corresponding antibiotic (No 121), the given fungus did not grow, and the growth of the plant was normal. There are many other observations which demonstrate the antimicrobial action of antibiotics inside plant tissues.

The antibiotics used should not be toxic to the plants. It is known that among the antibiotics there are various preparations, some of which are very toxic and cause the poisoning of certain tissues or of the entire plant. To such antibiotics belong: gramicidin, mycetin, clavacin, catenulin, magnamicyn, etc. Clavacinan antibiotic produced by the fungus *Asp. clavatus,* inhibits the growth of cereal roots at a dilution of 1:1,000,000 (Wang 1948). Other antibiotics--penicillin, streptomycin, grisein, terramycin, etc--may for all practical purposes be considered nontoxic. They may accumulate in tissues in large quantities, do not cause any disturbances, and, at certain concentrations, even stimulate the growth of plants (Barton and Mac Nab, 1954; Askarova, 1951), Scheffer and Kloke (1954) introduced antibiotics into soil in which they later cultivated plants. Only very high concentrations of the antibiotics caused the inhibition of the growth of barley and rye.

There is no need to introduce over-large amounts of antibiotics in treatment. The antimicrobial doses of the antibiotics of this group are considerably lower than the doses which cause the poisoning of the plants. For example, penicillin inhibits the growth of bacteria in wheat tissues at a concentration of 3-10 units/g, while the plant can withstand a dose of more than 1, 000 u/ g. Streptomycin and globisporin inhibit growth of bacteria in plants at concentrations of 5-10 units/g while the plants can withstand a dose of more than 500 units/g, etc.

There are many antibiotics that occupy an intermediate position in relation to their toxicity. Such antibiotics can also be successfully used in the healing practice. Griseofulvin is one of them. Its therapeutic dose is 5-10 μ g/g. A dose of 20 μ g/g is toxic for wheat and causes a burn and the swelling of the roots (Stokes, 1954).

One should also emphasize another feature of the action of antibiotics, namely their ability to inactivate toxins formed by fungi and bacteria. Inactivation of toxic substances by products of microbial metabolism was mentioned before (Krasil'nikov, 1947 a). It has been shown, that the toxic effect of gramicidin can be eliminated by neutralizing it with the metabolic products of bacteria. Actinomycetes inactivated the toxin which was formed by the sporeforming bacteria, *Bac. subtilis* and *Bac. mesentericus.*

It is known, that in many infections (if not in all) the plants suffer from poisoning by the toxins which are produced by microbes developing in the affected tissues.

Therefore, by selecting appropriate microbes these toxins can be rendered harmless and thus the poisoning of the plant can be prevented. Clover seeds saturated with bacterial toxin did not germinate in our experiments, or germinated to a limited extent; their seedlings lagged in growth and soon perished. When the poisoned seeds were treated at once with antitoxin, the percentage of germinating seeds increased considerably and the growth of these seedlings was only slightly less than that of the controls (Figure 97). Toxin, which was mixed with antitoxin in a certain ratio had no toxic effect at all.

Figure 97. Antitoxic effect of metabolites (antitoxin) of bacteria on clover seedlings, poisoned with bacterial toxin: a--control; seeds of clover were not treated with toxin before sowing; growth normal; b--seeds treated with toxic product, formed by bacterial inhibitors--there is no germination or a weak one; seedlings soon perish; c--seeds treated with the same toxin and then with the antitoxin substance produced by bacteria.

The inactivating effect of certain actinomycetes was observed by us in experiments with toxins formed by the fungus *Botrytis cinerea*. This fungus grows abundantly on leaves of certain plants--birch, oak, etc which are wetted by the carbohydrate excretions of aphids. The toxic substances formed by the fungus affect the tissue of birch leaves. Toxins of the given fungus were obtained by us in a culture in a nutrient medium with honey produced from insects. Its application to the surface of a birch leaf causes burn and necrosis of the tissue with subsequent yellowing and death of the leaf.

An antitoxin against this toxin which was formed by an actinomycete antagonistic to *Botrytis cinerea* was found. The addition of a certain amount of antitoxin to the toxin neutralizes it, and the mixture obtained is rendered nontoxic for birch. This antitoxin is also beneficial when first symptoms of poisoning appear. The process of wilting and formation of necrotic patches stops and the leaves recover from the injury and continue to develop normally.

The antitoxic action of actinomycetes, or more exactly, the action of their metabolic products, was observed by us in experiments with toxins formed by the fungi *Fusarium vasinfecturn, Fusarium* sp., *Trichothecium* etc.

The phytopathogenic fungus--*Deuterophoma tracheiphila* causes the poisoning of citrus plants by its metabolic products. An antagonist *(A. griseus)* of this fungus was found among the actinomycetes, which produces antitoxic substances and inhibits the growth of the fungus. The antibiotic grisein, obtained from a certain actinomycete, also had suppressive effects on this pathogen, which is the cause of a disease of citrus plants.

It may be assumed that for any toxin of microbial origin an antitoxin can be found.

Among the microbes there are many species which form strong poisons not only for plants but also for animals and human beings (botulin, tetanus toxin, etc). Under natural conditions these toxins are inactivated by other microbes, which produce antitoxins. It is tempting to use these antitoxins against food poisoning and other toxicoses of man and animals.

As is seen from the above-mentioned data, antibiotic substances fulfill all the requirements demanded of healing substances in plant breeding.

The possibility of using antibiotics for curative purposes was also proven by indirect laboratory and field experiments.

The first experiments in this direction were performed with crude antibiotic substances, obtained from bacteria and actinomycetes (Krasil'nikov, 1947 a). More fundamental and systematic studies were performed with chemically purified substances. Antibiotics were employed in the struggle against infections of woody and grassy plants (Krasil'nikov, Mirzabekyan and Askarova, 1951; Askarova, 1951; Mirzabekyan, 1952). Mirzabekyan employed the specially chosen antibiotic, grisein, in treatment of apricot and peach trees suffering, from "bacterial wilt." This disease is caused by *Bact. armentaca.* It is expressed in the withering of the crown first, and then of the whole tree.

At first the experiments were performed on young one- to two-year-old saplings of apricots and peaches. They were artificially infected with a culture of *Bact. armeniaca* and a few days later, treated with antibiotics.

An aqueous solution of the preparation was introduced into the leaf surface by wetting it. In all cases where the plants were treated with antibiotic immediately after inoculation, the disease did not appear. In cases where the treatment was started after a delay and when the symptoms of the disease, the wilting of leaves, were already apparent the disease stopped developing, leaves and branches recovered, and the plant continued to develop normally.

A hundred per cent of plants, that were not subjected to treatment, were infected and perished (Figure 98).

Figure 98. Curative effect of antibiotics. Apricot seedlings infected with *Bact. armeniaca:* a and b--plants not treated with antibiotic; c--plants treated with antibiotic; d--control plants (noninfected).

Experiments with fruit-bearing, 15- to 20-year-old plants were performed in the experimental fruit section of the Academy of Sciences of the Armenian SSR and in one of the fruit farms of Armenia.

The antibiotic grisein was introduced into the stem and was also sprayed on the crown. The process of drying ceased after the treatment. The leaves and branches recovered and continued to grow normally (Figure 99). In cases where the injury was a severe one and where there were dying branches, an effect was also noticed. The antibiotic stopped the further spread of the disease and new sprouts appeared on the still living parts.

Figure 99. Curative effect of the antibiotic grisein in "bacterial wilting" of apricots: a--a tree not treated with the antibiotic; b--a treated tree.

335

Positive results were obtained upon the use of antibiotic substances in the struggle with "malsecco" of citrus plants under laboratory conditions. The disease known as "malsecco" is caused by the fungus *Deuterophoma tracheiphila.*

Young lemon trees artificially infected with the fungus easily succumbed to the "malsecco" disease. In the struggle with this disease antibiotics were selected which were later tested on experimental plants. The antibiotic solutions were introduced into the trunk and through the leaf surface.

Among the preparations tested, grisein had a curative effect. The plants recovered quickly or did not get sick at all, while the control plants which were not subjected to the treatment, died,

In order to sterilize the grafts of lemons which were used as grafting material Mirzabekyan (1955) saturated them with an antibiotic solution. Specialists have found, that the infection is introduced with the grafting material, In the laboratory experiment it was shown quite feasible to employ antibiotics in the practice of grafting plants; grafts treated with streptomycin or grisein were sterile.

Among the grassy plants, antibiotics were most frequently used with cotton cultures, affected by gummosis. This is a widespread disease and causes great losses to agriculture. It is caused by the nonsporeforming *Ps. malvacearum,* which is distributed and carried by the seeds. These often harbor the bacteria internally, thereby considerably complicating the struggle against infection.

Askarova (1951-52) preliminarily chose a number of antibiotics which inhibit *Ps. malvacearum* and she discovered that some of them (Nos 73/20, 160, 114, etc) penetrate the seeds of cotton freely and kill the bacteria which cause gummosis. These substances therein saturate the tissues of the entire seed and embryo but do not harm them. The ability of such seeds to germinate does not decrease and in some cases (preparations 160, 265) it even increases. At first the experiments were performed in growth containers and later on experimental plots and on industrial fields. Only crude antibiotics in the form of culture fluid were employed. The activity of this fluid was 1,000-2,000 units/ml.

The results were positive. Seeds treated with antibiotics germinated better and their shoots were taller and healthier. There were fewer diseased plants at the early stage of growth and at the end of vegetation, than in the control (Table 135).

Table 135
Effect of antibiotics on the appearance of gummosis in the cotton plant
(experiment under field conditions on small plots)
(after Askarova 1951)

Conditions of experiment	Morbidity, % in the cotyledon stage	Morbidity, % in stage of boll formation
Control (no treatment of seeds)	72.4	19.1
Seeds treated with preparation No 114	8.1	0.0
Seeds treated with preparation No 117	10.2	0.4
Seeds treated with preparation No 86	18.4	2.0
	Yield of seed cotton	
From control plot	1,640 kg	
From experimental plot, prep. No 114	3,040 kg	
From experimental plot, prep No 160	3,480 kg	
From experimental plot, prep No 86	2,250 kg	

In the experiments, seeds treated with antibiotics germinated earlier which was quite distinctly reflected in the subsequent development of the cotton plant (bud formation, flowering, opening of the bolls); the vegetation time of the plants was shortened by 8-10 days.

Similar results were obtained in experiments performed under industrial conditions. Pretreatment of seeds with antibiotics 105 and 114 before sowing lowered the disease incidence of cotton-plant gummosis 5-6 times and because of this a higher cotton crop was obtained. For example, in the kolkhoz "Kzyl-Argin" (Tashkent Oblast') twenty-five per cent of the control cotton plants were affected by gummosis and on the treated fields the morbidity decreased to 5.3%. The control cotton plants yielded 24 centners/ hectare of seed cotton and the experimental plants--30.6 c/ha (Askarova, 1951-1952).

Similar experiments were performed by Mirzabekyan (1952-1953) in Armenia. She employed grisein No 15 in the form of partially purified raw material in treatment of seeds. The incidence of the gummosis disease of the cotton plant decreased by 67-84%,

Of interest were the experiments of Askarova (1952) against the secondary gummosis infection of the cotton plant. In cotton-growing practice one often observes a secondary massive infection of a given culture by gummosis during vegetation.

This is often facilitated by rainfall. Under conditions of a field experiment and on the kolkhoz fields, application of antibiotic substances during this secondary infection also yields positive results.

Bel'tyukova (1951) disinfected seeds with an antibiotic microcide before sowing. The incidence of infection of the plants with pathogenic bacteria decreased considerably after such treatment.

Bushes of garden rose affected by mildew were specially treated by us with an especially chosen antibiotic of actinomycetal origin. After the leaves were washed three times with a solution of the crude antibiotic, the symptoms of mildew began to disappear and after some time the leaves acquired a normal or almost normal appearance. (Figure 100).

Figure 100. Treatment of bushes of garden roses, affected by mildew: A--bushes not treated with antibiotic; the leaves are covered with a white coating of fungus; B--bushes treated with antibiotic; green leaves, free of fungus.

A positive effect of antibiotics was observed in all cases where treatment was started in the early stages of the disease.

In strongly affected bushes the treatment had only partial results: the mildew cover disappeared, but the leaves acquired a brownish-green color, which either disappeared later, or as more often occurred, remained until the end of vegetation.

Positive results with antibiotics were obtained by Protsenko in floriculture, Gurinovich in market gardening and Afrikan and others in experiments with field crops and vegetable cultures.

Recently, foreign investigators extensively investigated antibiotics in their a struggle against plant diseases under conditions of production, in orchards, gardens and fields. For this purpose special preparations of antibiotics are produced--agrimycin 100, agristrep, phytomycin, acco-streptomycin, etc. They are crude preparations of streptomycin together with terramycin or some other antibiotic. Agrimycin contains 15%

streptomycin sulfate and 1.5% oxytetracycline (terramycin) in a powderlike filling. Agristrep is similar to agrimycin but differs from the latter in that it has a higher content of commercial streptomycin (37%). Phytomycin--a liquid preparation, contains 20 % streptomycin and is the most stable upon at storage. Accostreptomycin contains 45% streptomycin is also qulte at stable, and can be preserved for 2 years. Of wide use in plant growing is the antibiotic, actidione.

Treatment of fruit trees. The best results in the application of antibiotics were obtained in horticulture in the treatment of fruit and nut trees infected with bacteria. Goodman (1954 a, b, c) in Missouri, Young and Winter (1953) in Ohio, Hueberger and Poulos (1953) in Delaware, Ark (1953 a, 1954) and Dunegan (1954) in California, Kienholz (1954) in Oregon, Clayton (1955) in North Caroline, Kirby (1954) in Pensilvania and Mills (1955) in New York obtained good results, upon spraying and dusting antibiotics on plants infected with bacteria. Wherever such treatment was performed the morbidity of apple and pear trees decreased or stopped altogether. According to Goodman, Dunegan, Ark and others, 3-4 sprayings of agrimycin at a concentration of 30-100 μ g/ml completely eliminated the infection of woody plants. Ark sprayed powderlike crude streptomycin and a solution of a purified preparation. In treatment of apple and pear trees afflicted by *Bact. amylovorum* or of walnut afflicted by *Bact. juglandis* the pure preparation was superior. In treating nut trees, two sprayings of streptomycin sulfate solution at a concentration of 10 μ g/ml were applied. Dye D. and Dye M. (1954) successfully treated the seedlings of pear trees, infected by *Bact. juglandis* with a solution of streptomycin sulfate and dihydrostreptomycin at a concentration of 100 μ g/ml.

Good results are obtained upon treating fruit trees (cherry, etc) infected by *Bact. syringae,* with actidione. One or two sprayings with a solution containing 1-2 μ g/ml is sufficient to stop the disease. The characteristic spots on leaves cease to appear and those already existing disappear. Today, actidione is used before fruit formation and after the cherries ripened, although investigations show that this antibiotic does not poison the fruits and may be used during fruit bearing (Hamilton and Szkolnik, 1953; Cation, 1953).

In Germany, Klinkowski and Keller (1956) in their struggle against mildew of fruit trees used crude antibiotics (filtrates of culture fluids) obtained from specially selected actinomycetes. The preparations were applied to the trunks of the apple trees infected with the disease "white ripening", The experiments performed in orchards on a large scale yielded positive results.

Treating leguminous plants. Mitchel et al., (1952) tested antibiotics--streptomycin, terramycin, neomycin, aureomycin, patulin, subtilin, etc, (in all 12 preparations) in treatment of leguminous plants artificially infected with bacteria. He introduced these bacteria into the plants by the use of a paste, which he smeared on the stems. Under laboratory experimental conditions, kidney beans and soy were totally protected against bacterial wilt by treatment with streptomycin or dihydrostreptomycin. All control plants perished.

After the successful experiments of Mitchell and co-workers, large-scale field experiments were started for testing antibiotics on leguminous plants. On the experimental fields of Beltsville in Maryland streptomycin was used in the struggle against the fungal disease of Chilean beans, caused by *Phytophthora phaseoli.* Spraying with an antibiotic solution at a concentration of 100 μ g/ml the morbidity decreased markedly or even disappeared entirely. The application of crude streptomycin in concentrations of 50 μ g/ml gave better results than the use of chemically pure preparations even at higher concentrations. It is assumed that the crude preparation of streptomycin contains some other substance with antifungal properties.

In fighting mildew of beans, agrimycin at a concentration of 25 μ g/ml in a mixture with copper preparations of the same concentration, was used. This preparation was more effective than the copper preparation and agrimycin applied separately (Zaumeyer and Fisher, 1953; Zaumeyer, 1955).

Dekker (195 5) tested the action of a culture fluid of the actinomycetes-antagonists--*A. rimosus* on pea seeds infected with *Ascochyta pisi* and *Mycosphaerella pinodes.* The antibiotic substance penetrated the seeds and killed the disease agent therein. Such seeds germinated normally and gave rise to healthy seedlings, while in the control massive infection was observed.

Klinkowski, Kohler and Shroedter (1955) treated bean seeds with crude antibiotics formed by *Penicillium chrysogenum,* and *A. griseus* in order to prevent infection of the sprouts by the bacteria *Ps. phaseolicola.* The seeds were soaked in antibiotic substances and thus relieved of infection.

Treatment of vegetable cultures. Brian et al., (1951) treated infected lettuce and tomatoes with griseofulvin. This drug possesses strong antibiotic properties and inhibits numerous species of fungi, including phytopathogenic ones. It does not affect bacteria. Spraying lettuce, infected by the fungus *Botrytis cinerea,* with a solution of the antibiotic yields quite satisfactory results.

Similar results were obtained in experiments with tomatoes infected by the fungus *Alternaria solani.* Griseofulvin was either introduced into the substrate under the root system, or it was sprayed on the leaves. In the control containers the morbidity incidence was 100%, while upon treatment the disease was not apparent at all, or only a small percentage of the plants contracted the disease.

In one of the experiments the number of spots that appeared as a result of the disease an the leaves of the tomatoes, was counted. In cases where griseofulvin was introduced in a dose of 10- 2 0 µ g/ml per 1 g of substrate there were no spots on the leaves or only very few; in plants not treated with the antibiotic there were more than 1,250 spots on a single plant.

The antiblotic thiolutin (obtained from *A. albus)* was used by Gopalkrishmann and Jump (1952) against fusarium wilt of tomatoes (caused by *Fusarium oxysporium lycopersici.* The authors treated tomato seedlings by dipping their roots in the antibiotic solution before planting them in the soil. This procedure completely protected the plants against the disease. The control plants all succumbed to the disease. Microbiological analysis of the tissues of the plant showed that with small doses (10 µ g/ml) of the antibiotic there were no outward signs of the disease but the mycelium of the fungus could be found Iin the tissues. After treatment with large doses of the antibiotic (40-80 µ g/ml) the tissues of the plants were sterile and no mycelia were observed in them.

In another series of experiments the authors soaked tomato seeds in the antibiotic and planted them in the soil. After 12 days 100% of the control plants had fusariosis, while the treated plants showed no signs of the disease or only a very small number of them showed its symptoms.

No less effective results were obtained when antibiotics were used against bacterial infections of tomatoes or other cultures. Conover (1954, 1955) reported good results in treating tomatoes and pepper infected with *Bac. vesicatorium,* with agrimycin and streptomycin. After 5 sprayings with a solution of the antibiotic at a concentration of 200 µ g/ml, 74% of the plants were completely healthy and only 0.4%, were seriously affected. Among the control plants 12% were healthy and 34% were very sick. Ninety-five per cent of the treated plants were suitable for replanting and among the untreated plants only 27% could be replanted.

Cox et al., (1953) completely cured diseased pepper by spraying it 3 times with streptomycin at a concentration of 500 µ g/ml. Similar data are given by Crossan and Krupka (1955), who found that upon treatment with the antibiotic the disease agent in the leaves of pepper is totally eradicated. Higher concentrations of the antibiotic, according to Cox (1955), yielded a smaller effect and sometimes even caused an increase in morbidity. The author noted the beneficial effect of a mixture of streptomycin (100-200 µ g/ml) and copper preparations.

No less effective results are obtained upon treatment of celery infected with *Pseudomonas apii,* a disease which is very common in Florida. The use of agrimycin (300-600 µ g/ml) almost completely eradicates the disease. As in the case of treating tomatoes and peppers, a mixture of streptomycin and a copper preparation gives better results.

Sutton and Bell (1954) treated turnips infected by *Pseudomonas campestris.* They treated the seeds with aureomycin solutions diluted 1:2,500 and 1:1,000 before sowing. Short exposure of the seeds to such a solution completely eradicated the disease agent; the germination of the seeds was normal and even a stimulation of growth of the seedlings was noted. The plants were healthy, while in the control they were infected to an extent of 30-76%.

The disease of the eyes of potato tubers caused by the bacteria *Bact. atrosepticum* and *Pseudomonas fluorescens* is often the cause of serious injury to potatoes in the field and storage. The application of antibiotics in such cases gives very good results. Bonds et al., (1953-1955), at first in greenhouse experiments and later under field conditions, found that treatment of cut tubers of diseased potatoes with a streptomycin sulfate solution (25 µ g/ml) prevents the disease in 80-100% of plants. Dipping of the tubers for a short time in the solution of the preparation not only diminishes the morbidity but also increases the viability of the sprouts.

According to Webb (1955), treatment of potato eyes with agrimycin had little effect, however treatment with phytomycin had very good results. The tubers produce more viable plants with abundant flowering and the tuber crop was 15% higher that of the control. Heggested and Clayton (1954) had good treatment results wit tobacco infected with *Pseudomonas tabaci*. The authors used a streptomycin sulfate solution followed by agrimycin and agristrep in 200 μ g/ml concentrations. These solutions were sprayed on the plants 2-3 times during the summer. As a result, there was almost no plant morbidity, while more than 30% of the control plants perished. The effectiveness of the antibiotic was higher than that of the copper preparations. Beach and Engle (1955) in Pennsylvania achieved excellent results in the treatment of tobacco with antibiotics. They used phytomycin at a 100 μ g/ml concentration. Kirby (1955) treated diseased tobacco plants with agrimycin (100 μ g/ml) mixed with febram. All the authors noted a decrease in morbidity, improvement of growth, increase in number of leaves and also a greater development of the root system.

Antibiotics were successfully used in treating decorative plants. Robinson, Starkey and Davidson (1954) treated chrysanthemums infected with bacteria with solutions of streptomycin, terramycin, neomycin, chloromycetin and other substances. The best results were obtained with the first three preparations. The antibiotics were introduced into the roots or into the grafts. The treated plants either did not succumb to the disease at all, or the incidence of disease was very low, while the control plants all perished. The antibiotics eradicated the infection in the plant tissues (Pramer, 1955).

Treatment of grain cultures. Attempts have been made to use antibiotics in diseases of cereals. Wallen (1955) used various antibiotics against wheat rust, caused by *Puccinia graminis var. tritici.* The rust-effective antibiotic in his experiments was actidione. Concentrations of 50-500 μ g/ml, although toxic to the plants, lowered the morbidity to 0-5 %. At a lower concentration (2 5 μ g/ml) no toxicosis manifested itself and the percentage of morbid plants was within the range of 50-60%, while the morbidity in the control was 100%.

The crop of the wheat treated with the antibiotic was higher than that of the untreated plots. The percentage of germinating seeds was higher among the experimental plants (more than 90%) than among the controls.

Leben, Army and Keitt (1953) applied the antibiotic helixin "B" against the disease of oats, caused by *Helminthosporium victoriae* and against the disease of barley, caused by *Helminthosporium sativum.* The antibiotic considerably lowered the incidence of disease in the plants. In the control, under conditions of an experiment in growth containers, there was 28-29% of diseased plants and under field conditions--2%; after treatment with the preparation no diseased plant was observed. In the growth containers and under field conditions diseased plants did not exceed 1%. Positive results were also obtained in laboratory experiments, using antibiotics against wheat rust (caused by *Tilletia foctens),* oat rust (caused by *Ustilago avanae),* and barley rust (caused by *Ustilago hordei).*

Henry et al., (1952 -1953) achieved an almost complete eradication of the rust disease of wheat by treatment with actidione mixed with a preparation called "Dixie clay." Even better results were obtained when actidione was used as a dust mixed with "Dixie clay" and a preparation called "Captan", which alone did not give good results in the struggle against the mentioned diseases.

Antibiotics have not so far been effective in combating virus diseases of plants. Attempts were made to use various antibiotic substances against tobacco mosaic (Schlegel. David and Rawlins (1954) and certain other virus diseases (Leben and Fulton, 1952); the small positive effect sometimes observed was not caused by suppression of the virus particles but by the action of the antibiotics on the host plant, which increased its growth and enhanced the resistance of the tissues (Zaumeyer, 1955).

The given data on the use of antibiotics in plant growing are as yet scanty. However, there is a basis for hope that these substances will prove to be no less effective in the treatment of plants than in the treatment of animals and human beings. Experiments show that a number of antibiotics may already be widely applied in agriculture, in the struggle against fungal and bacterial diseases of woody and grassy plants.

Antibiotics are in many cases not inferior in their action to modern antiseptics, and often surpass them. It is possible that in the future, with the study of conditions and the mechanism of action, and with improvement of the methods of their application, antibiotics will become even more effective.

If the antibiotics, after their introduction into the plant stems are directed into the root system, and accumulate there at a higher or lower concentration, then these preparations would probably be effective against root diseases; this might be of great importance.

One also should not overlook the economic side of this enterprise.

At first one should assume that it will be advisable to apply antibiotics to the most costly cultures, mainly in horticulture, in the struggle against diseases of fruit and decorative plants.

In these cases, not only the cost of a given treatment of the tree is of importance, but also the time necessary for growing a fruit-bearing plant, should be considered. Antibiotics are less harmful to the health of man than antiseptics. A certain portion of the chemical substances which enter the plant tissues, when they are treated with antiseptics, concentrate there, and to some degree lower the nutrient qualities of the plants both as food and fodder. The possibility is not excluded that certain elements may prove to be harmful to man and animals.

Antibiotics are less dangerous in this respect. They cannot accumulate in concentrations which are toxic to animals and man. One may choose antibiotics which are altogether harmless. Certain antibiotics have an activating effect on plants and increase their growth. All this illustrates the necessity of devoting more attention to the antibiotics than has been done until the present.

CONCERNING THE EPIPHYTIC MICROFLORA

In discussing the problems of interaction between microorganisms and higher plants one cannot ignore the epiphytic microflora. Microbial epiphytes are the organisms which concentrate on the surface of the green parts of vegetating plants and are nourished by the excretions of the latter.

The epiphytic microflora has been but little studied, especially its quantitative and qualitative composition on various plants.

On the surface of the aerial parts of plants one finds different microorganisms--bacteria, actinomycetes, fungi, yeasts, algae and protozoa. Their number may by very high. Duggeli (1904) counted many thousands of microorganisms on the surface of cereal seeds. From 80,000 to 25 million bacterial cells and from 4,000 to 7,200 fungi in one gram of wheat seeds have been detected by Morgentaller (1918). The author points out that on healthy seeds there are almost no fungi.

In germinating wheat seeds there are 60,000 bacterial cells per gram of grains, and in nongerminating seeds-- 13 millions. Mack (1936), Kent-Jones and Amos (1930), inspected 21 samples of wheat seeds from different countries, and found from 8,000 to eight million bacterial cells on the surface of one gram of seeds. Gustafson and Parfeitt (1933) in a similar study counted from 46,000 to 3,260,000 bacterial cells in one gram of wheat seeds.

Rautenshtein (1939) studied the microflora of wheat seeds in the various stages of ripening: milky, waxy, and full maturity stage. The results of his observations are given in Table 136.

Table 136
Quantitative and group composition of microorganisms on ripening seeds of wheat
(number of cells in thousands in one gram of seeds)

Wheat variety	Maturation stage	Total No of microbes	Bacteria	Fungi	Actino-mycetes	Yeasts
Cesium 0111, second class	Milky	8,050	7,250	150	650	0
	Waxy	5,525	5,275	225	25	0
	Full	17,650	17,050	500	100	0
Cesium 0111, first class	Milky	33,375	32,500	625	250	0

Waxy	72,000	70,900	500	600	0
Full	41,500	41,250	250	0	0

Bacteria are the most numerous among the microorganism groups which were found. Yeasts are not always encountered.

With ripening of the seeds the number of microbes on their surface increases. James, Wilson and Stark (1946) counted from 280,000 to 164 million microorganisms in one gram of wheat seeds. The numbers of microbes found on the surface of the green parts of plants are not smaller. Many investigators counted from 49,000 to 6,300,000 epiphytes in one gram of tissue (Khudiakov, 1953; Thomas and Hendricks, 1950; Stirling, 1951; James, 1955, and others). According to Kroulik, Burkey and Wiseman (1955), the number of epiphytes in one gram of tissue of green plants of corn, oats, clover, lucerne, garden grass, and other plants varies from 1,540,000 to 99,200,000. These numbers vary from species to species, in relation to the age of the plants, and also with the soil-climatic conditions. As a rule, the number of epiphytes on the surface of young plants is larger than the number on ripening ones. In relation to seeds, the opposite picture was observed. Bacteria form the greatest part of the epiphytic microflora. The species composition of the bacteria is quite diverse, but the dominating part of it is considerably small. Almost all the investigators noted the predominance of bacteria with a yellow pigment, classified as *Ps. herbicola* on plants. This bacterial species was described by Duggeli (1904), He found that these bacteria were the dominant species. Their total number reached 380,000 and more in one gram of tissue.

According to Weller (1929), *Ps. herbicola* comprises 90-100% of the total bacterial flora on seeds of wheat and rye. Upon germination of seeds in the soil, these bacteria soon disappear, reappearing toward the end of ripening. On growing plants, according to the author, there is abundant growth of lactic-acid bacteria. On seeds of barley and oats Weller found sporeforming bacteria. Rautenshtein (1939, a, b) found 75-98% of *Ps. herbicola* among the bacteria populating the surface of wheat seeds. Cocci and Sarcina are encountered in a few cases. He also found a great number of lactic-acid bacteria. Among the fungal flora, Rautenshtein found the fungi *Cladosporium herbarum*, *Trichoderma koninglii* and less often *Dematium*, *Asperigillus*, *Penicillium*, *Oospora* and also the species *A. globisporus* and *A. griseus* and the yeast species of the genus *Torulopsis*.Many heat-resistant and thermophilic bacterial forms were found. The majority of these forms belonged to the sporeforming species of the type *Bac. mesentericus*. They also grow at a temperature of 17-20° C.

James, Wilson and Stark (1946) distinguish between two types of epiphytic bacteria--type A and type B. Type-A bacteria form yellow colonies and are all considered to be cultures of the same species--*Ps. herbicola*.The other type belongs to the colorless *Pseudomonas* species. Of the fungal flora these authors found the following on plants: *Acrostalagmus*, *Alternaria*, *Penicillium*, *Aspergillus*, *Botrytis*, *Cephalosporium*, *Fusarium*, *Torula*, *Monilia* and other fungi. There were also phytopathogenic species among them *Helminthosporium sativum*, *Hormodendron pallidum*, *H. viride*, *Alternaria tennis*, *Fusarium culmorum*, *Cladosporium herbarum*, *Septoria nodorum*, etc.

James (1955) gives the following data on the extent of distribution and accumulation of *Ps. herbicola*. Out of 200 plants of oats, barley, and flax, growing in different regions of Canada which were studied, more than half contained this microbe in amounts of 100,000 and more, about 15 % of the plants contained from 10,000 to 100,000 bacteria and some samples were free of this bacterial species altogether.

Clark (1947) and others observed the yellow bacteria in great numbers on the green parts of the cotton plant.

The predominance of these bacteria on other plants was noticed by many investigators (Burri, 1903; Mack, 1936; Thomas and Hendriks, 1950; and others).

Wallace and Lochhead (1951) point out the connection between the epiphytic and rhizosphere microflora. The latter, according to these authors, is intermediate between the microflora of the soil and the epiphytic microflora.

Khudiakov studied a great number of plants (1953a). He investigated the migration of bacterial epiphytes from the surface of the seeds during the germination of the latter, to the seedlings and later to all the plant organs including fully formed and mature seeds. More than 20 microbial species were isolated from different wheat varieties by this author. Among these microbes there were three species of yeasts, two species of fungi and the remaining were bacteria, among which three,. cultures produced a yellow pigment, three others orange or red, one green, and all the others colorless. According to the author's observations, each of these species is the predominant form on some seed variety.

Upon analysis of preharvest seeds of wheat (Moscow variety, 2411) grown in the Moscow Oblast', 97% of the bacteria found were Ps. herbicola; no yeasts or colorless bacteria of the Ps. fluorescens group were observed. On another wheat variety which was grown alongside the former and under the same conditions, there were 60% yeasts and no Ps. herbicola bacteria were observed.

Reciprocal cross infection of the wheat varieties by the isolated cultures has shown that the latter were not specific. Epiphytes from one wheat variety, when transmitted to another variety, grew as well as they did on seedlings of their own host plant. Khudiakov has shown that epiphytic microflora can be changed at will by treating the seeds before sowing with the corresponding microflora. He treated sterilized oat seeds with cultures of epiphytic yeasts and bacteria which he isolated, and sowed them in open ground.

Those microbes that had been artificially introduced (Table 137) were found on these plants. On the control plants the bacteria which usually concentrate on this species predominated.

Table 137
Effect of bacterial inoculation of seeds on the composition of the epiphytic microflora of oats

Organisms introduced with the seeds	Number of colonies on plate	% yeasts, No 1 red	% yeasts, No 2 mycelian	% yeasts, White	% Bacteria, Ps. herbicola	% bacteria, yellow green	% bacteria, others
Red yeasts No 1	80	80.8	0	6.2	0	0	5.0
Mycelial yeasts No 2	107	1.8	90.6	2.8	0	0	4.6
Bacterium sp. yellow-green	132	5.3	4.5	14.4	7.5	65.1	3.0
Control seeds (not inoculated with bacteria)	84	2.3	2.3	9.5	80.9	0	4.6

Kvasnikov and Sumnevich (1953) investigated woody plants in Central Asia--poplar, apple, pear, cherry, maple, Greek nut, etc and grassy cultures: corn, lucerne, cotton, Sorghum cernium, milo, potato, sugar beet, cabbage, etc. In all cases, lactic-acid bacteria were found on all the plants in great quantities. According to the authors, on wild plants the number of these bacteria is considerably smaller. The nearer the plants are to places of human habitation, the more lactic-acid bacteria one finds on the surface of plants.

Kroulik, Burkey and Viseman (1955) divide the bacteria into chromogenic and colorless groups. Among the former, the Ps. herbicola type predominates and among the latter--lactic-acid bacteria Lactobacterium plantarum.

In our studies we investigated various species of grassy and woody plants growing in the central belt of the USSR. The number of bacteria and fungi on the surface of leaves and branches has been determined. Similarly to other investigators, we also detected hundreds of thousands and millions of bacterial cells per gram of tissue. In the different plants the predominant epiphytic microflora varies in its species composition. In some plants Ps. herbicola predominates and in others--other species of the genus Pseudomonas and Bacterium, and sometimes-- lactobacilli. Quite often one finds large numbers of yeasts of the genus Torula (T. rosea, or T. alba) and of the genera Sporobolomyces and Mycotorula.

Large numbers of microorganisms are found on the surface of fruits. Studies show that on berries and fruits there are bacteria, fungi and yeasts, actinomycetes and even protozoa. Epiphytes are encountered on wild as well as on cultivated fruits.

On berries as on other parts of the plants the most numerous group of microbes are the bacteria with fungi and yeasts following. Often there are as many as hundreds of thousands or even millions of them on 1 g of berries.

The number of microbes varies with the variety and species of the berries, with the degree of naturity, and with climatic and other external conditions. As a rule, their number increases with the ripening of the berries.

The quantitative ratio between bacteria, yeasts and fungi also change.

The microflora of the vine grapes has been the most thoroughly studied. According to the data of Akhinyan (1952) the total number of microorganisms on the surface of vine grapes of the "Kakhet" variety ranges between 3,000 and 4,000,000 per 1 g, depending on the region and where the vine was grown (Table 138).

Table 138
Distribution of microflora on the surface of vine grapes
(according to Akhinyan, 1952) (number of cells in 1 g of berry)

Region (Armenian SSR)	Yeasts	Fungi	Bacteria
Ashtarak region			
Oshakan village	20,850	4,100	650,000
Voskevaz village	75,000	--	296,000
Artashat region			
Aizestan village	4,008,600	123,000	7,500
Yuva village	3,500	--	60,000

The presence of such a large microflora on the surface of plants cannot be explained by their being carried over mechanically from the air. The accumulation of certain specific species speaks against it. The latter evidently grow and multiply on the plant surface. Consequently, they must find there sufficient quantities of food substances necessary for mass reproduction.

Plants, as has been pointed out above, excrete various volatile and nonvolatile substances--with the aid of special glands or by guttation. In the drops formed by the guttation of rye grass glutamine was found (Chibnall, 1939), Genkell (1946) observed the excretion of mineral salts together with the fluid excreted by plants of salty marshes. Vigorov (1954) found in the guttation drops of 7-to 9-day-old wheat seedlings 1.8 mg/ml of dry substance, containing 5-10 mg/ml ammonium-nitrogen and 40-45 mg/ml phosphorous compounds. The author observed, that the intensity of guttation depends on illumination, soil humidity, on the presence of nitrogen and other nutrient elements. Introduction of ammonium salts into the soil elevates the excretion of nitrogen compounds in the guttation drop. One of the tests for the presence of organic substances in the fluids is the growth of microorganisms in them. According to the author, fungi grow abundantly in guttation drops.

According to the data of Kholodny (1944 a,b,c) all or many of the organic substances excreted by plants are used by microbes as sources of nutrition.

The significance of the epiphytic microflora in the life of the plant is many-faceted. Among the epiphytic microflora there are many activators (*Ps. herbicola,* yeasts, etc), which form biotic substances--vitamins, auxins, folic acid, thiamine, riboflavin and other compounds, and also organisms forming antibiotic substances with strong antimicrobial properties.

On the surface of leaves and stems of plants there are microorganisms forming toxic substances. In the epiphytic group there are also parasitic and phytopathogenic forms.

One should assume that the metabolic products of the epiphytic microflora behave in a certain manner in the plant tissue, having a definite effect on them. The ability of leaves to absorb various substances was known for a long time. On this basis methods of extrarhizal feeding have been elaborated and also methods of introduction of substances with the aim of changing certain physiological functions shedding of leaves, arresting of flowering, etc.

It was also established that plants can absorb various microbial metabolites, vitamins, antibiotics and other compounds through the leaf surface. As was indicated above, these substances not only enter the plant through the leaves but they can be introduced by this route in large quantities for the purpose of feeding as well as for fighting bacterial and fungal infections. Among the epiphytic microflora there are numerous antagonists which produce antibiotic substances which suppress their competitors, and among them also phytopathogenic microbes. Growing abundantly on plants, such organisms may fulfill a protective role removing or suppressing infectious agents originating from without. If we were to change the composition of the epiphytic microflora on the surface of the green parts of plants at will, and form certain coenoses of antagonists there, this would prove to be of great value to plant and fruit growing.